Molecular Politics

Molecular Politics

Developing American and British Regulatory
Policy for Genetic Engineering, 1972–1982

SUSAN WRIGHT

The University of Chicago Press
Chicago and London

Susan Wright is lecturer in History of Science and in Political Science at the University of Michigan. She is the coauthor and editor of *Preventing a Biological Arms Race*, published by MIT Press.

The University of Chicago Press, Chicago 60637
The University of Chicago Press, Ltd., London
© 1994 by The University of Chicago
All rights reserved. Published 1994
Printed in the United States of America

03 02 01 00 99 98 97 96 95 94   5 4 3 2 1

ISBN (cloth): 0-226-91065-2
ISBN (paper): 0-226-91066-0

Library of Congress Cataloging-in-Publication Data

Wright, Susan, date
    Molecular politics : developing American and British regulatory policy for genetic engineering, 1972–1982 / Susan Wright.
        p.   cm.
    Includes bibliographical references and index.
    ISBN 0-226-91065-2 (cloth). — ISBN 0-226-91066-0 (paper)
    1. Genetic engineering—Research—Government policy—United States.
    2. Genetic engineering—Research—Government policy—Great Britain.   I. Title.
QH442.W75   1994
363.17′9—dc20   93-47054
                                                                       CIP

⊗ The paper used in this publication meets the
minimum requirements of the American National Standard
for Information Sciences—Permanence of Paper for
Printed Library Materials, ANSI Z39.48-1984.

This book is printed on acid-free paper.

For Jonathan

It's molecular politics, not molecular biology, and I think we have to consider both, because a lot of science is at stake.

Anonymous speaker, Enteric Bacteria Meeting, National Institutes of Health, Bethesda, Maryland, 31 August 1976

Rosencrantz: What's the game?
Guildenstern: What are the rules?

Tom Stoppard, *Rosencrantz and Guildenstern Are Dead*

CONTENTS

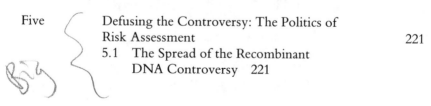

ILLUSTRATIONS

Figures

Tables

PREFACE

Science in the late twentieth century has become an intensely contested material and ideological resource over which nations, corporations, universities, scientists, and nongovernmental organizations struggle at many levels—local, national, international, and transnational. The subject of this book—the rise and fall of controls for genetic engineering in the United States and the United Kingdom—provides a window on these struggles.

In 1975, the controversy surrounding the emerging field of genetic engineering and the conditions under which it should be pursued was just beginning to heat up. Fresh out of graduate school with a doctorate in the history of science and sensitized by the antiwar and environmental movements to the complicity of science in militarism and environmental degradation, I saw the genetic engineering issue as a case study in social negotiations over the promotion and control of new and possibly risky fields of science and technology. In this my views were shaped by those scientists who anticipated the social and ethical problems posed by military, commercial, and other uses of genetic engineering and saw their responsibility to society in terms of raising these issues publicly. Some took risks in their careers for doing so and they deserve thanks beyond their influence on this book.

My effort to understand the dynamic of scientific development through the lens of the evolution of genetic engineering policy has been at the same time a search for appropriate methods of analysis. It was clear to many observers of the early struggles over the control of the field that science and politics were in some sense intertwined. But, initially, few were aware of the potential global scope of the interests being expressed in these struggles. Certainly my early orientation, which focused on issues and procedures at the national and local levels, now seems myopic. In addition, I came to recognize the essentially static nature of the categories "science" and "politics" and the need to use methods that dissolved the boundary between them.

A first step toward broadening my approach came in 1976–77, which was spent in Cambridge, England. Conversations with David Dickson, Bob Young, and Edward Yoxen alerted me to the fact that the British response to the problem, while it generally resembled that of the United States in content, differed significantly in terms of the form adopted for policymaking. An exploratory round of interviews with those engaged in formulating and participating in the policy adopted in the United Kingdom and a long conversation stretching into the early hours of the morning with Sydney Brenner confirmed these differences and, in addition, pointed to the important role of the British trade unions locally, in the politics of individual laboratories, and nationally, in policy formation. The idea of pursuing a comparative study of the two countries, each of which was in the forefront of promoting and controlling genetic engineering, took root at this stage.

In 1977, the struggles over genetic engineering policy erupting in the United States in local communities and in Congress and, more quietly, within the British Genetic Manipulation Advisory Group, made clear that policy developments in the two countries involved deep conflict over the nature of the issues being addressed and who should address them. At this point, University of Michigan political scientist Jack Walker crucially influenced my thinking about analyzing policy formation. In the quiet and unassuming manner that is so missed by his colleagues, he encouraged me to examine the long and generative debate in political theory between pluralists, who assumed that policy formation could be understood by examining decisions in formal policy arenas, and postpluralists, who claimed that investigating the "mobilization of bias" of the policy arena itself is essential for understanding who exerts a decisive influence on policy formation and how. The influence of Peter Bachrach and Morton Baratz, Frederick Frey, Steven Lukes and other postpluralists on my approach to analyzing policy processes will be evident in what follows.

The postpluralist approach to policy formation posed a further theoretical problem for my analysis: that of conceptualizing the social "interests" assumed to be expressed within and around the policy arena. As a historian, I found the tendency of some political theorists to conceive of interests in absolute or rational terms problematic. Discussions with Jonathan King and Edward Yoxen on the formation and evolution of the interests of the biomedical research sector as well as the visible transformation of the interests of genetic engineers and their institutions in the late 1970s reinforced my sense that interests should be seen as contingent, formed by particular historical circumstances.

In 1979, I embarked on several years of research on the evolution of the British and American controls. This included interviews with scientists, government officials, politicians, representatives of corporations, members of public interest organizations, trade unionists, and others, and archival research on both sides of the Atlantic; observation of the quarterly meetings of the NIH Recombinant DNA Advisory Committee; and research on the principal stages in the development and refinement of the genetic engineering techniques.

It is not possible to name all of those who contributed to developing my understanding of the issues but I hope that the extent of my indebtedness is evident in the references given in the notes. I am deeply grateful to those who granted interviews and provided access to relevant documentation. Especially for the early period of policy formation, I also want to acknowledge the importance of the materials drawn together by Charles Weiner and deposited in the MIT Institute Archives. In using this rich collection of documents, interviews, and tapes, I have benefited from his foresight in documenting the genetic engineering issue and related policy decisions as these evolved in the 1970s. I am also grateful to colleagues who took the time to initiate a historian into the complexities of their field, especially to John Valentine, for guiding me through a series of key papers in genetic engineering, and to Ellen Elliott, Richard Goldstein, Stuart Ketcham, Jonathan King, Bruce Levin, Stuart Newman, Bob Rohwer, and Barbara Rosenberg, for responding to numerous technical queries.

In the 1980s, two developments contributed to enlarging and modifying the postpluralist framework that guided my initial research. First, industrial and military applications of genetic engineering drew attention to the international and transnational character of the interests being exerted in struggles over promotion and control of the field and underscored the restrictive character of early definitions of the genetic engineering issue. My engagement in the 1980s in addressing the problem of military development of this field and collaboration with members of the Committee on Military Use of Biological Research of the Council for Responsible Genetics enlarged my understanding of the global scope of the politics of technological development, the subtle ways in which seemingly neutral scientific endeavors contribute to that development, and the moral and social choices involved.

The second development had a different character. This was the challenge to the modernist ideal of universal, objective knowledge posed by the many strands of postmodern thought that swept through history, social studies of science, and many other academic fields in the 1970s and 1980s. I confess to an early resistance based on an inaccurate impression that all postmodern analysis was nihilistic in character and as such obstructed constructive political change. Rethinking that resistance owes much to the influence of Richard Falk, who persuaded me to take a long second look at those strands of postmodernism that question colonizing forms of knowledge, open awareness to the perspectives of other cultures, and yet recognize the need for normative judgment.

In particular, Foucault's critique of the social evolution of power, knowledge, discourse, and practice has been influential both positively and negatively. Positively, Foucault's recognition of the relations among power, discourse, and practice provided theoretical support for my sense that the discourses emerging from policy arenas expressed the power relations of these arenas. Negatively, Foucault's disinclination to analyze subjects and agency posed an important problem: while I accepted the claim that the effects of power are disseminated through social institutions, I could not accept the tendency to deny that power is also expressed by responsible groups and institutions. My own experience observing the struggles of various institutions and groups over genetic engineering policy indicated otherwise. The challenge posed by Foucault's position was to synthesize methods of analysis that incorporated his insights into the power/knowledge connection but did not relinquish the postpluralist insight into the essential role of individuals and organizations in mobilizing bias. In responding to this challenge, I have benefited from Steven Lukes' appraisal of postpluralism and Nancy Hartsock's feminist critique of the limitations of Foucauldian discourse analysis. By contextualizing interests and extending the postpluralist analysis of power to address the mobilization of power in discursive practice, I hope to have shown the benefits of denying the academic boundary between voluntarism and poststructuralism.

Throughout the period in which ideas have taken shape and have been transformed into written form, I have been sustained and stimulated by friends and colleagues. I thank Ditta Bartels, Carola Carlier, Michael Cohen, David Dickson, Richard Falk, Richard Goldstein, Donna Haraway, Jonathan King, Sheldon Krimsky, Bruce Levin, Everett Mendelsohn, Dorothy Nelkin, Stuart Newman, David Ozonoff, Marc Ross, Arthur Schwartz, Robert Sinsheimer, Charles Weiner, Tom Weisskopf, Bob Young, and Edward Yoxen for discussion of issues and reactions to drafts. I have a special debt to Paul Grams, who took the responsibility for editing the manuscript far beyond the usual limits, generating through acute observation creative tensions that provoked development of my general position. I also wish to acknowledge the contribution of the three anonymous readers for the University of Chicago Press, in history of science, politics, and biology, for detailed and incisive responses to the manuscript.

It is pleasant to thank those who aided my research and the preparation of the book. Staffs of archives and government offices in the United States and Britain provided friendly assistance with source materials. I have also been fortunate in doing research with the resources of a great library close at hand. I thank the reference staff of the University of Michigan Library, to whom I turned on countless occasions, and Grace York, for her expert guidance through the library's documents collection.

I am greatly indebted to the following people, not only for their high standards but also for their warm support for the project: Sue Meyer, for wordprocessing early drafts and for facilitating my transition from the typewriter to the word processor; Sandra DuMonde for transcribing interviews and helping in many essential ways to prepare the manuscript for publication; Dan Hayes, Christine Skwiot, Jenna Didier, Rachel Meyer, Kristin Blackburn, Jim Herron, Jeannette Lim, and Tiffany Espinosa, for bringing joy as well as accuracy to the tasks of library research and proof reading; Jo Ann Berg, for preparing the index under huge time constraints; Nancy Myers, for editing down an extravagantly long early chapter; David Oliver for preparing the figures; Jaye Schlesinger for the cover illustration; and the staff of the University of Chicago Press for their excellent assistance throughout.

During the years of research and writing, the Residential College has provided a supportive context for interdisciplinary scholarship at the University of Michigan. I have benefited from the encouragement of its directors, John Mersereau and Herbert Eagle, and the interest of my colleagues. I also thank the School of the History and Philosophy of Science of the University of New South Wales for a renewing term in the Antipodes and the Department of Political Science for encouraging the more global dimensions of my interests over the past few years. My students in Ann Arbor and Sydney in courses on the nature of science, the history of biotechnology, and new approaches to global security have played an essential role in asking tough questions and insisting on going beyond easy answers.

For the funding that made the book possible, I am grateful to the Rockefeller Foundation and to the National Science Foundation for several years of support in the 1980s, and to the Office of the Vice President for Research and the Rackham Graduate School of the University of Michigan, for assistance that enabled me to bring the book to completion.

In addition to those I have already named, many friends have cheered me on in countless ways. But for them, I doubt that a manuscript would have emerged. Special thanks to Buzz Alexander, Dan Axelrod, Walter and Francelia Clark, Charles and Sally Doak, Ellen Elliott, Matthew Evangelista, Barbara Fields, Joan Filler, Cecilia Green, Margot Halsted, Rosemary and Thomas Hill, Nicolaus Mills, Eliana Moya-Raggio, Yvette Perfecto, Mitch and Carole Rycus, Bob Rohwer, Joan Ross, John Vandermeer, and John Wright.

I have left my greatest debt until last. My son Jonathan Wright, to whom this book is dedicated, grew up as it evolved, tolerated the vagaries of a distracted parent, and always believed in her project of attempting to understand the role of science in a turbulent world.

Ann Arbor, Michigan
March 19, 1994

ACKNOWLEDGMENTS

The introduction, chapter 2, and chapter 5 are substantially revised versions of previously published essays: "Science and the Social Warp: Writing the History of Genetic Engineering Policy," *Science, Technology, and Human Values* 18 (Winter 1993): 79–101; "Recombinant DNA Technology and Its Social Transformation, 1972–1982," *Osiris* 2 (1986): 303–60; "Molecular Biology or Molecular Politics? The Production of Scientific Consensus on the Hazards of Recombinant DNA Technology," *Social Studies of Science* 16 (1986): 593–620.

INTRODUCTION
Exploring the Boundary between Politics and Science

That humans could intervene in the genetic constitution of living things was almost uniformly hailed as a breathtaking development in the field of molecular biology, pregnant with possibilities both for science and for society. It was also observed that the advance marked the end of biology as an analytic form of inquiry and the beginning of a synthetic science. Equally momentous was the pause molecular biologists took to assess how genetic engineering should proceed: for researchers collectively to examine the future impact of a new field of science and technology and to attempt to forestall adverse effects was a nearly unprecedented event in the history of science. The making of policy during this unusual period is the subject of this book.

The broad contour of events that followed the first genetic engineering experiments in several California laboratories is well known. The demonstrated power to modify DNA—and thereby the organisms it controls—proved a magnet for worldwide attention. Because viruses and bacteria were the initial loci for the early experimentation, some of that attention quickly focused on the possible health and environmental hazards of the new techniques. Astronauts returning from the lifeless moon had just been sequestered to prevent any introduction of novel pathogens to the planet, and the newly created microorganisms unshielded by quarantine of some sort seemed potentially dangerous—to both scientists and the public. Broader concerns addressed the social and ethical implications of reshaping the genetic constitution of living things: By whom would this new technology be used? To what ends?

Following ground-laying discussions of these questions, held largely within the community of biomedical researchers in whose midst the techniques had emerged, a period of debate and policy development produced national controls for research and development in the United States, the United Kingdom, and most other countries where research was under way. These controls specified safety precautions designed to contain possible hazards.

The promulgated controls did not remain in place for long, however, despite intense public controversy, especially in the United States. Within a decade, fears about the dangers of recombinant DNA technology largely vanished. Within the scientific community, voices calling for caution in genetic engineering experimentation were all but swamped by those convinced that there was little or no reason to impede research; some scientists decried their own earlier apprehensions. For the general public, the controversy all but faded away. The burden of proof shifted from science and industry to show that their experiments and procedures were safe, to the public to show that they were dangerous. Controls grew progressively weaker, and policies of crisis intervention replaced the earlier policies of prevention, first in the United States and eventually in most other countries. At present, some twenty years after the origination of the techniques, most restrictions on research in genetic engineering and on industrial applications have been abandoned.

This book provides a fine-grained comparative analysis of the evolution of policy in the United Kingdom and the United States, two countries that were engaged in the field from the outset and were foremost in establishing controls. Part 1 examines the formation of the principal institutions and interests that played important roles in shaping genetic engineering policy. Chapter 1 examines the evolution of government institutions responsible for promoting and controlling science and technology, and of public and private interests in government policy. Chapter 2 examines the formation of new interests in the field of genetic engineering itself, as it was transformed in the 1970s and early 1980s from an academic area of research to one with strong corporate connections.

Part 2 analyzes the early events that led up to the establishment of government oversight in both countries: the decision of the committee assembled by molecular biologist Paul Berg to publicize its concerns about the new field, the assessment given by the British committee chaired by Lord Ashby, the historic Asilomar conference held in February 1975, the development of guidelines in the United States and a code of practice in the United Kingdom, the appointment of committees to implement these controls in each country, and the concurrent formation of a discourse concerning the nature of the problem and its solution.

Part 3 explores the response of the biomedical research community and other interested parties to the strong challenge mounted against the American policy in the U.S. Congress as well as in various communities. Chapter 5 examines scientific meetings and other contributions from scientists that were influential in producing a change not only in the discourse concerning the possible hazards of genetic engineering but also in the practice of assessing these hazards. Chapter 6 examines the response of the biomedical research sector to congressional intervention, the role of a new discourse about biohazards in justifying the positions taken, and the factors that ultimately derailed legislative efforts.

Part 4 investigates the implementation of policy in each country. At this stage (1976–78) American and British practice diverged. The National Institutes of Health (NIH) guidelines were voluntary and applied only to NIH grantees; they were implemented by an expert committee appointed by the NIH director. The British controls were eventually mandatory, were applied uniformly to all sectors, and were implemented by a broadly constituted committee—the Genetic Manipulation Advisory Group (GMAG)—with representation of the scientific, business, and labor sectors as well as of a vaguely defined "public." While the NIH committee moved quickly toward revision and relaxation of the original guidelines (chapter 7), GMAG's first two years were mainly taken up with applying the controls established in 1976 (chapter 8). Moreover, the discourses concerning hazards in the two countries diverged in an important way. While the American committee focused predominantly on hazards to surrounding communities, the British committee focused primarily on hazards within the workplace.

Part 5 examines the processes through which controls in both countries were weakened to the point of being virtually dismantled. In the United States in the period 1979–82, several major revisions, superimposed on a series of incremental decisions, produced a reversal of the legacy of the Asilomar conference, opening the entire spectrum of living things to genetic manipulation, with controls remaining for only a few limited classes of experiment (chapters 9 and 10). The United Kingdom, though lagging behind the United States, followed suit (chapter 11). Part 5 analyzes how this reversal was achieved in each country, how it was made politically acceptable to various attentive sectors, and how the discourse concerning the hazards of genetic engineering evolved in the process. The final chapter appraises the implications of the rise and fall of the controls, especially what this history reveals about the political and economic interests affecting the disposition of science in the late twentieth century.

Because attempts to anticipate and control the future consequences of science and technology are rare, the genetic engineering case deserves special attention: how and why did the scientific community collectively pause to debate the direction of biogenetic science and technology? And why and how was it decided to proceed with "business as usual?" How did a consensus within the field emerge, and how were policymakers persuaded? Since many scientists actually slowed down or even stopped their research entirely, the decisions taken in the United States and the United Kingdom that ultimately led to generally uncontrolled exploration also raise fundamental questions about the dynamics of the development of science and technology.

This book attempts to interpret and explain the evolution of British and American genetic engineering policies. Today, however, with a broad debate about "the objectivity question in history" in progress, such an attempt requires some preliminary justification.[1] In what sense, and according to what criteria, can a historical account claim to interpret and explain its subject matter?

In the 1960s, traditional empiricist beliefs that science progressively discovers the unknown and that scientific development is governed by an internal logic were broadly challenged, notably by Thomas Kuhn's *Structure of Scientific Revolutions*.[2] Rich historical work in the 1970s reinforced the Kuhnian challenge, revealing science as the product of specific communities characterized by shared frameworks of metaphysical assumptions, methodological norms, and conceptual commitments.[3] In contrast to traditional claims that science was autonomous, unaffected by its social and cultural context, science was shown to be contingent, as much a product of its social environment as an account of natural phenomena.

The fading of traditional empiricism also opened the study of scientific development to the winds of change blowing from other fields. In history, studies of the roles and experiences of women and ethnic minorities did not simply amplify the existing scholarly corpus to include hitherto missing dimensions but also challenged dominant assumptions about the writing of history, especially its assumption that historical truths were universal.[4] In both history and literature, poststructuralist theory undermined the traditional opposition between representation and reality, proclaiming the textuality of each and the impossibility of escaping the "prison-house of language."[5] In anthropology, a hermeneutic approach to knowledge rejected the empiricist distinction between experience and interpretation, contending that humans are "fundamentally self-interpreting and self-defining, living always in a cultural environment, inside a 'web of signification we ourselves have spun.'" There was, therefore, "no outside, detached standpoint from which [the analyst could] gather and present brute data." In attempting to understand culture or society, the analyst was limited to "interpretations and interpretations of interpretations."[6] Although differing in important respects, all of these intellectual currents broke with the ideal of history as a politically neutral, objective account of the past that assumed an absolute distinction between subject and object and aimed at universal knowledge.

In the intellectual climate of the 1970s and 1980s, the "black box" of science, previously seen by empiricists as an authoritative system of knowledge that developed independently of its social and cultural contexts, was opened to examination. Four main approaches to these new problems, incorporating social constructivist, ethnographic, radical, and feminist positions, were pursued.[7]

Social constructivism has sought to understand the nature and development of science and technology in terms of the professional—and occasionally the more general political and ideological—interests of scientists and engineers that are assumed to shape scientific and technological choices. This approach has resulted in an impressive variety of empirical studies addressing the effects of social interests on the development of science and on the closure of scientific controversies. The philosophical basis for this approach, known as the "strong program," asserted that all knowledge claims in science should be opened to historical and sociological investigation, regardless of their truth value.[8] A key assumption is that scientific results are underdetermined by evidence, with the consequence that there will always be considerable flexibility in their production, and hence scope for social negotiation. The goal of social constructivism, guided by the strong program, has been to "show why particular scientific accounts were produced and why particular evaluations were rendered . . . by displaying the historically contingent connections between knowledge and the concerns of various social groups in their intellectual and social settings."[9] This goal has also been extended to technology, by assuming a parallel flexibility in the design of technological artifacts and systems.[10] In sum, social constructivists have sought to explain the contents of the black box of science in terms of the concerns and interests of actors.

Ethnographic studies of science have focused on the culture of science, producing important studies of the laboratory as the site of construction of scientific knowledge, experiments and experimental practice, scientific discourse, and the mobilization of discursive and visual resources in the pursuit and promotion of science. The crucial move, distinguishing this approach from social constructivism, is its insistence on the reflexivity of knowledge claims: analysts of science must be seen as in society, not outside it, and their assumptions and concepts must be opened to the same kind of analysis as science itself. "Society" can be no more of a black box than "science."

In making this move, ethnographers of science have tended to shun "causes" and "interests," arguing that while analysts of science may follow scientists around, enter their laboratories, watch them at work, and listen to their conversations, they cannot explain why scientists make the choices they do or why controversies are settled in one way rather than another.[11] Such insistence on intractable particularity has left little room for generalized understanding beyond the heightened self-awareness of the observer—their protests to the contrary notwithstanding. And some have embraced an extreme form of phenomenalism, claiming that to talk of real natural or social worlds is to fool ourselves. On that view, the objects that we claim to encounter cannot survive that claim on them;

they dissolve into the verbal images constructed of them. In general, ethnographers have insisted that opening the black box of science does not provide a warrant for trying to explain what is found inside.

Independently of the social constructivist and ethnographic approaches but occasionally interacting with them, radical and feminist treatments of science and technology have sought to understand how the values and goals of capitalism on the one hand and of patriarchy on the other are expressed in the development and effects of science and technology. In addition to formulating programmatic goals,[12] radical approaches have studied science as an expression of class-based ideologies,[13] the formation of engineering under capitalism,[14] and the shaping of science policy by military, political, and economic interests.[15] Feminist analysts have produced an impressive literature examining the processes through which women have been excluded from science and analyzing the gendered character of its dominant metaphysical assumptions, methods, and concepts as well as the concept of gender itself.[16]

With some cross-fertilization from the other two approaches, social constructivism and ethnography have dominated the academic mainstream.[17] These positions, however, have left major questions concerning the disposition of the contents of the black box of science unanswered or even unasked. Social constructivists have rarely pressed the question of why some interests play influential roles in shaping science while others do not. Many social constructivists apparently assume that the social shaping of science and technology happens in a world in which interests compete on equal terms for influence in a democratic arena. By failing to register that some groups gain access to policy arenas while others are excluded, that some interests are recognized while others are ignored, their analysis is implicitly conservative, accepting the established arrangements for science policymaking as if these were natural and inevitable. Ethnographers, on the other hand, refuse to explain why science develops as it does. Either by rejecting explanations and causes or by ignoring bias in policymaking, both approaches embrace forms of political quietism, swerving from engaging the question of why science is goal-oriented and how its goals are defined.

In summary, while social constructivist and ethnographic approaches open the black box of science/technology, in general, both allow the black box of political decisions concerning science to stay closed. Consequently, both have produced strong temptations to focus on a single level of scientific activity rather than on relations between the microlevel of the scientist, the laboratory, and the experiment and the macrolevel of the state, the economy, and the institutions of science. While radical and feminist analysts recognize the importance of opening the policy box, so far their efforts to do this have been limited. Feminist analysts have concentrated on questions about the nature of the metaphysics, methods, concepts, and discourses of science rather than on social and political questions about development.[18] With some exceptions, radical science analysts, have had few resources to pursue broad political and social answers to the questions that they have raised about the roles of science and technology under capitalism.

In contrast to the approaches that now characterize the social studies of science, most other treatments of science still embrace traditional empiricist conceptions of its history; the assumption that science and its methods are isolated from social influence is deeply embedded in the natural sciences and the general culture. Science purportedly achieves truth, eliminates error, and disseminates objective understanding, and its history merely recounts that process.

The received view of the history of genetic engineering policy is a case in point. On this view, the problems posed by the new field are defined as essentially technical in nature, and the history of policy formation is portrayed as the history of how those problems were resolved. Thus, according to the retrospective account of Donald Fredrickson, the NIH director under whom genetic engineering controls were first developed and later dismantled, the main parameters of this history were "science and its practitioners," assumed to be responsible for achieving the understanding necessary to form a rational policy, and "the public," assumed to be immersed in a prevailing "antiscientific" culture. The dismantling of the American controls is portrayed as rationally directed by technical experts with privileged access to specialized knowledge in the face of resistance from an irrational public.[19] The genetic engineering episode was cast as a "war" between truth and error, rational analysis and irrational alarmism, in which rationality and truth ultimately won.

Considerable evidence concerning the evolution of genetic engineering policy throws such views into question, however. In the first place, the sociopolitical context of policymaking is far too important to be ignored. It is notable that nowhere in Fredrickson's account is there mention of the relevance of the social, economic, and political climate in which recombinant DNA policy evolved, or of the growing interests in industry, scientific, and government circles in developing the new field rapidly. Yet there is abundant evidence indicating that these affected the perceptions and actions of key participants. The received view does not attempt to explain why such evidence can be set aside.

To assume that the formation of genetic engineering policy was a dispassionate, politically neutral process of technical assessment excludes consideration of the complex motives affecting policymakers or of the possibility that policies were shaped by social and cultural conditions. A portrait of scientists, government representatives, and corporate executives objectively assessing the implications of recombinant DNA technology on purely technical grounds can be nothing more than a silhouette. Moreover, the received view takes the discourse about the issues posed by genetic engineering as self-evident: it cannot explain why the terms and categories of debate changed over time or why those in the United States and in the United Kingdom differed significantly.

Bringing Power Back In

Two main temptations are thus involved in writing the history of genetic engineering policy: the impulse of traditional empiricism to exclude social influences entirely, offering the writer the satisfaction of producing an account not rendered ambiguous or problematic by consideration of the (by no means clear and unambiguous) society in which policy was made; and the present impulse in the social studies of science to insist that both science and its history are shot through with social and cultural influences but to deny the possibility of a privileged view of either. The crucial element that these approaches exclude by their own theoretical premises is an investigation of the political economy of science—the power relations that affect the direction and growth of research and that crystallize in government policy. The received view, in claiming that policy rests on a technical logic, would deny that power relations are relevant; the broad tendency of contemporary science studies is to ignore them.

Yet genetic engineering—from its inception an obviously major scientific and technological resource—provoked in those responsible for its development and control actions and arguments that demand political and economic analysis. Examining this dimension of genetic engineering policy is crucial to explaining its development. The history of policy formation needs to be investigated in ways that are sensitive not only to technical and cultural characteristics of the issues raised by genetic engineering but also to the nature and operation of the power relations characterizing the actors.

An important challenge posed by the rich evidence associated with the evolution of the policies guiding genetic engineering is to develop a theoretical framework that not only encompasses what transpired at a particular level of analysis—for example, the state institutions responsible for science, legislatures, corporations, or scientists working at the bench—but also addresses the power relations among actors and reveals the political processes that linked those levels.

In the background of the choice of a theoretical framework are several issues concerning the nature, location, and operation of power, assumed in its most general form to be a capacity attributed either to human agents or to systems in which they act, to bring about effects on other actors (who, it is assumed, would act differently in its absence).[20] First, there is the question of who exercises power. Is power possessed by agents (either individuals or social groups) or is it an attribute of systems or structures? A voluntarist tradition that runs from Hobbes and Locke to Mills, Dahl, and Lukes assumes that the historical subject has an "ineradicable and perhaps crucial explanatory role"[21] that is rooted in the complementary ideas of human agency and human responsibility; in contrast, various forms of structuralism posit systemic relations as fundamental and in the limiting case, as determining human action. Debates in the 1950s between Mills and Parsons and in the 1970s between Marxists Miliband and Poulantzas exemplify these polarities.[22]

Some choices must be made but these do not need to be polarized. Since various social groups were visibly struggling over genetic engineering policy, it seems reasonable to assume that these actors were (voluntarily) pursuing their interests and that they did so within limits set by possible structural constraints. It also seems important to leave open the possibility that agents may act to change and modify structural constraints.

But the problem of identifying interests raises further theoretical issues.[23] There are three main conceptual moves to be considered. Behavioralists have argued that interests are only revealed through observable conflict. The problem with this position is that it is easy to think of situations in which actors do not pursue their interests on account of some constraint: they are deterred, they believe their action cannot affect a given situation, and so on.[24] Alternatively, some have proposed a realist position according to which interests are goals and values revealed under hypothetical conditions in which participants are unconstrained.[25] The problem here is reliance on a counterfactual condition that can never be directly observed and the meaning of which could generate endless debate. A third constructivist move—and the one adopted in this study— is to argue that interests should be determined by investigating their historical formation. This position assumes that interests are shaped by social and cultural circumstances, and that they may be defined independently of their expression in any given political arena by investigating their formation and the values, beliefs, and practices established during that process. Such an approach has been underutilized in political science and sociology, perhaps because, in requiring a historical study of interest formation, it entails crossing traditional disciplinary boundaries.

A second issue concerning power is the epistemological question of how power should be located—a question that gave rise to extended debate in the 1960s and 1970s. Pluralists assumed not only that power is distributed throughout democratic societies but also that its operation should be determined by investigating observable behavior, specifically, the decisions taken in formal policy arenas.[26] This position was countered by postpluralists Schattschneider, Bachrach, and Baratz, who argued that power in democratic societies was far more concentrated, operating most crucially and influentially out of sight of those who attend only to the formal policy process.[27] Postpluralists therefore argued that it was not sufficient to confine attention to concrete decisions made in the formal policy process, since that approach excludes any consideration of the informal processes taking place elsewhere, which might affect the scope of the issues placed on the formal agenda. These might ultimately be far more influential for policy formation than formal decisions are.

The nature of the debate between pluralists and postpluralists is nicely represented by Saunders in a diagram (fig. 1). If B wants A to address a demand, the matter may be taken up in a formal policy arena and either accepted or rejected (level I). Or A may take steps to ensure that the matter is kept off the formal agenda through a variety of tactics (level II). Or B may fail to articulate the demand because he or she anticipates that it will be rejected or ignored (level III). Finally, B may even fail to formulate his or her demand because A is able to influence or even determine B's very desires (level IV). Schattschneider memorably registered the nonneutrality of organizations at all of these levels: "All forms of political organization have a bias in favor of the exploitation of some kinds of conflict and the suppression of others, because *organization is the mobilization of bias*. Some issues are organized into politics while others are organized out."[28]

While pluralists held that only events at level I could be observed, postpluralists Bachrach and Baratz argued that what they called "nondecisions" at levels II, III, and IV were equally if not more important: "Of course power is exercised when A participates in the making of decisions that affect B [level I]. But power is also exercised when A devotes his energies to creating or reinforcing social and political values and institutional practices that limit the scope of the political process to public consideration of only those issues that are comparatively innocuous to A [levels II–IV]."[29] In other words, interests act and power is exercised not only in votes "on stage" but also in determining such matters as the selection of a policy arena, the appointment of decision makers, the organization of agendas, and the dissemination of decisions. Such actions, they claimed, might never be registered formally but could mark the outcome significantly. Thus, they represented a second face of power that had been ignored by pluralists.

Pluralists responded by arguing positivistically that, because only decisions in a formal arena were observable, the second face of power did not exist.[30] Bachrach and Baratz defended their position by claiming that

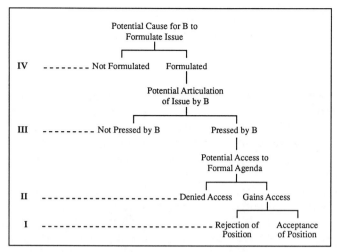

Fig. 1 Levels of non–decision making. Adapted from Peter Saunders, *Urban Politics: A Sociological Interpretation* (London: Hutchinson, 1979), 29.

nondecisions *could* be observed whenever there was conflict, since this would allow covert as well as overt grievances to surface.[31] For Lukes, such a defense conceded far too much epistemological ground. To assume that conflict was necessary to register the operation of power neglected its "most effective and insidious use," namely, to prevent conflict from ever arising. According to Lukes, "to assume that the absence of grievance equals genuine consensus is simply to rule out the possibility of false or manipulated consensus by definitional fiat." The most fundamental operation of power is to manipulate "perceptions, cognitions and preferences" in such a way that people fail even to formulate demands.[32]

This study adopts a framework that is both postpluralist and constructivist in orientation. I assume that policy arenas generally bear the imprint of their creators in the form of structural bias that unevenly distributes influence, access, and control of the agenda, and that the effects of bias may be investigated at all the levels of policy formation identified above. Moreover, in contrast to the pluralist position, I assume that interests shaping the policy arena can be defined independently of those effects, by examining their historical formation. This approach can be used to elucidate the most problematic level of "non–decision making" (level IV) identified by postpluralists—the shaping of values and preferences so that some issues are not even formulated. Lukes notes that "bias . . . is not sustained simply by a series of individually chosen acts, but also, most importantly, by the socially structured and culturally patterned behaviour of groups, and practices of institutions."[33] How such behaviors and practices are related to the historical formation of interests is an important concern of this book.

Power and Discourse

Postpluralist analyses of power opened up the examination of policy formation to a wide range of evidence concerning interests and their effects on the policy arena. However, they largely omitted consideration of a third issue: the relation between power and language. It is here that the linguistic turn in general, and the analysis of poststructuralist Michel Foucault in particular, is most valuable.

Foucault placed "discursive practice" at the center of his analysis of social systems, arguing that such practices embody power by supporting a "normalizing gaze"—a socially defined system of rules that permits certain statements to be made, orders these statements, and allows us to identify some statements as true, others as false, and still others as irrelevant.[34] His achievement was to demonstrate the organic, systemic relations between discursive practices, disciplinary techniques, and social institutions, challenging in the process the liberal separation of truth and power. As he expressed this challenge, "truth isn't outside power or lacking in power: . . . truth isn't the reward of free spirits, the child of protracted solitude, nor the privilege of those who have succeeded in liberating themselves. Truth is a thing of this world: it is produced only by virtue of multiple forms of constraint" (131). Consequently, "'truth' is linked in a circular relation with systems of power which produce and sustain it, and to effects of power which it induces and which extend it" (133).

In claiming that power achieves its effects through disciplinary practice, Foucault refocused the problem of power from the voluntarist concern with "regulated and legitimate forms of power in their central locations" to their "ultimate destinations with those points where it becomes capillary [completely dispersed]" (96). In contrast to a voluntarist conception that associates power with actors or agents, Foucault argued that the analysis of power should not "concern itself with . . . conscious intention or decision," and that it should refrain from attempting to locate agents or sources of power. Instead, the analysis should work at the level of "on-going subjugation, at the level of those continuous and uninterrupted processes which subject our bodies, govern our gestures, dictate our behaviors, etc." (97).

Within the sociology of science, a move rather parallel to that of Foucault was made by Callon, Latour, and others in transferring the focus of analysis from the interests of scientists and of other agents to "networks" or "actor-worlds"—alliances of human and nonhuman elements such as

scientific and technical institutions, scientists, and the objects and procedures that scientists create. The development of science and technology
is conceived as the process of formation of such networks, in particular
the formation of relatively stable ensembles of procedures, instruments,
theories, results, and products to which various actors give their allegiance. Like Foucault's approach to discursive practice, this approach dis-
solves voluntarist conceptions of interests into merely the "temporarily
stabilized outcomes of previous processes of enrollment."[35]

But the poststructuralist move to place disembodied discursive practices (Foucault) or networks (Callon and Latour) at the center of analysis
tends to obscure if not obliterate questions of agency and cause. This
move has occasioned vigorous criticism from those who wish to preserve
clarity both in normative issues concerning the use of power and in explanation of development and change. Nancy Hartsock argues that Foucault's exclusive emphasis on the "capillary" nature of power—the sense
that power is dispersed through networks and hence is everywhere—
makes any concept of accountable exercise of power disappear. Hartsock
argues further that, in replacing the traditional idea of sovereign power
with the conception of power as existing in local material institutions,
methods that analyze the effects of power exerted by large institutions
are also replaced by methods that focus exclusively on local, individual
levels: "Power [for Foucault] is everywhere and so ultimately nowhere."
Thus Foucault ultimately produces a disorienting sense of ungroundedness: in rejecting attempts to define the sources of power or reasons for
transformations of discursive practice, he finally "stands on no ground
at all."[36]

Steven Shapin has pursued a roughly parallel line of argument with
respect to Latour's dismissal of interests and rejection of explanation as a
goal in accounts of the development of science. Shapin notes that, although Latour appears to ban interests, they are assumed in the background of his account, since Latour's "technoscience" is goal-directed.
But why science should be goal-oriented and who is generating and defining these goals is not addressed. Shapin, like Hartsock, is uneasy with
the ungroundedness of Latour's position: "This is a world in which anything and anybody can be an actant or an actor. . . . It is the world of the
seamless web, a world in which everything is connected to everything
else. . . . Ultimately, those that truly inhabit the seamless web can say
nothing intelligible about its nature, even, if they are consistent, that it is
seamless and that it is a web."[37]

Such criticisms do not mean, however, that the poststructuralist conception of the dispersal of power through discursive practices is necessarily incompatible with a voluntarist concept of agency. There is no reason
not to assume that power may be expressed at the center as well as at the

periphery, by agents as well as through institutions and discursive practices. The historical problem of revealing the processes through which power is diffused—from the center to the periphery and possibly vice versa—should not be foreclosed. Furthermore, the postpluralist treatment of power can yield important methods for analyzing those processes by opening the achievement of structural bias to examination. A key component of this analysis, missed by postpluralists, is the role of discursive practices in establishing and consolidating structural bias.

An important purpose of this study is to reveal the links between power, structural bias, and discursive practice in the evolution of genetic engineering policy. Discourse, understood in a Foucauldian sense, is treated (in a non-Foucauldian, causal manner) as linked not only to preferred practice but also to the power to control the policy arena. The analysis below draws attention to the effects of the choice of discourse and to the close relation between discourse and practice at all levels of policymaking (fig. 1) but in contrast to a poststructuralist analysis, it does so in relation to the evolution of interests in shaping the policy arena. The process of establishing a specific discourse may be expected to reveal much about the politics of a given policy arena because discourses achieve currency for political reasons.

The techniques described above are used throughout this book to show how interests were formed, policy arenas selected, participants chosen, scopes of decision making restricted, and results of proceedings disseminated—how, to use Schattschneider's words, "bias" was "mobilized" as the result of the operation of certain interests and the exclusion of others. In particular, these methods are applied to the intriguing problem of why the controversy that surrounded the question of genetic engineering hazards in the United States disappeared so rapidly. Intensely debated in 1975–77, the hazard question was almost a nonissue by 1979. It will be shown how interests acted to reduce the virtually unlimited complexity of the hazard problem, to change the discursive practices associated not only with hazards but also with hazard assessment, and ultimately to achieve closure of the controversy (chapters 5 and 6). This approach investigates the fine structure of the operation of interests in genetic engineering and, in doing so, aims to establish the links between the mechanisms that limited and ultimately closed debate on the genetic engineering controversy and the broad social and political context in which the debate occurred.

In summary, an examination of the roles of interests in shaping policy requires four interrelated levels of analysis: the social and economic contexts in which governmental and sectoral interests in the field were formed; the formation of those interests, including their characteristic beliefs, commitments, and practices; the operation and expression of those interests in the development of policy; and the evolution of discursive practices in contributing to and being shaped by policy struggles and negotiations.

One further methodological point is important. In this study, comparative analysis of policy formation in two countries that developed distinct systems of government control for genetic engineering is used for two main reasons: first, because differences in either conception or implementation underscore the arbitrary nature of decisions that might otherwise be seen as natural or logical; second, because interactions between two national systems provide important indications of the operation of transnational influences (especially, in this case, the influence of corporations and international scientific organizations and the interests of national governments in supporting the ability of their scientists and corporations to compete internationally).

Underdetermined but Not Unconstrained

To return to the question raised initially about the possibility of historical interpretation and explanation: this treatment of the development of genetic engineering policy is located neither in the traditional empiricist camp of those who see history as simply "uncovering" the past nor in the various poststructuralist camps of those who insist that texts are all we can know. History reduces neither to the sum of the facts nor to the sum of the texts. The empiricist position is unsatisfying because it does not respond to the major strength of poststructuralist criticism that "human inquiry is necessarily engaged in understanding the human world from within a specific situation" and that "this situation is always and at once historical, moral, and political."[38] Poststructuralist positions, on the other hand, while they have been responsible for developing that understanding in many ways, have often turned away from the project of accounting for development and change.

The position taken in this study is that, although experience is never unambiguous (contrary to the positivists)—that is, events never speak for themselves—historical interpretation is constrained in various ways. First, the general purposes and interests of the historian map the terrain of events in different ways, depending on her purpose and focus.[39] For example, an account of the unfolding of recombinant DNA technology that focused on the experiences of the scientists involved would tell a different (although not necessarily conflicting) story from the one told here.

Second, historical interpretation is further constrained by the methods of inquiry that it adopts. These may be more or less constricting, more or less open. For example, as noted above, to account for the formation of genetic engineering policy on merely technical grounds excludes the whole social realm from investigation. Similarly, an account that presupposed that a specific social group—biomedical researchers, or corporate executives, or government bureaucrats—dominated policy formation would close off other possibilities from investigation.

Finally, historical interpretation is constrained by the cumulative impact of the evidence on which it draws. Events and discourses, what happened or what was said, limit in essential ways the historian's analysis. Actions cumulate and discourses persist in ways that cannot be ignored. It is here that traditional criteria for evaluating history—consistency, inclusiveness, accuracy, attention to context—come into play.

Thus the objects of history, while they may be underdetermined, are never unconstrained. Further, they are accommodated to both the political values and commitments of their examiners and the cumulative impact of evidence. As such, historical interpretation enters a larger process of debate and dialogue in which its claims and the relations, categories, and values it uses in establishing them are tested against competing positions.[40] Thus one of the essential needs of a democratic society is to ensure the openness of that larger debate. As Joan Scott has written in connection with the pluralist and poststructuralist challenges to objectivism in history,

> Written history both reflects and creates relations of power. Its standards of inclusion and exclusion, measures of importance, and rules of evaluation are not objective criteria but politically produced conventions. . . . [H]istory is inherently political. There is no single standard by which we can identify 'true' historical knowledge, however well-trained we have been in graduate seminars on methods and historiography. Rather, there are contests, more or less conflictual, more or less explicit, about the substance, uses, and meanings of the knowledge we call history.[41]

My purpose in the following account is to illustrate how the evolution of recombinant DNA policy can be interpreted and explained through contesting readings of historical evidence, sensitive to the categories of analysis, the scope of interpretation, and the evidence to be encompassed.

PART ONE
Institutions and Interests

ONE

Social Interests in Promoting and Controlling Science and Technology

Recombinant DNA technology emerged during a period that may be seen as a watershed with respect to societal promotion and control of science and technology in both Britain and the United States. After World War II, the science policies of each country evolved through three roughly parallel phases. In the first phase, long periods of sustained growth in government funding of research and development were followed by a tightening of government expenditures in the late 1960s and early 1970s. In the second phase, starting in the mid-1960s and ending in the 1970s, concern on the part of various sectors of the public about the impact of science and technology on the environment, the workplace, and consumers resulted in challenges to unregulated technological development and in broadly supported efforts to control and even to anticipate the impact of new technology. The third phase, beginning in the 1970s and continuing into the 1980s, saw a broad reversal of the regulatory impulse and vigorous efforts to direct scientific research toward industrial application.

In the background, and casting a crucial influence on these developments, were the changing economic fortunes and interactions of the advanced capitalist countries: the long postwar economic boom, the increasing globalization of capital and integration of world markets, and the deepening economic strains of the 1970s and 1980s. The expansion of the American and British economies (along with those of all other capitalist countries) in the two decades following the Second World War provided the base for the huge growth in scientific research and development in that period, with the resulting emergence of "big" science; acceleration in technological innovation; spiralling gains in productivity and the development of the powerful postwar technologies; the ever more powerful weapons of mass destruction, nuclear power, and synthetic chemicals.

The parallel declines of the American and British economies that occurred in the 1970s and 1980s also influenced important changes in science policy. The eclipse of American domination of the global economy in the 1960s and 1970s, experienced in terms of persistent inflation, high levels of unemployment, recession, and growing trade deficits, produced a tightening of government expenditure for scientific research as well as an influential campaign on the part of the private sector against regulation. At the same time, the progressive breakdown of British manufacturing capacity and the erosion of British markets (reflected in the lurching of the British economy from crisis to crisis throughout the 1950s, 1960s, and 1970s) produced far more serious constraints on British government support for science.

Finally, the major economic problems faced by both countries in the 1980s—the closures, bankruptcies, and deindustrialization of the United Kingdom, accelerated by the savage monetary policies of the Thatcher government, and the sluggish economic growth and increasing national debt of the United States—contributed to a transition to more selective support for science, deregulation, privatization, and the formation of close ties between academic science and private industry. In addition, the uncertainties accompanying the generally declining rate of profit increased pressures to exploit technological advances. The foreign-trade imbalances troubling both nations also meant that these commercial pressures manifested themselves regularly in calls for the maintenance of each nation's leadership in scientific research and intensified international competitiveness with respect to new technology.

The purpose of this chapter is to examine the formation, in these changing economic contexts, of institutions and interests that played major roles in shaping genetic engineering policy in the United States and the United Kingdom. Of particular importance are, first, those institutions that, in the postwar boom provided abundant funding (especially in the United States) for biomedical research, producing specific social relations between government patrons and scientist clients; second, the institutions formed in the late 1960s and early 1970s to protect society and the environment from undesirable effects of technological development; third, the commercial, financial, and scientific interests that reacted to the earlier regulatory impulse, producing new alignments among science, industry, and government in the late 1970s and early 1980s. A major purpose of this book is to show how these institutions and interests shaped the policies for the control of recombinant DNA technology that evolved in each country.

1.1 Expansion of Government Support for Science,
1945 to the Late 1960s: The United States

From the end of the Second World War until the late 1960s, all of the capitalist countries participated in a process of sustained and rapid economic expansion dominated by the United States, with its huge economy and advanced technological base intact at the end of the war. The growing resources of the U.S. government made possible enormous changes both in the sponsorship and in the scale of scientific research and development.

Before World War II, government support for research was modest and largely limited to government institutions. The universities were mainly independent of government support and acted as autonomous, self-regulating entities, competing with one another for personnel and resources. In 1907, the American university was described as a "voluntary cooperative association of highly individualistic persons for teaching and for advancing knowledge."[1] Research was seen as a "relatively minor though important adjunct" to the process of higher education; research funds were limited, coming largely from private foundations and general operating funds.[2] For the biological sciences, the Rockefeller Foundation played a highly influential role, initiating and conceptualizing a program of research that shifted attention to the molecular level.[3]

The patronage of scientific research underwent substantial changes during and after World War II. During the war, the nation's scientific and technical expertise was mobilized to serve military and defense needs. War-related research in the universities, funded through contracts awarded by the Office of Scientific Research and Development, produced a new alignment between research in the universities and the goals of the federal government. Scientists were drawn into science policymaking at the highest levels of government. The effectiveness of generously funded government research programs was demonstrated by the results: the atomic bomb, radar, the proximity fuze, penicillin. Whether it was science that won the war is highly debatable, but the *perception* of science as a crucial factor in the nation's power and prestige was firmly established by the war effort.[4] The nation "emerged from the war disposed toward generous support of science and its institutions."[5]

In the years immediately following the war, the paradigm of government-sponsored, mission-oriented research became firmly entrenched, dislodging and marginalizing the prewar roles of private foundations as patrons of science.[6] The Department of Defense (DoD), the Atomic Energy Commission, and the National Aeronautics and Space Administration (NASA) assumed responsibility for sponsoring research in the physical sciences.[7] In a move described by one of its former directors, James Shannon, as "the most influential and far-reaching in the history of federal support for university research," the NIH, created in 1930, took responsibility for all the remaining wartime contracts for biomedical research, thus becoming the primary funding agency for the biological sciences.[8] By 1950, five new disease-oriented institutes had been established, and broad legislation empowered the Surgeon General to establish further institutes.[9] Only the National Science Foundation (NSF), established in 1950, diverged from the pattern of linking research goals to perceived national needs.[10]

The expanding postwar economy supported spectacular growth in the federal science budget. In 1946–56, federal research and development expenditures grew from $917.8 million to $3.45 billion, an increase in real terms of 166 percent; in the following decade, research and development expenditures grew even more rapidly, to $16 billion, an increase in real terms of 282 percent.[11] Generally, this growth reflected changing national priorities. Initially, the principal impetus for growth came from the intensification of the Cold War and the growing momentum of an arms race, charged by the discovery of 1949 that the Soviet Union was developing its own nuclear weapons. As Alex Roland has observed, the arms race "proved more important for science, for it opened the government purse wider than ever before in peacetime and set off the mad scramble for weaponry that President Eisenhower would come to call the military-industrial complex."[12] The launching of Sputnik in 1957 symbolized the growing threat to the scientific and technological underpinnings of U.S. military superiority. The overall result of Sputnik, therefore, was to accelerate the mobilization of science behind an all-out effort to protect national security. The growth of federal funding of science continued on its steep upward curve (figs. 1.1 and 1.2).

While the requirements of the arms race and the space race dominated the picture (together they accounted for 55.7 percent of the federal expenditure for basic and applied research in 1966), federal support for biomedical research also grew vigorously in this period, powerfully shaped by several political forces. Improvement of the nation's health was widely perceived as an important need and strongly endorsed by President Truman. The main components of Truman's program—national health insurance, medical education, hospital construction, and expansion of biomedical research—were radically altered by two powerful lobbies. In the first place, a strong private lobby headed by philanthropist Mary Lasker pressed for increased funding for research aimed at the cure of certain diseases, especially cancer. At the same time, the American Medical Association (AMA) mounted a multimillion dollar campaign against

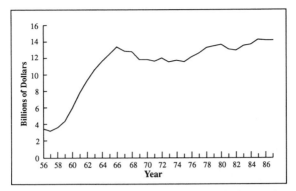

Fig. 1.1 Total U.S. government obligations for basic and applied research, 1956–87, constant 1982 dollars (NSF fiscal year GNP deflator). From National Science Foundation, *Federal Funds for Research and Development: Detailed Historical Tables, Fiscal Years 1955–1987* (n.d.).

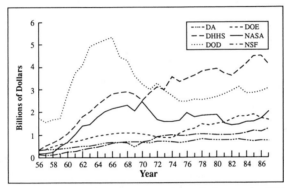

Fig. 1.2 U.S. government obligations for basic and applied research, 1956–87, constant 1982 dollars, by agency (NSF fiscal year GNP deflator). From National Science Foundation, *Federal Funds for Research and Development: Detailed Historical Tables, Fiscal Years 1955–1987* (n.d.).

the health insurance, medical education, and hospital construction proposals in Truman's program.[13]

Spurred by the cancer lobby, pushed by the powerful chairs of the health appropriations committees with the support of the NIH leadership, and deterred from doing anything else with health dollars by the AMA, Congress voted ever-increasing appropriations for biomedical research.[14] As a former NIH administrator noted, "Support of research was the only respectable way in which legislators could simultaneously respond to the desire of people to do something about disease and their aversion to anything smacking of—to use a quaint phrase—'socialized

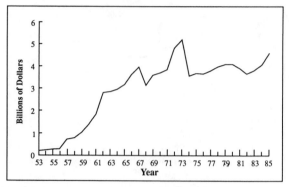

Fig. 1.3 NIH appropriations, 1953–85, constant 1982 dollars (NSF fiscal year GNP deflator). From Department of Health and Human Services, National Institutes of Health, *NIH
Almanac, 1986–1987* (Bethesda, Md.: NIH Office of Information, n.d.).

medicine.'"[15] For over two decades, the NIH budget grew roughly
exponentially, from $2.8 million in 1945 to over $1 billion dollars
(over $3 billion in 1982 dollars) in 1968 (fig. 1.3). Biomedical research
funding also represented an increasing share of the total federal budget for research, growing in the period 1956–66 from 8.8 percent to
21 percent. In 1961, the Department of Health, Education, and Welfare
(DHEW) assumed the lead role for federal support of basic research in
universities.[16]

The long period of rapidly expanding American support for science
came to an end in the late 1960s for several reasons. Most important, the war in Vietnam drained dollars away from all domestic programs. In addition, the funding of President Lyndon Johnson's Great
Society programs in housing, health, and education competed with other
components of the federal budget. By 1963, the total U.S. research and
development budget had reached almost $12.5 billion and was growing
at roughly 20 percent per year.[17] Continuing such a rate of growth
would have swamped the federal budget. Both Congress and the administration began to question whether the nation was seeing an adequate
return on its ample investment in science. In addition, congressional
committees began to probe federal research programs for unnecessary
duplication of research and other forms of waste in federal programs.
Neither the belief that basic research was the underpinning of technological development nor the policy that scientists alone should direct
spending of federal funds survived this scrutiny. "Accountability" became the new rallying cry.[18]

Biomedical research in particular received sustained scrutiny from politicians. Hearings presided over by Congressman L. H. Fountain created
a climate of criticism and a focus on extravagance in research, raising

questions about the adequacy of peer review for ensuring that funds were used effectively.[19] President Johnson, spurred by the Lasker lobby, increased the pressure and, in a direct confrontation with researchers' claims for the fruitfulness of undirected, "pure" research, sounded a clarion call for concrete results. In July 1966, the president asked the assembled directors of the NIH, along with the Surgeon General and the Secretary of Health, Education, and Welfare, whether "too much energy was being spent on basic research and not enough on translating laboratory findings into tangible benefits for the American people."[20] After uproar in the scientific community, Johnson helicoptered to Bethesda to praise the NIH for its "billion dollar success story."[21] Despite that gesture, the tide had turned. Federal support for biomedical research leveled off, and the NIH entered what a later NIH director called an era of "selective growth."[22]

The campaign to make biomedical research accountable continued under Richard Nixon. Nixon, under the influence of close friends in the cancer lobby and well aware that cancer was fast achieving visibility as a public issue, endorsed the idea of targeted research. Spurred by the report of a blue-ribbon panel appointed by members of Congress, which came out in favor of an all-out attack on "the implacable foe," Nixon, in his State of the Union address in January 1971, announced a "moon-shot for cancer": the concentrated effort that had "split the atom and [taken] men to the moon" would now be "turned toward conquering this dread disease." After considerable maneuvering in Congress to secure leadership with respect to the cancer issue, key members of Congress— Edward Kennedy, who became chair of the health subcommittee of the Senate Committee on Labor and Public Welfare, and Paul Rogers, Kennedy's counterpart in the House—eventually endorsed legislation. In 1971, Congress passed the National Cancer Act, giving the National Cancer Institute considerable autonomy, making its directorship a presidential appointment, and establishing a new, presidentially appointed advisory board. The federal coffers for biomedical research were opened once again—this time, however, selectively.[23]

These political pressures were reflected in trends in federal support for science. When the long upward trend in total federal funding of research came to a sudden halt in 1966, cuts in spending were most keenly felt by the DoD and NASA (fig. 1.2). Of the leading federal agencies responsible for research, only support for the DHEW and the NSF ran counter to the overall trend. The increase in DHEW support, mainly reflecting the NIH budget, was the most dramatic. From 1973 onward, the DHEW became the leading federal sponsor of research (fig. 1.2). The agency's support for research in the universities far surpassed that of any other source (fig. 1.4).[24]

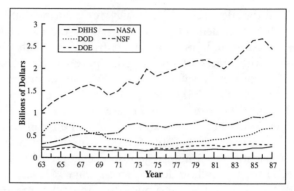

Fig. 1.4 U.S. government obligations for basic and applied research in the universities, 1963–87, constant 1982 dollars, by agency (NSF fiscal year GNP deflator). From National Science Foundation, *Federal Funds for Research, Development, and Other Scientific Activities*, vols. 13–15; National Science Foundation, *Federal Funds for Research and Development: Detailed Historical Tables, Fiscal Years 1955–1987* (n.d.).

Impact of Postwar Federal Support on the Organization and Patronage of Scientific Research

The massive infusion of federal funds into scientific research and the subsequent leveling-off of funding had far-reaching effects on the organization and nature of scientific research in general and biomedical research in particular. Perhaps the most important effect was that the government replaced the university as the leading patron of research. As federal support for research grew in importance, decisions on priorities for research became increasingly centralized under the control of the principal government agencies responsible for support: the DoD, NASA, DHEW, and NSF.

In the biomedical area, the great influx of federal funds in the 1950s and 1960s meant that the NIH, as the immediate donor of funds, emerged as the center of a vast research system unmatched in size and scope throughout the world. The funds that flowed so freely into the nation's universities and medical schools in the 1950s and 1960s, in addition to supporting research, provided thousands of faculty members with released time from teaching duties, new facilities and equipment, training programs for pre- and postdoctoral students, and career-enrichment programs for junior faculty members. Between 1950 and 1967, NIH research grant appropriations grew from $6.4 to $225 million.[25] The impact was felt in terms of a rapid expansion of life science departments and medical schools; new buildings, facilities, and equipment; and a growing supply of graduate and postdoctoral students. By 1967, when NIH appropriations peaked temporarily after twenty years of rapid growth, NIH funds were supporting 67,000 senior research investigators, academic programs and projects in more than two thousand universities and medical schools, and advanced training in basic science and clinical specialties for more than thirty-five thousand students.[26] In the medical schools, production of Ph.D.'s soared, while production of M.D.s increased only modestly. As Strickland notes in his detailed study of biomedical research policy after World War II, federal funds were used rather precisely for the purposes for which they were intended.[27]

A major feature of American science in the 1960s, then, was a biomedical research network with components in every substantial university and medical school in the country, with the NIH at its center. Without doubt, the scientific standards achieved under NIH patronage were high. Studies of NIH, even those inclined to be critical, uniformly concluded that the scientific achievements of the research carried out under NIH sponsorship were great. Nevertheless, as a government agency existing by virtue of appropriations decided jointly by the president and Congress, the NIH's patronage of science was certainly subject, like that of other federal agencies, to economic and political pressures.

One source of pressure on the agency was reflected in the terms of its mission. The legislation that established the original institute in 1930 stated its purpose as "study, investigation, and research into the fundamental problem of the diseases of man." The prevailing view within the biomedical research community was that its primary mission was the development of a scientific basis for modern medicine through progress in understanding basic biological processes.[28] On the other hand, politicians mindful of the political mileage to be made from "breakthroughs" in the control of disease, and the pharmaceutical industry, hoping to be given new technologies and new products, emphasized the rapid achievement of cures of major diseases. Thus political support for biomedical research involved a strong utilitarian interest that was easily accommodated only as long as funding was generous and hopes for the conquest of disease through basic research ran high.

A second pressure on the NIH originated in the atmosphere of intense international rivalry in which the American scientific and technological effort evolved after the Second World War. As the authors of a survey of national research systems observed in 1968, "Almost all [U.S.] science and technology activities are part of planetary competition, and are expected to reinforce not only the country's military position but also its economic potential, social equilibrium, and international prestige."[29] The biomedical sciences, no less than other fields, were part of the international race for scientific supremacy. References to "American world leadership in biology" and to scientific successes that placed "the United States at the forefront of biomedical research in the entire world in record time" were characteristic of official reports on the NIH track record.[30]

Thus considerable pressure devolved on the NIH leadership: failure to capitalize quickly on a promising new technique entailed loss of prestige both directly and indirectly through a "biological brain drain." And that, in turn, involved the possibility that the president and Congress would initiate changes in personnel at the top of the NIH hierarchy.

In summary, research priorities of the NIH were shaped, first, by the need to demonstrate the utility of biomedical research to the political community, and second, by the need to maintain its global leadership in the biomedical area. Inevitably, these pressures were transmitted from the organizational center of the system to the universities and scientists at the periphery. When controversy in the 1970s threatened to limit or abrogate biogenetic research, the scientists involved founded their arguments against restriction of their work on precisely these two grounds of practical utility and international competition.

Impact of Postwar Federal Support for Research on Universities

As a result of the large postwar influx of federal funds for research, university incomes from this source grew substantially, and research acquired a major and often dominant role in the universities.[31] But although the universities flourished financially during this period, their earlier independence and autonomy were substantially eroded. In the new order, funds for research went not to the university or department head but to the individual researcher, in the form of grants for specific projects. The allocation of funds was determined by direct negotiation between the researcher and an agency. Projects were screened and evaluated by review panels generally composed exclusively of scientists in the same field as that of the project under consideration. Thus research directions came to be determined principally by federal agencies and leading researchers on review panels rather than by universities. And the primary ties of university scientists shifted from their departments and universities to their peer groups across the country who would have the greatest influence in the evaluation of their work.[32] In the words of a report from the Organization for Economic Cooperation and Development, the price of the university's prestige and flow of material resources was that it became "the medium for a scientific activity whose development it did not control."[33]

The nature of the constraint became clearer when the overall decline in federal funding in the late 1960s pinched the universities. Federal support for basic and applied research leveled off in real terms after 1967 and did not begin to rise again until the late 1970s (fig. 1.5). The leading research universities that had grown so dependent on federal support in the 1950s and early 1960s were particularly affected.[34] As a result of this downward

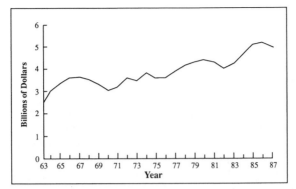

Fig. 1.5 Total U.S. government obligations for basic and applied research in the universities, 1963–87, constant 1982 dollars (NSF fiscal year GNP deflator). From National Science Foundation, *Federal Funds for Research, Development, and Other Scientific Activities*, vols. 13–15; National Science Foundation, *Federal Funds for Research and Development: Detailed Historical Tables, Fiscal Years 1955–1987* (n.d.)

trend in federal support, university administrations sought alternative sources of funding and emphasized research areas for which federal funding continued to grow. The biomedical sciences were one such area. Although biomedical research funds no longer flowed as freely as they had in the period before 1968, growth continued, particularly in targeted areas such as cancer research. Between 1967 and 1974, NIH support for basic and applied research in the universities grew by 38 percent, to $0.9 billion.[35]

These funds now came with strings attached, however. As politicians began to demand a payoff from basic research and the need to justify additional increments in the NIH budget became more acute, the utilitarian dimension of biomedical research support became increasingly influential,[36] and undoubtedly this emphasis filtered through to university administrators whose resources were strained by cutbacks and who were eager to cultivate new sources of support. Interest shifted from research areas that had no immediate connection with practical applications to areas where the potential for practical application was clear. Genetic engineering—a field founded on advances in molecular biology and funded by the NIH, and one for which the practical implications were immediately evident (chapter 2)—was likely to be selected for special attention.

Impact of Postwar Federal Support for Research on Scientists

Patterns of federal funding of science also had important consequences for the nature and practice of research. The conventional wisdom of the 1960s was that the American system of support, based on review by peers who were members of the national review panels organized by each government agency, was fairer and more objective than systems in which the distribution of support occurred at the institutional level and that the system fostered a spirit of freedom in scientific inquiry.[37] But while American science was not rigidly constrained, it nevertheless was limited in less obvious ways by the goals of the government agencies that supported it. Science journalist Daniel Greenberg noted, "Recognition of the freedom that does prevail in science [must be joined to] the realization that the laisser-faire system takes place within the bounds of an intricately constructed, subtly functioning system of government that, in effect, defines the possibilities of science by governing the availability and use of resources."[38] Specifically, the form of support for science in the United States affected not only the direction and goals of research but also the work roles of scientists, the criteria for recognition for their work, and even the nature of research, thereby imposing a subtle (and occasionally not so subtle) structure on their activities.

Most evident were changes in the organization of research, from the typically independent, small-scale work of the prewar era to increasingly complex projects that required, in the words of one observer, "access to large-scale, complex resources, diverse technical skills, and collaborative scientific relationships."[39] The changes were most striking in fields such as high-energy physics, but similar changes occurred in many of the life sciences, where it became common for professors to develop teams of doctoral and postdoctoral students and technicians to work on aspects of a large problem. This meant that, in addition to the qualities of rigor and creative problem-solving ability that were the traditional hallmarks of good science, scientists in the postwar era were required to develop the qualities of managers and entrepreneurs: the ability to spot talent in students, to foster team spirit, to promote a scientific project in terms attractive to government sponsors, and to maintain it thereafter by producing results on time.[40]

The switch to federal patronage of science also generated significant changes in the criteria for recognition of scientists' work. As the proportion of university research funded by the government grew, so did the importance attached by universities to that source of income. In addition to the quality of research, a new criterion for recognition was the ability to keep the scientific enterprise thriving by bringing in grants. By the 1960s, it was widely felt that scientists who were able to obtain research grants were more likely to be promoted than those who were not.[41]

Federal funding also affected the nature of research pursued under its sponsorship by encouraging a mixing of the attitudes, ethics, and goals of research at the "basic" and "applied" ends of the scientific spectrum.

The need to obtain federal support for research fostered an adaptation of basic science to the goals of the federal agencies. Cell and molecular biologists, for example, began to justify their work in terms of the national goal of finding a cure for cancer and other diseases. Some deliberately switched to the investigation of tumor-causing viruses, knowing that this work would be more readily funded. One observer has noted, "Scientists found it both interesting and financially advantageous to pursue their investigations of 'basic research' questions within a potentially applied context, or else to move back and forth quite regularly between 'basic' and 'applied' foci for their investigations." [42]

In addition, federal mechanisms for awarding grants created subtle but important shifts in the structure of research. Serial-competitive grant renewal practices tended to set limits on the kinds of research that scientists, particularly younger, less-established scientists, could safely undertake. It became important to select problems with a short-term payoff, or problems that could be broken into smaller units with payoffs that were clear in light of the larger problems being pursued. [43]

Generally, the earlier independence of researchers was replaced, after World War II, by the interdependency of the federal patron-client relationship. Scientists lost their earlier autonomy: the utilitarian goals of federal sponsors were now established in arenas largely beyond their control. But as recipients of federal support, and as members of the peer review panels that implemented this support, they still had considerable influence on how the general goals of the federal sponsor would be met. Furthermore, heads of federal agencies were, more often than not, former clients, drawn from the same scientific culture they were expected to support and sometimes to control. Federal agencies were able to exert substantial influence over the direction as well as the conditions of research (through the distribution of support and the sanction of its withdrawal), although they needed the collective acquiescence of their scientist clients in order to exercise such authority. Thus when questions of the control of research arose, scientist-clients turned readily to their federal patron as a "natural" arena for policymaking.

1.2 Expansion of Government Support for Science, 1945 to the Late 1960s: The United Kingdom

The United Kingdom, like the United States, participated in the postwar economic boom, although, as several analysts have shown, because of the location of much of its capital abroad and its attempt to maintain an international strategic role, the British domestic economy grew more slowly and more hesitantly than the economies of its major competitors. [44] Nevertheless, at least until the early 1960s, the expanding British economic base supported increasing government funding of research and development.

In contrast to the United States, where federal support for research and development was being strongly influenced by national priorities, in the United Kingdom, the autonomy of scientific research was protected by an entrenched system of support in which university funds for research were supplemented by grants awarded by government-supported research councils. Principal support for the universities after 1919 came from large block grants administered by the University Grants Committee (UGC). Composed mainly of representatives of the universities, the UGC was conceived not only as an organization responsible for dispensing funds but also as a buffer or shock absorber that would insulate the universities from direct government interference.[45] Through these funds, the universities themselves supplied basic support for their researchers. For the next fifty years, the UGC was the direct concern of the treasury, whose permanent civil servants provided a secretariat for the committee and generally shared the committee's belief in university freedom. Through Conservative and Labour governments alike, this administrative structure persisted—a circumstance that two analysts of the British university system attribute to the "extraordinary stability of the British system of elite recruitment to positions of political, industrial or bureaucratic power."[46]

More specific direction came from the Research Councils, whose principles of operation were established in 1918 by the Haldane report and, like those governing the operation of the UGC, persisted virtually unchanged until the 1960s. By then, four councils existed: the Department of Scientific and Industrial Research, coordinating major defense and civilian development efforts deemed important for the national interest, the Medical Research Council (MRC), the Agricultural Research Council (ARC) and the Nature Conservancy (NC), responsible for medical, agricultural, and environmental research. An Advisory Council for Scientific Policy (ACSP), composed largely of prominent researchers, in principle formulated overall research policy for the British government, but in practice the Research Councils firmly asserted their independence.[47]

This system oversaw steady increases in spending on research and development as the British economy shared in the postwar economic boom. Government support for research and development increased in real terms by 44 percent between 1955/56 and 1964/65, growing from £222 million to £427 million (£1.5 to £2 billion, in constant 1980 pounds).[48] As in the United States, the British government emphasized research and development for military purposes, which consumed 74 percent of the nation's total research and development support in 1958/59.

Within the civilian sector, much research spending backed glamorous, high-prestige products such as aircraft and nuclear power plants rather than chemicals, vehicles, or machinery (the products that were to prove so important for increasing productivity and expanding markets in the 1960s and 1970s).[49]

By the early 1960s, the costs of Britain's efforts to maintain itself as one of the world's leading powers were becoming evident. The stagnation of the country's older industries as well as evidence revealing its decline relative to other European economies triggered widespread reassessment of government policy for scientific and technological development, both within the incumbent Conservative government and among the major political parties.[50] Three main issues came to the fore in this debate. First, the great expense of ambitious military and civilian projects and the perception that military development did little to improve industrial productivity called official research priorities into question. Second, shortages of skilled personnel, especially in applied science and engineering, and a "brain drain" to the United States and other countries provoked calls to expand higher education, particularly in technical fields. Third, enhancing governmental control over science and technology policy was widely accepted as desirable.

Moves to achieve these goals were made by both the Conservative and the Labour governments of 1960s and early 1970s, each with its own emphasis. In December 1963, under Alec Douglas-Home's Conservative government, the UGC was transferred out of the Treasury and placed under the Minister of Science, Lord Hailsham.[51] From that point onward, the UGC not only financed the universities but also began to shape priorities for science and education. Harold Wilson's Labour government, elected in 1964 with a manifesto to forge links between science and socialism "in the white heat of [scientific] revolution,"[52] made other moves both to promote science and to control it. The science policy machinery was reorganized, with the Department of Education and Science given responsibility for education and civilian research, including the Research Councils and the UGC. Technology, on the other hand, came under the purview of a new Ministry of Technology, which included the Atomic Energy Authority, the National Research and Development Corporation, and industrial research establishments. Wilson's government also attempted to scale down spending on large defense and aerospace "prestige" projects, although institutional inertia allowed many of these to survive.[53]

The 1960s in the United Kingdom are often portrayed as a period of expansion in support for science, and in some important respects they were. During the 1960s, following a call for university expansion by a government committee (the Robbins committee), the universities grew

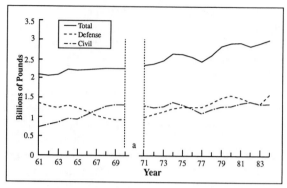

Fig. 1.6 U.K. government expenditures for research and development by the Research Councils and by the universities, 1961–84, constant 1980 pounds (Central Statistical Office deflator [*Economic Trends,* annual supplement, 1987]). From Central Statistical Office, *Research and Development: Expenditure and Employment,* Studies in Official Statistics, no. 21 (London: HMSO, 1973, tables 9, 9A; Central Statistical Office, *Research and Development: Expenditure and Employment,* Studies in Official Statistics, no. 27 (London: HMSO, 1976), table 6; *Economic Trends* no. 309 (July 1979), table 6, 111; no. 334 (August 1981), table 6, 102; no. 359 (September 1981), table 3, 114; no. 382 (August 1985), table 9, 89; no. 394 (August 1986), table 3, 84.

very rapidly. Eight new ones were created, and ten former colleges of advanced technology acquired university standing. The number of university teachers doubled, from 14,000 in 1960 to 30,000 in 1968.[54] Spending on research and development in the universities through the block grant to the UGC kept pace with this expansion, increasing by 95 percent in real terms between 1960/61 and 1969/70. Spending on the other main source of support for civilian research, the Research Councils, also expanded, growing by roughly 130 percent in real terms from £30 million in 1961/62 to £95 million in 1969/70 (fig. 1.6).

However, the United Kingdom's deep economic difficulties, the relentless erosion of markets for its manufactured goods, persistent balance-of-payment problems, and a "stop-go" economy bordering on crisis meant that the steady growth of the science budget of the 1950s could not be maintained.[55] Increased civilian research and development was achieved by the Labour government only through substantial cuts in military research and development (fig. 1.7). The growth in the British government's total support for research and development was to be substantially curbed in the 1960s, averaging less than 2 percent per year. A gloomy assessment by the Council for Scientific Policy in its first report in May 1966 invited the Research Councils to "consider their long term programmes in order to assess the effect of tapering-off of growth rates particularly in relation to new and scientifically desirable projects which might be excluded."[56] Hopes of salvaging "new and scientifically de-

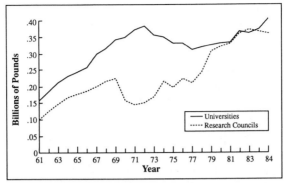

Fig. 1.7 Total U.K. government expenditures for research and development, and for military and civilian research and development, 1961–84, constant 1980 pounds (Central Statistical Office deflator [*Economic Trends,* annual supplement, 1987]). From Central Statistical Office, *Research and Development: Expenditure and Employment,* Studies in Official Statistics, no. 21 (London: HMSO, 1973), tables 9, 9A; Central Statistical Office, *Research and Development: Expenditure and Employment,* Studies in Official Statistics, no. 27 (London: HMSO, 1976), table 6; *Economic Trends* no. 309 (July 1979), table 6, 111; no. 334 (August 1981), table 6, 102; no. 359 (September 1981), table 3, 114; no. 382 (August 1985), table 9, 89; no. 394 (August 1986), table 3, 84.

sirable projects" were the smoldering embers of the "white heat of revolution."

In 1970, the boom in spending on civil research and development came to an end when Edward Heath's Conservative government revived military research and development and, to balance the budget, slowed the growth of civilian science. Government support for the Research Councils from 1969/70 to 1973/74 increased by only 2 percent in real terms (compared to 14 percent in the previous decade), while that for university research declined by 24 percent (fig. 1.6).

A major indication of the government's desire to achieve greater selectivity in the funding of research was a discussion paper issued in 1971 by Lord Rothschild, head of the Central Policy Review Staff, a government think tank providing a high-level review of government policy.[57] These proposals injected into British science policy a utilitarian emphasis on "accountability" that had been in the background of science policy discussions for some time and that opposed the traditional emphasis of the Research Councils on the protection of scientists' autonomy in research. The report proposed that much research previously initiated by scientists should be carried out on a customer-contract basis, the customer being the government department seeking certain results and the contractors being the scientists undertaking research. In the blunt words of the report, "The customer says what he wants; the contractor does it (if he can); and the customer pays." The traditional defense, that research initiated by scientists following their own interests was ultimately the most productive, was dismissed with the icy comment, "The country's needs are not so trivial as to be left to the mercies of a form of scientific roulette."[58]

The report raised howls of protest from scientists committed to "pure" research. Nevertheless, the central principle of accountability was accepted by the government, and a substantial portion of the Research Councils' budget was turned over to various ministries with the understanding that these funds would be used to commission research. Although it was subsequently argued that these changes made little difference in practice, particularly on research undertaken by the ARC and the MRC,[59] they certainly had a generalized impact on the policies of all the British agencies involved in the funding of research. They symbolized an end to an era of science policy committed to the principle of scientific autonomy and the onset of a new era in which the goals of research and development were directly defined by the state and progress toward their realization was used as a criterion for support.

In summary, in both the United Kingdom and the United States, the twenty years following World War II were marked by unprecedentedly high levels of government support for biomedical research. The distribution of these funds was at first largely under the control of semi-autonomous agencies—the UGC and the Research Councils in the United Kingdom, the NIH in the United States—in which scientists and academicians had considerable power. The UGC was in principle designed to protect scientists from government control, but the NIH had for a long time practically the same effect. When in the late 1960s the growth of government spending on scientific research was curtailed, the notion that the direction of such research was a matter of public policy and that the scientists involved were accountable for practical results came to the fore. In the United States particularly, this meant that various lobbying interests acquired considerable power in the funding-allocation process, but in both countries the ebbing of public spending meant that scientific research had increasingly to justify its support on utilitarian grounds. That research would provide practical benefits as well as theoretical advances became a regular feature of the justifications given by scientists for government support.

1.3 Reassessing Science and Technology, 1965–1975

At the same time that politicians began to demand accountability from scientists, both the effects and the direction of science and technology were challenged more vigorously than ever before. Growing numbers of people—including many scientists—began finding the social role and impact of modern science and technology increasingly problematic. This reassessment of science had two principal sources: the antiwar movement, which challenged the militarization of science, and growing public concern about consumer safety, environmental degradation, and occupational hazards, which brought into prominence a sense that the price of unrestrained technological progress was too high. What emerged from all of these movements was a strong perception that the vastly increased power to intervene in natural processes provided modern science and technology with vastly increased power to disrupt or destroy those processes. Public confidence in technological development as the key to social progress gave way to disenchantment. These challenges to accepted forms of deployment of science and technology took different forms in the United States and the United Kingdom but were influential in both countries.

The Antiwar Movement

The 1960s and 1970s saw widespread efforts among scientists to disengage science from the Vietnam War and from its use by the military in general. This movement was accompanied by the articulation of important challenges to traditional views of science as an autonomous, morally neutral pursuit whose fruits would provide the basis for social progress.

Of course, opposition to the militarization of science was not new. The bombing of Hiroshima and Nagasaki had touched the consciences of many scientists who shared with Robert Oppenheimer a deep sense of the loss of innocence of science.[60] In general, however, concerned scientists in the postwar years expressed their concern through established channels and dissented from government policy in ways that did not fundamentally challenge society's growing demand that scientific research produce knowledge for military purposes.

The majority of scientists involved in the movement against the Vietnam War seem to have acted in much the same spirit. Like the atomic scientists, they focused on questions concerning the use and abuse of science rather than on the economic and political structures that had evolved after the Second World War to support scientific research. At the same time, a more radical critique of science and its social role was developed by other scientists, particularly members of such new organizations as Scientists and Engineers for Social and Political Action (SESPA) in the United States, later renamed Science for the People, and the British Society for Social Responsibility in Science (BSSRS), both of which were committed to publicizing the misuse of science and technology and to promoting public involvement in science and technology policy.[61] These scientists doubted that the misuses of science they delineated were merely accidents or perversions and asked how these misuses were generated. The announcement of the formation of SESPA stated:

> We reject the old credo that "research means progress and progress is good." Reliance on such simplistic ethical codes has led to mistaken or even perverted uses of our scientific talents. As an antidote we shall establish a forum where all concerned scientists—and especially students and younger members of the profession—may explore the questions, Why are we scientists? For whose benefit do we work? What is the full measure of our moral and social responsibility?[62]

The attempt to answer those questions resulted in a shift of emphasis away from antiwar issues taken as isolated problems to the relations between those issues and the goals of the public and private institutions that supported scientific research, away from seeing the social responsibility of the scientist as one of publicizing and opposing misuses of scientific knowledge to a broader vision of working to change the social systems that appeared to cause such misuse. The BSSRS similarly moved from action on particular issues of use and abuse to an attack on the supposed neutrality of science and to a theoretical analysis of the role of science as a productive force and as a form of ideology, incorporating interpretations of the external world that legitimized dominant groups in society.[63] Members of both organizations began to ascribe the abuses of science to its management within political and economic systems whose goals were likely to be profit or social control rather than the public good. Harvard molecular geneticist Jonathan Beckwith and two colleagues warned of the dangers of misuse of their work in bacterial genetics, and he went on to say:

> We had seen research, sometimes deriving from basic bacterial genetics, used to create weapons of biological warfare. We had seen drug companies exploiting antibiotic research to enrich a few and exploit the many. We had seen a misdirection of health priorities—where super research is carried out on heart transplants and cancer and little attention paid to major health problems such as malnutrition, lead poisoning, infant mortality and delivery of health care in general. . . . We concluded that unless the present system in the United States is changed there will always be the danger and the likelihood of the use of science for the exploitation and oppression of people all over the world. We don't pretend that the problem is peculiar to the United States or to capitalist countries; but at least a prerequisite for a change to a society where there can be much greater participation in decision-making is the elimination of a system which is based on the exploitation of man by man.[64]

In summary, in response to the militarization of science and particularly to the most evocative symbol of that process in the 1960s, Vietnam, this new and radical critique of science and its social role emphasized the contradiction between the traditional image of the scientist as a disinterested seeker after truth and the actual conditions under which much scientific work was managed. Science was no longer seen as a politically or morally neutral pursuit but as an integral part of a larger social process informed by political and economic goals that were not necessarily in harmony with the goal of improving the human condition. From this perspective, the responsibility of individual scientists came to be seen more broadly as necessitating an effort to reform the social systems supporting science.

Public Awareness of the Impact of Science and Technology on Consumers, the Environment, and the Workplace

The 1960s also saw a surge of interest in a wide range of consumer, environmental, and occupational health and safety issues that directed public attention to the impact of science and technology. Several material and political conditions helped to push the protection against the effects of modern technology to a relatively high place on the political agenda by the end of the decade.[65] First, as a result of the enormous postwar growth in productive capacity and the progressive integration of scientific advances into industrial processes and products, consumers in the 1960s were faced with a bewildering array of products that were often aggressively and misleadingly marketed. Consumers' need for guidance on the quality, effectiveness, and safety of their purchases provoked soaring sales of subscriptions to consumer magazines.[66] Second, the effects of modern technologies on the natural environment were noticeably severe. Radioactive fallout from nuclear weapons tests, the concentration of persistent pesticides in food chains, oil pollution from massive spills, the discharge of industrial and domestic wastes into the air, rivers, and oceans—all produced effects that extended far beyond local ecosystems. In the 1960s, it became clear that the impact of modern technology was "far larger (greater biomass removal), more permanent (species extinction, severe depletion by direct removal and habitat removal), more vital (habitat and species removal, reproductive impairment through pollution), and far broader (multiple ecosystems drawn on and disrupted for production needs)" than that of earlier eras.[67] Moreover, these effects were dramatized by a series of disasters in both countries.

Finally, although less obvious, the hazards of the workplace had become, for those exposed, far more threatening than those of the general environment. Workplace hazards had always been serious, reflecting employers' lack of concern and resistance to change, trade union apathy, and the absence of government controls. In the United States in 1969, it was estimated that 14,500 workers were killed on the job and 2.2 million were injured. Furthermore, in the 1960s, the accident rate rose significantly and, by the early 1970s, exceeded postwar levels.[68] Similar trends were cited in the United Kingdom. Cases of occupational disease caused by silicon, asbestos, beryllium, and organic chemicals drew attention to the potentially deadly nature of many commonly used industrial materials.[69] These hazards were multiplying as hundreds of new chemicals, the products of the petrochemical revolution, were being introduced into the workplace each year.[70] The idea that new toxic substances or pathogenic organisms might slip into use through the reckless application of modern technology—a major concern of some who were to take part in the debate over genetic engineering—was thus not novel by the time the debate broke out.

Several books published in the 1960s helped to generate a stark sense of the dimensions of these problems and, perhaps just as important, a sense that remedial action was both possible and urgently needed. Perhaps more than any other work, Rachel Carson's *Silent Spring,* published in 1962, raised public alarm over the long-term environmental disruption likely to result from uncontrolled deployment of advanced technology. *Silent Spring* provided an accessible analysis of the impact of persistent pesticides, brilliantly demonstrating how the interconnectedness of biological systems could mean that the introduction of even low levels of pesticides in one place could disrupt life processes elsewhere, possibly thousands of miles away. Barry Commoner in *Science and Survival,* published in 1966, generalized this alarm, warning that the long-term effects on the biosphere of radioactive fallout from nuclear tests, water pollution from use of phosphates in detergents and nitrates in fertilizer, and air pollution from auto exhaust would likely be permanent and irreversible. "We have been massively intervening in the environment without being aware of many of the harmful consequences of our acts until they have been performed and the effects—which are difficult to understand and sometimes irreversible—are upon us. Like the sorcerer's apprentice, we are acting upon dangerously incomplete knowledge. We are, in effect, conducting a huge experiment *on ourselves*" (28).

Furthermore, these authors proclaimed that the public had the right to challenge unsound uses of technology and to decide how it should be controlled. "It is the public that is being asked to assume the risks that the insect controllers calculate," wrote Carson. "The public must decide whether it wishes to continue on the present road and it can do so only when in full possession of the facts" (13). Thus in the 1960s, there was a growing sense not only that new technologies were having far-reaching and serious impacts but also that changes in traditional political structures were needed to address these problems.

The Response in the United States

In the United States, an important factor that opened up the possibility of redirecting public science policy in the 1960s and 1970s was a change in the political complexion of Congress. The Democratic victories of 1958 and 1964 shifted the balance of power in Congress from a conservative coalition of Republicans and Southern Democrats to a new group of liberal Democratic reformists. The first priorities of the liberal Democratic congresses after 1958 were major social reforms in areas such as civil rights, poverty, aid to elementary and secondary education, and health insurance for the elderly and the poor. The costs of implementing these reforms, combined with the inflationary pressures generated by the war in Vietnam, meant that, by 1965, there was little slack in the federal budget for further expensive domestic programs. Protective legislation to regulate consumer products, the workplace, and industry—legislation that passed on most of its costs to business and the consumer—appealed to legislators who remained eager to take on social reform issues. Congress in the midsixties provided a political environment in which safety was likely to be an attractive issue.

That it could also be a winner was demonstrated by the success of auto safety legislation, championed by Ralph Nader, author of the influential *Unsafe at Any Speed,* and Senator Abraham Ribicoff. The auto industry had opposed legislation. The passage of the bill showed that a powerful industry could be defeated. Moreover, passage also transformed consumer protection into a popular and highly visible issue.[71] Within Congress, liberal reformists acted swiftly to capitalize on the new public receptivity to safety issues by proposing a whole range of new legislation. President Johnson, sensing the pressure to attend to these issues and aware of growing disaffection within organized labor for new and costly social programs, developed new "quality of life" proposals that included pollution control and consumer protection. In 1968, Johnson added occupational health and safety to this agenda. As Martin Noble observed, Johnson's action was an "opportune response" both to "the [Democratic] party's commitment to the new politics of middle-class reform and the traditional concerns of organized labor."[72] No less sensitive to popular issues, President Richard Nixon also endorsed a workplace safety program. In 1970, Nixon made the need for environmental controls a major focus of his State of the Union address.[73]

Meanwhile, outside Congress, the consumer safety and environmental movements snowballed, and memberships in environmental organizations grew substantially from the midsixties onward; environmental issues received increasing attention in the news media, and sales of *Silent Spring* and other books on the environmental crisis soared.[74] What had happened was the formation, primarily in the middle class, of a new mass consciousness of environmental issues. The widespread perception of the image of the earth beamed from space in 1969 as a tiny, lonely, vulnerable ball of life reflected the emergence of a new sensitivity to the possibility of a planet endangered by the ravages of technology.[75] The time was ripe for action.

For a few years, the politicians and the attentive public formed a resonating system in which pressure from the public received a ready response from politicians and political action reinforced the public movement, and a broad range of reform legislation was enacted to mitigate the effects of technological development on the consumer, the workplace, and the environment. Many of these new statutes established standards for consumer goods, effluents, and exposure levels for harmful substances. Others established goals, institutions, and procedures for assessing and responding to technological impacts. The National Environmental Policy Act (NEPA), passed in 1969, set the provision of "safe, healthful, productive, and aesthetically and culturally pleasing surroundings" as a national goal. It also required all federal agencies to prepare "environmental impact statements" for actions "significantly affecting" the quality of the environment. The Occupational Safety and Health Act of 1970 imposed on virtually all private employers a duty to provide working conditions free of serious recognized hazards. The act gave the Secretary of Labor the authority to promulgate and enforce standards, to inspect any workplace without advance notice, and to propose penalties upon discovery of violations.[76]

Generally, the legislation of the late 1960s and early 1970s codified three methods of public control over new technology. First, it established a principle of public review of decisions on the regulation of science and technology. NEPA's environmental impact statements, for example, were designed to facilitate and even encourage public review. Such procedures were complemented by new "sunshine" laws, such as the Freedom of Information Act passed in 1966 and amended in 1974, which gave the public broad access to government documents, and the Federal Advisory Committee Act, passed in 1972, which provided public access to government decision making.[77] Second, the new legislation significantly expanded the concept of legal standing, enabling citizens and groups to take legal action against government agencies or private industry not only on the narrow grounds of invasion of property or legal rights but also on the broader grounds of violation of the public interest. The Occupational Safety and Health Act established the right of workers to bring action to require the Secretary of Labor to abate an imminent danger. Finally, the legislation emphasized preventive measures designed to forestall undesirable impacts of new technology. NEPA, for example, was seen as a means to force federal agencies to address the question of how to minimize or avoid damage to the environment.[78] Generally, the aim of such legislation was not to stop technological development but to attempt to avoid its destructive tendencies or, in the words of one observer, "to pick [the technological] roses without getting cut by the thorns."[79]

It is also important to recognize the limits of the new wave of protective legislation, particularly with respect to the distribution of influence and access to policymaking arenas. While the new legislation provided a variety of ways in which the public could influence government regulation of technology and opened up governmental processes to public scrutiny to an unprecedented extent, it did not facilitate direct participation in the policymaking process. Responsibility for the appointment of advisory boards to government agencies, for example, remained firmly in the hands of those agencies. Representation of citizen groups on these boards was limited.[80] The formulation of policy for the development and control of science and technology remained the prerogative of government and private industry.

The Response in the United Kingdom

Although interest in environmental issues may not have reached the same levels of intensity and saliency in the United Kingdom as in the United States, the available evidence suggests that it was nevertheless widespread and increasing steadily in the 1960s. Moreover, a series of major disasters—the mining tip calamity that engulfed the town of Aberfan in 1966, the *Torrey Canyon* wreck that dumped 117,000 tons of crude oil off the coast of Cornwall in 1967, the explosion at a nylon plant at Flixborough that killed twenty-eight people—served to focus and dramatize the concern and to catalyze a public outcry that moved Harold Wilson's Labour government and Edward Heath's Conservative government to action. The Labour government was poorly prepared for the environmental issue: that party's commitment, after all, had been toward mobilizing science and technology for industry, not protecting the environment against them. But the government moved hastily to include these questions on its political agenda, creating in 1969 a central scientific unit on pollution and, in 1970, a Royal Commission on Environmental Pollution. After the 1970 election, the new Conservative government established a Department of the Environment, with broad authority for environmental protection. Several new pieces of environmental legislation were adopted in the 1970s, including the 1974 Control of Pollution Act, which was intended to strengthen the powers and responsibilities of local authorities for the control of air and water pollution and toxic wastes, as well as to make information on pollution control available to the public.[81]

These actions largely continued a tradition of regulation that differed greatly from the adversarial, legalistic approach characteristic of the United States. Environmental and workplace conditions in the United Kingdom fell under the aegis of numerous statutory provisions accumulated over roughly a century of regulation, applying to air pollution, factory and mine safety, river pollution, pesticides, and nuclear installations, and enforced by various inspectorates. In contrast to the United States, where absolute standards were set and litigation was expected to deal with violations, the British inspectorates had evolved a pragmatic, informal approach to regulating industry, based primarily on personal contact between inspectors and company representatives and with business conducted in an atmosphere of persuasion, conciliation, and cooperation. The chief inspector of the Alkali and Clean Air Inspectorate stated in 1972: "We look on our job as educating industry, persuading it, cajoling it. We achieve far more that way. The Americans take a big stick and threaten to 'solve your problem.' We say to industry, 'Look, lads, we've got a problem.' In this way, we've got industry well and truly tamed." [82]

The criminal sanctions available were rarely used: in the fifty years before 1967, the Alkali and Clean Air Inspectorate had prosecuted only three cases for air pollution violations. Prescribed penalties were in any case very small. [83]

At the center of the practice of pollution control in the United Kingdom was a concept known as "best practical means," which in many ways encapsulated the major differences between regulation in the United States, based on promulgated standards, and the flexible approach used in the United Kingdom. Best practical means, a concept elusive to those used to precise standards, was the product of the process of consultation between inspectorates and the regulated industries. The chief inspector of the Alkali and Clean Air Inspectorate wrote in 1973, "One needs to live with [best practical means] and use it regularly, like a good system of contract bridge played with a cooperative partner, in order to properly understand all its nuances." [84] To assess best practical means, inspectors took into consideration the state of pollution technology, local (i.e., physical/geographical) conditions, and the economic costs to the industry. There was, therefore, considerable scope for negotiation. As one analyst concluded, "Almost anything could in principle be regarded as coming within the scope of best practical means." In practice, this meant that economic costs to the polluter were balanced against damage to the surrounding environment. [85]

The regulation of emissions and effluents was shielded from public scrutiny by stringent legal restrictions on the release of information to the public. The Official Secrets Act of 1911 prohibited public disclosure by government officials of virtually all information collected in government offices other than documents expressly approved for public release. This broad restraint on the free flow of information about pollution of the U.K. environment was reinforced by the restrictions of specific statutes. For example, the Rivers Act of 1961 prohibited the disclosure of information on discharges of effluent and the limits set on such discharges. Inspectors were placed in a position in which penalties to individuals for disclosure of information could be far more severe than those for pollution.[86]

Although its supporters defended the British system of pollution control as effective, the various disasters of the 1960s and media coverage of unacceptable air and water pollution levels in various parts of the country combined to support a new perception that the system was not working well enough. A government White Paper stated in 1970: "The British system of law in this, as in related fields, does not traditionally rely on the very heavy penalty as the main deterrent. It relies rather on persuasion and the belief that, especially to industrial firms, it is the disgrace that counts and not the fine. The weapon of prosecution has in the past been sparingly used. But the government now believe that the present penalties are both incoherent and generally too lax."[87]

In this context of criticism both within the government and outside it, the Royal Commission on Environmental Pollution began its work. The commission was given broad powers to advise the government on pollution and to recommend policy. Its first chair was biologist Sir Eric Ashby, master of Clare College, Cambridge, who would later become the chair of the first British committee on genetic engineering. In 1972, Ashby was succeeded by Sir Brian Flowers, rector of the Imperial College of Science and Technology, London, who chaired the commission until 1976.

The royal commission's reports on environmental pollution (the first three under Ashby) addressed both substantive issues relating to the magnitude and the nature of Britain's environmental problems and procedural issues concerning approaches to control. With respect to pollution control procedures, the commission made a break with tradition by rejecting the statutory confidentiality of British environmental legislation. The industry's defense that secrecy was necessary to protect commercially useful information from competitors no longer seemed valid. As the commission stated, "Its only value is to protect industry against the risk of common law actions or against misconceived or ill-informed allegations that the environment is being dangerously polluted." It was time for this "needless cloak of secrecy" to be withdrawn.[88]

In most other respects, however, the royal commission called for gradual modification and adjustment of the British regulatory system rather than radical change. Although its fifth report acknowledged that the public needed greater input into decisions on best practical means, the concept was still defended as "inherently superior to control by nationally fixed and rigid emission standards," which were judged to be "generally impracticable and unnecessary." Furthermore, and no doubt with the environmental battles under way in the United States in mind, the commission, admitting that fines imposed on offenders were meager deterrents, nonetheless argued that an "aggressive policy of confrontation, involving prosecution for every lapse" would "destroy the basis of cooperation [between industry and the inspectorates]" and would be "counterproductive" and "inappropriate." [89] Finally, in no case did the commission propose opening actual policymaking to include the public as well as the inspectorates and the regulated industries.

In contrast to its response to environmental problems, the British government eventually made important changes in policymaking structures for addressing occupational health problems. In a political climate characterized by growing union concern over health and safety issues, the Labour government appointed a committee chaired by former National Coal Board chair Lord Robens and charged with investigating occupational health and safety standards and procedures. The committee's report, issued to the Conservative government in 1972, gave rise to a broad national debate not only on the need for more stringent controls but also on the need for changes in governance.

The report recognized the severity of workplace hazards and proposed a thorough overhaul of the "tangle of jurisdictions" applying to workplace regulation and their replacement by a unified regulatory framework. [90] At the same time, the committee came down emphatically on the side of voluntary control and self-regulation. There was, its report insisted, too much law rather than too little, creating the impression that the responsibility for safety in the workplace should rest with the state rather than with employees and management. The resulting "apathy" was, in the committee's opinion, "the most important single reason for accidents at work." The best approach to safety was to impose on employers a general duty to provide a safe working environment and to impose on employees a general duty to observe safety and health provisions, but voluntary standards and codes of practice were to be preferred to statutory regulations. For similar reasons, the Robens committee accepted the concept of best practical means and opposed a statutory provision requiring the appointment of safety representatives and safety committees as "too rigid and, more importantly, rather too narrow in concept." [91]

The debate that followed was predictable. The Confederation of British Industry (CBI), representing employers, voiced approval of the report, commenting only that "it has come down . . . remarkably close to the line [we] suggested."[92] The trade unions and the Labour party, on the other hand, expressed strong criticism. Labour M.P. Neil Kinnock stated in the House of Commons debate on the Robens report: "To suggest that the law is the main cause of apathy is a distortion of reality. It is like saying the crutch has made the cripple."[93]

The Conservative government generally accepted the Robens recommendations and prepared legislation in accordance with them, but before it could act, its administration came to an end: the general election of February 1974 returned the Labour party to office. In the same year and in a climate of strong union commitment to strengthening workplace regulation,[94] the new government passed the Health and Safety at Work Act of 1974, a piece of legislation that departed significantly both from tradition and from the Robens recommendations. The act brought virtually all workplaces—including scientific laboratories—under its jurisdiction, extending coverage to five million new employees. It defined a general regulatory framework as well as new institutions for controlling workplace hazards. Administration of the act was made the responsibility of a nine-member Health and Safety Commission representing employees, employers, and local authorities. For the detailed implementation of policy, the act also created the Health and Safety Executive, comprising the former inspectorates and the former Employment Medical Advisory Service.[95]

The Health and Safety Commission was granted broad powers to develop two types of controls: statutory regulations enforceable through the courts and criminal law, and codes of practice that were not legally binding. Penalties for violation of statutory requirements were made somewhat stiffer. But whether health and safety policy would move in the direction of increased voluntarism or increased enforcement was left to be decided through further interpretation of the act.

The most novel feature of the act—as well as a significant departure from the philosophy of the Robens committee—was its creation of statutory rights of employees to participate in decision making on safety issues. At the national level, trade unions were now able to participate fully in the development of health and safety policy through their representation on the Health and Safety Commission. At the local level, the act created a statutory right for employees to appoint their own safety representatives.[96] The powers of the latter were further defined in a later piece of legislation, the Safety Representative and Safety Committee Regulation of 1977, which gave safety representatives extensive rights to be informed about workplace hazards, to inspect the workplace, to receive training in safety policy and procedures, to investigate accidents, and to establish safety committees.[97]

American and British Responses Compared

The wide publicity given to the effects of uncontrolled technology on a burdened planet and its inhabitants meant that virtually all participants in the debate over genetic engineering policy would be sensitive to the possibility of environmental and health hazards. When molecular biologists began to realize the constructive and synthetic powers their field possessed, they did so in a social context where such abilities had become widely seen as subject to public scrutiny and at least to the possibility of governmental control.

This examination of the evolution of environmental and workplace regulatory policy in the United States and Britain suggests that governmental and public responses would take different forms in the two countries as a result of the differing types of access to formal policy arenas. In the United States, the impact of new protective legislation was felt largely in terms of stricter standards for effluents and emissions, laws encouraging the anticipation of the effects of new technology, and greatly increased interaction between the public and the government on environmental and industrial safety issues. The public acquired greater access to the policy process and more powerful legal instruments for challenging government decisions. That a citizen had a right to a healthy environment, which could be asserted in the courts, was a new feature of the U.S. political landscape that was obviously on the minds of the biomedical researchers who came to decide and debate the use of recombinant DNA technology. Yet while environmentalists and others were given a voice in policymaking, they were not given a formal policy arena. That American biomedical researchers would be able to assert a right and even exclusive responsibility for developing national controls for genetic engineering probably owed something to the general pattern of public exclusion from *direct* participation in formulating environmental and industrial safety policy.

In the United Kingdom, in contrast to the United States, environmental controls remained to a great extent voluntary in nature and largely shielded from public scrutiny by the Official Secrets Act. The reforms recommended by the Royal Commission on Environmental Pollution and adopted by the British government, although they extended environmental protection and probably tightened the application of best practical means, had little effect on the basic form of policy. Implementation of environmental law remained a matter of negotiation between the regulated industries and the inspectorates responsible for control of effluents and emissions. With respect to industrial safety, however, reforms were more radical. The new levels of participation achieved through the Health and Safety at Work Act brought an important change in employee-employer relations. In this respect, the 1974 act was a major break with tradition, one that would provide trade unions with a formal arena and an effective voice in policymaking.

1.4 Deregulation and Selective Growth: 1970s and 1980s

The impulse to protect society against the destructive tendencies of modern science and technology turned out to be a relatively brief moment in the history of regulation. Toward the end of the 1970s, government efforts to accommodate public pressures for controlling the impacts of science and technology began to give way to pressures in both countries to roll back regulation and, beyond this, to develop closer links between academia and commerce.

Two general conditions, one economic, the other political, encouraged these trends. Beginning in the mid-1960s, U.S. markets were increasingly penetrated by the faster-growing economies of Japan, Germany, and some other European countries, and U.S. domination of the global economic system was gradually eclipsed. In 1971, the United States registered its first merchandise trade deficit since World War II, a reflection of the erosion of its older industries such as textiles, iron and steel, automobiles, and clothing as a result of more efficient technology and cheaper labor pools elsewhere. In the mid-1970s, the new phenomenon of "stagflation"—high inflation combined with high unemployment— made its appearance. All of these problems were intensified by the rising cost of raw materials, particularly oil.[98]

The declining competitiveness of U.S. industry generated pressures on corporations to find ways to cut costs and to develop new export markets. Key growth areas in the early 1970s were the newer, research-intensive industries such as computers, semiconductors, telecommunications equipment, aircraft, engines, chemicals, and pharmaceuticals. Throughout the 1970s, the trade balance for these products grew rapidly in favor of the United States. Between 1972 and 1979, it almost quadrupled, from $11 billion to $39.3 billion. (The trade balance for manufactured products from industries that were not research-intensive, however, dropped from − $15.0 billion to − $34.5 billion in the same period.)[99] The pressures on corporations to move quickly to develop these research-intensive, high-technology fields therefore increased.

If the economic problems of the United States in the 1970s were serious, those in the United Kingdom were reaching crisis proportions. The British economy, like that of America, was deeply troubled by stagflation and trade deficits. In addition, the British industrial base, long in decline relative to those of other nations, now began to contract absolutely. Only the growth of North Sea oil production in this period prevented an absolute fall in industrial production.[100] For reasons similar to those in the United States, corporate and government leaders looked to new forms of technology to provide a remedy for Britain's ailing economy.

Second, the political fortunes of the scientific establishments changed significantly in the late 1960s and early 1970s. In the United States, scientists lost the privileged treatment they had come to expect, and the ready flow of federal support for research ceased. Symptomatic of this political reversal was President Nixon's abolition of the post of presidential science adviser and the Science Advisory Committee. Scientific leaders saw these moves as signs of a new "disillusionment" with science among politicians and the American public. Similarly, the former confidence of the British scientific establishment was jolted by spending cuts in the late 1960s and early 1970s and the move by Lord Rothschild to replace the former protection of the autonomy of British science with a utilitarian emphasis on accountability. At this point, therefore, scientific leaders began to search for ways to regain their political standing, to try to shape a more positive image of science, and to cultivate new sources of support.

The Response of the American and British Scientific Establishments

Some of the earliest—and most intense—responses to the environmental and antiwar movements' critiques of the social role of science in technologically advanced societies came from those most closely connected with the support of science in Britain and America. In general, leaders of the scientific establishments in both countries perceived these movements as a serious threat to scientific development. They reacted by characterizing their targets in sweeping and derogatory terms. In an address to celebrate the five-hundredth meeting of the Biochemical Society in London in December 1969, Philip Handler, president of the U.S. National Academy of Sciences, described the antiwar and environmental movements as a "violent world reaction" against science that would severely jeopardize its future.[101] Alvin Weinberg, director of Oak Ridge National Laboratory, characterized left-wing critics of the use of science for the war in Vietnam as "scientific abolitionists."[102] Sir Peter Medawar, president of the British Association for the Advancement of Science in 1969, proclaimed with reference to those who "wring their hands over the miscarriages of technology and took their benefactions for granted" that "to deride the hope of progress is the ultimate fatuity, the last word in poverty of spirit and meanness of mind."[103]

Other and more substantive concerns appear to have been behind these attacks. As science budgets declined in the late 1960s, leaders of the British and American research establishments feared that criticism of science would generate an atmosphere in which politicians might withdraw even more support from science. Philip Handler expressed this concern: "The [manner] of directing attention to these problems has turned both our decision-makers and our youth against science. This nation may yet pay a dreadful price for the public behavior of scientists who depart from fact and reason to indulge themselves in hyperbole."[104] John Maddox editorialized: "There is a danger that the committees which decide how money should be spent will also limit the uses which can be made of science in the years ahead. This, in the long run, is where the penury now being forced on science will cause the most serious damage." Worse still, governments might attempt to eliminate branches of research entirely because of fear of the problems they might generate.[105]

Two principal themes characterized the warnings of scientific leaders in this period. On the one hand, the "excesses" of modern science and technology—the ammunition of environmentalists, antiwar activists, and left-wing scientists alike—had to be contained. As Philip Handler proclaimed in an address in 1970, "We [must] attempt to manage our technological civilization yet more successfully, remedying the errors of the past, building the glorious world that only science-based technology could make possible." On the other hand, it was equally necessary to contain what were seen as the excesses of public discontent with science and technology if the "disenchanted" public were not to undermine government support for science or even to bring certain lines of inquiry to a full stop. To achieve this, it was necessary to rebuild science's image and to replace the focus on the destructive dimensions of science and technology with an optimistic perception of its constructive potential. Handler noted in the same speech, "Until the terms 'science' and 'education' again evoke warm, positive associations in the public mind, we cannot expect significant growth of the fundamental endeavor."[106]

In 1967, the National Academy of Sciences established a committee chaired by Harvey Brooks, dean of engineering and applied physics at Harvard, to respond to the growing demand that scientific research achieve greater social relevance. The report acknowledged that scientists needed to become more responsive to social requirements and interests. However, it also warned, "Real dangers are involved when the nonscientist attempts to impose his own value system on what should be largely scientific decisions."[107] In 1969, Brooks chaired a second National Academy of Sciences committee, commissioned by Congress to undertake a study of technology assessment. This committee issued strong calls for the anticipation and prevention of the side effects of technology, for the neutrality and detachment of the assessment process, and for the development of a pluralistic decision process.

At the same time, the report made clear that the authors did not envisage a radical reorganization of the existing forms of promotion and control of science and technology: "The needed climate requires that private industry be encouraged to find its own solutions—not compelled to follow solutions formulated from above." By contrast, a "Luddite response" in which "people with much power and little wisdom lashed out against scientific . . . activity, attempting to destroy what they found themselves unable to control," would be "tragic." Public participation in the assessment process, while it was to be encouraged, was also to be carefully confined to "well-defined channels," with mechanisms provided to "filter out for summary treatment truly frivolous or irresponsible claims."[108] Genetic engineering, appearing only three years after these proposals, would be a test case for the strategies envisioned for coping with the problematic public response to new science and technology.

The Response of Private Industry

As the U.S. economy deteriorated in the late 1960s and early 1970s and competition for the development of new, more profitable technologies increased, leaders of private industry began to perceive the environmental, consumer safety, and occupational health and safety movements—which were still commanding congressional attention and action—as serious threats to the future of American business.[109] As David Rockefeller, chair of Chase Manhattan Bank, observed in a speech to the Advertising Council in December 1971: "It is scarcely an exaggeration to say that right now American business is facing its most severe public disfavor since the 1930s. We are assailed for demeaning the worker, deceiving the consumer, destroying the environment, and disillusioning the younger generation." All of industry was under attack: "Consumerism is equated in the public mind with the idea of the individual against business: *all* business."[110] According to one executive, "My industry is regulated up to its neck. You are regulated up to your knees. And the tide is coming in." Public demand for safety had gone too far. Industry would be "engulfed by a rising tide of entitlement."[111]

Two main goals began to be aired and pursued by corporate leaders in the 1970s. First, they aimed at participating more actively in the formation of government regulatory policy in order to oppose from within the twin excesses of regulation and democracy.[112] Second, they sought to promote rapid technological development. In this regard, the formation of closer ties between American industry and American science was seen as essential, and it was anticipated that the universities, with their impressive manpower and research capabilities, could play a crucial role.

In the 1960s, the universities and industry had grown apart, partly because of public opposition to the role that certain industries had played for the military, but perhaps more fundamentally because industry support for university research had become almost insignificant as a result of the huge growth of federal support in the 1950s and 1960s. (In 1968, industry's share of total academic expenditures on research and development was a mere 2.5 percent.)[113] As the importance of scientific research for the pace of technological development became apparent, however, corporate leaders began to see the universities as radically underused sources of knowledge for generating technological innovation. In 1974, Frank Cary, chair of IBM, hosted a dinner for the chairs of the largest U.S. high-technology corporations, thirteen Nobel laureates, and several academic leaders, at which he emphasized the need to initiate a closer collaboration between academic science and industry. The importance of this offer for the universities was immediately recognized. Cary's message was soon relayed on the editorial page of the leading U.S. science journal, when Philip Abelson, editor of *Science,* wrote: "If anything is clear, it is that we cannot depend solely on the wisdom of politicians in the solution of long-range problems. We must find better ways. A closer cooperation of academic scientists and dynamic elements of industry could lead to effective actions."[114]

These goals were energetically promoted in the 1970s by corporate leaders, with the active support of scientific leaders. New organizations were created and established organizations were revived to provide industry with a more effective voice both in Washington and in the courts.[115] Calling for "free enterprise," "innovation," "efficiency," and "rationalization," these organizations redefined the "public interest" in terms of the primacy of economic growth for the improvement of the social order.[116] Arguments that progress based on technological innovation was being stifled by regulation and excessive public demands for participation in decision making were framed in terms of this new discourse. Samuel P. Huntington, professor of political science at Harvard University, proclaimed in a report published by the Trilateral Commission, an organization whose members represented the commercial, banking, and political establishments of the United States, Japan, and Europe: "The arenas where democratic procedures are appropriate are, in short, limited. . . . In many situations, the claims of expertise, seniority, experience and special talents may override the claim of democracy as a way of constituting authority."[117]

The idea of a new alliance between industry and academia proposed by Frank Cary at the IBM dinner in 1974 was also vigorously encouraged. In 1978, the American Council on Education, an organization representing the nation's largest universities, established the Business–Higher Education Forum, "dedicated to closer understanding and cooperation

between business and higher education on problems of mutual concern."
The same year, six leading professional societies established the National
Commission on Research, whose members included several representatives of corporate research and whose support came largely from foundations with strong industrial ties. It soon became evident that the
commission's political function was to reinforce the university-industry
alliance. In addition, many smaller interactions—for example, execu-
tives-in-residence programs in the universities, joint university-industry
meetings, and workshops on innovation and regulation—began to cement the alliance.[118] By the time Edward David, president of Exxon Research and Engineering, addressed the annual meeting of the American
Association for the Advancement of Science in 1979 on the need for a
"synergistic relationship between the scientific community and industry"
that would encourage joint research programs, consulting arrangements,
and collaborative study of public policy issues affecting science and industry, this relationship was actually well on its way to being realized.[119]
Recombinant DNA technology would prove to be a pacesetter in that
respect.

Science Policy in the United States

The history of U.S. science and technology policy since 1975 is in important respects the history of the realization of the goals framed by U.S.
corporations in the 1970s: the containment of the regulatory and democratic impulses of the late 1960s and the promotion of rapid technological
development, which was perceived to be essential for the growth of the
leading American industries. The Ford, Carter, and Reagan administrations proved responsive to those goals in varying degrees that reflected
the interests of the coalitions supporting them.

The sensitivity of President Ford and Vice President Nelson Rockefeller to the joint interests of the scientific and business communities was
expressed in the reestablishment of the Office of Science and Technology
Policy in 1976, Ford's establishment of a group of prominent industrial
and scientific leaders to serve as advisers on science and technology
policy, and his executive order calling for "inflation impact statements"
for all regulatory decisions.[120] Also significant was Ford's reversal of the
previous downward trend in federal support for research (fig. 1.1). At
the same time, the Democratic congresses of that period continued to be
responsive to continuing public pressure for controls on technological
development, passing legislation such as the Toxic Substances Control
Act in 1976, which required for the first time premarket testing of potentially dangerous chemicals.

Not surprisingly, in view of Carter's Trilateral Commission membership and his support by multinational corporations, the Carter administration went a considerable way toward articulating, if not entirely executing, a new science and technology policy that responded to the platform of corporate and scientific leaders. But these policies were also constrained by the need to respond to the environmentalists and unions that had also supported Carter's bid for office. In the first place, and in response to industry emphasis on limiting regulatory costs, various moves were made to subject regulatory policy to economic analysis. For example, Executive Order 12044, issued in March 1978, required regulatory agencies to integrate cost-benefit analyses into regulatory decisions. Also, a Regulatory Analysis and Review Group, headed by William Nordhaus and with the active involvement of presidential science adviser Frank Press, was set up in the White House to oversee regulatory policy and to ensure that economic effects of regulations were taken into account.[121] Although the need for these new reviews of regulatory policy was phrased in terms of the need for "efficiency" and "consistency," considerable evidence indicates that their fundamental purpose was to make the social control of technology subject to market constraints. However, there were limits to how far those who favored this path could go. Those parts of the Carter administration committed to maintaining and extending the protective policies initiated in the early 1970s resisted the idea that economic benefits might justify exposure to significant—or unknown—risks. As Eula Bingham, head of the Occupational Safety and Health Administration, stated in 1978: "Worker health and safety are to be heavily favored over the economic burdens of compliance."[122] Such resistance produced a more moderate outcome than might otherwise have been the case.[123]

A second emphasis of the Carter administration's science policy was on support for industrial innovation and stronger links between universities and the private sector. As the U.S. economy slid further into decline, inhibiting large budget increases for basic research, the administration instituted selectivity in funding, using relevance to industrial and military development as a criterion.[124] An extensive domestic policy review on industrial innovation, completed in October 1979, recommended legislation to establish a new patent policy to encourage industrial development of government-sponsored research, expansion of an NSF program to foster university-industry cooperation, expansion of NSF support for small businesses, and further efforts to reduce the costs of regulation. Some of these measures were acted on by Congress. For example, the Patent and Trademark Amendment Act of 1980 gave universities, small businesses, and nonprofit organizations rights to patents arising from federally supported research—a major government action that facilitated industry access to university research.

The Carter administration's initiatives paved the way for further policy developments by the Reagan administration that would prove even more favorable to industry. Brought into office by a coalition representing the interests of the far right, multinational corporations, and the Trilateral Commission, Reagan was able to pursue the Carter administration's goal of rapid technological development with little obligation to attend to criticism from labor unions, environmentalists, or consumer advocates. But in contrast to Carter's acceptance of government intervention, Reagan's development of science policy was shaped by a strong commitment to scaling back what he saw as the unnecessary intrusion of the federal government into the lives of citizens and the activities of private business. Given these interests and influences, the Reagan administration's science and technology policy was characterized by three principal goals: rapid deregulation, selectivity in support for scientific research, and stimulation of technological innovation through tax incentives. In pursuing these goals. Reagan drew upon many of those who had been influential in forming the corporate consensus on the need for deregulation and stimulation of technological innovation in the 1970s.[125]

Commitment to deregulation in all spheres was a key element of Reagan's election campaign. Once in office, he pursued this goal with zeal. Protective legislation for the environment and the workplace was vigorously attacked through the appointment to key agencies of officials with strong commitments to deregulation and privatization, through major cuts in funding for the regulatory agencies, and through the agencies' interpretations of regulatory statutes.[126] A crucial condition for all regulatory action was set by Executive Order 12291, signed by Reagan in February 1981, requiring regulatory agencies to conduct "regulatory impact analyses" for new rules, to demonstrate that expected benefits outweighed costs and that the course of action that entailed the least cost to society had been chosen, and to submit all proposals to the Office of Management and Budget for review. At one stroke, regulation was made subject to economic imperatives.[127]

The Reagan administration, like its predecessor, was also committed to increasing federal support for science, since basic research was seen as providing a base of knowledge essential for later industrial development. Support for research rose steadily during Reagan's first term: from FY 1982 to FY 1985, the increase in support for basic research was 27.3 percent; the increase in support for basic and applied research was 10.6 percent (fig. 1.1). Selectivity in support for science continued to be emphasized. As presidential science adviser George Keyworth emphasized, beyond the intrinsic merit of a project, "pertinence" to the economic and social needs of the nation would provide the basis for selection.[128] In practice, this translated into support for areas of science that were seen as particularly important for national defense and high-technology industry, such as advanced computer technology, biotechnology, and materials research. In contrast to the Carter administration's commitment to underwriting the costs of certain kinds of civilian technology, the Reagan administration generally took an anti-interventionist position with respect to applied research and development, holding that these were best left to the private sector.[129] There were, however, major exceptions to the Reagan philosophy: areas such as defense and space where the federal government was seen as the primary user were generously supported.

Finally, in the climate of fierce competition for development of new technology in the 1980s, the Reagan administration, like its predecessor, actively pursued mechanisms to spur innovation and to promote the transfer of knowledge from the universities to private industry. In keeping with its commitment to limiting government intervention in private industry, the Reagan administration emphasized patent and tax reform rather than direct government outlays for research and development in industry. The most important element of this policy was a set of provisions in the 1981 Economic Recovery Tax Act that provided substantial tax credits for research and development as well as general incentives for new capital investment. In addition, in 1982, the Department of Commerce began to promote the use of tax shelters for joint research and development ventures for investors and corporations.[130]

These measures, combined with those already established during Carter's term, had three main effects. First, they encouraged a huge flow of capital into the small high-technology firms specializing in such fields as computer technology and genetic engineering. Venture capital investment in small firms had declined from 1970 to 1975, but from 1975 onward, it skyrocketed—from $10 million in 1975 to $4.5 billion in 1983 (a real increase of almost 25,000 percent), with the greatest proportion of this funding flowing into high-technology firms. Second, investment in public stock offerings of small high-technology firms increased dramatically, particularly after 1980. (The capital raised in this way increased by almost 8,000 percent in real terms, to $1.5 billion in 1983.) Hundreds of these firms were established in this period. Third, hundreds of tax-sheltered "research and development limited partnerships" were formed between small high-technology firms and private investors in the 1980s. (In 1982–83, over $200 million was raised for biotechnology firms alone.)[131]

Changes in tax law were also instrumental in opening up university research for corporate investment, and generally in nurturing a synergistic relation between universities and corporations. The universities, eager to find new sources of support, opened their laboratories to corporate research projects. Corporations, eager to tap a vast new source of knowledge and expertise, soon produced tempting proposals. A path-breaking agreement was made between Monsanto and Harvard Medical School in 1975. Harvard Medical School received $23 million over ten years for research on an antitumor agent. In return, Monsanto received first rights to all patentable results of the project.[132] From the late 1970s onward, arrangements of this kind were made with increasing frequency, and industry support for research and development in the universities rose rapidly. By 1982, *Business Week* was claiming that the "trickle of industrial support of science at universities a few years ago is turning into a torrent."[133] (The amount spent by private industry on the universities that year was $326 million, and such investment was rising at the rate of roughly 11 percent per year.)[134]

Generally, what these new funds bought for corporations, in addition to rights to any patentable research, were windows on new fields of research with potential industrial application. At the same time, university researchers were encouraged to become more responsive to commercial interests. Basic research, seen traditionally as disinterested, now became infused with commercial meaning. As one observer put it, "The dynamics of the international marketplace have given Bacon's claim that knowledge is power a new twist: scientific knowledge has become a strategic tool for company planners, a way of guiding them toward future products and markets."[135]

While these developments were resisted by some scientists and generated controversy on some campuses, the general trend toward closer alignment of university research with industrial interests continued. The strengthening of these links was reflected in the increasing number and variety of university-industry contractual arrangements: university centers and institutes supported by industry, jointly owned or operated laboratories, research consortia, cooperative programs, and jointly owned research companies.[136] Scientific knowledge, traditionally seen as a public resource, was quickly being transformed into a commodity.

With the deregulatory, tax, patent, and budget allocation policies of the Carter and Reagan administrations in place, the governmental response to the corporate goals formulated in the early 1970s was virtually complete. That the biomedical researchers who were the majority of the policymakers on the genetic engineering issue were to prove mainly compliant with expressed corporate objectives during this period should be seen as part of a general social trend.

Science Policy in the United Kingdom

Despite important differences in rhetoric and ideology, the response of the British Labour government to the United Kingdom's serious economic problems anticipated in certain respects the more extreme response of the Conservative government that succeeded it in 1979. Although the Labour party was not fully converted from Keynesianism to the monetarist philosophy pursued so vigorously by the Conservatives, it went a good part of the way toward adopting policies that reduced government spending, squeezed credit, and allowed unemployment to rise.[137] At the same time, however, the Labour government attempted to respond to the claims of the trade unions that formed its base of support. It accommodated demands for pay increases and generously supported the new agency—the Health and Safety Executive—and the commission responsible for it, created by the Health and Safety at Work Act of 1974. While many other government expenditures were cut back, the government's grant in aid to the Health and Safety Commission increased in real terms by 3 to 4 percent between 1977/78 and 1979/80.[138]

Science, however, was one of the areas subject to budget cuts. In the early 1970s, government support for science had grown steadily, at about 4 percent per year—a far cry from the heady days of the 1960s, when growth averaged about 12 percent per year, but nevertheless substantial enough to enable British science agencies to continue to support a wide range of research and to operate on the assumption that excellent proposals would be funded. The turning point came in December 1973, when major cuts were announced by the Conservative government and were carried out the following year by the Labour government. From 1973 until 1978, support for civil science declined substantially—by some estimates as much as 21 percent, or over 4 percent per year.[139] By 1976, the Advisory Board for the Research Councils (ABRC)—the successor to the Council for Scientific Policy and the committee responsible for advising the Secretary of State for Education and Science on general policy and support for the Research Councils—anticipated a trend toward level funding.[140]

Both sides of the dual support system were hurt by the cuts. Government support for research and development in British universities through the UGC's block grant grew by about 7 percent from 1971/72 to 1978/79, although it was estimated that the amount spent for research per staff member actually decreased by 28 percent in this period as a result of the expansion of the universities and the increasing seniority of their faculties.[141] The Research Councils fared no better. Support for all of the councils fell by 11 percent in real terms between 1974/75 and 1978/79.[142]

This decrease in funding combined with the utilitarian emphasis of the Rothschild reforms of 1971 to shift the policies of the ABRC and of the individual councils away from a responsive mode of operation, in which the ideas and interests of individual scientists provided the primary impetus of scientific development, to an interventionist mode, in which scientific development was actively shaped by government priorities. The 1970s may be seen as a transitional stage in which policymaking bodies came slowly to support a need for a change in policy but at the same time attempted to continue the practices of the old regime. The tension is evident in the MRC's annual report for 1973/74. On the one hand, the MRC recognized that "the changes taking place . . . are inevitably away from the extreme laissez faire view which holds that the good of society is best achieved by the sum of the independent interests of individual scientists." On the other hand, the MRC insisted that "long-term fundamental research for which the application is not immediately apparent should be adequately safeguarded" and that, "while it is right for society to state priorities and objectives, decisions on matters of scientific feasibility, scientific strategy and management are matters for scientists."[143] A similar tension was reflected in the ABRC's 1979 recommendation to cut back funding for fields of "big" science such as high-energy physics, astronomy, space, and radio research in order to free resources to support new initiatives, and its warning in the same report against selecting fields for support solely on the grounds that they offered an immediate economic return.[144]

Despite efforts by the Research Councils to continue traditional commitments to undirected research, there were indications in the 1970s that a transition to interventionist policymaking was likely, and virtually inevitable if funding was held level or cut. Although it was later claimed that the Rothschild reforms had little immediate impact on the work of the Research Councils,[145] they almost certainly injected a new utilitarian emphasis into British science policy.

Such an emphasis was reinforced in 1976 when the government appointed a new committee, the Advisory Committee for Applied Research and Development (ACARD), with strong representation from private industry, to advise the government on the application of research and development in industry.[146] In 1979, ACARD produced a report on industrial innovation that called on the government to make publicly funded research and development more relevant to the needs of industry. Looking at the policies and strategies adopted by Britain's competitors abroad, the report concluded that too much attention had been paid to research and development for its own sake and not enough to the relevance of these activities to manufacturing and industry. It called for a "major effort . . . to place a greater emphasis in Britain on the sciences related to manufacture with special attention to the relationship between production processes and the design, quality and reliability of products, as well as to marketing." Many of the report's specific proposals were similar to those being considered by the Carter administration at roughly the same time: extra funds for research and development to support manufacturing technology, incentives for business investment in new plants and machinery, support for research and development by small businesses, and the strengthening of relationships between universities and private industry.[147]

Thus at the end of the 1970s the British government, like the U.S. government, had begun to create an interventionist agenda for science policy, in which scientific research was to be increasingly directed toward industrial goals. But it was Margaret Thatcher's Conservative government, elected in May 1979, that fully accomplished the change from a responsive to an interventionist science policy. Thatcher, like her ideological counterpart across the Atlantic, came into office with strong commitments to pursue cuts in government spending, privatization, deregulation, and the weakening of the trade unions. Perhaps most important of all, since it struck at the economic base, Thatcher responded to the United Kingdom's economic problems by instituting a form of economic triage rationalized by a philosophy of monetarism: inflation and the inefficiencies of British industry were to be cured by a stringent squeeze on credit. In John Kenneth Galbraith's words, the United Kingdom became a "Friedmanite guinea pig." The effect was to weaken an already ailing economy, bringing on a major recession and allowing British markets for manufactured goods to be increasingly penetrated by other nations.[148]

Under these increasingly difficult economic conditions, the Conservative government developed its science policy along two principal lines.[149] First, the science budget was held roughly constant, forcing funding agencies to define priorities and adopt selective policies. Both sides of the dual support system were squeezed. The Research Councils, after receiving an initial injection of funds honoring the commitment of the previous government, faced a long era of level funding. It became impossible to support both new initiatives and a full range of basic research. A study conducted by the Science and Engineering Research Council (formerly the Science Research Council) found that only half of the grant proposals in its top category were supported. Margaret Thatcher's response—that "people are not prepared to make decisions"—indicated that conditions would not change.[150] The universities fared no better. In November 1980, the government announced a policy of level funding for them as well. This was followed in March 1981 by the worse news of a 10 percent cut between 1981 and 1984.[151]

With this sustained squeeze on resources for research and development, the Thatcher government compelled implementation of the selective policies so long resisted by the Research Councils. A joint ABRC/
UGC report, which warned in 1982 of the damage to the research system
inflicted by cuts in support, also implicitly endorsed the need for selectivity by proposing the concentration of research into selected fields chosen by research committees in each university. Central oversight of such
plans would be provided by the UGC, which would review each university's selection strategy. Similarly, a report by the ABRC in the same
year, entitled *The Science Budget: A Forward Look,* argued that the "comprehensive coverage" of research in the 1970s in which funding agencies
attempted to maintain "a quality presence in virtually every field of science and technology where there were researchers willing and able to
work in it" was no longer possible. "We are faced," the report stated,
"with a situation in which gaps will and should appear." [152]

The second main emphasis of the Thatcher government's science
policy was to make research institutions more responsive to the needs of
U.K. industries. The opening lines of the ACARD report *Technological
Change: Threats and Opportunities for the United Kingdom,* published in
December 1979, bluntly stated: "The rate of technological innovation
in United Kingdom industry will need to increase if its products and
manufacturing processes are to match those of our major competitors.
This is a necessary condition of our future survival as a trading nation."
A series of official reports issued in the 1980s articulated the new interventionist science policy. Some of these urged development of new high-
technology fields. Others formulated general goals for British science.
The first joint report of the chairs of ACARD and ABRC, published in
1983, called for "improvement of linkages between sources of new ideas
and their potential exploiters" and urged development of "an educational
system that is geared to provide the country with the right skills at the
right time"; selectivity within as well as among research fields; priority
support for "generic technologies," such as computer technology and
genetic engineering, that could be applied to a variety of production
processes and products; concentration of work in "centres of excellence";
formation of the "right fiscal, financial, and regulatory climate for innovation"; and closer links between the universities and the private sector. [153] A second ACARD/ABRC report entitled *Improving Research Links
between Higher Education and Industry* recommended that the balance between basic and applied research in higher education be shifted toward
applied work, that industrial relevance be used explicitly as a criterion for
assessing research, and that "industrially-oriented research" be encouraged with additional support.

Despite strong rhetoric about the noninterference of government in the affairs of commerce, the Thatcher government swallowed its scruples when it came to shaping science to the needs of industry. The government's strong emphasis on development of research fields with industrial relevance meant that these fields received priority funding and that links between private industry and universities and Research Councils starved of support were greatly strengthened. The field of genetic engineering was apparently and often explicitly one of those to benefit.

Changes in the control and regulation of science and technology also occurred. Like the Reagan administration, the Thatcher government aimed at providing a social environment that placed minimal restraints on the activities of private industry. Although environmental regulation continued with little change,[154] the Thatcher government's impact on regulation of the workplace was substantial, for two main reasons. First, immediately following its election in 1979, the government launched a campaign to reduce government spending on "quangos" (quasi-autonomous nongovernmental organizations), bodies appointed by the government for a variety of advisory and regulatory tasks. Two of the most prominent quangos were the Health and Safety Commission and the Health and Safety Executive. A report issued by the prime minister in January 1980 announced the elimination of 246 quangos, with further cuts anticipated in the future. Although the Health and Safety Commission and the Health and Safety Executive could hardly have been eliminated by similar fiat, the report emphasized that their budgets were under scrutiny, drawing attention to the "rapid expansion" of these agencies and to their "unusually complex organization" and calling for "more explicit attention . . . to the costs and benefits of new health and safety measures."[155] Cuts in the commission budget duly came down, and substantial reductions were made in the inspectorate of the beleaguered agency in the 1980s.[156] At the same time, the power of the trade unions was being weakened both by growing unemployment and by the tough antiunion measures taken by the government. As jobs and pay increasingly dominated union concerns, the unions no longer were in a strong position to press for improvements in industrial safety.[157]

1.5 The Shaping of American and British Science Policy

In summary, the parallel declines of the British and American economies were accompanied by remarkably similar changes in the science policies of the two countries that greatly intensified competition among scientists for support for their work, facilitated a transition to selectivity in government funding, and channeled research toward the goals of government and private industry, particularly toward the provision of knowledge bases for new areas such as information technology and biotechnology.

Each country pursued these changes in its own distinct national style and in response to specific institutions and interest groups that were seen as obstacles to change. In the United States, some scientists and university administrators, accustomed to the liberal funding policies of an earlier era that respected scientists' autonomy, initially resisted pressures to gear research to industry needs. However, the huge influx of funds for research in high-technology fields made available by private investors and large corporations in the late 1970s and early 1980s (partly as a result of government tax and patent policies) proved too attractive, particularly in an era in which government support was harder to get; as chapter 2 shows, the 1980s saw ties between science and industry cemented as scientists and universities accepted industry support. In the United Kingdom, where the autonomy of scientific research had long been protected by institutional barriers and where funds for high-technology research were far more scarce than in the United States, the alignment between science and industry was achieved only by active state intervention. But the results were remarkably similar. Edward David's claim about the United States in 1982, that "autonomy is giving way in industry, government and academia alike to a close integration with the innovation system,"[158] applied equally well to the United Kingdom.

Each country also moved to weaken protective regulation of science and technology. In the United States, government commitment to regulation was undermined by the campaign launched by private industry to promote the notion of regulation as a barrier to economic growth and technological progress. Rampant deregulation was achieved under Reagan as a result. In the United Kingdom, the trends were more subtle. Environmental controls were not changed in any obvious way, but they were in any case considerably more flexible than those of the United States. In the area of industrial safety, the weakening of the trade unions and the cuts in support for the Health and Safety Commission in the Thatcher years considerably undermined workplace controls, without any change in the basic statute governing workplace safety.

In conclusion, the science policies of the United States and the United Kingdom in the 1980s were shaped to provide settings in which strong links between science and industry could be easily formed and in which the rapid development of new technology could be pursued without the restraints and efforts to anticipate side effects characteristic of the late 1960s and early 1970s. Genetic engineering was to be an initial testing ground for these policies.

TWO

The Social Transformation
of Recombinant DNA Technology,
1972–1982

While scientific, commercial, and other social interests were shaping government policies for the promotion and control of science and technology in the 1970s and 1980s, the field of genetic engineering emerged and came to maturity. This chapter examines the development of the new field from the early 1970s, when the possibility of genetic manipulation by recombinant DNA techniques was first demonstrated, to the early 1980s, when the field exploded commercially. In particular, I focus on the effects of its changing institutional backing, which transformed it from an almost exclusively academic field of research to one characterized by a network of corporate connections.

To explain why these changes occurred so abruptly and met so little resistance from members of the discipline, I shall argue in part that certain technical and social characteristics of molecular biology in the early 1970s made it ripe for practical application. However, these conditions alone cannot account for the intense pace of development that occurred later in the decade and into the 1980s. A comprehensive interpretation must also take into account the economic and political environment of the period, particularly the powerful incentives for commercialization generated by the state.

Throughout this analysis, it has been convenient for purposes of organization to refer to "technical" and "social" developments, but the distinction should not mislead the reader. The technical and social dimensions of molecular biology were closely intertwined: technical advances were often motivated by societal interests, and societal interests were shaped by technical advances.

In what follows, successive stages in the development of recombinant DNA technology are defined in relation both to technical innovations and to the response of the industrial and financial sectors that invested in the field. The data forming the basis of the account are accordingly diverse. The description of technical developments is based on scientific papers, reviews, and news articles of the period, as well as on interviews with those working in the recombinant DNA field and on archival sources such as grant proposals.[1] Undoubtedly a more detailed picture of the development of the recombinant DNA field will emerge as archival resources grow. Nevertheless, these data show substantial convergence for the issues under consideration. Although the materials revealed disputes about priorities and, in one case, a disagreement about the weight of a scientist's contribution to a major result, there is a high degree of consensus about the importance of specific techniques, experimental results, and conceptual advances for the development of the recombinant DNA field.

The account of the commercial response to recombinant DNA technology is based on evidence drawn from industry analyses, corporate prospectuses and reports, government studies, interviews with industry executives, and the abundant coverage of biotechnology in newspapers, scientific journals, and trade journals. Since industrial strategies are rarely revealed directly, this account is necessarily less complete and more tentative than that of the technical developments. The chapter concentrates on the impact of commercial interests on molecular biology rather than on the microstructure of those interests.

The evolving social and technical conditions of genetic engineering provided a further context in which government regulatory policy evolved. In particular, it will be argued in later chapters that emerging commercial and international competitive interests were responsible for shaping government policy in important respects. It might be argued that the reverse is true—that government policies shaped corporate responses to the possible hazards of genetic engineering—and in a restricted sense, this argument is persuasive: there appear to be few cases of corporate violations or circumventions of government policy, although the exceptions are revealing. On the other hand, there is much evidence to support the view that corporate and government interests in ensuring the international economic competitiveness of the field shaped government controls. Government controls were rarely a significant impediment to development: usually, they receded as the technology advanced. In general, the development of government policies for the control of recombinant DNA technology should be seen in the context of the commercial development of the field rather than the other way around.

2.1 Anticipations of Genetic Engineering, 1952–1970

Molecular biology in 1970 represented the culmination of a research tradition that originated in the 1950s in the merging of structural, biochemical, and informational approaches to the central problems of classical genetics. The main concepts underlying this tradition—that genes consist of DNA and that DNA encodes information determining the processes of replication and protein synthesis—were of course embodied in the model of DNA proposed by James Watson and Francis Crick in 1953. Two decades of research based on the Watson-Crick model produced dramatic theoretical and technical achievements. The genetic code was deciphered; the cellular machinery responsible for replication of DNA and protein synthesis was described in considerable detail; the biochemical pathways involved in replication, expression, and natural recombination were defined and the enzymes responsible for catalyzing these processes isolated; and a model for the regulation of protein synthesis in bacteria was developed and confirmed. These theoretical advances were reflected in new and impressive capacities to manipulate DNA. Viral DNA was replicated in the test tube; the lactose operon of the bacterium *Escherichia coli* was isolated and characterized; and the first complete chemical synthesis of a gene was achieved.[2]

This success in describing cellular processes at the molecular level had an immediate effect on related areas of research. The DNA model led to explanations for the genetic behavior of the simplest forms of life, the prokaryotes. It was now possible to define the main mechanisms of transfer of genetic information among bacteria in molecular terms. One of these mechanisms, conjugation, was shown to account for the transfer of drug resistance among bacteria through the transmission of small circular pieces of DNA called plasmids. A new class of enzymes, known as restriction enzymes, was discovered in the 1960s, and their function in protecting bacteria from foreign DNA inserted by bacterial viruses or by other means was clarified.[3]

How the DNA in the multiple chromosomes of higher organisms, the eukaryotes, functioned, how expression of this DNA was regulated, and how such processes could account for cellular differentiation and development were understood to be far more complex and difficult problems. But here too, major programs of research were under way. Many molecular biologists in the 1960s turned their attention to the study of animal tumor viruses, which, it was hoped, could be used as probes into the genetic functioning of the cells of higher organisms. As such, tumor viruses were seen as simultaneously offering a technique for elucidating the mechanisms of gene expression in higher organisms and a way of discovering how a normal cell was turned into a cancerous one. In spite of the complexity of such research problems, there was a strong sense in the 1960s that the DNA model would remain the basis for a coherent interpretation of biological processes and that ultimately the details of the functioning of DNA would be filled in.[4]

A notable feature of the molecular biology of 1970 was that, despite its achievements and the promise of further theoretical progress, the field had produced no major practical applications. A report issued in 1975 by one of the genetic engineering companies stated: "The 'practical' applications of molecular biology have been slow in coming. At the applied level, it is at this time still difficult to find any really important medical or industrial capability for which it matters at all that we know the genetic code or that DNA occurs in nature as a double helix."[5]

Two conditions, however, made practical application likely. First, new forms of technology were inherent in the very constructs of molecular biology; this was recognized by many practitioners in the 1950s and 1960s. If the living cell could be reduced to an information-processing machine, and mechanisms were found to alter at will the information fed to the cell, it followed that new forms of control and hence new practical capabilities would emerge.[6] Biochemist E. L. Tatum stated in his Nobel Prize lecture of 1958:

> With a more complete understanding of the functioning and regulation of gene activity in development and differentiation, these processes may be more efficiently controlled and regulated, not only to avoid structural or metabolic errors in developing organisms, but also to produce better organisms.
>
> Perhaps within the lifetime of some of us here, the code of life processes tied up in the molecular structure of proteins and nucleic acids will be broken. This may permit the improvement of all living organisms by processes which we might call biological engineering.
>
> This might proceed in stages, from the *in vitro* biosynthesis of better and more efficient enzymes, to the biosynthesis of the corresponding nucleic acid molecules, and to the introduction of these molecules into the genome of organisms, whether via injection, viral introduction into germ cells, or via a process analogous to transformation. Alternatively, it may be possible to reach the same goal by a process involving directed mutation.[7]

Second, research in molecular biology, while practiced chiefly in academia, was funded in a manner and on a scale conducive to practical results. As chapter 1 documents, in the United States, support for the field between 1950 and 1970 came largely from government funding of biomedical research, which grew exponentially in the postwar years. By the 1960s, support from molecular biology's primary sponsor was generous. As molecular biologist Sydney Brenner observed in 1974, "Watson and Crick may have invented [molecular biology], but Uncle Sam certainly fueled it."[8]

Uncle Sam did more than that. While research in the field was largely undirected in a specific, product-oriented sense, it was guided by the priorities and goals of its government sponsors. Work sponsored by the NIH was pursued with the general goal of understanding and controlling disease, and a major assumption underlying this support was that progress in basic research was a necessary condition for progress in medicine. A duality of purpose thus characterized molecular biology in the United States: the research itself was aimed at basic problems in biology, but the support for it was justified in terms of solving problems in medicine. Although most molecular biologists at this point probably saw their work primarily in terms of the first purpose, the ultimate rationale was practical. It is interesting and revealing that the term *genetic engineering,* meaning the "deliberate and controlled modification of the genetic makeup" of living things, gained currency at this time and that debate about the ethical and social implications of genetic engineering of human beings continued throughout the decade.[9]

It was also recognized that genetic engineering techniques would apply to animals, plants, and microorganisms. Tatum's former student, Nobel laureate Joshua Lederberg, foresaw "dramatic applications": viruses might be used to carry new genes into human or plant chromosomes, providing therapy for human genetic disease and the means to improve crop plants; animals might be used to generate specific genetic material as "spare parts" for human use.[10] James Danielli, head of the Center for Theoretical Biology at the State University of New York at Buffalo, anticipated an "age of synthesis" in which molecular biology would be applied in the creation of organisms tailored to carry out specific tasks in industry and agriculture—for example, bacteria to manufacture chemicals or digest pollutants and plants to fix nitrogen from the air.[11] Lederberg also saw a darker side to the possible applications. In 1970, he warned the U.N. Conference of the Committee on Disarmament that the techniques of molecular biology might be used to develop new weapons—novel infective agents against which there might be no defense.[12]

An indication that the practical potential of molecular biology was beginning to be taken seriously in the private sector was the establishment in 1967 of a lavishly supported molecular biology research institute in New Jersey by the giant Swiss pharmaceutical company, Hoffmann–La Roche. According to a report in *Science,* company literature represented the basic program of the institute in terms of a "long-range commitment to fundamental research designed to yield substantial benefits to humanity in terms of scientific progress." But it was also apparent that the company expected a return on its investment in "fundamental research." "Obviously, we hope for products," stated the company's president at a dedication ceremony for the institute in 1971. He went on to discuss the company's social responsibilities if and when advances in "fields such as

genetics, antiviral agents, and other areas provocative of social controversy" occurred. As the report in *Science* noted, "the belief that entirely
new forms of marketable therapy . . . may emerge from basic molecular
biology lies at the heart of the Roche approach to basic research."[13]

Across the continent in Berkeley, California, similar prospects guided
the establishment of a new firm known as Cetus. From its founding in
1971 by Ronald Cape, a biochemist and molecular biologist, and Peter
Farley, a physician, both of whom held MBA degrees, the ultimate goal
of the company was to tap the practical potential of molecular biology.
Cetus's initial business was based on a rapid screening process for selecting new strains of organisms with desirable commercial properties, such
as microorganisms producing high yields of antibiotics.[14] But this aspect
of the firm's activities was seen as a short-term capital-raising effort.
The long-term technological base, it was assumed, would develop out of
the "fundamental knowledge developed over twenty years in molecular
biology."[15] With this goal foremost, Farley and Cape recruited Joshua
Lederberg and microbiologist Arnold Demain, an expert in industrial
fermentation, as scientific advisers.[16]

Thus by the 1960s and 1970s, considerable efforts were being directed
toward the types of genetic engineering that Tatum, Lederberg, Danielli,
and others had anticipated. In general, the aim of this work was to develop the ability to insert into a recipient cell segments of foreign or
exogenous DNA (i.e., DNA from a source other than the chromosome
of the recipient cell). A review published in 1971 listed over five hundred
references to experiments that involved the insertion of foreign DNA
into mammalian cells.[17] Ways of genetically modifying higher organisms
by transduction and transformation were developed. Other work attempted to transform cells in tissue culture using selected DNA. Further
efforts were aimed at using animal viruses to carry foreign DNA into
human cells; a highly controversial procedure used an animal virus
known to make arginase to treat two children suffering from a hereditary
deficiency of this enzyme.[18]

Before 1970, important technical difficulties impeded these genetic engineering projects. Little or no control could be exerted over foreign
DNA transferred into a cell by transformation or transduction, or over
the mode of insertion into the cell's chromosomes. Moreover, both bacterial cells and cells of higher organisms possessed natural mechanisms
for protecting themselves against DNA recognized as foreign and tended
to degrade it. As a result, the procedures used in these early experiments
lacked precision, and the processes by which gene expression was occasionally obtained remained obscure. Uncertainties of this kind doubtless
accounted for inability to reproduce many experimental results.[19]

2.2 The First Gene-Splicing Experiments, 1969–1973

With hindsight, it is possible to see that by 1970 all the techniques necessary for genetic engineering with bacteria were available. Plasmids and
bacterial viruses provided vehicles to carry foreign DNA into living cells;
techniques that enabled bacterial cells to take up relatively large pieces of
DNA without being killed in the process were developed in 1970; methods for synthesizing DNA and for making DNA copies from messenger
RNA (mRNA) provided a limited means of making pure sources of
DNA; various enzymes discovered and isolated in the 1960s provided a
way to join DNA from different sources (an application demonstrated
by H. Gobind Khorana and his coworkers in the synthesis of the gene
for alanine transfer RNA, reported in 1970); and finally, the site-specific
restriction enzymes that were becoming available in the late 1960s made
it possible to cut DNA at exact locations.[20]

The potential use of these enzymes for cutting and joining diverse
pieces of DNA from different sources was soon recognized. Robert Helling, a participant in one of the first successful gene-splicing experiments,
recalled that in the late 1960s this idea was in the air.

> By 1968 or 1969, I was interested in thinking about some way to start
> manipulating genes—putting them together in new ways and so forth.
> There was somehow a feeling in the field—no one talked about it ex-
> plicitly—that the field had matured. Ligases were discovered, joining of
> ends, circularization, so there were ways you could put DNA together.
> So it was a natural progression of one's thought to start thinking of
> tying different kinds of DNA molecules together. It was something I
> had on my mind for a long time. And others did.[21]

Molecular biologists began to pursue these ideas experimentally. In
1967, for example, in a grant application to the NIH, Joshua Lederberg
proposed what must have been one of the earliest schemes for genetic
engineering in bacteria. Through procedures that are in some respects
strikingly similar to those used in the first successful gene-splicing experiments, Lederberg envisioned joining DNA from different sources
enzymatically. The hybrid DNA would then be introduced into the bacterium *Bacillus subtilis*.[22]

Typical of the duality of purpose that had come to characterize research
in molecular biology, Lederberg's goals were at once scientific and practical. On the one hand, he justified the work in terms of developing an
increased ability to examine the regulation of genes in bacteria: "The
research utility of freely moving genes from another species into bacteria
needs no elaboration, e.g., for the study of transcription-control, and to
facilitate analyzing the genetic competence of DNA from differentiated

tissues of higher organisms." On the other hand, he also pointed to the practical applications that might follow the development of a successful genetic transfer process: "Important practical utilities would follow from the incorporation of human genes into suitable cryptic virus DNA for kinds of transductional therapy for human genetic disease." [23]

Genetic Engineering Achieved

The first steps toward an effective method of genetic engineering were taken independently by Peter Lobban, a graduate student working under A. Dale Kaiser, a phage geneticist in the biochemistry department at Stanford University Medical School, and by Paul Berg, a biochemist in and chairman of the same department, together with two postdoctoral fellows, David Jackson and Robert Symons. Lobban and Berg conceived of the same type of procedure at roughly the same time. In essence, both groups envisaged, first, obtaining a well-defined piece of DNA; second, attaching this piece of DNA to a virus; and third, using the virus as a "vector" to carry the foreign DNA into a living cell. [24]

There were, of course, important differences between the scope of Lobban's research debut as a graduate student and that of Berg's research program, which was financed by a major grant from the NIH and staffed by a team of postdoctoral assistants, graduate students, and technicians. Berg's work was more ambitious, and as a recipient of National Cancer Institute funding, he had the long-term goal of understanding cancer at the molecular level through the study of the action of tumor viruses. But there were also striking similarities in the general aims and interests that informed the two sets of initial experiments: each project incorporated the dual theoretical and utilitarian concern that, as I have indicated, characterized American biomedical research as a whole.

Lobban and the Berg team each understood that techniques effective in genetic engineering might both serve as powerful tools for studying gene structure and regulation in higher organisms and permit industrial application. Lobban noted in his thesis proposal that the end product of a successful gene-splicing process would be bacteria capable of "synthesizing the products of genes of higher organisms." [25] David Jackson (an associate of Berg's) later recalled: "The concept of using bacteria as factories to produce mammalian proteins was a very current one in 1970 and 1971. That was one of the major motivations for wanting to do this piece of research, because as you thought of it some more, you could see more and more ways in which it was going to be very broadly applicable." [26]

The Berg group, whose procedures will be described here, used DNA from the bacterial virus lambda and from a monkey tumor virus known as SV40. Both of these pieces of DNA occur in the form of closed loops. In the first step of the process, each loop of DNA was cut with a restriction enzyme. The Berg group used the enzyme Eco RI, which had been isolated in the laboratory of Herbert Boyer at the University of California, San Francisco. Eco RI cut each loop in just one place, converting it into a linear molecule. Next, the ends of each piece of DNA were enzymatically modified using lambda exonuclease and terminal transferase. In a way similar to that proposed by Lederberg, adenine nucleotides were added to the ends of one type of DNA and thymine nucleotides to the ends of the other type. When the two types of DNA molecules were mixed together, their ends joined, forming circular hybrid molecules containing the SV40 and lambda virus DNA.[27]

Lobban's procedure, completed shortly before the Berg group's, was similar except that he used two molecules of circular DNA from the virus known as P22, which attacks the *Salmonella* bacillus, and converted them into linear pieces using detergent rather than restriction enzymes.[28]

Berg, with David Jackson and Robert Symons, completed the cutting and splicing of the SV40 and lambda virus DNA in 1971. He intended to continue by first using the SV40 virus as a *vector,* in the terminology that quickly came into general usage, to insert lambda virus DNA into animal cells and then using the lambda virus as a vector to insert SV40 DNA into cells of the bacterium *Escherichia coli.* This experiment was suspended in 1972, however, when news of Berg's intentions triggered concern about possible hazards, particularly the possibility that *E. coli* might inadvertently be transformed into a novel carrier of tumor virus DNA in human populations (see chapter 3). Other types of genetic engineering work continued, however.

By the fall of 1971, Berg and his colleagues were giving seminars on their work, and the idea of splicing DNA from different sources was becoming widespread.[29] The next step in the development of the techniques—the insertion of foreign DNA into bacteria in such a way that it would be replicated—was taken jointly by Herbert Boyer; Robert Helling of the University of Michigan, who spent the academic year 1972/73 working with Boyer; Stanley Cohen of Stanford University Medical School; and Annie Chang, Cohen's technician.

Prior to their collaboration, Boyer, Cohen, and Helling were each pursuing quite distinct problems in basic research. Boyer was exploring the molecular mechanisms involved in restriction and modification, the processes by which bacteria secrete restriction enzymes to protect themselves from "foreign" DNA and, through methylation, protect their own DNA against degradation by these enzymes. Cohen's and Chang's research focused on bacterial drug resistance, particularly the mechanisms governing the replication and transfer of plasmids. Helling was examining transduction, the process by which bacterial viruses transfer genes between bacterial cells, in *E. coli.*

All recognized the value of controlled genetic transfer for their research and at the same time were aware of the practical potential of the technique. In a grant application to the National Institute of General Medical Sciences in 1968, Boyer stated, "The applications of this area of research to the possible future of 'genetic surgery' are obvious."[30] This dual appreciation is also plain in Helling's recollection of his perceptions of the implications of gene splicing at this point.

> Probably the most important thing was that [controlled gene splicing] opened up the study of genes from other organisms to the same techniques available with viruses and *E. coli.* It meant that you could do anything with the genes of any organism. . . . It doesn't matter where a gene comes from. You can move it from the chromosome up to a plasmid, from one in three thousand genes to one in three. . . . You can amplify it on the plasmid. You can make enormous amounts of enzymes. . . . I think that the potential for affecting agriculture and medicine was obvious to us—I know it was.[31]

In the spring of 1972, the controlled transfer of foreign DNA into bacteria was brought closer by important observations on the action of the restriction enzyme Eco RI. Vittorio Sgaramella, a Stanford researcher, noticed that circular DNA molecules from the bacterial virus P2, after being cut by the enzyme, would reform spontaneously.[32] Shortly afterward Janet Mertz and Ronald Davis, of the biochemistry department at Stanford, suggested an explanation for this phenomena. They showed that the cuts made by Eco RI were not blunt, as had been assumed, but staggered, so that the break on one strand of the double helix was several nucleotides away from the break on the other. Like Sgaramella, they showed that the two DNA fragments created by the enzyme would spontaneously join, or recombine, with one another.[33] Their finding was confirmed when Boyer and his colleagues determined the exact composition of the nucleotide sequences at the "sticky" ends and showed that the enzyme cut the DNA so as to yield complementary sequences on the two projecting ends.[34]

This result, which became widely known in the summer of 1972, immediately suggested a way to cut and splice DNA that was much easier than the procedures used by Lobban and Berg, for the projecting ends produced by Eco RI were already able to link up with the ends of other DNA fragments produced by the action of the same enzyme. As Helling recalled: "Herb Boyer told me that they had just found out that the ends were 'sticky.' This instantaneously told me and presumably others in the field how we could tie DNA from different sources together."[35] Mertz and Davis expressed the same view in their published paper: "Quite possibly, . . . one may, in this simple way, be able to generate specifically oriented recombinant DNA molecules *in vitro*."[36]

Boyer and Helling immediately embarked on a research program that aimed at splicing a piece of foreign DNA into the bacterial virus lambda and using the virus to insert foreign DNA into the bacterial host *E. coli*. When problems arose with the use of the lambda DNA, the researchers joined forces with Cohen and Chang to search for a suitable plasmid to use instead. The plasmid they selected—later called pSC101—proved extremely useful, not only for this experiment but for many later ones. The plasmid, which could be cut open in a single place with the restriction enzyme Eco RI, carried the gene for resistance to the antibiotic tetracycline. Genes for resistance to a second antibiotic, kanamycin, were spliced into pSC101, and the hybrid plasmid was introduced into *E. coli* bacteria. Cohen, Chang, Boyer, and Helling showed that the bacteria had indeed taken up the hybrid molecule and that the new genes were being processed by the bacteria as if they were their own.[37] Successful experiments by Cohen and Chang using foreign DNA derived from *Salmonella* and *Staphylococcus* bacteria suggested that "interspecies genetic recombination [might] be generally attainable."[38]

The next step was immediately clear: the same procedures would be tried with genes from higher organisms. By August 1973, Cohen, Boyer, Helling, and Chang, together with John Morrow, a graduate student in Paul Berg's laboratory, and Howard Goodman, another member of Boyer's department, had shown that DNA from the toad *Xenopus laevis,* which normally codes for a specific type of RNA, could be introduced into *E. coli* and replicated in the bacterium. Further, they showed that the *Xenopus* DNA was transcribed into RNA.[39] This suggested, although it did not prove, that bacteria might also be induced to express genes from higher organisms. The bacterial synthesis of proteins made by higher organisms was a step closer to realization.

The Social Response to the First Gene-Splicing Experiments

What was remarkable about the *Xenopus* experiment was not the linking of DNA from different sources per se: the recombination of DNA in nature was a well-known phenomenon. This process normally occurred, however, only between the genes of related species and their viruses. The deliberate, controlled linking of genes from different species was seen, by contrast, as a novel and remarkable achievement. As Sydney Brenner of the Cambridge Laboratory for Molecular Biology wrote in a report to the first British committee to assess the implications of recombinant DNA technology (the Ashby committee): "It cannot be argued that this is simply another, perhaps easier way to do what we have been doing for a long time with less direct methods. For the first time, there is now available a method which allows us to cross very large evolutionary barriers and to move genes between organisms which have never had genetic contact."[40] A similar view was aired in a report circulated by Cetus Corporation in 1975.

> The significance of this power cannot be exaggerated. Perhaps the most important breakthrough has been the capability to transfer genes from one species to another.
>
> Classically, the definition of a "species" has been that an organism in such a group could *not* breed (i.e., exchange genetic material) with a member of another such group. Nature, through evolution, has created barriers to the exchange of genes *between* species. *Within* a species, breeding for improvement is possible and has long been practiced. However, it is not possible to create an animal with the combined characteristics of a dog and a cat by mating them. These species barriers have so long been accepted as logical and almost absolute that it is only within the past months that scientists have seriously contemplated the ramifications of breaking these species barriers.[41]

The impact of this new capacity to move pieces of DNA between species at will was understood by molecular biologists in the dual scientific and practical terms that characterized the perceptions of its inventors. On the one hand, they saw the ability to do controlled genetic engineering experiments as a powerful means to open up new lines of inquiry into the structure and function of DNA. "The scientific importance of what can be done with these methods cannot be overemphasized," Brenner impressed upon members of the Ashby committee. On the other hand, they also anticipated industrial applications. Brenner's report discussed the possibility that bacteria could be genetically reprogrammed to synthesize proteins normally made by higher organisms. Although engineering the bacterial regulatory apparatus to decode the genes of higher organisms was problematic, he outlined a technique for its accomplishment (section 2.3) and predicted that the industrial production of drugs such as insulin and the genetic engineering of agricultural plants would be achieved.[42] At the same time, the power of this capacity for moving pieces of DNA between species evoked growing concern that the new technology might generate significant health, environmental, and social problems.

As molecular biologists pondered the implications of their invention, news of its potential importance for industry began to filter into the rest of society. The tone and emphasis of press coverage depended crucially on the source and was, in general, mixed, for reasons to be addressed in chapter 3. But a strong theme throughout was the industrial potential of the technology. Many of the scientists close to the work stressed not only the potential for advancing the field of molecular biology but also its commercial significance. These uses were invariably portrayed as promising and beneficial. In justifying their research to funding agencies, scientists were used to equating its practical application with social benefits. It seemed natural, no doubt, to justify genetic engineering to the press in a similar way.

The business magazine *Fortune,* in an article run in early 1974 on the prospects for genetic engineering, featured Cetus and its scientific advisers Joshua Lederberg, Arnold Demain, and Donald Glazer and predicted that "transfers of DNA, those master molecules of life, [would be employed] to improve production capabilities of microbes."[43] Press coverage of the *Xenopus* experiment in May 1974, which was based on a press release from Stanford University Medical Center and interviews with Lederberg and Stanley Cohen, duly emphasized the "promise" of the new gene-splicing techniques for medicine, industry, and agriculture. The techniques could "revolutionize" the pharmaceutical industry, according to the *San Francisco Chronicle.*[44] The new methods will "meet some of the most fundamental needs of both medicine and agriculture such as supplies of now scarce hormones and nitrogen-fixing microorganisms," according to the *New York Times.*[45] Bacteria reprogrammed through the insertion of foreign genes to make commercially valuable substances such as insulin, growth hormone, and interferon began to be referred to as "factories." Profitable uses of gene splicing came to be counted prominently among its "benefits."[46]

While media images of the commercial potential of genetic engineering were optimistic, the corporate world seems to have responded slowly to them. Ronald Cape, founder and chair of Cetus, recalled in 1982 that, although he saw the commercial implications of gene splicing "immediately," most U.S. corporations failed to perceive its potential at this stage. Even at Cetus, an active program of research and development in genetic engineering did not begin until 1976.[47] Nevertheless, there were signs that some institutions were taking notice. The giant British chemical company Imperial Chemical Industries (ICI) initiated the monitoring of recombinant DNA techniques at their primary source, the universities. In the fall of 1974, ICI launched a joint research program with the Department of Molecular Biology at the University of Edinburgh. Forty thousand pounds was allocated to support a research group headed by Kenneth Murray, who had developed a bacterial virus for use as a vector in gene splicing.[48] In the same year Brian Hartley of Imperial College, London, began outlining a research and development program for the commercial production of enzymes for industrial use.[49] In November

1974, Stanford University filed for patents on behalf of Cohen and Boyer to cover both the basic processes they had developed with Helling and Chang and the plasmid cells, pSC1O1, used in the procedures.[50]

2.3 Visions of a Commercial Future, 1974–1976

Technical Advances

After the first experiments demonstrated the feasibility of gene splicing, research entered a new stage in which use of the techniques spread rapidly and problems associated with their development were attacked simultaneously on many fronts. By 1976, the NIH alone was sponsoring 123 projects, at an approximate cost of $15 million—up from two projects, at an approximate cost of $20,000, in 1975 (table 2.1).

Practical and commercial relevance was the lure of much of the work, and, as described in chapter 1, science policy in the United States and elsewhere encouraged this trend. It was easier to defend proposed scientific research to a funding agency if the result might advance the understanding and use of genes seen as relevant to health, agriculture, or industry. As work in the field progressed, an increasingly close blend of scientific and practical interests came to characterize research.

DNA from a wide range of sources began to be used in gene-splicing work.[51] Increasingly, in addition to genes derived from the cells of organisms traditionally studied in genetics—the fruit fly, frog, and mouse, for example—genes of practical and commercial importance were being selected. In the microbiology department at the University of California, San Francisco, work was launched in May 1977 on the cloning of the insulin gene. About a year later, at the Institute of Molecular Biology in Zurich, work began on the cloning of the gene for interferon, an antiviral

Table 2.1   NIH Support for Recombinant DNA Research

Fiscal Year	No. of Projects	Support (millions of dollars)	Fiscal Year	No. of Projects	Support (millions of dollars)
1975	2	0.02[a]	1979	847	103
1976	123	15[a]	1980	1,061	131
1977	349	42[a]	1981	1,400	164
1978	546	61	1982	1,588	185

Sources: National Institutes of Health; Office of Technology Assessment, Commercial Biotechnology: An International Analysis (Washington, D.C.: U.S. Government Printing Office, 1984), table 57, 310.

Note: Recombinant DNA techniques were often employed in only a part of given project.

[a]These estimates may be low since the data were collected before the NIH guidelines requiring notification of the use of recombinant DNA techniques went into effect.

agent released in very small amounts by human cells infected by viruses. Broad programs of plant genetics were initiated with the ultimate goal of tailoring food plants for increased yields, resistance to pests, and the ability to fix their own nitrogen. Plant viruses such as the cauliflower mosaic virus and the tumor-producing bacterium *Agrobacterium tumefaciens,* which act by transferring genes into plant cells, became the objects of intensive study as potential carriers of foreign DNA into plants.[52]

Most research during this period was guided by one or more of four general goals.

> *Control:* the development of the means to define and control precisely the piece of DNA to be inserted into a living cell and the manner of its insertion
>
> *Amplification:* the development of the means to increase the efficiency of replication of a piece of foreign DNA
>
> *Generalization:* the provision of a wide variety of sources of pure DNA so that gene-splicing techniques might be applied to any gene from any source
>
> *Expression:* the manipulation of bacteria so that they would not only replicate foreign DNA but also express genes from higher organisms, for example, by synthesizing specific proteins

Five developments were particularly important for advancing these goals. First, new vectors were constructed that replicated more prolifically in a bacterial cell. The plasmid used by Cohen, Boyer, Helling, and Chang, pSC1O1, made only one or two copies of itself in each cell. In 1974, Boyer, Charles Yanofsky of Stanford, and Donald Helinski of the University of California, San Diego, and their collaborators constructed a plasmid, designated Co1 E1, that could have as many as several thousand copies per cell. Construction of composite plasmids was aimed at combining the capacity to yield many copies within the cell, the presence of several sites that could be cut by different restriction enzymes, and the capacity to link with very large pieces of DNA. One versatile product was pBR322, a plasmid developed in Boyer's laboratory and later widely used in research and industry.[53] Convenient strains of the bacterial virus lambda were also developed in the same period.[54]

Second, techniques for preparing DNA for cloning were developed. Early methods used a source of DNA enriched in a particular gene, but this approach was limited by the availability of such sources. For genes coding for small proteins, it was possible to synthesize DNA using the methods developed earlier by Khorana.[55] But at the time, this technique could be used only for DNA whose structure was known.

How was it possible, in general, to locate and delimit a gene of interest on a long strand of DNA containing many genes? The approach taken in 1975 and 1976 was to use mRNA as a starting point. While DNA occurs in a roughly uniform concentration in all cells, a high concentration of mRNA corresponding to a particular gene product occurs in cells that specialize in synthesizing that product. (For example, the beta cells of the pancreas contain high concentrations of insulin mRNA, which is in the process of being translated into insulin.) mRNA corresponding to a particular gene was converted into DNA using reverse transcriptase, an enzyme discovered in 1970 that can copy a single strand of mRNA into a complementary strand of DNA, known as copy DNA (cDNA). The disadvantage of this method was that only the parts of DNA that are transcribed into mRNA were obtained; other parts, such as the surrounding control regions, were not produced.[56]

An alternative approach to gene preparation was to take a whole strand of DNA with all of its unknown genes, cut it into many pieces with a restriction enzyme, and clone all of the fragments. This became known as the shotgun method and was widely used from 1974 on.[57]

Third, methods for recognizing and selecting colonies of bacteria that contained a specific type of DNA were developed. Typically, restriction enzymes cut DNA into many fragments (and in shotgun experiments that number was large), thus presenting researchers with the problem of picking out the bacteria containing the target gene. Early experiments solved this problem in various ways: for example, genes that supplied functions necessary to the survival of the bacterial host could serve as tools for selection. These methods could not be generalized, however.[58]

By 1975, new selection methods made it possible to screen clones of bacteria containing many different DNAs directly on the plates on which they were cultured and to select a clone containing a desired gene. These methods used as a probe a minute amount of radioactively labeled DNA complementary to the gene of interest and able to bind only to that gene. When the bacteria on the culture plates were exposed to this probe, it could pinpoint the gene and clone of interest.[59]

Fourth, major advances occurred in the ability to define the genes chosen for cloning in molecular terms. Specifying the precise nucleotide sequences of a gene (sequencing) had been an extremely laborious procedure, possible only with a small number of units. In 1975 and 1976, new methods, introduced by Frederick Sanger at Cambridge University and by Allan Maxam and Walter Gilbert at Harvard, made it possible to sequence DNA with relative ease. These techniques greatly advanced the study of DNA structure, especially of those segments that do not code for proteins; they also made it possible to define exactly the DNA segments used for cloning purposes.[60]

Fifth, the new capacities for sequencing DNA and advances in its chemical synthesis produced further refinements. It became possible to synthesize "sticky" DNA ends and to add these to blunt-ended pieces of

DNA that experimenters wished to clone. These artificial sticky ends, or linkers, became the snap couplers of gene splicing, allowing researchers to insert DNA fragments into precise sites in vectors almost at will.[61]

Of the four main goals that guided work at this stage, the most ambitious was that of expression (the goal that had been anticipated even before the first successful recombinant DNA experiments). From the outset, this goal embodied both scientific and practical interests. Expression was seen as the means to make large amounts of gene products for further study and to probe the differences between gene regulation in higher and lower organisms. At the same time, it was recognized that expression could make possible the commercial production of substances normally made only by higher organisms (often in minute amounts). This blend of interests is clearly reflected in an account of developments in the field in July 1976.

> The possible expression of eukaryotic DNA [i.e., that from higher organisms] to produce functional proteins when cloned in E. coli has been a key question. A positive answer would greatly facilitate [study of the genes of higher organisms]. . . . Another main reason for the interest in whether eukaryotic genes can be expressed at the protein level is industrial. Will the vectors which have been developed for E. coli hosts enable certain valuable eukaryotic proteins to be produced in organisms such as E. coli when grown in fermenters?[62]

In 1974, molecular biologists had no clear answer to that question, but they were beginning to explore possible experimental approaches to it. As noted above, one of the earliest analyses of the problem was given by Sydney Brenner in a memorandum to the Ashby committee in 1974. Brenner argued that bacteria probably would not automatically process the genes of higher organisms, but he anticipated a solution to the problem. He proposed linking a chosen animal gene, such as the gene for insulin, directly to a bacterium's control element, the operon. He predicted that this would guarantee that the product would be synthesized by the bacterium, although he acknowledged that there were questions about whether the protein produced would fold in the normal way. He also predicted that, once expression was achieved, the yield (a critical parameter once commercial considerations entered the picture) could be increased using "standard genetic tricks."[63]

By 1975, several research groups had begun to explore the expression problem. Chang and Cohen examined the processing of mouse genes in bacterial cells; the proteins produced differed from those produced by the same genes in mouse cells.[64] Ronald Davis and his coworkers produced evidence suggesting that E. coli could process a yeast gene.[65]

In general, most eukaryotic genes inserted into bacteria before 1977 were not automatically processed and expressed; whether they could be remained an open question. Some of those involved in developing the techniques, however, believed that it was just a matter of time—perhaps only one or two years—before the problem would be solved. Lederberg, for example, recalled that "many people made a lot of fuss at that time about not [yet] getting faithful expression of eukaryotic DNA in bacteria. . . . I was sure that [the problem] would soon be surmounted."[66]

Commercial Responses

In contrast to the intense business interest in genetic engineering that would emerge only a few years later, the response from the commercial sector at this point was generally cautious. An exception was the small firm Cetus, which enthusiastically embraced a vision of a commercial future based on gene splicing. In a report circulated to potential investors in 1975, Cetus described the "revolution" to come in glowing terms. Contrasting the results to be achieved by genetic engineering with those due to natural evolution and conventional breeding, the authors wrote:

> No process of mutation or evolution will ever cause a micro-organism to manufacture insulin, human growth hormone, or an antibody to save a patient dying of encephalitis. The changes in DNA necessary to do that are so complicated that it is statistically valid to say that they will *never* happen randomly. Gene splicing can and will make these things possible. Gene stitching finesses evolution; it breaks nature's species barriers. We propose to do no less than to stitch, into the DNA of industrial micro-organisms, the genes to render them capable of producing vast quantities of vitally-needed human proteins. . . . Cetus proposes to make these proteins in virtually unlimited amounts, in industrial fermenters. . . . We are proposing to create an entire new industry, with the ambitious aim of manufacturing a vast and important spectrum of wholly new microbial products using industrial micro-organisms.[67]

An extensive list of potential products followed, ranging from hormones, vaccines, single-cell proteins, and amino acids to enzymes, antibiotics, and interferon.

This vision of the future was not entirely an effect of public relations hype but was, to some extent, shared by other organizations with commercial interests. ICI's grant to the Department of Molecular Biology at the University of Edinburgh and Stanford University's application for patents on the results of the work by Cohen, Boyer, and Helling have been noted. In addition, early in 1975, Cohen was recruited by Cetus as a member of its board of scientific advisers. In April 1976, Herbert Boyer was persuaded to join Robert Swanson, a thirty-two-year-old businessman with a background in venture capital, to form Genentech, a company specifically committed to developing commercial applications of gene manipulation. By the fall of 1976, at least six multinational drug companies—Hoffmann–La Roche, Upjohn, Eli Lilly, Smith Kline and French, Merck, and Miles Laboratories—had initiated small research programs in genetic engineering.[68]

While these events indicated interest in the commercial development of genetic engineering, they did not necessarily reflect strong commitments on the part of major corporations. The amount of capital invested in gene manipulation by private sources was small. Boyer and Swanson started Genentech with $1,000 of their own money. Cetus found it difficult to interest large drug and chemical companies in forming contracts for research and development projects.[69] (In comparison, the main supporter of genetic engineering, the U.S. government, had invested roughly $15 million in research through fiscal year 1976; see table 2.1.) The attitude of the large multinational companies that dominated the pharmaceutical and chemical industries, if they were interested at all, seems to have been to establish a means of following developments in molecular biology and to wait and see what emerged. As one ICI scientist put it, "We see our job as poaching on all the academic work now going on."[70] The same point of view was expressed more bluntly in an internal ICI memo: "To invest money in [a genetic engineering company] to achieve expression of mammalian genes in *E. coli* would be worthless because the likelihood is that this will first be done by one of the many academics working in this area; then we will all have access to the information."[71]

2.4 Genetic Engineering Enters the Business Arena, 1976–1979

The refinements in gene splicing achieved in the years immediately following the first experiments maintained the rapid growth of the field. The number of projects using these techniques that were supported by the NIH alone grew nearly fivefold from 1976 to 1978 (table 2.1). A broad spectrum of research programs was initiated, and the techniques were applied to a wide range of hosts other than *E. coli: Bacillus subtilis;* the bread mold *Neurospora:* yeast, plant, and animal cells in culture; and even whole plants and animals. The following account focuses on research using *E. coli,* but it is important to realize that this work was only part of a complex of research programs undertaken at this time.

Technical Advances

A blend of basic and applied, of analytic and synthetic interests characteristic of the earlier phases of development, continued to guide research.

Research aimed at understanding the structure and function of genes at the molecular level and research aimed at practical results, particularly bacterial synthesis of the gene products of higher organisms, progressed simultaneously, often in the same laboratory.[72]

In 1976–79, analytic work using cloning and sequencing techniques produced a wealth of new data on the molecular structure and organization of DNA. These data began to suggest problems with the accepted model of DNA, particularly its assumptions that the transfer of genetic information in the cell proceeded linearly from gene to product and that the genes themselves occupied defined, stable sites on chromosomes.

Even before the advent of recombinant DNA technology, there was good reason to suspect that the organization and regulation of genes in higher organisms would be more complex than those in bacteria. The differentiation of cells in higher organisms suggested the existence of intricate control mechanisms. Further, evidence on the variation of genome size with species suggested that the proportion of DNA in higher organisms that coded for proteins (known as structural DNA) was low; it thus raised questions about the function and organization of the remaining DNA in these organisms.[73] Even so, the new data surprised scientists accustomed to thinking in terms of the models that had worked so well for bacteria and their viruses.

A major surprise came in 1977, when the first evidence that the structural genes of higher organisms were discontinuous, unlike those of their bacterial counterparts, was presented at a conference at the Cold Spring Harbor Tumor Virus Laboratory. According to *Nature,* the scientific audience was "amazed, fascinated, and not a little bewildered." As data accumulated, it became apparent that most structural genes in higher organisms are broken up into stretches of nucleotides that code for amino acids and stretches that do not code for anything. The "silent," noncoding regions were called intervening sequences or introns.[74]

Subsequently, much evidence suggested that in expression both coding and noncoding DNA is transcribed into a long precursor RNA molecule and that this molecule is then somehow "edited" by the cell to remove the stretches corresponding to introns, leaving the shorter mRNA molecules that carry translatable parts of the code spliced together in order. The reason for the existence of introns and their processing mechanism remained mysterious, however.[75]

In the same period, the practical problem of getting bacteria to synthesize the gene products of higher organisms became the object of intense research activity in the United States and Europe. The focus of research was not the expression of genes in general, but the expression of genes coding for commercially important proteins. The substances anticipated in the Cetus report—hormones, vaccines, interferon, insulin—were now seen as immediate goals.[76]

The discovery of introns introduced a new problem in realizing bacterial synthesis of the gene products of higher organisms since, as was soon recognized, bacteria do not have the machinery to remove introns. As it happened, however, two ways around this difficulty already existed. For small proteins whose amino acid sequence was known, intron-free DNA molecules could be made synthetically. For larger genes, experimenters could start with mRNA (if available), which contained the information encoding a protein in continuous form, since introns in the code had already been removed. Intron-free cDNA could then be made from the mRNA with the enzyme reverse transcriptase.

But the use of intron-free DNA did not overcome the incompatibility between the bacterial apparatus for expression and animal structural genes. The strategy pursued by several laboratories to circumvent this problem was to attempt to trick a bacterium into processing a gene from a higher organism as if it were its own, by placing the gene immediately after the bacterial operon—essentially the scheme proposed by Brenner in 1974.

Confirmation of the feasibility of this approach came in the fall of 1977 when teams headed by Herbert Boyer at the University of California, San Francisco, and by Keiichi Itakura at the City of Hope National Medical Center in Duarte, California, used this method to achieve the first complete bacterial synthesis of an animal protein. The animal gene used was a chemically synthesized gene for a small peptide hormone, somatostatin, produced in the human brain. The plasmid employed in the experiment contained the entire regulatory part of the lactose operon and a part of the structural gene coding for the enzyme beta-galactosidase, followed immediately by the somatostatin gene. (Just how much of the beta-galactosidase gene was needed for the procedure to work was determined by trial and error.) The composite plasmid was inserted into *E. coli*. When the lactose "switch" was "turned on" by exposing the bacteria to lactose, the somatostatin DNA was expressed along with the beta-galactosidase DNA as if it were a bacterial gene. The outcome was a hybrid, or "fused," protein, consisting of beta-galactosidase and somatostatin. After this protein was extracted from the bacteria, the somatostatin part was separated chemically from the beta-galactosidase part.[77] What had been from the beginning a primary goal of gene splicing had been accomplished.

Over the next two years, this technique for obtaining the bacterial expression of fused proteins as well as methods of gene synthesis, sequencing, cloning, and detection were energetically developed by laboratories

Table 2.2 Announcements of Bacterial Expression of Commercially Important Substances, 1977–80

Date of Announcement	Journal and Date of Publication	Substance	Company or University	Support	Source for Announcement
2 November 1977[a]	S 12/9/77	Somatostatin	City of Hope Medical Center; University of California, San Francisco	Genentech	NYT 11/12/77
June 1978	PNAS 8/78	Rat proinsulin	Harvard; Joslin Diabetes Foundation	NIH	WSJ 6/12/78
September 1978	PNAS 1/79	Human insulin	Genentech; City of Hope Medical Center	Genentech	WSJ 9/7/78
December 1978	PNAS 12/78	Chicken ovalbumin	Upjohn	Upjohn	BG 12/25/78
January 1979	N 12/21/78	Rat growth hormone	Howard Hughes Medical Institute; University of California, San Francisco	NIH; NSF; Lilly	CEN 1/22/79
July 1979	S 8/10/79	Human growth hormone	Howard Hughes Medical Institute; University of California, San Francisco	Lilly	NYT 7/17/79
July 1979	N 10/18/79	Human growth hormone	Genentech, City of Hope Medical Center	Genentech, under contract with KabiGen AB	NYT 7/22/79
January 1980	N 3/27/80	Human leukocyte interferon	Biogen	Biogen	WSJ 1/17
March 1980	B 12/80	Thymosin alpha-1	Genentech	Genentech; National Cancer Institute; Hoffmann–La Roche; Battelle Memorial Institute	WSJ 3/10/80

Sources: B = *Biochemistry;* BG = *Boston Globe;* CEN = *Chemical and Engineering News;* N = *Nature;* NYT = *New York Times;* PNAS = *Proceedings of the National Academy of Sciences of the United States of America;* S = *Science;* WSJ = *Wall Street Journal.*

[a] Announcement was made at a congressional hearing.

in the United States and Europe, several of which were engaged in races to establish priority for the production of particular substances. Expression of a variety of animal genes, including those for growth hormone, insulin, and interferon, followed in quick succession (table 2.2).

Commercial Responses

Following the somatostatin experiment, a moderate supply of capital from corporate and venture capital sources began to flow into the genetic engineering field. Standard Oil of Indiana seems to have initiated this trend when it bought $10 million of shares in Cetus in October 1977 and January 1978. Other investments of a similar nature followed (tables 2.3 and 2.4).

At least two new companies appeared at this time. Genex Corporation was founded in Maryland in July 1977 by J. Leslie Glick, chairman of Associated Biomedic Systems, a firm specializing in research and development in the life sciences and in the manufacture of products for biological research, and Robert Johnston, a venture capitalist, with initial funding from the Koppers Company and from InnoVen, a venture capital firm. Paul Berg's former collaborator David Jackson was hired as chairman of Genex's scientific advisory board in the same year.[78] In Europe, Biogen was founded by two businessmen, Dan Adams and Ray Schafer, and nine scientists, including the genetic engineers Walter Gilbert (who would resign his position at Harvard to become Biogen's chief executive officer in 1981), Kenneth Murray, Brian Hartley, and Charles Weissmann. The company received substantial backing from International Nickel, a Canadian metals corporation.[79]

These investments in the new genetic engineering firms in 1977 and 1978 marked the beginning of interest in genetic technology on the part of multinational corporations. In addition to buying shares in the new genetic engineering firms, between ten and fifteen such corporations in the United States had initiated in-house work by the end of 1977. Further, in August 1978, Eli Lilly teamed up with Genentech in a joint venture to develop an insulin production process. Lilly also signed a similar contract with a competing research group at the University of California, San Francisco. In Europe, Unilever started a genetic engineering program for cross-breeding plants. And Hoechst, the huge German chemical and pharmaceutical company, established a research program aimed at the production of insulin and antibiotics.[80]

At this stage, three distinct forms of commercial interest in genetic engineering emerged. First, venture capital companies such as Kleiner and Perkins, Wilmington Securities, and InnoVen (a concern established by Monsanto and Emerson Electric) began to see the new bioengineering

Table 2.3 Equity Investments in Genetic Engineering Firms, 1976–80

Year	Investing Corporation	Genetic Engineering Firm	Amount (millions of dollars)	Source
1976	International Nickel	Cetus	0.5	Dickson
1977	Standard Oil of Indiana[a]	Cetus	8.5	Cape
	International Nickel / Kleiner and Perkins / Mayfield Fund / Monsanto	Genentech	1.0	Debut; Lewin
1978	National Distillers[a]	Cetus	5.0	Cape
	Standard Oil of Indiana[a]	Cetus	4.5	Cape
	Standard Oil of California	Cetus	13.0	Cetus
	Wilmington Securities	Genentech	0.5	Genentech
	InnoVen[a]	Genex	0.475	Glick
	International Nickel[a]	Biogen	0.35	Biogen
1979	Lubrizol	Genentech	10.0	Genentech
	Schering-Plough[a]	Biogen	8.0	OTA[b]
	International Nickel	Biogen	1.25	Biogen
	Koppers[a]	Genex	3.0	Genex
1980	Abbott Labs	Amgen	5.0	OTA
	International Nickel	Biogen	4.61	Biogen
	Koppers	Genex	12.0	Genex
	Lubrizol	Genentech	15.0	OTA
	Monsanto	Biogen	20.0	OTA
	National Patent Development Corp.	Interferon Sciences	0.6	OTA
	Nuclear Medical Systems	Genetic Replication Technologies	0.95	OTA
	Schering-Plough	Biogen	4.0	OTA
	Tosco	Amgen	3.5	OTA

1981	Bendix	Engenics[c]	1.75	OTA
	General Foods	Engenics[c]	.5	OTA
	McLaren Power & Paper Co.	Engenics[c]	1.25	OTA
	Mead Co.	Engenics[c]	1.25	OTA
	Dennison Manufacturing Corp.	Biological Technology Corp.	2.0	OTA
	Allied Corp.	Calgene	2.5	OTA
	American Cyanamid	Molecular Genetics	5.5	OTA
	Campbell Soup	DNA Plant Technologies	10.0	OTA
	Continental Grain	Calgene	1.0	OTA
	Dow	Collaborative Research	5.0	OTA
	Ethyl	Biotech Research Laboratories	.95	OTA
	Fluor	Genentech	9.0	OTA
	International Nickel	Biogen	2.5	OTA
	International Nickel	Plant Genetics	N.A.[d]	OTA
	Koppers	Engenics[c]	1.25	OTA
	Koppers	DNA Plant Technologies	1.7	OTA
	Phillips Petroleum	Salk Institute Biotechnology/ Industrial Associates	10.00	OTA
1982	Baxter-Travenol	Genetics Institute	5.0	OTA
	Corning/Genentech	Genencor	20.0	OTA
	DeKalb	Bethesda Research Laboratories	0.6	OTA
	Getty Scientific Corp.	Synergen	4.0	OTA
	Gillette	Repligen	N.A.[d]	OTA
	Johnson & Johnson	Enzo Biochem	14.0	OTA
	Kellogg	Agrigenetics	10.0	OTA
	Lilly	International Plant Research Institute	5.0	OTA
	Lubrizol	Sungene	4.0	OTA
	Martin Marietta	Molecular Genetics	9.7	OTA
	Martin Marietta	Native Plants	5.0	OTA
	Martin Marietta	Chiron	5.0	OTA
	Schering-Plough	DNAX	29.0	OTA

(continued)

Table 2.3 (continued)

Year	Investing Corporation	Genetic Engineering Firm	Amount (millions of dollars)	Source
1983	Allied Corp.	Genetics Institute	10.0	OTA
	BioRad	International Plant Research Institute	1.0	OTA
	Martin Marietta	Chiron	2.0	OTA

Sources: Biogen = Biogen N.V., prospectus, 22 March 1983, 37; Cape = Ronald Cape, personal communication, July 1984; Cetus = Cetus Corporation, prospectus, 6 March 1981, 32–33; Debut = "A Commercial Debut for DNA Technology," *Business Week*, 12 December 1977, 128; Dickson = Dickson, "Recombinant DNA Research: Private Actions Raise Public Eyebrows," *Nature* 278 (1979): 494–95; Genentech = Genentech, Inc., prospectus, 14 October 1980, 29; Genex = Genex Corporation, prospectus, 29 September 1982, 10; Glick = J. Leslie Glick, personal communication, 27 July 1984; Lewin = Roger Lewin, "Profile of a Genetic Engineer," *New Scientist*, 28 September 1978, 925; OTA = Office of Technology Assessment, *Commercial Biotechnology: An International Analysis* (Washington, D.C.: U.S. Government Printing Office, 1984), table 13, 100–101[b]

[a]Investing corporation represented on the board of directors of genetic engineering firm. (This information is noted only for Biogen, Cetus, Genentech, and Genex.)

[b]This source shows equity investments in firms involved in a wide range of new biotechnologies. Of these firms, only those listed as using recombinant DNA technology in the *Genetic Engineering News* annual guides to biotechnology companies are included in this table; see *Genetic Engineering News* 2, no. 6 (1982); 3, no. 6 (1983); 4, no. 8 (1984).

[c]Company established by Stanford University and the University of California, Berkeley.

Table 2.4   Aggregate Equity Investments in Genetic Engineering Firms,
1976–83

Year	Amount (millions of dollars)	Year	Amount (millions of dollars)
1976	0.5	1980	72.4
1977	14.5	1981	49.4
1978	18.8	1982	111.3
1979	22.3	1983	13.0

Sources: See table 2.3.

firms as attractive possibilities for high-risk investments. The primary goal was to obtain extremely high rates of return—possibly as high as 100 or even 1,000 percent—from the early rapid growth of successful new industries or businesses. The risk, of course, lay in the placement of capital.

Venture capital companies were gambling that no unforeseen problems, either social or technical, would interfere with the first few years of growth of the new genetic engineering firms. (They were not to be disappointed: the return for Wilmington Securities on its 1978 investment of $0.5 million in Genentech was about 1,600 percent by 1980.) These companies were encouraged, too, by the cuts approved by Congress in the 1970s in taxes on capital gains. By 1978, Congress had reduced such taxes to a rate of 28 percent, down from 48 percent a decade earlier.[81]

Second, and for rather different reasons, multinational corporations also began to invest in the new firms. At this stage, multinationals saw the prospects for genetic engineering largely in defensive terms. Their primary objective was to protect existing products and markets from being undermined by the new technology, by securing a dominant position with respect to its development and application. Agreements such as Eli Lilly's insulin contracts with Genentech and the University of California, San Francisco, were designed to preserve domination of world markets for specific products. A second concern related to the first was to position themselves to dominate new routes of diversification. Small-scale, in-house research on recombinant DNA provided a source of expertise for evaluating developments in the field. Investments in genetic engineering firms bought not simply shares but also representation on the boards of directors of these companies, providing, as it were, an early-warning system regarding the prospects of the new biotechnologies. As a corporate planning manager for Standard Oil of California stated in 1979, the company wanted "a window on this technology."[82]

Third, the new firms themselves had nothing to protect and everything to gain from a swift entry into the new field. Their primary interest was to survive, grow, produce, and market their own products in competition with one another and eventually with the corporations as well. Biogen's first prospectus made these goals clear. "It is not Biogen's strategy," the company's directors wrote, "to conduct research on a cost-plus contract basis. Instead, its goals are to become a research-based production and marketing company and to participate in the eventual commercialization of its research. . . . Increasingly, Biogen seeks to retain certain markets, production rights, joint ventures, and other interests." [83]

In order to survive, new firms in the field not only had to acquire sufficient capital to finance the expensive early years of research and development before marketable products emerged but also to lure top-ranking scientists and skilled technicians with competitive salaries and attractive fringe benefits. Top-quality staff were essential both for carrying out fast-moving, competitive research and development and for attracting capital. As a report of the U.S. Office of Technology Assessment later observed, the number of Ph.D.'s employed became as important for rating the quality of new genetic engineering companies as more traditional indicators such as earnings and rate of growth. It was no accident, then, that these firms sprang up close to the campuses responsible for much of the pioneering work in the field and that a very high proportion of their scientific personnel came from university laboratories. [84]

From the beginning, the new firms used aggressive research and development, proprietary, and communications strategies to advance themselves. The owners of Genentech noted that "the Company plans to establish a strong proprietary position in its technology through the diligent pursuit of its patent claims and the protection of its knowhow [through trade secrets]." [85]

The firms promoted an exuberantly optimistic picture of the future for genetic engineering, based on images of a high-growth, high-technology field. Press coverage in 1977 and 1978, for example, featured the following statements made by executives and scientific advisers for these companies:

> "There's an explosive new industry here, no doubt about it." (Ronald Cape, Cetus)
>
> "I expect that genetic engineering in its broadest sense will be involved with industry within the next few years." (Arnold Demain, Cetus)
>
> "Molecular biology has reached the point where it can become involved in industrial applications. . . . The field is opening up rapidly." (Herbert Boyer, Genentech)
>
> "We're building another IBM here. . . . The opportunities are incredible." (Peter Farley, Cetus) [86]

These small companies tempted investors with a vast array of potential applications covering a broad spectrum of production technologies: in medicine, new drugs such as hormones, vaccines, antibiotics, and anti-viral agents (e.g., interferon); in agriculture, new plants that would fix their own nitrogen; in food production, synthesis of amino acids and vitamins; and in the chemical and mining industries, enzymes for use in manufacturing, pollution control, medical diagnostics, and food process-ing and bacteria for refining rare minerals and cleaning up pollution.[87]

In summary, large multinational corporations, venture capital compa-nies, and fledgling genetic engineering firms followed distinct strategies toward the development of the recombinant DNA field. By the end of 1978, increasing competition for entry into and dominance of the new field had clearly marked the discourse surrounding its potential with a new sense of urgency. As the *Economist* noted in December 1978, an "industrial cornucopia" of "prizes" from genetic engineering was on its way, and the "list of companies climbing on the genetic engineering bandwagon" was growing. And as headlines in the American *Boston Globe* and the British *Evening Standard* ran, "Clone business: It's growing fast. It's growing fast" and "All aboard the gene machine."[88]

2.5 The "Cloning Gold Rush," 1979–1982

The Economic and Political Environment

In 1975, the authors of the Cetus Corporation's "Special Report" had stated that "it is only a matter of time before the race begins to [exploit recombinant DNA] developments commercially," but they could not have foreseen the rapid acceleration of this race or the extent to which sectors other than private industry would become involved. It is unlikely that these changes would have occurred so rapidly in the United States without a major shift in the economic and social priorities that sur-rounded technological development. When the new techniques first ap-peared in 1973, public interest in federal controls for regulating the development of new technologies was strong, and the wave of protective legislation initiated by Congress in the 1960s had not yet passed. The sensitivity of the general public, members of Congress, and government agencies to issues bearing on the protection of the consumer, workers, and the environment from the side effects of science and technology re-mained high. The following years saw a reversal of regulatory policy in general, as the business sector actively sought a reduction of protective controls and the initiation of new forms of government support for tech-nological development in the private sector (chapter 1). In the late 1970s, this agenda began to be translated into government policy, producing progressive weakening of regulatory controls (transformed into an inten-sive assault after the election of Ronald Reagan) and the formation of new science and technology policies that promoted technological innovation and encouraged the formation of new links between the universities and the private sector. Government support for science and technology was selectively directed toward supporting those fields that promised to alleviate the decline of American industry by providing increases in productivity, new products, and new bases for industry. Recombinant DNA and other new biogenetic technologies satisfied all of these conditions and were specifically targeted. Furthermore, investment in genetic engineering was facilitated by the Supreme Court's landmark decision *Diamond v. Chakrabarty,* which allowed patent coverage not only on the processes for making novel organisms but also on the organisms themselves.[89]

Technical Advances

The level of laboratory activity and the support for it grew extremely rapidly after 1978. The leading source of government support, the NIH, increased funding for recombinant DNA research at a rate of about 34 percent per year in the period from 1978 to 1982. In that period, the number of projects jumped by 191 percent, from 546 to 1,588, and support itself jumped by 203 percent, from $61 million to $185 million (table 2.1). Investments in recombinant DNA activities from the private sector climbed even more rapidly.

The dual analytic and synthetic thrusts of research continued, yielding, as in the earlier periods of development, results pertinent to both science and commerce. With respect to analytic research, the use of rapid sequencing, cloning, and other techniques produced vast quantities of data on the structure and organization of DNA as well as a series of important discoveries and surprises. The discovery of introns has already been described; other theoretical puzzles—the discovery of anomalous forms of DNA, of extensive repetitions of DNA bases, of transposable genes, and of a variety of placements of control regions in relation to structural genes—further complicated the picture.[90]

The proliferation of anomalies and problems with respect to gene structure and regulation in higher organisms appeared symptomatic of a breakdown in the accepted theory of DNA structure and function. But whether these problems could be encompassed within the existing theory or indicated a need for a radically different theory of gene structure and regulation remained unresolved at this stage.[91]

While recombinant DNA techniques generated major new theoretical problems in molecular biology, the synthetic applied work using the technology continued apace. The "fused"-protein method for synthesizing somatostatin and insulin had shown that their production by genetically altered bacteria was possible, not that it was commercially viable. Research to maximize expression—by nucleotide-level tinkering, by increasing the purity, yield, and stability of protein products, and by varying the secondary structure of mRNA (the way in which the RNA molecules folded)—was actively pursued.[92]

This work on practical application, however, was limited by the absence of a general theory of the functioning of genes in higher organisms. The methods used were essentially "bootstrap" operations—empirical approaches that, if successful, became the basis for the next round of trial and error. As one molecular biologist employed as vice president for research in a genetic engineering firm put it: "We don't understand exactly how each of these subsystems [for expression] functions. At this point, [we use] a very empirical kind of approach. . . . People in the field are still in the data-gathering stage at the same time as they are trying to get high-level expression. . . . We have ideas as to what might work, what might not, but they don't have a very high predictive ability at this point."[93]

Even after efficient expression of a gene was obtained, the process of large-scale fermentation of the bacteria containing that gene presented many further problems. Huge cultures were susceptible to contamination by other microbes that could outcompete the genetically engineered host; production efficiency tended to decrease, sometimes substantially, with increases in culture volume; temperature, acidity, and other conditions in the fermenter had to be precisely controlled for effective production. It took ICI twelve years to move from laboratory fermentation of single-cell protein to full-scale production.[94]

Commercial Interests Intensify

Technical hurdles notwithstanding, as the power of recombinant DNA technology began to prove itself, competition to invest in it intensified. Several factors contributed to the heating up of the recombinant DNA race. Genetic engineering firms lured investors with prospects of high returns on investments and of potentially enormous markets if, indeed, inexpensive bacterial production techniques materialized. J. Leslie Glick, president of Genex, projected returns of 14.6 percent and higher and a total market for bacterially produced substances of $40 billion by the year 2000.[95] Markets for genetically engineered plants, development of which was seen as a more distant prospect, were estimated to be even larger.[96]

In the pharmaceutical field, prospects for new markets appeared not only large but possibly imminent, particularly for substances that were expensive or difficult to make by other means. Interferon figured prominently in such calculations. Made only by human cells in extremely small amounts, this substance had been hard to obtain in pure form and to test clinically. It was suspected, however, that interferon might be effective in treating a wide range of diseases in which viruses were implicated, including cancer and the common cold. Clearly, investment in research and development for its production with recombinant DNA technology was a gamble that could be worth billions of dollars if its potential was realized. In 1980, the brokerage house E. F. Hutton estimated that the annual world market for interferon would be $50 to $100 million.[97] A reporter for the *Los Angeles Times* described the product's commercial significance: "Interferon . . . is a glamour stock in the scientific market-place. Money and glory await whoever can find a way to manipulate bacteria to produce quantities of the substance, one of the most potent biological substances known to occur in man."[98]

By 1980, competition to produce substances such as interferon and insulin by genetic engineering and other new production technologies was intense. All four of the leading genetic engineering firms (Cetus, Genentech, Genex, and Biogen) and at least a dozen drug and chemical corporations in the United States, Europe, and Japan were pursuing re-search on efficient methods of interferon production. For many of these firms, genetic engineering technology for interferon production was a prime target.[99]

Similarly, developing new technology for the production of insulin spurred competition between the leading American and European pro-ducers: Lilly, which controlled 90 percent of the American market, and Novo Industri, which controlled 60 percent of the European market. By 1980, Lilly and Novo were engaged in a race to produce a pure, cheap, human insulin. Both companies had genetic engineering research pro-grams, and Novo was also exploring a chemical method for converting pig insulin into the human variety.[100]

Firms also competed for personnel, since, as the number of genetic engineering ventures increased, the supply of skilled scientists with ex-perience in the recombinant DNA field diminished. In 1981, an invest-ment company reportedly approached twenty researchers before it found one without a commercial tie.[101] Since knowledge and experience were as critical to recombinant DNA ventures as capital, it was important to hire outstanding scientists before the supply disappeared.

Commercial interest was also stimulated by a steady stream of an-nouncements, mainly from small genetic engineering firms, about tech-nical advances in the bacterial production of commercially important substances, pilot production plants, and clinical trials of substances made by bacteria (tables 2.2 and 2.5). Particularly eye-catching were the an-nouncements of the long-anticipated bacterial expression of human in-sulin by Genentech in September 1978 and of human interferon by Biogen in January 1980, both of which stimulated glowing, largely un-

Table 2.5   Announcements of Technical Advances Related to Bacterial Production of Commercially Important Substances, 1979–81

Date	Event	Company	Source
1979			
July	Animal tests of somatostatin	Genentech/Hoechst	E 7/14/79
July	Animal tests of human insulin	Genentech/Lilly	E 7/14/79
September	Proposals for pilot production of insulin and somatostatin	Genentech/Lilly	NYT 9/11/79
October	Plans for human trials of human insulin	Lilly	WSJ 10/21/79
1980			
April	Proposals for pilot production of five human hormones	Genentech	E 4/5/80
April	Plan for large-scale production of single-cell protein	ICI	E 4/5/80
June	Proposal for pilot production of interferon	Schering-Plough	WSJ 6/24/80
June	Successful tests on biological activity of human growth hormone, interferon	Genentech	N 6/26/80
July	Clinical trials of human insulin	Lilly	NYT 7/24/80, N 7/31/80
September	Safety of human insulin shown	Lilly	NS 9/11/80
October	Animal tests of interferon	Genentech/Hoffmann–La Roche	NS 10/9/80
December	Plans for clinical trials of human growth hormone	Genentech/KabiGen AB	N 12/11/80
1981			
March	Pilot production of bovine growth hormone	Genentech/Monsanto	WSJ 3/16/81
June	Vaccine against foot and mouth disease	Genentech/International Minerals and Chemicals Corp.	WSJ 6/19/81
September	Clinical trials for human growth hormone in United States	Genentech	WSJ 9/29/81

Sources: E = *Economist;* N = *Nature;* NS = *New Scientist;* NYT = *New York Times;* WSJ = *Wall Street Journal.*

critical press coverage worldwide. The fanfare surrounding the Biogen announcement, made at a press conference in Boston by company scientists Walter Gilbert and Charles Weissmann, conveyed the specific message that the firm had won the "interferon race" and, more generally, that companies like Biogen were the leading repositories of state-of-the-art knowledge in the field. In fact, the results achieved by Biogen were neither novel nor commercially significant. Similar conclusions were later reached about the Genentech announcement. "Cloning by press conference," as one observer called it, seemed to rival the importance of cloning advances achieved in the laboratory.[102]

A final, crucial factor in the intensification of competition is that, particularly after 1979, the United States offered a political and economic environment that was extremely supportive of the development of recombinant DNA technology (section 1.4). Attractive tax incentives promoted university-industry cooperation, and cuts in taxes on capital gains and the provision of tax shelters for research and development ventures stimulated a large flow of venture capital into new fields like genetic engineering. As a result, far more capital was available for investment in genetic engineering in the United States than in Japan and many European countries.

All of these factors were influential in shaping perceptions of the commercial significance of recombinant DNA technology and in fueling further competition for controlling interests in the new field. As a result, from 1979 on, investments in genetic engineering from various sectors began to accelerate. Theoretical and production-oriented work, in commercial and academic settings, was greatly expanded by this influx. The further development of the field demonstrates the pervasiveness of the effects of commercial investment, as the material power biologists had produced came to be translated into economic power through the infusion of capital. Three phases of this process can be distinguished.

Phase 1: Corporate Investments, 1979–82. The largest and most influential investors in genetic engineering were the huge pharmaceutical, chemical, and oil corporations, and the first phase of investment was marked by their investment in genetic engineering firms and by their contracts with both commercial and academic laboratories.

Equity investment in genetic engineering companies increased steadily through 1979, rose sharply in 1980 to $72.4 million and again in 1982, to a record high of $111.3 million, and thereafter, declined abruptly (tables 2.3 and 2.4). As noted earlier, these investments, in providing the means to follow developments in genetic engineering at close range, protected multinational corporations from technological erosion of their dominance of world markets. Many corporations spread their risks by buying simultaneously into several firms. The Koppers Company, for example, invested in Genex, Engenics, and DNA Plant Technologies in 1979–81. Schering-Plough invested in Biogen and acquired DNAX. International Nickel invested in Cetus, Genentech, Biogen, and International Plant Genetics.[103]

Competition for these investments drove up both the value and num-

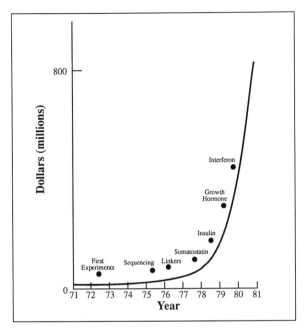

Fig. 2.1 Equity investment in biotechnology companies, 1971–81. Adapted from J. Leslie Glick, "The Biotechnology Industry," paper for the American Law Institute and the American Bar Association, Committee on Continuing Professional Education, October 1981.

ber of the bioengineering firms. Both variables rose roughly exponentially until 1982. In 1979 alone, the paper value of the four leading firms almost doubled.[104] The number of small firms specializing in the field rose from about twelve in 1978 to about fifty-two in 1982.[105] The cumulative equity investment in all types of biotechnology companies rose from $50 million to over $800 million between 1978 and 1981 (fig. 2.1).[106] Only in 1982 did equity investments in these firms slow. The reasons are not entirely clear. It is likely, however, that at this point large corporations began to invest more heavily in in-house activities. Also, perhaps the pool of qualified scientists needed for new firms had dried up.[107]

Multinational corporations also invested in genetic engineering through research and development contracts with biotechnology companies. Many such linkages were formed in this period: KabiGen AB with Genentech for the production of human growth hormone, Hoffmann–La Roche with Genentech and Shell Oil with Cetus for the production of interferon, American Cyanamid with Molecular Genetics for bovine growth hormone, Dow Chemical Company with Collaborative Research for rennin, Standard Oil of California with Cetus for fructose, and the Koppers Company with Genex for the biocatalytic transforma-

tion of aromatic chemicals from coal distillate derivatives.[108] Research
and development contracts provided many genetic engineering firms
with short-term capital. In 1979, for example, 86 percent of Cetus's total revenues, 63 percent of Genentech's total revenues, and 33 percent
of Genex's total revenues were derived from corporate-financed contracts.[109] In return, the corporations achieved access to state-of-the-art
knowledge of the field in a relatively inexpensive form, as well as board
representation and, typically, exclusive worldwide rights to market the
product of the venture and thus also to control market development.[110]

Despite the transfer of significant expertise to the genetic engineering
firms, the universities and other research institutions remained the primary source of theoretical knowledge and the site of the most important
advances in genetic engineering. From 1981 onward, the multinational
corporations also moved to tap these resources. The time was right: de-
creasing federal support for basic research in biology, resulting partly
from continued high inflation and partly from cuts in funding, pressured
academic institutions to find alternative sources of support. New government policies catalyzed university-industry collaboration.[111] The result
was a rapid flow of corporate funds into research contracts, research partnerships, and other contractual arrangements with the universities. One
analyst later described the trend: "The message [to investors] was clear:
biological science on campus became big business. Companies recognized this. Why on campus? Simply because the cutting edge which laid
the foundation for the technology was being carried out there."[112]

Through contracts with universities, large corporations committed a
total of about $250 million in 1981 and 1982 (table 2.6). The conditions
of support varied widely, but all of these grants were characterized by
the very fact of conditionality, a key element of which was the provision
of corporate access to state-of-the-art knowledge in university laboratories. The effect was to blur the distinction between the power of genetic
engineering to produce knowledge and its power to produce profits.[113]

Phase 2: Public Stock Offerings, 1980–83. The first phase of investment
in genetic engineering was overlapped by a second, in which genetic engineering firms raised capital through public stock offerings. These sales
required advance publicity, and this was generated through announcements of technical advances and commercial investments, with the significance of both tailored to shape the expectations of the audience. The
significance of technical developments was often magnified while problems were overlooked. And the essentially cautious, protective intent of
corporate investments was obscured as headlines announced that genetic
engineering was "warming up for big pay-off" (*Science*), that "the [bio-
technology] bandwagon begins to roll" (*Nature*), that the "cloning gold
rush turns basic biology into big business" (*Science*), and that "there's
gold in them thar recombinant genetic bits" (*New York Times*).[114]

Table 2.6   Investments in Biological Research in Universities and Research Institutes
by Multinational Corporations, 1981–82

University or Research Institute	Corporation	Amount (millions of dollars)	Contract Length
California Institute of Technology	Du Pont	0.15	—
Cold Spring Harbor Laboratory	Exxon	7.5	5 years
Cornell University	Procter and Gamble	0.119	—
Harvard Medical School	Du Pont	6.0	5 years
Harvard Medical School	Du Pont	23.0	12 years
Harvard Medical School	Seagram	6.0	—
Harvard University	FMC Corp.	0.57	3 years
John Hopkins Medical School	American Cyanamid	2.5	5 years
Massachusetts General Hospital	Hoechst	50.00	10 years
Massachusetts Institute of Technology	Whitehead Foundation	127.5	—
Rockefeller University	Monsanto	4.0	5 years
Salk Institute	Phillips Petroleum	10.00	—
Stanford University	Upjohn	0.1	—
Stanford University	Smith Kline Corp.	0.047	1.5 years
University of California, Davis	Allied Chemical Corp.	2.5	5 years
University of California, San Francisco	Lilly	0.780	5 years
University of California, San Francisco	Merck	0.226	—
University of Maryland	Du Pont	0.5	—
Washington University Medical School	Monsanto	4.0	5 years
Washington University	Mallinckrodt	3.9	3 years
Yale University	Celanese	1.1	3 years
Total		250.5	

Sources: Data on university-industry contracts for genetic engineering in Judith L. Teich, "Inventory of University-Industry Research Support Agreements in Biomedical Science and Technology," paper prepared for the National Institutes of Health, January 1982; Judith Johnson, "Biotechnology: Commercialization of Academic Research," Issue Brief IB81160, Congressional Research Service, Library of Congress, July 1982; Norton D. Zinder and Jackie Winn, "A Partial Summary of University-Industry Relationships in the United States," Recombinant DNA Technical Bulletin 7 (March 1984): 8–19, 52.

By 1980, public expectations of the future of biotechnology were well primed, and the scene was set for the first public stock offerings. The firms going public were not disappointed. When Genentech offered its shares on the stock market in October 1980, the event produced what the Wall Street Journal called "one of the most spectacular market debuts in recent history."[115] In the first twenty minutes of trading, shares soared from $35 to $89, settling by closing time on the first day at $71. Overnight the company's value increased from about $75 million to over $500 million, and the company's founders, Swanson and Boyer, became millionaires.[116] Six months later, in March 1981, Cetus followed Genentech's lead. Although by this point buyers' interest was not as intense—Cetus's shares sold for $23—the event was registered as the largest single initial stock offering in U.S. corporate history. The sale raised $115 million for the company and set its market value at about $600 million.[117]

After the first few public offerings, public response to the sale of stock in these firms tended to follow the overall rise and fall of the stock market. In 1982, when the market softened, the response to new offerings was unenthusiastic, but the long bull market from July 1982 to June 1983 supported a further boom. Thereafter, interest in new issues decreased as the 1983 boom was followed by a decline in 1984.[118]

Phase 3: Denouement, 1982–84. At the peak of investment interest in genetic engineering, in 1980, enthusiasts were prone to claim that "biotechnology was limited by only the imagination."[119] The reality, however, was that recombinant DNA technology was limited by commercial and technical feasibility, and difficult problems remained to be solved before the techniques could result in marketable products. By the middle of 1981, more sober appraisals began to replace the inflated expectations of the previous two years. An article in the *Economist* warned that there would be "many disappointments. The market forecasts are all guesses. To turn the base metal of biology into big profits will need not only a lot more basic research but also a lot more practical experience, a lot more process engineering and much bigger investments than most people are contemplating today. Risks will be high, patents hard to enforce, competition frenetic and most products (when they do come) rapidly obsolescent."[120]

With time, investors in genetic engineering companies became attentive to the continued absence of marketable products and to the long lead time between the initiation of research and development and the attaining of products. Price-earnings ratios—a favorite yardstick of investors—were abnormally high for these firms, reflecting low or nonexistent revenues from sales.[121]

The end of the genetic gold rush did not mean, however, that the commercial potential of genetic engineering would fail to be realized. Most analysts continued to believe that it would be, but that it would take longer and would perhaps be less sweeping than the glowing projections of 1980 had suggested.[122] The appearance of the first American genetic engineering product—human insulin from Genentech and Lilly—in October 1982 reinforced this sense. While the technical feasibility of industrial genetic engineering was confirmed, the initial cost of Humulin considerably exceeded that of insulin prepared by extraction from pig and cow pancreas glands—an outcome that contrasted with the rosy predictions of plentiful, low-cost products only a few years earlier.[123]

Large corporate investors began to shift their focus to their own in-house research and development programs. Du Pont established one of the largest in-house agricultural research laboratories in the United States, at a cost of $85 million; Eli Lilly built a $50 million biomedical research center for recombinant DNA technology and immunology and a $9 million pilot plant and laboratory for the production of genetically engineered products.[124] The acquisition by Schering-Plough of the new firm DNAX (and with it access to experienced recombinant DNA researchers such as Paul Berg and Arthur Kornberg of Stanford University and George Palade of Yale University) and by Occidental Oil of Zoecon, a plant genetic engineering company, similarly provided facilities and access to state-of-the-art expertise.[125]

Additionally, research and development limited partnerships (a tax-shelter scheme promoted by the Reagan administration) brought in considerable capital to smaller genetic engineering firms. Under revisions of the tax code that came into effect in January 1982, investors could shelter profits while using losses to offset other sources of income and deducting almost 100 percent of the initial investment from income, thereby almost doubling the potential rate of return. In 1982 and 1983 alone, nine such firms raised $367 million through these arrangements.[126]

Commercial Interests Expand

The urge to profit from genetic engineering was not restricted to private investors. Universities, their resources depleted, looked to patentable genetic engineering research as a means to generate new income; individual states in the United States saw the technology as a way to regenerate their industrial bases; and other countries hastened to catch up with the pace of development being set in the United States.

In the early 1980s, leading American research universities pursued diverse routes toward profiting from basic research: consulting arrangements, industrial associates programs, research contracts with corporations, and joint university-industry research institutes.[127] In the fall of 1980, Harvard proposed a plan to found a genetic engineering company. The plan deeply divided the faculty and was eventually dropped,[128] but similar schemes were pursued at Stanford; the University of California, Berkeley; Michigan State University; and Yale.[129] The tax and patent policies of the Carter and Reagan administrations provided powerful incentives for these new alliances (section 1.4).

From the inception of recombinant DNA technology, the United Kingdom, Japan, and most of the European countries had pursued research aimed at developing and using the techniques. However, the pace of development was significantly slower in these countries (with the pos-

sible exception of Japan) than in the United States. In the United Kingdom, this slower pace reflected the effects of several factors, particularly
the lack of a ready supply of venture capital, structural problems with
government promotion of new fields of technology, and less fluid relations between private industry and the universities where research was
carried out.

By 1980, the European countries were scrambling to catch up with the
United States. High-level committees were established in several of these
countries and by the European Economic Community (the trade affiliation of nine European countries) to plan strategies for supporting the
development of biotechnology. Official reports called for substantial
government support for genetic engineering and other forms of biotechnology, the training of a new work force of specialists and technicians,
and the establishment of government-sponsored genetic engineering
companies. Celltech, a company jointly owned by the British government and several financial and banking corporations, appeared in 1980,
as did Transgene, jointly owned by the French government and a
consortium of corporations including the giant oil company Elf Aqui-
taine. In 1982, the British started to plan the formation of an additional
company to capitalize on genetic engineering in agriculture.[130]

Comparable developments on a smaller scale also took place in some
states in the United States. Like the European countries, these states
wanted to compete with the geographical areas where the science and the
commercial development of genetic engineering were concentrated. In
the Midwest, for example, states hit hard by the long recession in the
automobile and steel industries began to compete to attract new industries, and genetic engineering was high on their list of priorities. By 1981,
North Carolina, Florida, and several midwestern states had begun to develop policies for attracting new genetic engineering industries.[131]

From 1979 to 1984, the race to control the commercial development
of genetic engineering intensified and ultimately extended far beyond
the university laboratories and genetic engineering firms where it had
started. Multinational oil, chemical, and pharmaceutical corporations,
venture capital firms, private investors, universities, national states, and
state governments in the United States entered this race, all lured by the
prospect of high returns on investments and encouraged by strong government incentives.

Press coverage in this period often portrayed the genetic engineering
firms as the pacesetters of recombinant DNA technology, and they
clearly played an important role in this regard. However, these firms
depended on the universities for expertise and on the multinational corporations for capital. By 1983, the long-term survival of many of the
smaller firms in competition with the corporations was doubtful, for

their short-term survival had been achieved largely through trading board seats, manufacturing rights, and marketing rights for equity capital and contracts. As one observer of these trade-offs put it, "These giant mid-wives [of the genetic engineering industry] just might smother the baby."[132]

The dominance of multinational corporations was manifested not only through their investments in the bioengineering firms but also through their support to leading universities and through the development of major research and production facilities of their own. Support by large corporations for research and development in genetic engineering in the United States was comparable to that of the federal government and seemed likely to exceed government levels as genetically engineered products were achieved.[133] By 1982, the corporations were advantageously positioned to shape further commercial development of the recombinant DNA field.

2.6 A New Commercial Ethos

Prior to the emergence of recombinant DNA technology, the practice of molecular biology was guided largely by the traditional ethos of academic research and its commitment to the development of knowledge through an essentially cooperative and communal effort. As appropriation of recombinant DNA techniques became a primary commercial goal, however, norms and practices shifted from those sanctioned by the universities to those dictated by the requirements of commerce.[134] Knowledge increasingly came to be seen as a form of "real estate," a valuable commodity to be appropriated for private profit, rather than as a communal resource. Consequently, norms and practices with respect to the treatment of scientific knowledge underwent a distinct transformation.

First, corporations, universities, and individual scientists alike began to pursue patents on techniques and organisms—a trend given impetus by the Supreme Court's decision on the patenting of life-forms in 1980.[135] Stanford University's early move to secure patents on the results of the gene-splicing work of Cohen, Boyer, Helling, and Chang was an important precedent for the field.

Second, secrecy began to inhibit the sharing of information and materials among researchers. The need of genetic engineering firms to secure patent coverage for their biogenetic inventions produced confidentiality arrangements under which employees agreed not to disclose proprietary information or to share materials. Universities implicitly supported this new norm by encouraging researchers to seek patent protection for their results. Symptomatic of these changes were the contradictions that now

began to embroil university research and teaching. Complaints of researchers' unwillingness to share ideas were aired;[136] legal struggles over
ownership of cell lines flared up.[137] To a degree, these issues were aired
within the universities and at congressional hearings. Some universities
issued guidelines to minimize conflicts of interest, but it must be noted
that these measures did not significantly hinder the development of corporate links with university research, which continued to be formed.[138]

Finally, experimental results were not seen simply as demonstrations
of biologists' power to advance science or apply it practically but were
also used as advertisements for the commercial viability of certain enterprises. Consequently, some of the techniques of business advertising
came to influence the dissemination of scientific results. Some results,
announced before they appeared in journals, were timed to prod investors or protect a patentable process; other results were kept out of the
public domain; and a modicum of unwarranted exaggeration, particularly premature claims for the industrial application of achievements, became frequent.

A case in point was the announcement of the expression of the human
insulin gene by teams of scientists at Genentech and the City of Hope
National Medical Center at a press conference in September 1978. The
presentation of the results made headlines as a major scientific and technological "breakthrough" that would make possible the commercial production of better and cheaper insulin for the world's hundred million
diabetics and, in the future, of a whole range of other medically and
industrially important substances as well. Yet as Andreopoulos, head of
the news bureau at Stanford University, pointed out, the biological activity of the molecules had not been demonstrated. Nor were the techniques used to achieve this result novel by that point: one month earlier
a team at Harvard had reported the expression of the rat insulin gene in
a similar experimental system.[139] Furthermore, many technical hurdles
had to be overcome before the manufacture of human insulin by bacteria
might become commercially viable: the amount of insulin produced in
the experiment was minute; the host organism, *E. coli,* had to be shown
to grow well in large culture volumes; and the cost of the product had to
compete with the cost of insulin made by more conventional means.

The most significant aspect of this event was commercial rather than
scientific. At the press conference, representatives of Genentech also
made public a multimillion-dollar agreement with Eli Lilly under which
the tiny company and the pharmaceutical giant would jointly develop the
insulin technology. Lilly would gain the rights to manufacture and market the hormone, while Genentech would receive cash and royalties on
its sale. At this point Genentech needed investments, and the announced
partnership with Lilly underscored the small firm's promise as an investment prospect.[140]

Biogen's press conference to announce cloning of the interferon gene similarly exemplifies the hyping of results for commercial reasons. That a leading scientist could candidly acknowledge after the event that the conference was orchestrated to create maximum impact for his company underscored the arrival of the new norm.[141] Of the spin on such events, a Wall Street analyst later warned: "Beware of 'hype.' News releases and grapevine information are meant to inform but they are not pure information: they are designed to entice investors."[142]

In summary, as genetic engineering became seen as a promising investment prospect, a turn from traditional scientific norms and practices toward a corporate standard took place. The dawn of synthetic biology coincided with the emergence of a new ethos, one radically shaped by commerce.

2.7 A Transformation of Interest

The social transformation of recombinant DNA technology from an academic field into a field with strong commercial ties was remarkable both in the extent and the rapidity of the change. In 1972, when the possibility of genetic manipulation by recombinant DNA techniques was first demonstrated, the parent field of molecular biology was an almost exclusively academic domain. The vast majority of its practitioners worked in universities and research institutions, and financial support came largely from government sources. But as recombinant DNA technology revealed its practical potential during the 1970s, the field was rapidly initiated into a new commercial role.

By the late 1970s, strong links had formed between the field and private industry at both the individual and the institutional levels. Practitioners, formerly cloistered in academe, became equity owners, executives, advisers, and consultants for the new genetic engineering firms, and were hired in increasing numbers by multinational corporations. By 1981, it was said to be difficult to find a recombinant DNA practitioner who did not have an industrial connection. Leading research universities accepted major corporate support for research institutes and programs, and established mechanisms to enable them to profit from the commercial application of research results. Traditional academic norms and practices were displaced by those dictated by the exigencies of commerce. Practical benefits—once offered as pro forma justification for grant applications—were now a principal rationale for commercial sponsorship of research. The formerly distinct endeavors of "basic" and "applied" research merged in the search for efficient methods of producing commercially important substances such as insulin and interferon.

Pondering the problems wrought by the commercialization of molecular biology, Donald Kennedy, president of Stanford University,

observed, "What is surprising and unique in the annals of scientific innovation so far, is the extent to which the commercial push involves the
scientists who are themselves responsible for the basic discoveries—and
often the academic institutions to which they belong."[143] The formation
of commercial ties between academic science and industry was not new,
but in the case of genetic engineering, the trend appears to have affected
a greater number of practitioners and to have had more intense effects.[144]

It was not accidental that molecular biology yielded a synthetic form
of inquiry that was highly adaptable to commercial application. The concepts of "program," "code," and "system" central to molecular biology
made the idea of the genetic reconstruction of nature immediately accessible, even natural. In addition, government funding policies generated
strong incentives for molecular biologists to recognize and exploit the
practical potential of research results. Thus there were powerful psycho-
social reasons for the readiness of researchers to perceive and pursue
practical application.

But these conditions alone do not explain the rapidity with which the
field became dominated by commercial considerations—particularly in
the light of other conditions that, under different political and economic
circumstances, might have acted as brakes on development. Significant
technical problems with the application of recombinant DNA techniques
for industrial production remained to be overcome prior to the "genetic
engineering gold rush" in 1979. Furthermore, as the following chapters
show, an important public debate on the hazards of recombinant DNA
technology continued throughout the initial period of development. Despite these technical and political problems, rapid commercial development proceeded.

This analysis shows that two conditions were important in accelerating
the assimilation of recombinant DNA technology by the industrial sector. First, economic pressures for the industrial development of new
technologies in the 1970s were high; particularly, they promised for petrochemical and pharmaceutical corporations a much-needed possibility
for new growth and above-average returns on investment. Second, after
1979, a supportive political environment fueled investment in new technologies, making it profitable to invest in research and development even
if risks were high. Both conditions encouraged rapid secularization of the
recombinant DNA field. The millions of dollars that changed hands as
contracts were consummated, as patents were sought on the results of
research, and as the sponsorship of research shifted from governments to
the private sector, particularly to multinational corporations, represented
a real change in the character of genetic engineering and the interests of
practitioners.

By the early 1980s, there was little indication that these changes would be merely transitory, or that they would be confined to recombinant DNA technology. On the contrary, there were many signs that the commercial development of recombinant DNA technology was only the beginning of a more sweeping transformation of the parent field of molecular biology. The discovery in 1975 of methods of producing monoclonal antibodies—extremely pure, uniform proteins that recognize and bind to specific substances—and the immediate anticipation of a wide range of medical and industrial applications strongly reinforced this perception. In the words of an industry analyst, what emerged was "a more generalized interest in biology as an industrial tool. . . . Genetic advances and biological advances were coupled early on because it was virtually axiomatic that where a useful biological process was found, genetics could improve it. Hence, just as in electronics the development of the transistor was a major advance in electronic amplification, manipulation of the gene became a major advance in biological amplification." Private industry was now sensitized to the commercial promise of molecular genetics, and the expectation was that the field would yield further commercially important developments: "Venture capitalists are keeping an eye out for the 'next recombinant DNA.' The sudden recognition that a research tool could turn into an industrially promising tool keeps everyone on the lookout. DNA was a good model, a good success story. There are bound to be others."[145] The political economy of science in this period ensured that the radical shift in the interests of molecular biologists exemplified in the social transformation of genetic engineering would endure.

PART TWO
Policy Initiation

THREE

The Emergence and Definition of the Genetic Engineering Issue, 1972–1975

3.1 Introduction

When the first techniques of genetic engineering appeared in 1972 and 1973, they were taken generally to signify a quantum leap in power to control the genetic constitution of living things. Sensitized by advances in other fields, many scientists saw this new power as double-edged, promising scientific and technological advances on the one hand and possibly serious social, ethical, and environmental problems on the other. Robert Sinsheimer, then chairman of the Division of Biology at the California Institute of Technology, later summarized this reaction.

> It is the success of science that has ended its pleasant isolation from the strident conflict of interests and the passionate clash of values. The great discoveries in molecular and cellular biology—in particular the elucidation of the structure and function of nucleic acids—have provided us with a definitive understanding of the nature of life. Earlier in this century, splendid discoveries in physics and chemistry provided us with a definitive understanding of the nature of matter. From that understanding has come the technology to reshape the inanimate world to human purpose. And many are less pleased with the consequences. Now the description of life in molecular terms provides the beginnings of a technology to reshape the living world to human purpose, to reconstruct our fellow life forms—each as are we, the product of three billion years of evolution—into projections of the human will. And many are profoundly troubled by the prospect.
>
> With the advent of synthetic biology we leave the scrutiny of that web of natural evolution that, blindly and strangely, bore us and all of our fellow creatures. With each step we will be increasingly on our own. The invention and introduction of new self-reproducing, living forms may well be irreversible. How do we prevent grievous missteps, inherently unretraceable?[1]

This chapter addresses the early response to Sinsheimer's question and the shaping of this response by various interests in the field—especially the interests of the community of scientists responsible for developing the techniques. In particular, the expression of these interests in the choice of policy arenas, selection of participants in those arenas, restriction of agendas, and definition of the general policy path that lay ahead are examined. An essential dimension of this early process was the progressive restriction of discourse surrounding genetic engineering—the formation of informal (often unspoken) rules about what was to be addressed and what was not to be addressed, what was central and what was peripheral for decision making, and the norms of truth and falsehood. A central argument of this chapter is that the reduction of discourse was an integral part of the process through which major interests in the new field affected the formation of policy.

3.2 Social Interests in Genetic Engineering

The advent of genetic engineering activated diverse and sometimes conflicting interests in the future development and use of the field: the scientific community, universities, governments, private industry, the military, labor unions, and pressure groups all responded in ways shaped by their own specific goals and needs.

Many of these responses were strongly influenced by changes in science policy from the late 1960s onward (see chapter 1). Scientists came under increasing pressure from governments to demonstrate the utility of their research. In the United States, the leveling off of the postwar exponential growth in government support and calls for "accountability" from politicians brought to the fore the utilitarian dimension that had always been present in the American research effort. Practical results from science assumed increasing importance in the eyes of politicians and the bureaucrats who oversaw the American research effort. Similar pressures developed in the United Kingdom. There too a decline in government support for civilian science set in, particularly after 1973, when both sides of the British dual support system—the universities and the Research Councils—were hurt. Commitment to protecting the autonomy of the universities and scientific research (previously a stronger tradition in the United Kingdom than in the United States) gave way before a growing emphasis on practical results. The prospect that genetic engineering would produce not only important progress in solving fundamental biological problems but also a wide array of practical applications fitted well with this new policy emphasis on both sides of the Atlantic. Consequently, the new field promised to fulfill important needs for several interest groups.

For scientists, the techniques promised not only the novel directions in research and the fast results required by serial-competitive funding arrangements, but also a powerful justification for continued support. Since molecular biology had yielded spectacular theoretical advances but almost nothing in terms of practical results, the prospect that genetic engineering would lead to the latter made it especially valuable to researchers. One British scientist commented in 1974: "We've been so much accused of wasting public money on this sort of rarified and abstruse new subject of molecular biology . . . that it's very encouraging to see . . . perhaps for the first time some really concrete indications that we're going to be able to offer something valuable to the human race. I think that's given us a boost, because we're all very conscious of the need to prove that we are worth the public money we are spending."[2]

For university administrators, the techniques offered a prospect of generous and diverse sources of funding in a period when support had become less assured and when the ability to demonstrate a practical payoff was becoming a significant criterion in assessing grant applications. As the chair of the Department of Microbiology at the University of Michigan Medical School pointed out to the university's board of regents in 1976, "Any institution prevented from using recombinant DNA techniques will be placed at a competitive disadvantage, and will eventually become dependent for knowledge and materials on other universities conducting this type of research. . . . It's entirely possible that [the] university will lose top-notch faculty whose work has been stymied, and we probably won't be able to attract their equal for the same reason."[3]

For the government agencies responsible for promoting biomedical research, genetic engineering offered both scientific advances of the type that would foster national preeminence in science and practical applications that would demonstrate the utility of the biomedical sciences to politicians. These prospects were immediately perceived and defined in terms of international competition. In the United States, it was feared that significant delays in development might hamper American researchers in what was seen as a race with other countries. A panel of eminent scientists drawn from the fields of biochemistry, molecular genetics, and cell biology stated in 1976, "If we were to lose momentum at this time, we would lose . . . our forefront [international] position in this field."[4] And the same fears and pressures were evident in the United Kingdom. A prominent British researcher stated on national television in 1977: "If there are to be economic benefits coming out of this work, it would be a tragedy if the United Kingdom was not there to reap some of these. We know what happened in the past; it would be a great pity if in the end this country found itself paying licensing fees and royalties to other countries for products."[5]

For private industry, the techniques signified potential sources of new products and, just as important, a new, energy-efficient mode of production. This meant at once a threat to existing sources of profit and new opportunities for obtaining a high rate of return on investment. As shown in chapter 2, although large multinational pharmaceutical, chemical, and energy corporations did not immediately take major action, they established ways to monitor closely the development of the field.

For each of these groups, the continuing development of recombinant DNA technology, and especially its practical applications, held powerful attractions; few among them were likely to bear delaying the work patiently.

While competitive interests in the new biology were being activated in academia and industry, the effects and the direction of Western science and technology were being challenged more vigorously than ever before. Growing numbers of people—including many scientists—were finding the power of science and technology to transform the environment increasingly problematic. These challenges to accepted forms of deployment of science and technology took different forms in the United States and the United Kingdom. In the United States, a diverse public interest movement had mobilized to secure legislation aimed at two main goals: prevention of the undesirable side effects of technology and expansion of public participation in regulatory policy. The response from the liberal U.S. Congress of the late 1960s and early 1970s was a mass of safety legislation (section 1.3). In the United Kingdom, the governmental response to calls to address environmental and workplace hazards was more muted. Generally, the reforms that were instituted did not change the character of British environmental and occupational health and safety policy; controls remained largely a matter of negotiation between private industry, local government officials, and government inspectorates and were mainly shielded from public scrutiny. An exception was the Health and Safety at Work Act of 1974, which gave employees extensive rights to representation in local safety committees as well as on the Health and Safety Commission, the body responsible for administering the act. The Health and Safety Commission was established in 1974, early in the evolution of genetic engineering, and supported by a Labour government committed to expanding the roles of trade unions in forming occupational health and safety policy.

The effectiveness of pressure groups and trade unions in influencing policies aimed at controlling the impact of science and technology varied. In the United States, environmentalists, trade unionists, and consumer-safety advocates, despite considerable prominence in the press and increased access to policy processes, had no direct means to participate in policymaking: such groups were given a voice in policymaking but no formal policy arena. In the United Kingdom, on the other hand, although pressure groups were generally less visible and vocal, at least one sector—the trade unions—had considerable means at its disposal to affect occupational health and safety policy through its representation on safety committees and on the Health and Safety Commission.

Between 1962 and 1979, British trade unions grew substantially, almost doubling their total membership. In 1979—the peak year for union membership—membership exceeded eight million, or roughly 55 percent of the British work force, compared with 25 percent in the United States.[6] Especially important for the future influence of the unions on genetic engineering policy was the unionization of scientific and technical workers in this period. British technicians, as well as many scientists, were represented by several unions, of which the strongest and most influential was the fast-growing and militant Association of Scientific, Technical, and Managerial Staffs (ASTMS). Formed in 1968 from a merger between the Association of Supervisory Staffs, Executives, and Technicians and the Association of Scientific Workers, the ASTMS membership grew rapidly under the guidance of its general secretary, Clive Jenkins, from 75,000 to 450,000 by 1978.[7] In the 1970s, the growing sensitivity to the industrial health hazards coupled with several serious industrial accidents served to place health and safety high on the list of union concerns. A union policy document issued in the 1970s stated:

> For too long we have accepted standards in our places of work which continually expose our members to unnecessary risk. The responsibility to provide a safe place of work rests with the employer, but too often we have accepted assurances from employers that these responsibilities have been met only subsequently and disastrously to find out the opposite. . . . We have seen in ASTMS the radical shift in attitude from pensions being a "gift" from the employer to the view that they are a deferred income and an investment. Similarly with health and safety, we need a shift from allowing health and safety to be controlled by the employer to a system of agreement by negotiation. Consultation on health and safety is no longer enough. . . . Health and safety must never be sold for short term benefits. No agreements on danger money, risk payments, etc. are acceptable.[8]

Such a document should not be read, however, as an indication of opposition to emerging technologies with unknown hazards. On the contrary, as a technical union, ASTMS, in harmony with the positions of both the leading political parties, supported the energetic development of new science-based industries in the United Kingdom. The emphasis was on safety *and* development of new technologies rather than on safety alone. In the late 1970s, ASTMS members would explore the possibility of developing new biotechnology-based industries with officials at the National Enterprise Board. The response of the ASTMS to the issues posed by genetic engineering differed from that of environmental organizations in the United States, which tended to focus exclusively on the risks posed by the new field.[9]

In political terms, ASTMS foresaw a two-pronged approach to the health and safety issue: locally, through negotiation with employers, and nationally, through pressure for higher safety standards. Thus ASTMS and other trade unions formed a powerful interest group whose political influence, at least while a Labour government was in office, was substantial. In the United States, in contrast, scientific workers in many industries and in virtually all university laboratories did not belong to unions and had no formal political voice on the question of health and safety in the laboratory.

The possible power of genetic engineering inevitably stimulated military interest. In the United States, such interest was associated primarily with the chemical and biological warfare programs, whose expansion in the 1950s and 1960s resulted in extensive use of the biological sciences and a research and development infrastructure encompassing at its height in the mid-1960s some three hundred universities, research institutes, and corporate laboratories.[10] Potential weapons applications of future advances in molecular genetics were noted by military officials in the 1960s. In 1962, the chief chemical officer, Major General Stubbs, noted in testimony before the House Subcommittee on Department of Defense Appropriations that developments in microbial genetics in the previous years had "advanced man's knowledge immeasurably about the basic ingredients of life and indicated ways in which these ingredients may be altered or transformed to suit his desires" and that genetic studies of microorganisms were receiving "ever-increasing attention" by the U.S. Army. Among the army's goals were the construction of novel viruses through the recombination of the RNA of different viruses, the development of methods of inducing and selecting mutations in populations of microorganisms, and the development of insects with increased resistance to insecticides and cold. Stubbs emphasized that "the major contributions to biological weaponry and defense will result from research and better understanding of the sciences of genetics." By 1969, the DoD was anticipating that progress in molecular biology would make it feasible to produce novel infective microorganisms with the capacity to be "refractive to the immunological and therapeutic processes on which we depend to maintain our relative freedom from infectious disease." From a military perspective, the biological sciences, especially molecular biology, offered a potential for overcoming existing defenses as well as the lack of battlefield utility that had previously caused bioweapons to be seen as militarily ineffective.[11]

At the same time, some pressures impinged on the U.S. government to limit biological warfare research. At a conference on the U.S. chemical and biological warfare policy, organized by the American Academy of Arts and Sciences and the Salk Institute and chaired by Harvard biochemists Paul Doty and Matthew Meselson in July 1969, a note of anxiety about the weapons implications of future military exploration of molecular genetics was sounded. Doty argued that "the ultimate consequences of developing successive generations of biological weapons probably go far beyond what we can expect out of the chemical side"—a view seconded by James Wiggins, a former U.S. representative to the United Nations, who argued that "ultimately, if pursued far down this road, some nation will set loose a contagion that will rage unchecked until the whole population of the world is destroyed."[12] Later that year, Nobel laureate and molecular biologist Joshua Lederberg, in public testimony to the House Foreign Affairs Committee, expressed similar concerns. Pointing to the rapid advances being made in his field, Lederberg warned that the military could develop "agents against which no reasonable defense can be mounted" and suggested that, given the relative inexpensiveness of biological weaponry, "our continued participation in BW development is akin to our arranging to make hydrogen bombs available at the supermarket."[13] The following year, in testimony to the Conference of the Committee on Disarmament in August 1970, he argued that the future development of molecular biology was double-edged, depending entirely on the manner of exploitation.

> From the very beginning, it was inescapable to me that these new approaches for the understanding and manipulation of living organisms had potential implications for human progress of very great significance. On the one hand molecular biology could increase man's knowledge about himself and lead to revolutionary change in medicine in such fields as cancer, aging, congenital disease, and virus infections. It might also play a vital role in industry and agriculture. On the other side it might be exploited for military purposes and eventuate in a biological weapons race whose aim could well become the most efficient means for removing man from the planet. As a student of evolution, and having studied it in the microcosmos with bacterial cultures, I knew man had no guaranteed place on our earth. . . . To these long-standing threats would now be added new ones, potentially of our own invention.[14]

Such positions provoked debate. Some biologists questioned positions such as Lederberg's, holding that known biological-warfare agents could pose an extreme threat that could be only marginally increased by future advances in technique.[15] In important respects, the situation was similar to that of the atomic physicists in 1939, when physicists differed in their assessment of the military implications of uranium fission.[16] Nevertheless, many members of the scientific community who were attentive to the question of military development of molecular biology feared that further advances might overcome the unreliability and delayed effects of biological weapons, taking them, as Bernard Feld, a participant at the American Academy of Arts and Sciences meeting, put it, out of "the category of useless weapons."[17]

Thus the scientific community pressed to inhibit military spending on using the biological sciences to develop weapons. As the Vietnam War dragged on, strong political forces at home and abroad supported these pressures.[18] Indeed, by the time that the first techniques of genetic engineering were emerging in the early 1970s, these pressures had helped to produce a major change in the U.S. chemical and biological warfare policy, especially unilateral renunciation of biological weapons, and achievement of the 1972 Biological Weapons Convention, a comprehensive international treaty requiring biological disarmament through prohibition of development, production, and stockpiling of biological and toxin weapons. In theory, the convention prohibited the use of further advances in molecular genetics for weapons purposes. However, vagueness in its provisions and the absence of machinery for verification of compliance, as well as the general weaknesses of international law, meant that the convention provided no iron-clad assurance that military establishments would not proceed to explore the weapons potential of genetic engineering. In particular, the limitations of the treaty's central prohibition on development, production, and stockpiling left a sizable gray area of research and development in which defensive and offensive interests coincided.[19]

This survey of the various interests that could be expected to be aroused by the emergence of new and powerful techniques in molecular biology underscores several important characteristics of the dynamics of policymaking. First, the analysis suggests that in the United States, individual researchers, university administrations, and the government agencies responsible for administering biomedical research policy would form a strongly bound, mutually reinforcing system in which interests in the scientific and industrial payoffs of the techniques would reinforce one another. In the United Kingdom, the same forces were also operating; however, the interests of technical unions in health and safety formed an important countervailing force.

Second, the power and novelty of the new field meant that two further sectors—industry and the military—would follow scientific advances attentively and could be expected to attempt to influence governmental policy wherever it impinged on their specific interests in development, for example, in the control of industrial secrets or in the pursuit of industrial applications in the case of industry, and in the pursuit of research and development in the "gray zones" of the Biological Weapons Convention in the case of the military.

Third, the influences of the critical science movements meant that the interests of scientists in this field would not be monolithic—at least, not initially. The wide influence and appeal of the environmental movement and the critical analysis of the use of science by the military, stimulated by the war in Vietnam, meant that many biologists on both sides of the Atlantic were deeply sensitized to the possibility of misuse of scientific knowledge. The diverse interests at work in the scientific community meant that initial perceptions of the potential effects of recombinant DNA technology would vary widely.

3.3 Precedents

By the time recombinant DNA technology made its appearance in 1972, the community of biomedical researchers was already aware of some types of problems the new field would pose. In the most general way, the genetic engineering controversy was prefigured in the larger debate over the direction and control of science and technology that began in the late 1960s. In addition, specific discussions of ethical issues associated with the possibility—at that time speculative—of genetic engineering in humans and of the hazards of new areas of biomedical research gave indications of the shape of the future response to the actual emergence of genetic engineering. Each of these debates established certain forms of discourse that would reappear later as the controversy over recombinant DNA technology waxed and waned throughout the 1970s.

The Control of Science

The widespread public disenchantment with the traditional goals of scientific and technological progress of the late 1960s brought not only protective legislation and moves to increase public involvement in science and technology policy but also a strong defensive reaction from leaders of science and industry (chapter 1). Representatives of the scientific establishments in the United States and Britain responded with a counterattack to critiques of science associated with the anti-war and environmental movements, denouncing them as threats to the future of science. Some who would later play key roles in the formation of genetic engineering policy spoke out. John Maddox, editor of the prestigious British journal *Nature,* referred to the "interlocking heresies" of the environmental and antiwar movements, roundly denouncing the group of scientists at Harvard who had linked their opposition to the Vietnam War to their concern about the future uses of genetic engineering.[20] Lord Eric Ashby, former vice-chancellor of Cambridge University and chair of Britain's Royal Commission on the Environment from 1971 to 1973, warned in a strong attack on those whom he called "the zealots of the

New Left" that cultivation of "an ideology of anti-science" was "the stuff of which Fascism is made." "Pessimism," "abolitionism," "anti-progress," "antiscience," and "Luddism" became code-words for the enemy inside and outside the scientific community that had to be, if not eliminated, neutralized.[21] Strategies of appeasement, through more effective "management" of science and through rebuilding the public image of science, and of containment of public disenchantment, through controlled "public participation" in policy decisions, were well aired in conferences and official reports in this period, including an influential report in 1969 of the U.S. National Academy of Sciences (see section 1.4). As one observer of these trends wrote, science had to be "*seen* to be harnessed more purposefully to human needs."[22]

The concerns of the scientific establishments on both sides of the Atlantic were clearly articulated at a conference held in London in 1971, Civilization and Science: In Conflict or Collaboration? and sponsored by the CIBA Foundation (a private foundation supported by the Swiss multinational pharmaceutical corporation, CIBA-Geigy). The participants included Alvin Weinberg, director of Oak Ridge National Laboratory; biologist Sir Peter Medawar; Hermann Bondi, the chief science adviser to the British Ministry of Defense; Sir Alan Bullock, vice-chancellor of Oxford University; and two people who would play important roles in the formation of British genetic engineering policy: Lord Eric Ashby and Sir Gordon Wolstenholme, director of the CIBA Foundation and the convener of the conference.

The ambivalence of this group of prestigious scientists toward the question of the public's role in the direction and management of science was a persistent theme of the meeting. On the one hand, it was recognized that the public had to be involved; otherwise, it would have no reason to view harmful impacts of science and technology with any sympathy. As Weinberg put it, "All of us, and certainly the technologists, should from now on make sure that we have a consensus of the whole public, and that the public know what we are doing. The entire society can share the guilt." On the other hand, the participants also feared that the public might well threaten the "Republic of Science": "If the public has the right to debate the use of pesticides in agriculture, then should it not also have the right to debate whether or not we should do experiments that might lead to asexual reproduction, or cloning, by human beings or to a racial basis for intelligence? And if the unaccredited public becomes involved in debate on matters as close to the boundary between science and trans-science as the direction of biological research, is there some danger that the integrity of the Republic of Science will be eroded?" Alan Bullock expressed this fear with respect to genetic engineering: "If this [critical] attitude continues, the public will say 'This is so much a Pandora's Box you are giving us, and it is so evident that it will be misused, that we had just better stop the fundamental research.' I can hear people saying: 'We just can't tolerate the problems that are going to be created by genetic engineering, and we will shut it down as a gift too destructive to the ordinary inventions and the ordinary *mores* of human life.' . . . This is the dilemma."[23]

Thus the dilemma facing leaders of science and technology on both sides of the Atlantic at the beginning of the 1970s: if the public were not involved in decisions on the control of science, it might very well choose to withdraw its support; if it were involved, it might move to limit or even close down branches of science. The proceedings of the CIBA conference and other discussions of this period show those close to the science establishments of both countries already beginning to formulate a response: public involvement would be accepted but carefully controlled. The assessment of the techniques of genetic engineering, which emerged only a year after the CIBA conference, would prove to be a critical test case.

The Debate on Genetic Technology in the 1960s

Throughout the 1960s, the rapid advances in molecular genetics generated speculation and debate about the implications of the application of genetic technology to humans. Biologists began to look beyond the existing (highly restricted) scope of genetic intervention to a future in which even complex human traits would come under genetic control.[24] Although there is no clear evidence that these have a genetic basis, this assumption was widely held. At a symposium, The Control of Human Heredity and Evolution, held in 1963, Hermann Muller anticipated "genetic surgery" not only for eliminating genetic diseases but also for improving traits such as "intelligence, the native strength of social predispositions, general vigor, and longevity."[25] In 1965, Rollin Hotchkiss introduced the term *genetic engineering* for such intervention, speculating that it would ultimately be available for purchase for enhancing mental and physical characteristics that were deemed desirable.[26] Robert Sinsheimer on the occasion of the California Institute of Technology's seventy-fifth anniversary in 1966 predicted that "eventually . . . man will have the power to alter, specifically and consciously, his very genes. . . . Intelligence will be applied to evolution."[27]

These visions of a new power to alter living things evoked for some the social, ethical, and environmental problems generated by scientific advances in other fields, particularly nuclear physics. Salvador Luria, responding to Muller in 1963, compared the possibilities opened up by biological research to the power thrust on the world by the atomic bomb.

"What has been foremost in my mind," he stated, "has not been a feeling of optimism but one of tremendous fear of the potential dangers that genetic surgery, once it becomes feasible, can create if misapplied."[28] Hotchkiss compared the "genetic dilemmas to come" to the major environmental problems of technology that were becoming evident in the 1960s, and he warned that, in directing new technologies for supposedly beneficial purposes, "the pathway will, like that leading to all of men's enterprise and mischief, be built from a combination of altruism, private profit, and ignorance."[29]

At the end of the 1960s and beginning of the 1970s, the continued prospect of nuclear annihilation, the destruction wrought in Vietnam, and serious environmental disruption were prominent concerns. The Boston-based organization Science for the People (section 1.3) was especially active and vocal in warning against the abuse of genetic technology. In November 1969, three Harvard scientists who were members of the organization used the occasion of their isolation of the lac gene of the bacterium *Escherichia coli* to warn of the dangers of misuse of genetics and to call for broad participation in its future direction.[30] Salvador Luria wrote in similar terms in an article published in the *Nation*. James Watson warned of the dangers of cloning humans in the *Atlantic*.[31] It is clear that many believed that the public needed to be informed about the possible future uses of biology and to participate in some way in making policy. As Watson wrote, "This is a matter far too important to be left solely in the hands of the scientific and medical communities."[32]

The essence of the arguments at this time for restraint, for controls over the deployment of biogenetic technology, and for public oversight was captured by bioethicist Leon Kass, at that time executive secretary of the Committee on Life Sciences and Public Policy of the National Academy of Sciences, in a seminal analysis published in 1971. Kass warned not only of the dangers of the coercive use of the new biology in eugenic programs but also of the most likely dangers of voluntary or passive acceptance of the forms in which it might be deployed in a free-enterprise society, propelled by self-interest. While he recognized that genetic engineering might be used in the future for benign purposes, he warned that self-interest would lead far beyond those goals to a transformation of humans from subjects into objects of technological control. "We are witnessing the erosion, perhaps the final erosion, of the idea of man as something splendid and divine, and its replacement with a view that sees man, no less than nature, as simply raw material for manipulation and homogenization."[33] He went on to note a "painful irony" in this prospect: "Our conquest of Nature has made us slaves of blind chance. We triumph over nature's unpredictabilities only to subject ourselves to the still greater unpredictability of our capricious wills and our fickle opinions" (786).

Faced with these prospects, Kass urged caution and even abstention: "When we lack the wisdom to do, wisdom consists in not doing. Caution, restraint, delay, abstention are what this second-best (and perhaps only) wisdom dictates with respect to the technology of human engineering" (786). This position presupposed a radical shift in the assumptions informing technological development. No longer did it suffice to permit unrestrained technological development until social costs emerged: the burden of proof should be shifted to the *proponents* of new biomedical technology; concepts of "risk" and "cost" should include social and ethical consequences; regulatory institutions should be encouraged to exercise restraint. Like Luria and others who spoke critically of the direction of biomedical technology, Kass saw the remedy for the uncontrolled advance of this technology in national and international controls and effective public education about the implications of biogenetics. Undoubtedly, he articulated a view held widely inside and outside the biomedical community, that it was time to slow down, to use restraint, to restrict laissez-faire development of biogenetics, and to say no to some of the technological options that such development would offer before it was too late.

Advocating restraint and public oversight, however, was also seen as a threat to science, not simply because it might slow down research in a specific field but more fundamentally because the airing of such negative opinion might lead politicians to restrict research or to withdraw support for it. For example, Philip Handler, president of the National Academy of Sciences, complained in an address in April 1970 to the assembled members of the Federation of American Societies for Experimental Biology of the "shrill voices playing to the top galleries" that were "registering both in the public consciousness and in the budget process." Specifically referring to the statement of the Harvard researchers on the dangers of genetic engineering, he argued that "if scientists indicate that the doing of science is dangerous to society, one can hardly expect the Federal government to pay for it. The price of irresponsible hyperbole is great indeed."[34] Much the same view was aired by Philip Abelson, editor of *Science,* in July 1971,[35] and by Bernard Davis, professor of bacterial physiology at Harvard Medical School, in an address at Boston University in December 1969.[36] Warning that "irresponsible hyperbole" had already jeopardized support for research, Davis went on to claim that if genetic research were to be restricted, society would "pay a huge price" in forgoing "benefits" from that research to human health, agriculture, industry, and to the scientific understanding of life. He ended with an attack on scientific members of the "New Left" who, he claimed, were using science as a "scapegoat for social problems" and in the process contributing to "widespread public anxiety concerning gene technology." In June 1971, Joshua Lederberg took issue with Watson's call for state intervention to control application of biomedical research. Decisions on control, he argued, should be left in the hands of individual scientists and physicians and governed by the "moral sanctions of an informed community."[37] Although consensus was lacking, these exchanges reveal parameters of the future debate on the control of recombinant DNA technology. The difference would be that in the latter, the drive to reach consensus would be intense.

Laboratory Hazards in the Early 1970s

The second issue that provided indications of the future response to the emergence of recombinant DNA technology was the question of the safety of microbiological research; in particular, doubts were raised about the adequacy of procedures for controlling use of dangerous or potentially dangerous biological agents.

Several accounts have shown that the safety of microbiological laboratories was emerging as a concern in the early 1970s.[38] The abundant support for biomedical research in the 1950s and 1960s greatly increased the number of laboratories using microorganisms. Of course, much of this work did not involve dangerous pathogens and was of low hazard. However, the increase in government support for cancer research in the early 1970s (section 1.1) produced a major expansion of research on cancer-causing viruses, particularly on the question of the role of viruses in human cancer. Funding for the National Cancer Institute's viral oncology program between 1970 (before passage of the National Cancer Act) and 1977 increased in real terms by 80 percent, to $60 million.[39] This growth in support produced expansion of the use of viruses known to cause malignancy in animals, as well as an influx of biochemists and molecular biologists with no training in the handling of dangerous pathogens into this area of research. A casual attitude toward laboratory safety was often apparent, and unsafe practices appeared to be on the rise. Edwin Lennette of the California State Department of Public Health observed in 1973 that graduates from microbiology departments in the universities had excellent theoretical training but "little or no perception of the hazards involved in handling pathogenic agents or how to protect themselves and innocent bystanders against such pathogens."[40]

Moreover, the investigation of modified oncogenic viruses and other types of animal virus with properties that were difficult to predict was rapidly expanding and intensely competitive. A memo written by James Rose, head of the molecular-structure section of the Laboratory of Biology of Viruses at the National Institute of Allergy and Infectious Diseases (NIAID), in 1972 noted the following areas of activity where he believed potential hazards were being generated:

The distribution for research purposes of novel hybrid viruses formed
by linking SV40, a monkey virus known to be oncogenic in other
animals, and human adenovirus, a virus responsible for respiratory
disease, especially among children and young adults

The "ever-increasing" production in research laboratories of other
mutant animal viruses

The adaptation in the laboratory of oncogenic viruses to human cells,
possibly providing circumstances under which natural restraints on
human infection would be broken down

The construction of hybrid DNAs through the biochemical linking of
DNA strands from different organisms, with "unknown biological
potential" (Rose was referring to the results of Peter Lobban and of
Paul Berg and his colleagues, achieved in 1972 [section 2.2].)

In particular, Rose drew attention to the SV40-adenoviral hybrids and
called for restraints on their use.[41] The concern here was that if the hybrids escaped and produced an outbreak of adeno infections, they might
trigger a cancer epidemic. No one knew the probability of such a scenario, but to some it appeared realistic. Robert Pollack, a cancer researcher at the Cold Spring Harbor Laboratory, stated in 1973: "We're
in a pre-Hiroshima condition. It would be a real disaster if one of the
agents now being handled in research should in fact be a real human
cancer agent."[42] Moreover, Rose believed that the conditions of research
were likely to exacerbate the problem: "The production of [viral hy-
brids] could even be stimulated by the fact that they provide new subjects for study in a field which is already crowded and in need of new
problems."[43]

In the United Kingdom, the problem of laboratory safety was brought
home dramatically in March 1973 when a technician at the London
School of Hygiene and Tropical Medicine contracted smallpox from an
experiment using the virus, passing on the disease to two others before
being diagnosed. The event highlighted the possibility that mishandling
of dangerous pathogens could result in casualties even outside the laboratory and raised the question of whether exposure to cancer viruses was
obscured by the long lead time in producing symptoms of the disease.

An official inquiry into the smallpox outbreak revealed almost casual
handling of the virus at the London School of Hygiene and Tropical
Medicine.[44] A working party appointed to examine use of dangerous
pathogens and chaired by Sir George Godber, a former chief medical
officer for the Department of Health and Social Security, recommended
tightening standards and limiting use of the most dangerous pathogens
to registered laboratories. The Godber committee also recommended
legislation for such purposes and noted that the Health and Safety at
Work Act might be broadly applied to cover use of dangerous pathogens.

Cancer viruses were not addressed in the report, which confined itself to organisms capable of producing immediate, life-threatening infections in humans and animals.[45] In the United States, such dangerous pathogens were regulated by the Center for Disease Control, under the Public Health Service Interstate and Foreign Quarantine Regulations. Cancer viruses, however, were not covered by the Center for Disease Control regulations.

In the United States, the problem of controlling research with hazardous viruses was raised by Andrew Lewis, a young researcher at the National Cancer Institute who grew concerned about the possible hazards of work on the SV40-adeno hybrids that he and others were pursuing. In the late 1960s, Lewis discovered a type of virus that contained part of the SV40 genome responsible for the malignant transformation of cells and that was able to replicate effectively in human cells.[46] In addition, the virus coat responsible for triggering an immune response did not raise the usual response for SV40—suggesting that humans might not make antibodies to this hybrid. Lewis began to worry about the possibility that this novel organism could infect humans and that the SV40 DNA it contained would cause cancer. He stated in a paper published in 1969: "The pathogenicity of non-defective viruses which are hybrids between human pathogens and other viral agents such as SV40 must be considered. Such viruses could be maintained in human and sub-human populations, and as pathogens would represent unknown hazards."[47]

Lewis initially responded to what he saw as a serious gap in controls on research in an informal way, by issuing his own memorandum with guidance about safety procedures whenever he distributed samples of the hybrids and by asking recipients of the samples to do likewise. As demands became more numerous, however, he raised the problem of control with officials at NIAID and with the head of his laboratory, Wallace Rowe. Virtually everyone consulted at this point agreed that there was a problem. Three virologists whose advice was solicited agreed that researchers using SV40-adeno hybrids needed to be competent and that facilities needed to be specially equipped to provide containment, and they all proposed further measures for control—the strongest position being discontinuance of research until the risks could be carefully assessed. Lacking formal authority for regulating research, the director of NIAID continued to depend on Lewis's memo and forwarded the problem to the Center for Disease Control, and the NIH Biohazards Committee.

While the Center for Disease Control considered amending its regulations to cover "non-naturally occurring agents," no action was taken. One difficulty—although presumably it was not insurmountable—was that of defining appropriate precautions for the large variety of hybrid viruses that might be produced. A novel regulatory problem was that the organisms presented *possible* rather than actual hazards. Finally, there is considerable evidence of resistance to external regulation in the scientific community. The Center for Disease Control's unexplained inaction meant that the only form of control used for the hybrid cancer viruses was an institutional version of Lewis's informal memorandum, voluntarily committing researchers to follow safety precautions for the hybrids. This eventually became NIH policy.[48]

But Lewis, for one, eventually concluded that a voluntary mechanism would never suffice. At least two directors of major research laboratories—James Watson and Daniel Nathans—initially refused to sign the memorandum of understanding, holding that it would constitute an unjustifiable restriction of their freedom of inquiry. As the debate over controls for genetic engineering unfolded, Lewis would write in 1974, "It is unlikely that in the competitive atmosphere in which science functions that broad unenforceable requests for voluntary restraint will contain the potentially hazardous replicating agents which arise from the widespread application of the plasmid recombinant technology."[49]

Such competition explains the reluctance of researchers to impose external controls that might slow any particular line of inquiry involving cancer viruses. A private meeting in January 1973 on the hazards of cancer viruses arranged by Paul Berg at the Asilomar Conference Center at Pacific Grove, California, exemplifies this reluctance. Many researchers who would subsequently play important roles in the debate over genetic engineering controls attended. (In addition to Berg, David Baltimore, Andrew Lewis, Wallace Rowe, James Watson, and Norton Zinder were present.) Despite recognition at this meeting of the problems posed by tumor virus research, little happened: those present raised concerns about the safety of cancer virus work, called for "informed consent," noted low standards in laboratories, and urged an epidemiological study of laboratory personnel, but this private, intraprofessional discussion yielded nothing. The informal memorandum of understanding for research with the viral hybrids remained the scientific community's sole response to the laboratory hazards issue.

In summary, the scientific community's inaction with respect to the hazards of cancer virus research demonstrates the competitive pressures that inhibited researchers from even trying to control widely acknowledged hazards. The safety of research with cancer viruses was handled internally and quietly, and moves toward external regulation were headed off.

3.4 Emergence of the Recombinant DNA Issue, 1972–1974

The immediate societal context in which Peter Lobban, Paul Berg and his colleagues, and Herbert Boyer, Stanley Cohen, Robert Helling, and

Annie Chang were pursuing the first recombinant DNA experiments was thus already charged with professional competition, with laboratory and environmental safety issues, and with the sensitivity of the public, including many scientists, to the hazards of powerful new technology. As a result, the early recombinant DNA experiments ignited questions that extended far beyond their immediate scientific and technical implications. Indeed, the initial excitement of the scientific community in response to the early experiments was tinged with concern. *Nature* published an anonymous response to Berg's initial success:

> There is a fly in the ointment which no doubt Berg and his colleagues are well aware of and which may cause them misgivings about proceeding further and could result in their reagent being left permanently in the deep freeze. . . .
>
> What would be the consequences if the reagent Berg and his colleagues have made somehow infected and lysogenized *E. coli* in someone's gut as the result of an accident? This possibility, remote though it may seem to be, can hardly be ignored, and it will be most interesting to learn what criteria the group adopts when it decides whether or not the scientific information that might be obtained by continuing the experiment justifies the risk. Perhaps those involved will decide that the game is not worth the candle.[50]

Privately, Berg's colleagues raised similar questions, producing sufficient doubt in his mind about the implications of inserting SV40 DNA into *E. coli* that he decided in 1972 to postpone the experiment.[51] Some thought that was the end of the problem, assuming that only Berg's laboratory had the expertise to pursue gene splicing. But research in other laboratories was proceeding, and concerns about possible hazards persisted. They erupted again, at a Gordon research conference in New Hampshire in June 1973, where Herbert Boyer presented some of the results of his first experiment with Cohen, Helling, and Chang, showing that it was possible to splice two types of plasmid, carrying genes coding for resistance to different antibiotics. This time, questions raised about the wisdom of continuing the work propelled the conference's two chairs, Maxine Singer of the NIH and Dieter Soll of Yale University, to send a letter noting the concerns expressed at the meeting to the National Academy of Sciences. The letter, which was published in *Science,* noted that the ability to combine DNA molecules from different sources offered "exciting and interesting potential both for advancing knowledge of fundamental biological processes and for alleviation of human health problems." But it also communicated fears that "such hybrid molecules might prove hazardous to laboratory workers and to the public" and proposed that the academy establish a committee to examine the problem and "recommend specific actions or guidelines should that seem appropriate." The letter also proposed that "related problems such as the risks involved in current large-scale preparation of animal viruses also be considered"—reflecting the sense of the meeting that, in addition to novel genetic engineering techniques, other emerging techniques posed special problems.[52]

The Gordon conference was the first in a long chain of events, surrounded by public controversy and debate, that eventually resulted in the formation of national controls for genetic engineering in the United States and the United Kingdom. Eventually, a broad consensus on the implications of genetic engineering, in which attention was restricted to the question of hazard, provided the conceptual basis for policy. At this stage, however, *before* pressures developed to reach agreement on whether and how to control the field, wide differences of opinion about the possible impacts of genetic engineering were expressed in private meetings on the issue and in the personal correspondence of scientists.

For many scientists, genetic engineering marked a radical departure from earlier techniques. Sydney Brenner testified in 1974 that genetic engineering allowed scientists for the first time "to cross very large evolutionary barriers and to move genes between organisms which have never before had genetic contact."[53] Similarly, Paul Berg, defending a position of caution to microbiologist Irving Crawford, argued that there was no evidence that the genes of SV40 or adenovirus had ever been linked to plasmids or phage vectors. If there had been, the environment of these combinations would have changed, resulting in a potential for the emergence of novel phenomena.[54]

Not everyone accepted such assessments. Writing to Paul Berg, Harvard microbiologist Bernard Davis argued that while "tumor viruses in bacteria might be a real danger . . . I doubt that we are likely to create any combinations of bacterial genes that have not already arisen in nature." According to Davis, the human gut was like a dense chemostat that was producing new combinations of DNA all the time; therefore, all combinations of bacterial and eukaryotic DNA that could have occurred probably had already done so: only those that had survived the process of Darwinian selection remained. Davis drew the conclusion that "new recombinants developed in the laboratory might be a menace only to those who work with them and thus have an immediate opportunity for infection. They would not spread into the community at large, and thus create a moral problem, unless they had a selective advantage; and if they had that advantage, and if I am correct in presuming that they are constantly being created in nature, they would already have become widespread."[55] For Davis, the fact that his argument implied that, at most, recombinant DNA technology would only pose hazards to laboratory personnel apparently eliminated the problem—or at least defused its ethical explosiveness. As will be seen, while this became the dominant position in the United States, the problem was conceived differently in the United Kingdom.

Eventually, Davis's skepticism about the existence of hazards associated with genetic engineering came to be widely accepted. At this point, however, most scientists saw recombinant DNA technology as capable of producing abrupt changes in the genetic constitution of living things. Given the heightened awareness in the early 1970s of the power of science and technology to transform the environment, scenarios for possible harm ranged from the escape and transmission of oncogenic DNA (the original basis for concern) to the possibility of creation of higher organisms with toxic properties. The principal possibilities contemplated at this time are categorized and listed here.

A. Hazards of genetic engineering of bacteria
  1. New combinations of drug resistance might be transferred into bacteria. As described in chapter 2, drug-resistance traits were being used as a way of identifying plasmids inserted into *E. coli* bacteria. The concern was that new combinations of drug resistance that did not already exist in nature would be conferred on *E. coli,* which would transmit these properties to other organisms.
  2. Bacteria might be constructed with other harmful attributes. Bacteria might be endowed with the ability to make a harmful product, such as a toxin. An example that was frequently cited was that of *E. coli* bacteria engineered to make botulinum toxin—one of the most potent toxins known. Another frequently mentioned example was that of *E. coli* bacteria endowed with the gene coding for the enzyme that degrades cellulose. The concern was that the activity of this enzyme would eliminate roughage from the human digestive tract, possibly causing chronic obesity and other diseases.[56]
  3. Bacteria engineered to make new proteins might trigger an immune response in a human host. Proteins similar but not identical to human proteins might cause a human host to make antibodies against its own proteins, initiating a novel form of auto-immune disease.
  4. Shotgun experiments in which a whole strand of DNA was cut into fragments with restriction enzymes and introduced into bacteria might result in the inadvertent cloning of viral DNA in *E. coli.*
B. Hazards of genetic engineering of viruses
  1. The construction of hybrid viruses through the linking of DNA from different viruses might produce an oncogenic virus with an extended host range, as with the SV40-adeno hybrids.
  2. The introduction of viral DNA into bacteria might provide a new route of exposure of humans to a variety of viruses.
C. Hazards of genetic engineering of plants and animals
  1. Plants endowed with the ability to make toxins or other harmful proteins would be introduced and might come to dominate ecosystems.

In each of these scenarios, not only the invention but also the viability of hazardous organisms was a question. As a result, much discussion focused on the ability of the bacterium contemplated for use as a host in genetic engineering experiments at this time, E. *coli* K12, to survive, to colonize animal (particularly human) hosts, and to transfer novel genes to other organisms in such a way as to cause a proliferation of damage— in other words, an epidemic.

Here again, discussion produced numerous uncertainties. Experiments conducted in the United Kingdom by plasmid experts H. Williams Smith and E. S. Anderson had shown that E. *coli* K12 did not survive effectively in the human gut under normal circumstances. However, the bacterium was able to survive for a few days, and it was also able to transfer plasmids to other E. *coli* bacteria resident in the gut at low frequency. Furthermore, the experiments had been performed in the absence of any selective pressure favoring the survival of the E. *coli* bacteria: in the presence of such a selective pressure (such as an antibiotic to which the bacteria were resistant), the probability of survival and transfer might be, in Anderson's words, "dramatically" enhanced.[57]

Moreover, the role of E. *coli* as a medically important pathogen was recognized. University of Alabama microbiologist Roy Curtiss noted in a widely circulated memo in 1974: "Infections with enteropathogenic strains of E. *coli* are probably responsible for the vast majority of diarrheal diseases and other enteric disorders among children and adults in the U.S.A. Furthermore, E. *coli* is one of the three main killers associated with patients dying of septicemia that secondarily arises because of diseases or states such as cancer, immune deficiency, transplantation, surgery, ulcers, appendicitis, peritonitis, etc." Curtiss also emphasized the considerable ability of E. *coli* to mate with and transfer genes to a wide range of other microorganisms, and in a great diversity of environments, including the intestinal tracts of humans and many other animals.[58]

Thus as the full power of genetic engineering became apparent, concerns about the pathogenicity of experimentally produced organisms were generalized. The possible impact of hybrid cancer viruses remained an important and unresolved issue. At the same time, the new technology seemed to generate the possibility of converting (with some low probability) the innocuous host organisms of genetic engineering experiments into agents of disease, possibly of epidemic proportions.

Correspondence among the scientists who became involved in the issue shows that these possibilities were being taken seriously. In 1974, Hans Kornberg, professor of biochemistry at the University of Leicester, wrote to Maxine Singer, conveying the news that British scientists had responded to the Singer-Soll letter somewhat skeptically and asking about further developments in the United States. In her response, Singer gave an assessment of the climate of opinion in the United States. This is an important statement, for she was probably in as good a position as anyone to gauge it.

> Here too there are those who consider the discussion "bad for science" for one reason or another. And there are many who, quite properly, worry about how their concerns for safety will get translated into restrictive rules. Overall, though, most serious people seem to view the problem as a significant one, especially as the biggest part of the problem is our ignorance as to whether or not we should even be worried. . . . If you want a scenario that is suggestive of the sorts of problems one worries about, the case of diphtheria is useful. The toxin is produced as the result of phage infection (phage coded) of an otherwise relatively harmless bacterium living in the throat. The toxin is harmless to the bacterium. But it gets out, circulates and causes a deadly disease. All you have to do is equate the phage with a plasmid to make the case.[59]

Such concern was not limited to the United States, despite Kornberg's sense of the skepticism of British scientists. With reference to the hazards of cloning tumor virus DNA, Sydney Brenner wrote:

> We know very little about the biology of these viruses and we cannot account for the almost universal occurrence of whole or defective [viral] genomes in the DNA of mammals and birds. Many experiments could be tried by "have a go" biologists who cannot control what they are doing with the present state of scientific knowledge. We would be justified in taking the view of fearing the worst. Thus one might accidentally produce microorganisms making proteins or fragments of proteins which could produce undesirable immunological reactions *in vivo* with unpredictable effects. The essence is that we now have the tools to speed up biological change and if this is carried out on a large enough scale then we can say that if anything can happen it certainly will.[60]

In addition to these potential health and environmental problems, there were also concerns about how and by whom genetic engineering would be applied. An editorial in *New Scientist* in 1973 warned of possible biological warfare applications.[61] In 1974, Brenner suggested that there might be problems of controlling "institutions which can, and probably do, practice secrecy in their activities, such as defense research laboratories and, more importantly, the major drug companies." The latter, he argued, could cause problems because of their capacity to find "off-shore 'havens' for this kind of work even if the major countries decided to control this research."[62] Robert Sinsheimer pointed out some years later that, in the absence of controls, genetic engineering would be available to "all sectors of society that have, or can purchase, the skills to use [these techniques]. They are available to entrepreneurs, to flower fanciers, to the military, to subversives. Their application does not require vast resources. . . . The potential for misuse is inherent in any scientific advance; it would seem to be particularly virulent here."[63]

Not everyone subscribed to concerns about genetic engineering or believed that they had overriding importance. A distinct counterpoint to expressions of concern about recombinant DNA hazards was expressions of concern about the restriction of scientific inquiry. Indeed, the latter was often portrayed as the bogey that might be raised if discussion of hazards became public. In 1974, Joshua Lederberg wrote to the NIH director expressing his deep concern that "carelessness about the details of enforcement and compliance [with possible controls] may result in a smothering blanket that goes far beyond any needs that are reasonable for the problem."[64] To Martin Kaplan of the World Health Organization, Lederberg wrote that "there really is great danger [of] the whole matter getting seriously out of hand and encumbering important research not only in molecular biology and in cancer biology but also in the study of pathogenic viruses generally."[65] Lederberg's junior colleague at Stanford, genetic engineer Stanley Cohen, echoed these sentiments in a memo to Paul Berg.[66] Bernard Davis, in the letter to Paul Berg cited earlier, noted the "danger of public over-reaction to the presumed menace, which could lead to renewal of fear of genetic engineering in man, and perhaps to fear of molecular genetics in general."[67] If hazards could arise in using new biotechnologies, in other words, many scientists were quick to insist that the general public should not deal with them; policy-making decisions were claimed to be the right and responsibility of scientists alone.

This analysis of early responses to the emergence of genetic engineering techniques shows that the issues associated with the new field were by no means self-evident. Genetic engineering did not exist in isolation either from society or from other aspects of biomedical science. First, its hazards might be seen as a subset of the laboratory hazards posed by the biogenetic techniques being developed in the early 1970s. Other biological hazards, including the increasingly widespread use of oncogenic viruses, also attracted attention and concern within the biomedical research community.[68] Second, genetic engineering provided the basis for a new class of technologies—a potential that was widely appreciated as soon as it made its appearance. The social and ethical implications of deployment of the field in industry, agriculture, medicine, and the military also exercised some scientists.

At this stage, no single discourse dominated public or private discussion of genetic engineering. Privately, in 1972–74, those close to the field raised a wide range of concerns. These concerns were progressively redefined and narrowed in both the United States and the United Kingdom—processes that occurred in tandem with a series of influential decisions concerning the forms of policy arenas, the selection of agendas, and the selection of participants.

3.5 Initiating Recombinant DNA Policy in the United States and the United Kingdom, 1972–1976

Given the initial diversity of the response to the emergence of recombinant DNA technology, the uncertainties involved, and the lack of empirical data about the hazards of the field, debate about how to respond might have continued indefinitely. In fact, sufficient consensus was so quickly achieved on the nature of the problem and the nature of an appropriate response, as well as on who should make the response, that initial controls were produced in a remarkably short time. In this section, the main decisions and events that produced an initial consensus and launched policies in the United States and the United Kingdom are described and analyzed.

The Gordon conference in June 1973 launched quasi-official policy-making on the genetic engineering issue, with the writing and publication of the Singer-Soll letter to the National Academy of Sciences—a step that met some resistance from leaders of the scientific community. According to one account, two leading biomedical researchers who attended the conference, Daniel Nathans of Johns Hopkins University and Sherman Weissman of Yale, questioned "whether an open debate might not lead to undesirable restriction of the traditional freedom of scientific research."[69] A majority voted to send the letter to the National Academy of Sciences, but a much smaller majority (forty-eight to forty-two) also voted to publish the letter in the leading U.S. science journal, *Science.* When the letter reached the editor of *Science,* there was resistance. "Do you really want to do this?" editor Philip Abelson is reported to have asked Maxine Singer. The letter, published in September 1973, framed the issue in technical terms: "Scientific developments over the past two years make it reasonable and convenient to generate overlapping sequence homologies at the termini of different DNA molecules. The sequence homologies can then be used to combine the molecules by Watson-Crick hydrogen bonding. Application of existing methods permits subsequent covalent linkage of such molecules." As one science writer observed, "The scientists were talking to themselves, over the heads of readers unversed in science, as parents do when they discuss adult subjects in the presence of children."[70] From a procedural perspective, the important feature of this process is that scientists facing a novel form of technology turned inward, to their own leading scientific organization, in their first effort to address the issues associated with it.

Philip Handler, president of the National Academy of Sciences, reinforced this move to keep the issue within the scientific community by asking Paul Berg to establish a committee to examine the problem and to propose short- and long-term actions. While Berg saw the committee as an informal group, Handler later affirmed that the Berg committee was acting on behalf of the academy.[71] More important than the formal status of the Berg committee, however, was its pivotal role in defining the genetic engineering issue and proposing mechanisms through which it would be addressed. Berg invited a number of leading molecular biologists and biochemists—David Baltimore, James Watson, Sherman Weissman, Daniel Nathans, and Norton Zinder (all previously involved in the question of the safety of cancer virus research)—to meet at MIT in April 1974. Invitations were also extended to Hermann Lewis of the National Science Foundation and Richard Roblin, a young microbiologist at Harvard University Medical School with an interest in ethical issues associated with the biological sciences. Roblin had also suggested Leon Kass and Jonathan Beckwith to represent "a more radical or skeptical [cautionary] view of the possibilities [of genetic engineering]," but neither was invited.[72] The resulting committee, with the possible exception of Roblin, was homogeneous: virtually all of its members and their views on the governance of science were well known to Berg.

Aiming at an open letter to the scientific community containing a series of proposals to deal with the recombinant DNA problem, the Berg committee focused narrowly on the question of the immediate hazards posed by the genetic engineering of viral and other types of DNA. Recollections of the meeting indicate that the committee members took the hazards issue seriously. David Baltimore later described the meeting: "We sat around for the day and said, 'How bad does the situation look?' And the answer that most of us came up with was, that the potentials for just simple scenarios that you could write down on paper were frightening enough for certain kinds of limited experiments using this technology, we didn't want to see them done at all. And these in fact were experiments that all of us would like to be doing."[73] (Baltimore's recollection dramatically illustrates how decisions on whether to slow research were being made by the very people on whom pressures to pursue genetic engineering were the strongest.) Some issues that almost certainly would have been confronted had the Berg committee been broader in composition seem either to have not been raised or to have dropped quickly out of consideration. Military use of genetic engineering technology, for example, was raised by Richard Roblin and addressed in an early draft of the Berg committee's letter but was later excised.[74]

The final version of the letter, published in July 1974 in the *Proceedings of the National Academy of Sciences, Science,* and *Nature,* constituted a crucial move in reducing and restricting discourse concerning the issues surrounding genetic engineering as well as defining mechanisms for its resolution. The letter addressed only the laboratory hazards of genetic engineering. Scientists were asked to defer two types of experiment—those involving plasmids that might confer novel resistances to antibiotics or the ability to make toxins and those involving the genetic engineering of animal virus DNAs (as in the original experiment proposed by Berg). They were warned that a third type of experiment, involving the linking of DNA from animals to plasmids or phage, "should not be undertaken lightly."[75]

The Berg committee's letter became best known in the press coverage that followed its publication for proposing a "moratorium" for genetic engineering, although in fact, as the details show and as Berg would later take pains to emphasize, more accurately it proposed a partial postponement of some experiments. But far more significant in the long run would be the proposals for further policymaking that accompanied its restriction of the issue to laboratory hazards. The letter called for an international conference of scientists involved in the new field "to review scientific progress in this area and to further discuss appropriate ways to deal with" the potential biohazards. At the same time, in a move that largely anticipated the outcome of such a meeting, the letter also called on the director of the NIH to establish immediately an advisory committee to oversee an experimental program of research exploring the hazards of the technology, to develop procedures for minimizing the spread of recombinant DNA molecules, and to draw up guidelines for work in this field.

The manner of dissemination of the Berg committee's letter influenced its reception and perceptions of its significance for both the scientific community and the general public. Prior to publication, the contents of the letter were made known to the scientific community at several scientific meetings. For example, David Baltimore discussed the letter with a group of potential users, tumor virologists, at the Cold Spring Harbor Tumor Virus Workshop in June 1974. Having ascertained that no journalists were present, Baltimore urged his scientific audience to emphasize the benefits of genetic engineering rather than the hazards, giving the impression to some members of his audience that what was really being proposed was a cover-up of current fears about the implications of the techniques that existed at this time.[76] Regardless of the accuracy of that perception, the emphasis on "benefits" would play an important role in the ensuing dissemination of the letter.

The contents of the letter were also made available to leading journalists prior to publication at a press conference at the National Academy of Sciences on 18 July 1974, given by Baltimore, Berg, and Roblin. Baltimore set the tone of this event in an initial statement that followed his own advice to the tumor virologists at Cold Spring Harbor. Highlighting the positive, Baltimore explained that "the technique holds the promise of generating new ways of making therapeutic compounds such as insulin. It might also be used to modify bacteria so that new strains can be developed which could turn nitrogen from the atmosphere into plant food." After describing the Berg committee's proposals, Baltimore ended with his own assessment of the hazards. In contrast to the kinds of statements that were being made privately at this time, Baltimore emphasized on behalf of the committee that the risk of hazard was "not enormous" but that "rather than accept the risk of hazard . . . the committee felt it more prudent to evaluate the hazard before carrying out the experiments."[77] This representation of the problem addressed by the Berg committee was generally reinforced by the statements of Berg and Roblin.

While many of the reporters appeared content essentially to accept Baltimore's formulation of the recombinant DNA issue, some took a noticeably more independent position in their questions. In particular, they pressed the three scientists to elaborate on possible military application of gene-splicing technology. While not denying the validity of the concern, Baltimore spoke for the committee in refusing to condemn in advance military use of genetic engineering and in indefinitely deferring its consideration, claiming that this was "a challenge we're going to have to meet when we know it exists."

The success of the Berg committee's letter in channeling debate is indicated by the press it received, which lauded the "rare" or "unprecedented" action of scientists' calling for a partial postponement of their own research, and depicted the issues raised by the letter largely in terms of an opposition between "benefits," framed in terms of scientific advances and technological applications, and "potential hazards."[78] But a few accounts, particularly that in *Science,* were more nuanced, noting the problem of applying recombinant DNA technology to biological warfare, the question of whether a voluntary moratorium on research could be made to stick, and possible procedural objections to the composition of the Berg committee. Nevertheless, coverage generally depicted the letter as an unusually responsible action taken with respect to a new field ripe with potential for great social benefit but facing a problem of laboratory hazards. From this point on in the public debate on genetic engineering, the future benefits—customarily used to justify funding— would be regularly enlisted to justify the possible hazards of research and development.

What was generally lost in the dissemination of the Berg committee's letter was any consideration of the significance of the committee's procedural and policy recommendations, especially its proposal that future control of the field should become the responsibility of the NIH and should take the form of guidelines for research.[79] These actions strongly reinforced the elimination of the social dimensions of the issues associated with genetic engineering and restriction of policymaking to technical dimensions. In a move indicative of the alacrity with which the scientific establishment assumed that it alone should address the implications of recombinant DNA research, NIH director Robert Stone, on the same day that the Berg committee's letter was released, announced that the NIH would establish the recommended advisory committee. In the following months, NIH officials drew up a charter for the committee, charging it "to investigate the current state of knowledge and technology regarding DNA recombinants, their survival in nature, and transferability to other organisms; to recommend programs of research to assess the possibility of spread of specific DNA recombinants and the possible hazards to public health and the environment; and to recommend guidelines on the basis of the research results." Significantly, the committee was seen by the NIH merely as "a technical committee, established to look at a specific problem." By the time of the international conference, which would be held in February 1975, the NIH advisory committee, known initially as the NIH Recombinant DNA Molecule Program Advisory Committee, was in place.[80]

The Initial Response in the United Kingdom

Waves from the Berg committee's activities reached the other side of the Atlantic well before the announcement of the committee's letter on 18 July. The correspondence of this period suggests that British and American scientists were in close contact. By June 1974, drafts of the letter were circulating in the United Kingdom, and a copy was in the hands of officials at the MRC.[81]

One of the British scientists who responded early on to the emerging debate in the United States was Hans Kornberg, professor of biochemistry at the University of Leicester and a friend of Maxine Singer and Paul Berg. After reading the Singer-Soll letter to *Science* in September 1973, Kornberg raised the question of the potential hazards of genetic engineering on the Biological Research Advisory Board of the Microbiological Research Establishment at Porton Down (the former British biological warfare facility). Although this committee apparently did not take up the issue, one of its members, the director of the Microbiological Research Establishment, Robert Harris, agreed that he would bring the issue before the Godber committee, of which he was a member and which was then addressing the question of controls for dangerous pathogens. Kornberg testified to the committee in July.[82]

Kornberg's own position at this time appears to have been somewhat circumspect. He was impressed by the concerns aired by Singer and Berg. (As noted earlier, Singer wrote to Kornberg in June 1974 indicating some of the scenarios that worried her.) At the same time, he recalled later that he "did not want . . . some kind of panic response which might cause the government to clamp down stringent restrictions on biological research without anybody being aware of what the possible hazards were." Kornberg also thought it important to "ventilate the possible benefits inherent in this technique."[83]

In response to Kornberg's evidence, the Godber committee agreed to address the hazards of genetic engineering further and to consider action similar to that proposed in the Berg committee's letter.[84] It is clear that by this point, however, other British scientists, especially potential users of genetic engineering, did not want the emerging field to be associated in the public mind with dangerous infectious pathogens like smallpox. Sydney Brenner expressed this sentiment: "It was expected that Godber would be rather noisy. I mean, it was [established as a result of] a small-pox epidemic and people had died. . . . Everybody . . . felt that [genetic manipulation] had better not be mixed up with the smallpox thing, which was really dangerous."[85]

Furthermore, if the Godber committee alone addressed genetic engineering, it would likely recommend that research with dangerous pathogens and with genetically engineered organisms be handled under the same administrative mechanism within the Department of Health. (As it indeed did when it reported in May 1975.) The available evidence suggests that prospective users of genetic engineering wanted to keep the control of genetic engineering under the purview of the Department of Education and Science and the Research Councils. Although the details of decision making within those government agencies remain obscure, the available evidence and the nature of later events point to the influence of that position.[86] By the time the Berg committee's letter appeared in *Nature,* the MRC had reported to the ABRC, the principal committee responsible for advising the British government on civil science policy, and by 27 July, the ABRC had proposed that a committee under the chairmanship of Eric Ashby be asked to assess the benefits and hazards of genetic engineering.[87] Brenner later recalled: "[The concern about not mixing GMAG with the dangerous pathogens advisory group] started . . . with Ashby. Ashby didn't want Godber to get involved with this kind of university science. . . . That is one of the reasons Ashby was set up."[88]

Like the function of the Berg committee, that of the Ashby committee was largely cast as a technical assessment of genetic engineering. According to its report, the committee would assess "the potential benefits and potential hazards" of genetic engineering, but it would not make "ethical

Table 3.1   Membership of the U.K. Working Party on the Experimental Manipulation of
the Genetic Composition of Microorganisms

Lord Ashby, Master, Clare College, Cambridge (chair)
P. M. Biggs, Director, Houghton Poultry Research Station
Professor Sir Douglas Black, Chief Scientist, Department of Health and Social Security,
London
Professor W. F. Bodmer, Professor of Genetics, University of Oxford
W. M. Henderson, Secretary, Agricultural Research Council, London
Professor H. L. Kornberg, Professor of Biochemistry, University of Leicester
Professor R. R. Porter, Professor of Biochemistry, University of Oxford, and Honorary
Director, MRC Immunochemistry Unit
Professor J. R. Postgate, Professor of Microbiology, University of Sussex, and Assistant
Director, ARC Unit of Nitrogen Fixation
Professor R. Riley, Director, Plant Breeding Institute, Cambridge
M. G. P. Stoker, Director, Imperial Cancer Research Fund Laboratories, London
Professor J. H. Subak-Sharpe, Professor of Virology, University of Glasgow, and
Honorary Director, MRC Virology Unit
Professor M. H. F. Wilkins, Professor of Biophysics, King's College, London, and
Director, MRC Cell Biophysics Unit
Professor R. E. O. Williams, Director, Public Health Laboratory Service, London

---

Source: United Kingdom, *Report of the Working Party on the Experimental Manipulation of the Genetic Composition of Micro-organisms,* Cmnd. 5880 (January 1975).

judgements about the use of the techniques." Again, the restriction of discourse was closely related to the choice of the policy advisers: like the Berg committee, the Ashby committee was composed of leading scientists and research directors (table 3.1).

Information to the Ashby committee flowed from two major sources. In the first place, evidence was received from some twenty-eight expert witnesses, all scientists, many of whom were using or were about to use genetic manipulation themselves. Much of this evidence is covered by the British Official Secrets Act and is not available. What has become public[89] suggests that, in the United Kingdom as in the United States, the response of scientists to the idea of controls on their research spanned a broad spectrum, including outright resistance, acceptance based on the assumption that controls would be voluntary, and concern about future social uses. Some dismissed the entire problem as "scaremongering." But the mainstream of British science seems to have accepted calling for a temporary postponement of research as a reasonable first step.[90] Arguments for this position were effectively summarized by Sydney Brenner in his written testimony to the committee. He observed that in many cases the hazards were completely unknown. In some cases, they might be trivial, but in others, they could be extremely serious. Whatever hazards existed, he pointed out, could be magnified immensely by scientists who were untrained in safe techniques or unaware of the possible problems. Brenner proposed as a technical solution to the problem an idea that had begun to circulate among biomedical researchers in 1974: using bacteria genetically modified to decrease the chances of their survival outside the laboratory or transfer of their DNA to some other organism. Like others who submitted evidence to the Ashby committee, Brenner envisaged the development of a system of voluntary controls administered by the organizations responsible for supporting research. What was needed, he suggested, was a code of practice that could be adjusted as knowledge of the area evolved.[91]

A second source of information to the Ashby committee was Paul Berg himself. Berg visited the United Kingdom in September 1974, debated the issues on BBC television with British scientists,[92] and made contact with Lord Ashby. Berg was apparently influential, coherently and reasonably soothing the concerns of those in the United Kingdom inclined to object to the moratorium on research. "I saw your programme," wrote Sydney Brenner to Berg, "and I thought you carried it off extremely well considering the number of morons that were participating."[93] Ashby wrote to Berg congratulating him on "a masterly handling of the discussion" and reported that "opinion in our committee is consistent with your views."[94] Berg wrote to Ashby that he was "most impressed and heartened by the response and actions in the U.K."[95] Thereafter, Berg and Ashby were in regular contact, and their correspondence shows substantial agreement about the form of policy needed for controlling genetic engineering.

The report of the Ashby committee appeared promptly in January 1975, constituting the first official report on the genetic engineering issue. Ashby later commented that the report was written "for the public to play for time while the experts got down to work" and that it was seen as important to "get the report out quickly to clear the field from sensational statements."[96] The overall tone of the report was one of optimism and reassurance. The potential of genetic manipulation for contributions to biomedical knowledge and industrial application was emphasized, and the benefits of the technology were confidently projected. In contrast, application of the technology for biological warfare was ruled beyond the remit of the committee and not addressed. The immediate costs that might result from laboratory or industrial hazards, on the other hand, were seen as more difficult to assess. The extent of the harm that might result should organisms containing new combinations of genes escape the laboratory or the factory and survive was deemed to be impossible to predict. Nevertheless, the report also suggested that the risks would probably be comparable to those presented by ordinary pathogens and that they could be controlled by the standard aseptic techniques and physical hardware used for those circumstances.

In addition, the Ashby committee, following Brenner's testimony, proposed to "disarm" the bacteria used as host organisms, rendering them incapable of survival outside the laboratory. And so the concept of biological containment (which was to prove highly influential in subsequent decisions on policies for the control of recombinant DNA technology) emerged for the first time in an official report.

In summary, the Ashby report, like the Berg committee's letter, claimed that, given suitable precautions, hazards would be far outweighed by benefits. Any suspension of work, its authors opined, could be "no more than a pause because the techniques open up exciting prospects both for science and for its potential applications to society, and the evidence we have received indicates that its potential hazards can be kept under control." The report proposed that the appropriate form of control would be a voluntary code of practice with the final judgment about appropriate procedures resting with researchers. The report was generally in harmony with the position of the Berg committee that research would be resumed as soon as voluntary controls were in place.[97] Ashby wrote to Berg in November 1974: "We are drafting our report now. . . . Briefly, our consensus (and it is a consensus) is that the pause has been valuable in prompting workers in this field to assess the situation. . . . All this seems consistent with your plans for the February conference."[98]

3.6 The Asilomar Conference, 24–27 February 1975

In the United States, meanwhile, plans were being made for the international conference proposed in the Berg committee's letter. The conference, held at the Asilomar Conference Center in Pacific Grove, California, in February 1975, proved to be a pivotal event in the history of the formation of policy for recombinant DNA technology. It produced a broad consensus among scientists in the United States and elsewhere about the nature of the genetic engineering problem and the contours of future policy that served to generate a highly influential public discourse concerning the problem and its solution. The following account analyzes the dynamics of the decision process that produced the final consensus: the distribution of influence in, and access to, the decision arena; the shaping of the formal agenda; and the dissemination of the resulting policy proposals.[99]

It is significant that the formation of informal policy continued to be sponsored by the National Academy of Sciences and was generated within an arena to which prominent molecular biologists had influence and access.[100] Paul Berg continued his responsibility for addressing the genetic engineering issue by chairing the organizing committee for the conference and by selecting its members, subject to academy approval.

Berg invited David Baltimore, Richard Roblin, and Maxine Singer to help him. In addition, Berg invited Sydney Brenner and Niels Jerne, chair of the European Molecular Biology Organization (EMBO), a major private organization for the support of molecular biology, to become the British and European representatives on the committee. (Jerne, however, played no active role on the committee and eventually resigned from it.) In the fall of 1974, the organizing committee carried out the tasks of developing the agenda, selecting the participants, and deciding on procedural ground rules.

The conference eventually recommended lifting the partial moratorium imposed since the previous July and replacing it with guidelines for genetic engineering research, many of which were quite strict. The image of this result, beamed across the world by largely positive news reports, was of an international community of scientists idealistically moving to restrict their own research and to anticipate its hazards. What was generally obscured by this image and its emphasis on the novelty of self-regulation was that the proceedings were also designed to enable research to move forward and that this goal was anticipated by the organizers from the beginning. Berg himself held that the moratorium should be lifted, that the way to gather information about the hazards was to proceed with research. Brenner's testimony to the Ashby committee indicates that, while acutely aware of the possible hazards of the field, he held a similar view. The formal goals of the meeting, set out in letters of invitation to participants, incorporated the assumption: "The purpose of the Asilomar meeting is to identify (a) the kinds of experiments one would like to do with joined molecules; (b) the important information that can be obtained from them; (c) the possible risks involved for the investigator and/or others; and (d) the measures that can be employed to test for and minimize the biohazards so that the work can go on."[101]

These goals, conveying a strong message that research would proceed, exerted a major influence both on the agenda for the meeting and on participation in it. Indeed, these key features of the meeting were closely interrelated. The organizing committee decided that the conference would have two main emphases. First, there would be reports on the state of genetic engineering and its use in research. Second, there would be reports from three panels that were set up to examine the potential hazards of genetic engineering, roughly according to the categories of hazard delineated in the Berg committee's letter: (1) plasmids and their use in cloning; (2) use of animal virus DNA; and (3) use of DNAs from higher organisms. In addition, Maxine Singer proposed an additional session on legal issues associated with genetic engineering. Early in the fall of 1974, chairs of the various panels were chosen and asked to select the members of their panel. Each technical panel was chaired, significantly, by specialists who were actively involved in or close to the research under consideration. Appointed by Berg, they naturally invited onto their panels colleagues who likewise were closely associated with the research (table 3.2).

That the Asilomar meeting was organized explicitly to recommend policy meant that particular care was taken in selecting participants. In contrast to scientific meetings, which generally follow a norm of openness, participation at Asilomar was by invitation only. A list of potential participants was drawn up by Berg by soliciting names from the organizing committee and the chairs of the panels, as well as from the NIH.[102] Clearly, Berg tapped the informal networks of scientists who were closely involved in the emerging field of genetic engineering. In addition, Berg drew on the advice of Maxine Singer and her husband, Daniel Singer, in selecting a panel of nonscientists to address ethical and legal dimensions of genetic engineering.

Invitees included several scientists from industry (from General Electric, the Merck Institute, the Roche Institute of Molecular Biology, and Searle) and one scientist with a military affiliation, Louis Baron of the Walter Reed Army Institute of Research.[103] In contrast, Berg and his coorganizers seem to have been cool to participation from scientists who might have questioned the future application of genetic engineering. Of the well-qualified scientists who were active in the public interest organization Science for the People, only Jonathan Beckwith, of Harvard University Medical School, was invited. When Beckwith was unable to attend, no formal invitation was made to any other member of the organization.

Press coverage of the meeting was also controlled. Initially, Berg planned to limit attendance at the meeting to eight reporters and to require that reports not be filed until after the end of the meeting. In addition, the organizers restricted access to the official tapes of the meeting until the year 2000. All of these conditions were arbitrary: the meeting was supported by government agencies, and as such there was a strong case for openness. Eventually, after considerable criticism and the threat of a lawsuit from the *Washington Post,* the conditions for press coverage were relaxed somewhat. (Twenty-one reporters eventually covered the meeting.) Nevertheless, the restrictions limited press coverage and subsequent analysis of the meeting significantly.[104]

Discussion of the implications of genetic engineering at Asilomar was constrained by the agenda. Most of the meeting was taken up with presentations on the state of the art of genetic engineering and reports from the three working groups. A single evening and a lunchtime talk were reserved for addressing broader social, ethical, and policy issues. This format itself carried an important message. The fact that reports on the

Table 3.2   Organizing Committee and Panels, Asilomar Conference

Organizing Committee

Paul Berg, Department of Biochemistry, Stanford University Medical Center, Stanford, California (chair)

David Baltimore, Center for Cancer Research, Massachusetts Institute of Technology, Cambridge

Sydney Brenner, Medical Research Council Laboratory of Molecular Biology, Cambridge, England

Richard Roblin, Harvard Medical School and Massachusetts General Hospital, Boston

Maxine Singer, Laboratory of Biochemistry and Metabolism, National Institute of Arthritis, Metabolism, and Digestive Diseases, National Institutes of Health, Bethesda, Maryland

Plasmid Working Group

Richard Novick, Department of Microbiology, Public Health Research Institute, New York (chair)

Royston Clowes, Department of Biology, Institute for Molecular Biology, University of Texas, Dallas

Stanley Cohen, Department of Medicine, Stanford University Medical Center, Stanford, California

Roy Curtiss, III, Department of Microbiology, University of Alabama Medical Center, Birmingham

Stanley Falkow, Department of Microbiology, University of Washington School of Medicine, Seattle

Eukaryotic DNA Working Group

Donald Brown, Department of Embryology, Carnegie Institution of Washington, Baltimore (chair)

Sydney Brenner, Medical Research Council Laboratory of Molecular Biology, Cambridge, England

Robert Burris, Department of Biochemistry, University of Wisconsin, Madison

Dana Carroll, Department of Embryology, Carnegie Institution of Washington, Baltimore

Ronald W. Davis, Department of Biochemistry, Stanford University Medical Center, Stanford, California

David S. Hogness, Department of Biochemistry, Stanford University Medical Center, Stanford, California

Kenneth Murray, Department of Molecular Biology, University of Edinburgh

Raymond C. Valentine, Department of Chemistry, University of California, San Diego, La Jolla

Animal Virus Working Group

Aaron J. Shatkin, Roche Institute of Molecular Biology, Nutley, New Jersey (chair)

J. Michael Bishop, Department of Microbiology, University of California Medical Center, San Francisco

David Jackson, Department of Microbiology, University of Michigan Medical School, Ann Arbor

Andrew M. Lewis, Laboratory of Viral Diseases, National Institute of Allergy and Infectious Diseases, National Institutes of Health, Bethesda, Maryland

Daniel Nathans, Department of Microbiology, Johns Hopkins University School of Medicine, Baltimore

Bernard Roizman, Department of Microbiology and Biophysics, University of Chicago

Joe Sambrook, Cold Spring Harbor Laboratory, Cold Spring Harbor, New York

Duard Walker, Department of Medical Microbiology, University of Wisconsin, Madison

Source: Based on documents from the Asilomar Conference, RDHC.

hazards of the research were sandwiched between sessions projecting visions of exciting new research approaches reinforced the assumption that had guided planning of the meeting—that research needed to proceed.[105]

In addition, discussion was shaped by informal, less explicit constraints. The goals of the conference organizers, the interests of the majority of the participants, and the form taken by the panel reports on hazards (see below) all suggest that the outcome—a proposal for guidelines for genetic engineering research—was widely anticipated. The Ashby committee had proposed that all research in the United Kingdom could proceed with precautions, and its report was circulated before the meeting.[106] The new NIH advisory committee was scheduled to meet immediately following the conference. Eyewitness reports of the science writers recorded that many of the participants were champing at the bit of the moratorium, pressing hard to resume their research. "The eagerness of the researchers to get on with the work in this field was most evident," one participant later observed.[107] One argument that was widely aired was, "If the dangers cannot be quantified, why put us out of business?" Guidelines and containment measures were seen by some as a means for returning to work, although there was a general reluctance to say as much.[108] Thus as a result of the restriction of participation at the meeting and the design of the agenda, discussion of the major issues faced at Asilomar was skewed. Few at this meeting were inclined to raise issues that might have continued the suspension of research in order to be adequately addressed.

When the 150 participants gathered at Pacific Grove on the morning of 24 February 1975, the cochair, David Baltimore, opened the proceedings and reinforced the prevailing bias by moving to restrict the scope of the meeting to the question of hazards and safety and by excluding broader social and ethical issues. As he stated in his introductory remarks,

> This meeting was conceived to lay out the existing technology, to consider what has been done, what might be done in the future, what benefits can come from the technology both in terms of knowledge and in more practical terms. But the impetus to call the conference was one of concern about potential hazards and about the safety of the techniques. So the ultimate focus of discussion must be on issues which are unusual for a scientific meeting. We must consider questions like: What should be done? What should we know before doing certain things? And basically, when do procedures in molecular biology become more of a hazard than a benefit?

Baltimore went on to rule out two further issues that he depicted as "peripheral to this meeting" and likely to "confuse it in a number of ways."

One . . . is the utilization of this technology in gene therapy or genetic engineering—which leads one into complicated questions of what's right and what's wrong—complicated questions of political motivation—and [of] which I do not think this is the right time [for discussion]. Second, an issue which I think is very serious and which many of us have cared about for a long time, which is the potentiality to utilize this technology for biological warfare. And again, although I think it is obvious that this technology is possibly the most potent technology for biological warfare, this meeting is not designed to deal with that question.

The issue that *does* bring us here is that a new technique of molecular biology appears to allow us to outdo the standard events of evolution by making combinations of genes which could be unique in natural history. These pose special potential hazards involving potential costs as well as enormous benefits. We are here in a sense to balance these costs and benefits right now, to design a strategy to maximize the benefits and minimize the hazards in the future.[109]

With the attention of conference participants confined to benefits (broadly defined) and costs (narrowly focused on laboratory hazards), the conference got under way.[110] Much of the meeting was taken up with sessions on the details of the new techniques and a morning on the ecology of the host bacteria and vectors contemplated as the basic tools of genetic engineering. Spliced in among these sessions were the reports of the three panels appointed by Paul Berg.[111]

The reports differed considerably in the level of detail of their analysis and in the level of concern expressed about possible new hazards. The plasmid working group's report was the most cautious and the most detailed. It emphasized the possibility that genetic engineering, "[like] other major technological and scientific advances . . . entail[ed] at least the potential of serious and unpredictable adverse consequences."[112] It mentioned as precedents the wide ecological damage caused by the accidental introduction of such organisms as fire ants, killer bees, and the carriers of chestnut blight and Dutch elm disease into the environment. The thirty-five-page, single-spaced report attempted to classify potential hazards of the introduction of foreign DNA into prokaryotes into six categories, from class I—experiments that were deemed to present an insignificant hazard and that could be performed on the laboratory bench—to class VI—experiments such as the cloning of the botulinum toxin gene in *E. coli,* which the group saw as holding such potential danger "as to preclude performance . . . at the present time under any circumstances."[113] The group used various theoretical principles in weighing hazards, conveying a sense, later challenged by other conference participants, that hazards might readily be classified and that research could then proceed on that basis.

The working group on the use of eukaryotic DNA similarly developed a classification of the hazards presented by the cloning of DNA from higher organisms, assigning experiments to classes II–V defined by the plasmid group. Two theoretical assumptions that would later be influential were used in the classification: first, the biohazard would not be less than that of the most hazardous component; and second, hazards could be ranked phylogenetically, with the hazard of the introduced DNA increasing with increasing proximity to humans. The group saw their classification as provisional and subject to modification depending on the outcome of hazard assessment experiments. It drew attention specifically to the possible hazards of large-scale applications of genetic engineering and proposed that such activities proceed only after demonstration that "the bacteria are 'safe'; that is, they will not be hazardous even if they escape the confines of their intended use."[114]

The animal virus working group produced only a single page proposing that the cloning of DNA from animal viruses should proceed according to the guidelines established by the National Cancer Institute for research on oncogenic viruses, which classified hazards as high-, moderate-, or low-risk. High-risk containment was proposed for high-risk oncogenic viruses and for highly pathogenic viruses. Most experiments were expected to fall into the moderate-risk category.[115]

On the question of implementation of safety precautions, the plasmid group provided the only detailed proposals, envisaging reviews of proposals for genetic engineering at the local level, with the principal investigator assuming the ultimate responsibility for safety. The other two groups did not elaborate on the question of implementation. All three reports seemed to converge in assuming that peer review and voluntary guidelines would provide an appropriate basis for control.

The fate of several proposals differing significantly from this recommendation is instructive. One came from virologist Andrew Lewis in a minority statement attached to the report of the animal virus working group. Lewis, sensitized to the problems of achieving compliance with voluntary controls for research with cancer viruses (section 3.3), stated that, given the uncertainties, "the risks exceed the rewards from the information to be obtained." Lewis called for a much more cautious approach to the manipulation of animal virus DNA, in which cloning of regions of DNA suspected to be oncogenic would be allowed only after the hazards had been fully assessed through a "slow, step by step approach."[116] The cold reception accorded Lewis's proposal by the conference participants was seen as symptomatic by a reporter from the *Rolling Stone:* "The source of the hostility is no mystery: This conference is *about* self-regulation—and [in regard to] Lewis's unfortunate experience, [the view is that] one bad apple need not spoil the barrel."[117]

A second nonconforming proposal came from the plasmid group, which was chaired by Richard Novick, a plasmid biologist at the Public Health Research Institute in New York and a veteran of antiwar efforts to dismantle the U.S. biological and chemical weapons programs of the 1960s. At Novick's urging, the group issued a statement warning about the potential application of genetic engineering to biological warfare and calling for a comprehensive prohibition on the military use of genetic engineering: "We believe that perhaps the greatest potential for biohazards involving genetic alteration of microorganisms relates to possible military applications. We believe strongly that construction of genetically altered microorganisms for any military purpose should be expressly prohibited by international treaty, and we urge that such prohibition be agreed upon as expeditiously as possible."[118] This proposal was not discussed at the meeting, nor did it appear in the final report. The majority of the conference members appear to have accepted the limits on the agenda proposed by the conference cochair.

Another untouched proposal came from nine Boston members of the Genetic Engineering Group of Science for the People. In an open letter circulated to the Asilomar scientists, the Boston group, who were not in attendance, criticized the structure of the Asilomar conference and called for broad public participation in the formulation of policy for genetic engineering: "We see even in the structure of this conference that a scientific elite is here alone trying to determine the direction that such regulation should take. . . . We do not believe that the molecular biology community, which is actively engaged in the development of these techniques, is capable of wisely regulating this development alone. This is like asking the tobacco industry to limit the manufacture of cigarettes." Specifically, the group recommended involvement of "those most immediately at risk—technicians, students, custodial staff, etc." in decision making on laboratory safety, social and environmental impact statements on the means and the goals of biological research, participation by nonscientists in the making of policy for genetic engineering, and continuation of the moratorium until these conditions were met.[119] These proposals received no consideration at Asilomar—even though they anticipated a legal analysis given on the last evening of the meeting (see below). It is not surprising that several of the reporters present concluded that Asilomar was intended to *avoid* public involvement rather than to encourage it. Challenges to the agenda or structure threatened to disrupt the conference's proceeding to authorize continuing genetic engineering research, and that goal was clearly preeminent in the minds of many.

Discussion of the various reports of the working groups exposed what one observer called the "central dilemma" of the meeting: "[N]o one had any real idea of what the risk might be or how to assess it."[120] In other words, the question of evaluating the hazards of genetic engineering—an evaluation fundamental to its rational control—simply had no answers. The problem was underscored in the following discussion:

> OLE MAALOE (University of Copenhagen): We are trying to give the public some assurance that we know what we are doing. To imagine that we can lay down even fairly general rules would be deceiving ourselves. . . .
>
> JOSHUA LEDERBERG (Stanford University): Either we tear up this piece of paper or we go into it very carefully. If it is likely to be crystallized into legislation, we had better be sure it is right. . . .
>
> PAUL BERG (Stanford University): We have to make some decisions. If you concede there is a graded set of risks, that is what you have to respond to.
>
> JAMES WATSON (Cold Spring Harbor Laboratory): But you can't measure the risk. So they want to put me out of business for something you can't measure.
>
> JOE SAMBROOK (Cold Spring Harbor Laboratory): As far as I am concerned, there is no absolute containment and all containment is inefficient.
>
> ROY CURTISS (University of Alabama): There is no way to decide right now what a safe organism is or what a safe cloning vehicle is.[121]

The pressures at Asilomar to produce a result that enabled work in the field to move forward were considerable. On the one hand, scientists such as James Watson and Joshua Lederberg were skeptical of any proposed controls for research. *Science* correspondent Nicholas Wade described their stance at the meeting: "Like a pair of *enfants terribles,* the two Nobel laureates were constantly discovering holes in the committee's positions and—although there seemed to be no concerted campaign— breaking the ice for the younger scientists who were eager to get the moratorium lifted on the easiest possible terms." At one point, Watson proposed simply that the moratorium be lifted. On the other hand, a larger group at the meeting saw guidelines and containment measures as the best means for returning to work. Wade quoted one of the working group chairs as saying: "The consensus here is that people want guidelines and containment so they can go and do their experiments, but no one will come out and say it—they're all chicken."[122]

At several points, the tensions at Asilomar seemed likely to produce an impasse. In the end, however, two factors seem to have been particularly influential in enabling the participants to reach a virtually unanimous conclusion. First, there was the promise of a technical solution to the hazard problem—or at least a part of it: the use as host organisms of strains of *E. coli* K12 with little ability to live or to multiply outside the laboratory. This idea, which Sydney Brenner had originally proposed to the Ashby committee, was elaborated in the plasmid working group's report.[123] It was energetically promoted by Brenner, who chaired a series of well-attended sessions on precisely how a series of mutations might be introduced into *E. coli* to ensure that it would die as required outside the test tube.[124] And the challenge of producing the strains was taken up by Roy Curtiss, who announced in a memo to the organizing committee that he was hopeful that his objective could be met in a matter of months.[125] Indeed, the expectation at the meeting—which was not to be borne out by experience—was that disabled strains would quickly be available.

Second, there was the very real possibility that legislators in the United States and perhaps elsewhere might act to control the new field if the conference produced nothing more than a controversy. Being regulated by one's colleagues troubled some researchers, but that it might preempt externally administered controls acted as a powerful pressure toward achieving consensus. At several points when the meeting seemed on the verge of disintegration, participants were effectively warned of the invisible threat of legislative intervention. Paul Berg reminded his audience: "If our recommendations look self-serving, we will run the risk of having standards imposed. We must start high and work down. We can't say that 150 scientists spent four days at Asilomar and all of them agreed that there was a hazard—and they still couldn't come up with a single suggestion. That's telling the government to do it for us."[126]

The specter of government action was also raised on the last night of the conference by two of the lawyers, Alexander Capron of the University of Pennsylvania Law School and Roger Dworkin of the University of Indiana Law School, who were invited to participate in a panel on the legal and ethical dimensions of genetic engineering, arranged by Daniel Singer. Eyewitness accounts describe the sobering impact of the lawyers' presentations, impressing on the scientists the discomforting truth that the public had a right to set limits to their freedom of inquiry whenever its pursuit impinged on others. In fact, this was the first time that the question was raised of the role of the public, through the legislature and the courts, in controlling the hazards of genetic engineering research. Capron spoke directly to the point: "Many people have talked about this research under the banner of 'academic freedom.' By overstating their case they risk provoking greater restriction. Freedom of thought does not encompass freedom to cause physical injury to others, and 'prior restraint' as one person has termed it, is absolutely an appropriate response where irreversible harm is threatened—which it may or may not be by various kinds of DNA recombinant work."[127]

The public has a right, Capron stated flatly, to become involved in assessing new technology. It also has a right to make an erroneous assessment. While the committee being established by the NIH might make an initial assessment of the technology, the policymaking process should also represent the public and should go beyond the narrow issue of safety to the broader questions of human impact.

Capron's insistence on the public's right to assess new technology was reinforced by Roger Dworkin, who reminded his attentive audience of the public's right to sue for damages and of the increasing severity of judgments on professional negligence. Even Dworkin's acknowledgment that the law also had a tradition of allowing expert groups to regulate themselves was of little consolation, for he noted that, if that discretion was abused, such groups were likely to be "massacred in court."[128]

Whether and how the public would be involved in the formation of genetic engineering policy was a crucial question. The dominant assumption at Asilomar that public involvement could only hamper researchers meant that those favoring speedy development of the field would naturally see assertions that the public could not be legally excluded from policymaking as recommendations that they act so as to obviate public involvement. Thus Sydney Brenner argued that the safety procedures recommended by the meeting should be sufficiently strict to be convincing to legislators: "We live at a time when I think there is a great antiscience attitude developing in society, well developed in some societies, and developing in government, and this is something we have to take into consideration. . . . Maybe this is the end of an era. It is very hard to tell in history where you really are. . . . I think people have got to realize there is no easy way out of this situation: we have not only to say we are going to act, but we must be seen to be acting."[129] From that point on, the theme of "being seen" to act, with its strong subtext of trying to fend off public intervention, was to be repeated frequently by scientists and research directors as they struggled to produce a solution to the genetic engineering problem.

The promise of the disarmed strains on the one hand and the threat of government intervention if scientists were not seen to be acting on the other were major influences on the consensus that emerged from the meeting. This consensus was not forged without intense debate. On the final morning, the organizing committee produced a provisional statement, which they had drafted during the night and early hours of the morning and which proposed, in essence, that the partial moratorium should end and that "work should proceed but with appropriate safeguards." Precautions for the experiments were to be classified as "high," "moderate," and "low," with the first two categories requiring physical containment in accordance with the National Cancer Institute standards and use of the disabled strains of $E. coli$ (table 3.3).[130]

Anticipating that dissent on the document might prevent any action

Table 3.3 Classification of Hazards in the Provisional Statement of the Asilomar Conference

Containment Level	Techniques/Practices	Types of Experiments
Low containment	Good microbiological techniques, encompassing use of lab coats, use of mechanical pipettes, a prohibition on eating in the lab, and use of biological safety cabinets for procedures that generate large aerosols.	Experiments involving organisms that normally exchange genetic information. Most experiments involving DNA from prokaryotes, lower eukaryotes, plants, invertebrates, and cold-blooded vertebrates.
Moderate containment	Physical containment required for handling moderate-risk oncogenic viruses, encompassing use of gloves and lab coats, use of biological safety cabinets for transfer operations, vacuum lines protected by filters, negative pressure that is maintained in limited-access laboratories.	Experiments with bacterial genes that affect pathogenicity or antibiotic resistance. Experiments with viral DNA involving the linkage of viral genomes in whole or in part to prokaryotic vectors and their introduction into prokaryotic cells (if "disarmed" vectors are used). Experiments with segments of oncogenic viral DNA if these segments are nontransforming. Experiments with segments of nononcogenic viruses. Experiments with DNA from warm-blooded vertebrates. Experiments with any DNA and animal vectors.
High containment	Use of facilities that are isolated from other areas by air locks, clothing exchanges, and shower rooms and that have treatment systems to inactivate or remove biological agents that may be contaminants in exhaust air and liquid and solid wastes. All agents are handled in biological safety cabinets. Personnel wear protective clothing and shower on leaving. Facility is maintained under negative air pressure.	Experiments with high-risk viruses. Experiments linking eukaryotic or prokaryotic DNA to prokaryotic vectors when the resulting organism might express a toxic or pharmacologically active product.

Source: Based on "Provisional Statement of the Conference Proceedings," 27 February 1975, RDHC.

being taken, the organizing committee decided not to put the matter to a vote. Instead, Berg, claiming that the provisional statement was "not a statement of the conference [but] a statement from the organizing committee, an attempt to pull together the views as we see them," proposed to receive reactions to the document and incorporate them, without further review, into a final version.[131] Asking the Asilomar scientists to accept a passive-reactive role, to relinquish control over the public statement that claimed to represent their views, immediately launched a major debate, if not rebellion in the ranks. When the issue was forced, it was decided to vote on the document, section by section. Reports of these final hours record the intensity of the discussions that followed—interactions that one observer described as "an episode of technological horse-trading conducted in a vocabulary that often seemed to fit the realm of science fiction rather better than that of hard science."[132] Frequently at issue were the high cost of moderate- and high-containment facilities and the delays involved in waiting for them and for the "disarmed" strains of the host organism.

Whatever the tensions and areas of disagreement apparent at the meeting, in the end most sections were approved by strong majorities. But at least one question—whether certain experiments should not be performed under any circumstances—produced prolonged debate. The tape of the final session shows some participants repeatedly trying to insert a statement to the effect that certain experiments were deemed so dangerous that they should not be performed under any circumstances. Berg and Baltimore initially resisted this proposal. When asked why the statement of the plasmid group had been omitted, Baltimore responded: "That was a conscious decision, based on the demonstrable split in opinion at this meeting about whether this is a philosophical question or a practical question. There were a number of people who told us at significant length that . . . this was a philosophical question of right of free inquiry." Finally, it was agreed to take a vote on the position—rephrased by Berg—that "there is a class of experiments that should not be done at all irrespective of the type of containment available today."[133] The proposal passed with only a few dissenting votes. Berg's formulation left unclear whether such experiments should never be done or only deferred. The organizing committee, in drafting the final version of the conference statement, opted for the latter position.

The following June, the final version of the Asilomar statement was published in *Science*. The report acknowledged great areas of uncertainty with respect to knowledge of the behavior of genetically modified organisms and their interactions with the environment.

> Little is known about the survival of laboratory strains of bacteria and bacteriophages in different ecological niches in the outside world. Even

less is known about whether recombinant DNA molecules will enhance or depress the survival of their vectors and hosts in nature. . . .

Nothing is known about the potential infectivity in higher organisms of phages or bacteria containing segments of eukaryotic DNA, and very little is known about the infectivity of the DNA molecules themselves. Genetic transformation of bacteria does occur in animals, suggesting that recombinant DNA molecules can retain their biological potency in this environment. There are many questions in this area, the answers to which are essential for our assessment of the biohazards of experiments with recombinant DNA molecules.[134]

Rather than wait on those answers, however, the final Asilomar statement maintained that hazards could be matched with suitable physical and biological containment precautions and that, under these conditions, most research should go forward. Berg's reformulation of the view of some participants that certain experiments were so dangerous that they should be absolutely prohibited reappeared in the final text as a qualification that certain risks were "of such a serious nature that they ought not to be done with presently available containment facilities."[135] In harmony with the position of the majority of the participants, it was also assumed that the appropriate form of control was a voluntary code of practice.[136] The Asilomar statement would provide the conceptual basis of the NIH guidelines for recombinant DNA research, whose development was initiated immediately after the conference.

3.7 The Asilomar Legacy

The initial stages of policymaking for recombinant DNA technology have often been described as "unprecedented" or "rare," and in some important respects they were. It was rare for scientists to act collectively to forestall problems associated with their research, and this action would cast a long shadow on the course of future policymaking. In general terms, the legacy of Asilomar was a sense that the burden of proof rested with scientists to show that the new biogenetic technology they wanted to develop was safe, and until they could do so, they should proceed with caution.

But an exclusive focus on this aspect of the early history of recombinant DNA policymaking distorts the whole picture. As emphasized earlier, policy formation is understood accurately only by examining how social forces both shape perceptions and govern the formal and informal processes that define the institutional environment, restrict participation in the decision process, and limit the scope of the issues addressed.

In the United States, leaders of the biomedical research community moved early on to transfer responsibility for an initial assessment of the issue to the National Academy of Sciences; in the United Kingdom, on the other hand, after an initial consideration of the issue by the God-ber committee on dangerous pathogens, the ABRC gave the Ashby committee the responsibility for making the initial assessment. But the obvious differences between the British and American institutions re-sponsible for forming an initial policy assessment—the one public and the other semiprivate—should not obscure their functional similarities: both were oriented to promote biomedical research, and their primary constituencies were the communities of British and American biomedical researchers. Members of the committees and groups appointed by these institutions—the Ashby committee, the Berg committee, the organizing committee for the Asilomar conference, and the Asilomar conference it-self—were drawn almost exclusively from their constituencies.

Given the basic similarities in the institutional positions of policy-makers in the United Kingdom and the United States and evidence of frequent exchanges of information about policy issues among scien-tists generally and among those with special responsibilities for forming policy, such as Paul Berg, Sydney Brenner, and Eric Ashby, it is hardly surprising that strong conceptual similarities characterized the arguments behind decisions taken in each country. Whether private or public, the bodies that assessed the issue and contributed to policy decisions were driven to adopt a single view of the problems posed by this emerging technology. It was in the interest of nearly all admitted to policymaking arenas that they not jeopardize control of policy by dissension, apparent inaction, or acknowledging as central dimensions of the issue that obvi-ously transcended the expertise of scientists.

Initially, scientists' responses to the possible implications of genetic engineering, shaped by the movements and countermovements of the late 1960s and early 1970s as well as by specific debates on the social and ethical dimensions of human genetic engineering and laboratory safety, were quite diverse. But the overwhelming tendency of the processes ex-amined in this chapter was to reduce the scope (and thereby the daunting complexity) of hazard evaluation by seeing broader social problems as falling outside the field of concern. Paul Berg, as an eminent represen-tative of American biomedical research and as chair of the Asilomar conference, insisted at a European meeting held shortly before the con-ference that the problems posed by genetic engineering techniques were not moral or ethical but only matters of public health.[137] Thus the genetic engineering "problem" was reprojected almost exclusively in terms of producing suitable safety precautions to contain hazards. In other words, it was reduced and redefined in terms that made it susceptible to a tech-nical "solution."

Only in one respect did the long-term impact of genetic engineering figure in the emerging policy discourse, namely, as "benefits." This configuration of genetic engineering's implications had roots in the utilitarian emphasis of biomedical research that stemmed from the political pressures on funding agencies to justify their appropriations (1.1, 1.2). Faced with questions about possible hazards, it seemed natural for scientists to evoke future "benefits" to justify continuation of their work. Thus, key reports and events disseminating the reductionist definition of the genetic engineering issue—the report of the Berg committee, the press conference on this report, the Ashby report, and the report of the Asilomar conference—opposed "benefits" of the field to its "potential hazards." In the press coverage of the issue that followed, hazards and benefits would be played off against one another countless times.

The most general finding of this chapter is that discourse and policy developed in synergy with one another. The proceedings of the Asilomar conference show that a reductionist discourse bearing within it the seeds of a technical solution expressed personal and economic interests in developing the field without external intervention and at the same time contributed powerfully to defining and reinforcing the central role of the biomedical research community in policymaking.

The power of the exclusion of the social from the definition of the genetic engineering problem and the virtual unanimity with which this definition was embraced meant that this perception of genetic engineering would quickly become dogma. Rarely in the future would policymakers deviate far from that basic position. On both sides of the Atlantic, those who contributed to the policy process would work largely within the boundaries of this discourse.

FOUR

Initiating Government Controls in the United States and the United Kingdom, 1975–1976

The achievement of the Ashby committee and the organizing committee of the Asilomar conference was to establish not only an influential reductionist discourse about the genetic engineering "problem" but also the central role of the biomedical research community in addressing it. This chapter examines the articulation of this policy paradigm in both countries and shows how the evolution of variant forms of the basic paradigm reflected the interplay of particular public and private interests. In particular, the chapter examines how the emergence of a new interest in the genetic engineering problem in the United Kingdom affected policymaking in that country, producing significant divergences in form and content from the American approach.

4.1 The Politics of the NIH Guidelines

Perhaps the most far-reaching proposal of the Berg committee in 1974 was to recommend that control of genetic engineering should fall under the aegis of the NIH and that controls should take the form of guidelines. In effect, the committee had already posited the paradigm under which work in the field could proceed. The NIH accepted this view and the participants at the Asilomar conference, sensitive to the need for rapid progress in the new technology, endorsed it (see chapter 3). By the time of the Asilomar conference, an NIH committee called the Recombinant DNA Molecule Program Advisory Committee was in place.

In assuming responsibility for developing policies to control genetic engineering hazards, the NIH became what political scientist E. E. Schattschneider calls a "pressure system," subject to several independent influences. Scientific interest in genetic engineering was intensifying, particularly in laboratories in the United States and Europe (see chapter 2). The number of projects involving genetic engineering supported by NIH grew from a handful in 1975 to over a hundred in 1976 (table 2.1). As the debate at Asilomar demonstrated, American scientists were keen to press ahead, aware that their colleagues elsewhere would do likewise, and fearful of being at a competitive disadvantage because of overly stringent controls. Thus considerable pressure impinged on the NIH from members of the biomedical research community to allow fast development of the field.

Fundamentally, speed in exploiting genetic engineering techniques appeared essential to most scientists because the first researchers would likely garner the greatest rewards. Whether to achieve advances in scientific understanding or to realize the "potential benefits" for producing "hormones, antibiotics and other drugs, vaccines, and enzymes," and for solving "problems of pollution, of energy and food shortages,"[1] or to reap a share of the profits accruing to commercial production of these benefits, those closest to the developing technology believed that success would attend immediate continuation of their efforts. Those already embarked on genetic engineering felt little desire to be significantly delayed (and many of the field's pioneers were prominent in the process of establishing controls); few public or private figures in the subsequent debate would wish to risk supporting stands that substantially obstructed progress.

The pressure for speed was expressed at two main levels. First, self-interest prompted some to demand relatively weak precautions for specific experiments and the rapid promulgation of controls. Rumors of "Saturday night" and "weekend" experiments, implying incipient rebellion within the ranks of biomedical researchers, were common in this period.[2] While most rumors were simply that, their very existence pressured the NIH leadership to establish controls with dispatch and to ensure that the controls satisfied client researchers. As the Recombinant DNA Molecule Program Advisory Committee chair, DeWitt Stetten, observed to a young scientist at Harvard Medical School in October 1975, "You appear to realize, and indeed to share, the sense of urgency which characterizes most good investigators. I should expect that were we to make regulations banning activity in this or any other field of science for a number of years, we should find these regulations very difficult or impossible to enforce."[3]

Second, genetic engineering immediately began to figure in international competition. By 1976, news of experiments conducted abroad under less stringent conditions than those contemplated by the NIH reached NIH officials.[4] As later interactions between scientists in other countries and NIH officials would reveal (section 7.1), if controls at home became significantly more stringent than those abroad, NIH officials risked putting American research at a competitive disadvantage in the international arena. In addition, commercial interest in genetic engineering was growing. Six multinational chemical and pharmaceutical corporations sent representatives to Asilomar. By 1976, several American firms were establishing in-house genetic engineering laboratories, Genentech was being launched by genetic engineer Herbert Boyer in collaboration with businessman Robert Swanson, and Cetus was circulating its first prospectus on genetic engineering to private investors (section 2.2). Unlike researchers supported by the NIH, industry could not be penalized for violating NIH controls. Industry officials could—and did (section 4.5)—respond to NIH policymaking with marked independence. The pressure such corporate independence exerted on NIH policymaking was indirect but telling. Any obvious flaunting of the NIH controls by the private sector would reflect badly on the NIH's authority to achieve compliance, thus inviting external intervention. Furthermore, that private industry represented a legal haven for scientists who felt overly constrained by the NIH controls would lend substance to the arguments of those who mooted revolt.

Moreover, the possibility of congressional intervention was quite real. A first manifestation was a hearing in the spring of 1975 conducted by Senator Edward Kennedy, the influential chair of the Senate Subcommittee on Health. The hearing, evidently stimulated by news of the Berg committee's letter and the Asilomar conference, addressed the roles of scientists and the public in the making of policy for genetic engineering. The hearing pitted arguments for self-regulation in science, put forward by molecular biologists Stanley Cohen of Stanford University and Donald Brown of the Carnegie Institution of Washington, against calls for the public to participate actively in the formation of science policy from Halsted Holman, an immunologist at Stanford University, and Willard Gaylin, a psychiatrist and president of the Hastings Institute (a private research institute established in 1969 to address bioethical issues). Cohen and Brown, maintaining that what was at issue in the genetic engineering case was controlling research, not technology, defended the position that policy decisions on genetic engineering were best left to the biomedical research community. Holman and Gaylin, on the other hand, took the position that scientific techniques such as genetic engineering were too powerful and their impact too pervasive for decisions on their deployment to be assigned exclusively to scientists.[5]

Kennedy himself voiced sympathy for public participation in science policy in a speech he gave shortly after the hearing at the Harvard School of Public Health. That scientists should regulate themselves was roundly denounced by Kennedy as "elitist and acutely parochial." The policy process for genetic engineering was, he said "inadequate because scientists alone decided to impose the moratorium and scientists alone decided to lift it. Yet the factors under consideration extend far beyond their technical competence. In fact they were making public policy. And they were making it in private. Regardless of the merits of their motivation, they reached their decisions in isolation from society." Borrowing a phrase from Holman's testimony, Kennedy argued that the public should be asked for its "informed consent" for developing new technologies such as genetic engineering.[6]

Considerable evidence shows that many early participants in developing genetic engineering controls were resistant, even hostile, toward public intervention. Gaylin's views provoked an angry response from Maxine Singer. In a stinging letter to Gaylin (with copies to Paul Berg and other members of the Berg committee, Stanley Cohen, Donald Brown, and Kennedy's assistant Larry Horowitz), Singer depicted his testimony as an "angry" and "provocative" attack on science, which could "well result in an undermining of Asilomar and its lessons." In her view, such "antiscientific" argument could well be counterproductive.[7] Correspondence within the RAC in this period finds Singer's view widely shared and shows an active interest in ensuring that congressional intervention did not materialize.[8]

In summary, the NIH was subject to pressures from several directions in the months following the Asilomar meeting: first, the threat of noncompliance by biomedical researchers and the risk of loss of American leadership in genetic engineering if controls were significantly more stringent than those being developed elsewhere; second, indications that private industry would not abide by controls it considered unreasonable; and third, the prospect of public intervention if controls were perceived to be too lax. The construction of controls by the NIH may be seen as a response to all of these pressures—as an attempt to enable research to proceed, to be "seen" to be acting responsibly, to demonstrate to members of Congress that the public *had* participated in the policy process, and to make no noise about regulating private industry.

4.2 Forming the NIH Recombinant DNA Advisory Committee

Following the acceptance by NIH director Robert Stone of the Berg committee's proposal that detailed guidelines should be drafted by an NIH advisory committee, planning for the formation of this committee proceeded. Stone appointed NIH administrators DeWitt Stetten and Leon Jacobs to chair and vice chair. Interviews with participants as well as correspondence from this period indicate that NIH officials and three members of the Berg committee—Paul Berg, David Baltimore, and Maxine Singer—consulted regularly, beginning at a National Academy of Sciences meeting in early July 1974, even before the release of the Berg committee's letter. According to Jacobs, there was "good agreement" about the role and constitution of the committee. Singer acted as a link between Berg and Baltimore and NIH officials. "I spoke to Leon Jacobs . . . and he seemed genuinely pleased to have me work as a sort of liaison between our planning group and himself regarding the establishment of the NIH Advisory Committee. He reported that the NIH proposals [for establishing the committee] had been expedited and were in the hands of the Secretary's office at HEW," Singer wrote to Berg and Baltimore in September 1974.[9]

In due course, the establishment of the committee was announced in the *Federal Register* with a rationale that played off potential benefits of recombinant DNA technology against possible hazards.

> Exploration of the genetics of microbial agents and of animal cells by use of the technology of study of DNA (deoxyribonucleic acid) recombinants offers tremendous promise of uncovering basic aspects of health and disease, and is appropriate for support by the National Institutes of Health.
> However, the use of this technology has various possible hazards because new types of organisms, some potentially pathogenic, can be introduced into the environment if there are no effective controls.

With this rationale, the advisory committee, to be known as the Recombinant DNA Molecule Program Advisory Committee (later shortened to Recombinant DNA Advisory Committee [RAC]), was charged as follows: "The goal of the Committee is to investigate the current state of knowledge and technology regarding DNA recombinants, their survival in nature, and transferability to other organisms; to recommend programs of research to assess the possibility of spread of specific DNA recombinants and the possible hazards to public health and the environment; and to recommend guidelines on the basis of the research results." The discourse established at Asilomar was in place: the focus was to be on recombinant DNA hazards; broader social and ethical issues were set aside. Within the limits imposed by this discourse, only scientists were qualified to deliver knowledgeable judgments: significantly, the committee was defined as "a technical committee, established to look at a specific problem."[10]

For some, NIH action on appointing the members of the RAC moved too slowly. The NIH had to follow DHEW rules and procedures and comply with the requirements of the Federal Advisory Committee Act of 1972; both involved delays. In addition, Robert Stone was retiring from office. In December 1974, Paul Berg wrote to Stone with an urgent appeal for action: "It is now nearly five months since the recommendations were sent forward and to my knowledge no committee has been named or approved, much less met to do business. Consequently I am writing to you in the hope that the same spirit that moved you so quickly on [the financing of the Asilomar conference] can be brought to bear on expediting the organizing and charging of the above advisory group." To "spur" Stone to action, Berg noted that the British had established a high-level committee chaired by Lord Ashby, which was already preparing an interim report. "Why should we be any less able to respond? Is our system incapable of acting in the same decisive way on an issue which you labeled urgent?" Berg closed by reminding the director of a motive for hurrying: "I believe that many investigators engaged in research on these matters will interpret the failure of the NIH to exert leadership in this area as a lack of concern or interest on the biohazard question. If there is nothing to show for our effort by the time of the Asilomar Conference (February 24–27, 1975), I fear it will be difficult to persuade scientists to continue with the voluntary suspension of the particularly worrisome experiments."[11] The prospect of a revolt of its clientele was naturally expected to provide the NIH leadership with a serious motive for action.

In fact, NIH efforts to select committee members were well under way. Shortly after Berg wrote to Stone, letters of invitation to prospective members of the RAC went out. The chair and the vice chair were NIH officials: Stetten, a biochemist by training, oversaw intramural research at the NIH; Leon Jacobs was associate director for collaborative research. Stetten, Jacobs, and other NIH officials consulted with key members of the biomedical research community such as Berg to fill out a committee of scientists (known to NIH administrators or to those they consulted).[12] A majority were actual or potential users of recombinant DNA technology, and many of the members were well known in their fields. From the perspective of the NIH administration, this was to be an "expert" committee (table 4.1).

Thus the NIH officials responsible for selection did not contemplate appointing members from occupations other than biomedical research: this idea, which originated among the RAC members at their first meeting, surprised Stetten, who thought the committee's charge made expertise requisite.[13] But moving to add a nonscientist did not stem from an interest in expanding the committee's agenda and addressing social and ethical questions associated with genetic engineering: the minutes of the first meeting record not only the recommendation for the appointment of a "lay" member but also that "the Committee cannot deal with ethical issues; it will have to restrict itself to considerations of safety and containment."[14] Rather, motives seem to have been frankly political and self-protective. Leon Jacobs recalled in 1980, "The real reason for the [nonscientist member] is to . . . counter the assertion that the committee was motivated by self-interest."[15] One RAC scientist put the matter more bluntly. "Like many other present members of the committee, I'm not sure this person could contribute to the deliberations, but I *am* sure that we need one for the purpose of being able to say we have one when there are complaints," Jane Setlow wrote to DeWitt Stetten in September 1975.[16] The first nonscientist, Emmett Redford, a professor of government at the University of Texas, Austin, eventually joined the committee in April 1976; a second nonscientist, LeRoy Walters, a bioethicist at the Kennedy Institute, Georgetown University, was added in September 1976.

One member of the RAC, Elizabeth Kutter, appointed in December

Table 4.1   Membership of the U.S. NIH Recombinant DNA
Advisory Committee, 1975–76

DeWitt Stetten, Jr., Deputy Director for Science, Office of the Director, National Institutes of Health, Bethesda, Maryland (chair)

Leon Jacobs, Associate Director for Collaborative Research, Office of the Director, National Institutes of Health, Bethesda, Maryland (vice chair)

Edward Adelberg, Professor, Department of Human Genetics, School of Medicine, Yale University, New Haven

Ernest Chu, Professor, Department of Human Genetics, Medical School, University of Michigan, Ann Arbor (left the committee in December 1975)

Roy Curtiss, III, Professor, Department of Microbiology, School of Medicine, University of Alabama, Birmingham

James Darnell, Professor, Department of Molecular Cell Biology, Rockefeller University, New York

Stanley Falkow, Professor, Department of Microbiology, School of Medicine, University of Washington, Seattle (left the committee in December 1975)

Donald Helinski, Professor, Department of Biology, University of California, San Diego, La Jolla

David Hogness, Professor, Department of Biochemistry, Stanford University, Stanford, California

Elizabeth Kutter, Member of the Faculty in Biophysics, Evergreen State College, Olympia, Washington (joined the committee in December 1975)

John Littlefield, Professor and Chairman, Department of Pediatrics, Children's Medical and Surgical Center, Johns Hopkins Hospital, Baltimore

Emmett Redford, Ashbel Smith Professor of Government and Public Affairs, University of Texas at Austin (joined the committee in April 1976)

Wallace Rowe, Chief, Laboratory of Viral Diseases, National Institute of Allergy and Infectious Diseases, National Institutes of Health, Bethesda, Maryland

Jane Setlow, Biologist, Brookhaven National Laboratory, Upton, New York

John Spizizen, Member and Chairman, Department of Microbiology, Scrips Clinic and Research Foundation, La Jolla

Waclaw Szybalski, Professor of Oncology, McArdle Laboratory, University of Wisconsin, Madison

Charles Thomas, Professor, Department of Biological Chemistry, Harvard Medical School, Boston

LeRoy Walters, Director, Center for Bioethics, Kennedy Institute, Georgetown University, Washington, D.C. (joined the committee in September 1976)

Source: National Institutes of Health, Recombinant DNA Advisory Committee, minutes for meetings held on 28 February, 12–13 May, 18–19 July 1975, 1–2 April, 13–14 September 1976, ORDAR.

1975, was both "expert" and "lay," in terms of the values that influenced RAC appointments. A biophysicist by training, Kutter taught at a small liberal arts college and therefore had embarked on a career track markedly different from that of other RAC scientists. Often disregarded by some members of the committee and actively resisted by others, Kutter's willingness to work hard outside the committee and her assertiveness within it meant that she played an active role in the proceedings, although her substantive contribution to the proceedings of this largely elite group of scientists was often limited to raising issues from a cautious perspective that were often subsequently set aside.

As the committee proceeded to draft guidelines, two main philosophies emerged among those who actively participated. On the one hand, some of the members—"hawks"—tended to urge controls that were somewhat less stringent than the Asilomar requirements and that would allow research to proceed at an accelerated pace. In general, hawks used the argument that the hazards of genetic engineering were unlikely to be great. On the other hand, other members—"doves"—tended to stress the committee's ignorance of the potential impact of genetic engineering and to urge adoption of containment levels at least as stringent as those proposed at Asilomar and in some situations more so. Most of the committee's battles over controls were fought along these lines. But regardless of where the views of RAC members fell along the spectrum from hawk to dove, they were largely united by a common acceptance of the basic terms of operation of the committee and its discourse on the genetic engineering problem.

4.3 Developing the NIH Guidelines, 1975–1976

From the initial meeting of the RAC in San Francisco immediately following the Asilomar meeting, to the release of the NIH guidelines in July 1976, hundreds of hours of convoluted discussion and debate inside and outside the committee were needed to produce the (foreshadowed) guidelines. In all, four meetings of the full RAC were held in 1975—in San Francisco (February), Bethesda, Maryland (May), Woods Hole, Massachusetts (July), and La Jolla, California (December). From these emerged recommendations on which the final controls were based.

At its second meeting, the committee established a subcommittee chaired by *Drosophila* geneticist David Hogness of Stanford University to draft guidelines. The Hogness draft was considered at the third meeting, in Woods Hole, where it was discussed, modified, and weakened in some respects. The Woods Hole draft evoked a storm of protest from biomedical researchers. In response, RAC chair DeWitt Stetten appointed a subcommittee, chaired by Elizabeth Kutter, to produce a new draft. At the fourth RAC meeting, in La Jolla, three distinct drafts of the guidelines, now labeled "Hogness," "Woods Hole," and "Kutter," were displayed side by side and paragraph by paragraph in a large variorum text. Before the La Jolla recommendations were adopted, the three versions were scrutinized paragraph by paragraph, debated, and voted on in proceedings that were often charged, and characterized by considerable maneuvering.[17]

From its first meeting onward, the RAC operated within the constraints of the discourse established at Asilomar. In particular, the committee assumed that genetic engineering research should proceed under suitable precautions developed by technical experts. This implicitly reversed the charge to the RAC, which specified that hazards should be assessed *before* guidelines were recommended, but the committee's agenda quickly became the drafting of guidelines, not the investigation of hazards. It was also assumed that responsibility for implementing the guidelines lay within the biomedical research community. Both assumptions came into play at the first meeting when the RAC agreed that the Asilomar provisional statement would be adopted as an interim measure for genetic engineering research, and that research would be peer-reviewed for compliance locally by institutional biohazard committees that would certify procedures and facilities, and nationally by the NIH committees, known as study sections, which were responsible for assessing the scientific merit of research grants. Appeals would be reviewed by the RAC.[18] Other issues would be either debated within this framework or set aside.

In essence the guidelines developed by the RAC conformed to two general principles taken as axiomatic at the Asilomar conference: (1) The hazards could be ranked on a scale. In general, it was believed that the closer the phylogenetic relation between the donor organism and the organism at risk, the greater the danger. For example, experiments involving mammalian DNA were assumed to jeopardize human health more than those involving DNA from prokaryotes.[19] (2) The hazards could be matched with a series of containment precautions of two kinds, namely, "physical" and "biological."

Physical containment was defined in terms of four levels, from P1, the standard techniques used in American microbiological laboratories, to P4, used to handle the most lethal pathogens such as anthrax, Lassa fever virus, and smallpox virus. Physical-containment levels P3 and P4 meant significant additional expenditures, estimated in 1977 to cost $50,000 and $200,000, respectively.[20]

Biological containment depended on the "disarmed" *E. coli* that had captivated the imaginations of the Asilomar participants. The RAC quickly defined three such categories. Using the standard laboratory strain of *E. coli* K12 as a host organism was claimed to offer a first level of containment (called EK1), since this strain is much less robust than wild ones and does not colonize the intestines of people and animals. EK1 containment also required that vectors with a low probability of transfer to other bacteria be used. The second level of biological containment, EK2, was predicated on the development of strains of *E. coli* viable only under artificial conditions—for example, ones needing certain chemicals for their culture. The EK3 level was defined as the use of EK2 strains whose inability to survive had been confirmed by actual tests in animals and humans. The several hosts eventually sanctioned for EK2 laboratory work (Roy Curtiss's was approved twenty-two months after Asilomar, in December 1976) were used for several years, but none was ever exhaustively tested for satisfying EK3-level requirements.

Thus the guidelines anticipated by the RAC would categorize experiments on a grid ranging from P1EK1 for the least hazardous to P4EK3 for experiments judged to be much more hazardous. Following the sense of the final statement of the Asilomar conference (in contrast to the sense of the conference itself, which favored certain absolute prohibitions), the guiding assumption was that almost all experiments and procedures could be performed safely under the proposed conditions and that those deemed most dangerous needed only to be deferred.

Virtually all of these principles were to be forcefully questioned as the RAC developed its guidelines, underscoring the fragility of the scientific consensus reached at Asilomar, and the variety of dissent the conference organizing committee had successfully outmaneuvered.

First, it was widely recognized that even the strictest physical-containment procedures assure only an imperfect barrier to protect the experimenter, and further, that this barrier can be undermined by carelessness or lack of training. A considerable body of evidence on laboratory safety showed that the higher levels of containment reduced but did not eliminate laboratory infections. When the RAC contemplated sliding many procedures into categories of lower physical containment than those prescribed at Asilomar, Paul Berg reminded the NIH that even P3 was imperfect; he cited the experience in his own P3 facility where those working with the SV40 virus commonly acquired antibodies to the virus—showing that they had been exposed and infected: "I'm convinced that physical containment is over-rated and that P3 containment, while reassuring to the psyche, is hardly the line of defense one would like to put the greatest reliance on. P3 physical containment is vulnerable to human error and, therefore, exposure of personnel within and out of the lab cannot be eliminated; reduced somewhat but not eliminated."[21]

The skill and scrupulousness of experimenters are obviously essential variables in physical containing of laboratory organisms. Those likely to pursue genetic engineering posed additional problems, because numerous molecular biologists were inadequately trained in safe microbiological technique and many had a distinct reputation for carefree procedure. Eating and smoking in the laboratory and pouring cultures down the drain were reportedly common. "The major difficulty is that a significant proportion of the investigators working with DNA recombinant molecules have had no prior training in microbiological procedures," RAC member Stanley Falkow, a microbiologist, wrote to a colleague.[22]

Second, although the Asilomar conference participants had hoped that biological containment would eliminate the problems posed by the unreliability of humans and the intrinsic leakiness of physical-containment measures, the assumption that *E. coli* K12 could be "disarmed" was not quickly borne out. Roy Curtiss, the microbiologist who took on the responsibility for constructing the weakened strain of *E. coli,* discovered that the task was far more difficult than anyone had anticipated. Significantly, Curtiss encountered a disconcerting habit among mutant strains of *E. coli* K12 he developed (which lacked the ability to make certain chemicals required for survival) to revert to type, reacquiring the power to synthesize these chemicals. Another problem was that breeding out such abilities was hampered by the bacterium's backup mechanisms, which meant that several mutations, rather than a single one, were necessary to destroy the capacity to make a particular chemical. "I now have far less confidence in biological containment than I had at the time of the Asilomar conference," Curtiss wrote to Stetten in August 1975. In the same letter, Curtiss emphasized how little was known about the chemical environments of the human gut or other places where *E. coli* bacteria would arrive in the event of an accident. Thus it was difficult to predict the probability of survival. "In sum, I will not be convinced about the biological protection afforded by any strain and/or vector until I have seen the appropriate testing data on survival and transmission that are obtained from experiments with animals and/or done in other relevant environments," Curtiss concluded. In other words, he judged that EK2 biological containment, for which the inability of the weakened strains of *E. coli* K12 to survive had been confirmed only by laboratory tests (rather than animal and environmental tests), would provide inadequate protection.[23]

Furthermore, other scientists emphasized how little was known about the ecology of *E. coli* and its plasmids and phages, particularly its ability to establish itself in the human intestinal tract. Feeding tests conducted by Anderson and Smith, which had been widely cited as evidence of the inability of *E. coli* K12 bacteria to survive for long in the human intestinal tract, were questioned by microbiologist Rolf Freter; he pointed to the essential *ambiguity* of these data: it was rare for *any* strain of *E. coli,* however robust, to establish itself in the intestinal tract in feeding experiments. The most that could be said of the Anderson and Smith experiments was that *E. coli* K12 had no *greater* capacity to establish itself than wild-type strains.[24]

Third, doubts were expressed at this time about the phylogenetic ordering principle: could it really be claimed to represent an accurate generalization about the hazards of cloning DNA, given the huge diversity of available organisms from which DNA might be derived? One skeptic was Sydney Brenner. At the La Jolla meeting in December 1975, he argued that shotgun experiments posed such a wide range of unknown hazards—for example, the possibility of cloning cryptic viruses, or genes encoding harmful proteins—that all such procedures were risk-laden.

> I think that one has to consider the hazards arising from the insertion of random DNA. What one must avoid is the creation of any selective advantage in any element of a microbe which in the world outside would be of adverse medical or economic significance. We don't know what the probability of that is. The essence of a shotgun experiment is that it explores a very large sample. That is the same whether you use *Bacillus subtilis, Drosophila* [fruit flies], or humans. So the production of the hazard is uniform, and should have one level of containment.[25]

In summary, the fundamental assumptions about the containment and classification of hazards that formed the basis of the controls contemplated by the RAC were fraught with great uncertainty—because of both the complexity of the relations between microorganisms and humans and the novelty and vastness of applied genetic engineering. As Falkow wrote in a draft preamble to the controls, "At present the hazards of many experiments may be guessed at, speculated about, or voted upon but they cannot be known absolutely in the absence of firm experimental data—and unfortunately, this needed data was, more often than not, unavailable."[26] Committee chair DeWitt Stetten expressed much the same thought.

> The problem, as I see it, will inevitably lead to different opinions since we are asking ourselves to forecast hazards of types of experiments of which there is no actuarial experience. The accidents which are envisioned have actually never occurred. Under these circumstances, each member of the committee . . . must rely upon . . . intuition, and the best that we can hope to achieve is a legislative type of resolution—namely, by vote. As men of science, we are fully aware that the legislative procedure is not designed to establish scientific truths.[27]

These uncertainties about the possible impacts of genetic engineering and about its physical and biological containment provided ample scope for debate as the RAC translated the general principles established at the

Asilomar conference into detailed containment requirements. Notably, the most intense, sustained debate within the RAC during its first year of operation concerned the levels of physical and biological containment to be required when using specific types of DNA in cloning experiments. At one level this debate was about the hazards posed by genetic engineering. But it often also reflected sociopolitical interests within the biomedical research community, either to move forward rapidly with research or to avoid external intervention.

In dealing with specific experimentation, the hawks on the RAC argued that the risks of genetic engineering were being overplayed; doves contended that too little was known to be confident that major hazards would never materialize. In practice, both factions accepted the general requirements defined at Asilomar. Proposed guidelines set too far above the Asilomar requirements produced an outcry from hawks and threats of "Saturday night" experiments; ones too lax produced protests from doves and threats of resignation from the RAC.

For example, the draft guidelines adopted by the RAC at Woods Hole in July 1975, when several of the more dovish members of the committee—Falkow, Helinski, and Rowe—were absent, not only specified some slightly lowered containment levels for experiments but also proposed several criteria for lowering requirements even further. These would have permitted experiments using DNA from cold-blooded animals and all "lower" eukaryotes (such as flies) at P2 physical containment and EK1 biological containment, that is, as the reporter for *Science*, Nicholas Wade, observed, under "standard laboratory techniques and materials, except for the Keep Out notice [on the laboratory door]."[28] Further, the maximum survival rate for *E. coli* bacteria that could be considered "disarmed" was raised from 1 in 10^8 to 1 in 10^6; and reductions in either physical or biological containment were allowed if the manipulated DNA could be shown to be "highly purified" and free of genes encoding harmful products.

The Woods Hole draft triggered an avalanche of criticism from dovish RAC members, much of it arguing that too little was known to relax requirements to the extent contemplated. Falkow, in a stinging letter to Stetten that ended with a threat of resignation, emphasized repeatedly the large areas of ignorance surrounding the ecology of *E. coli* K12: "We are totally ignorant of what happens outside the alimentary tract. . . . We are still ignorant of the ecology of *E. coli,* its plasmids and its phages. That is what we should say and is one major reason that cloning experiments are potentially dangerous. EK-1 host vectors are nothing more than the *status quo.* Obviously, no EK-2 vehicle has been tested for stability or to see if it lives up to its theoretical potential." Falkow attacked the lowered containment levels, particularly those for DNA from cold-blooded animals and lower eukaryotes, as "built on a foundation of quicksand." He continued: "It is incredible that without a single experiment and the limited information available that this Committee somehow comes up with the conclusion that these experiments are essentially safe and require minimum containment and existing host-vector systems. This is tantamount to a hunting license for any hack or high school student to do these experiments with the blessing of the NIH." "I wonder whose health and welfare is actually being protected," Falkow acidly concluded.[29]

Others took similar positions. Roy Curtiss described the Woods Hole draft as "a license to experiment with existing hosts and cloning vehicles and in the almost complete absence of controls and/or sanctions." He also recorded at this time his objections to the EK2 category of biological containment, maintaining that only the EK3 category (which was to employ "disarmed" strains already tested in humans and animals) provided sufficient protection; he acknowledged the unanimous opposition of those at Woods Hole to his position.[30]

Criticism also came from some fifty biologists at a conference at Cold Spring Harbor, who lamented to Stetten that "the [Woods Hole] draft appears to lower substantially the safety standards set and accepted by the scientific community as represented at the meeting at Asilomar in February 1975."[31] Paul Berg himself said much the same in a detailed six-page letter to Stetten, criticizing certain features of the Woods Hole draft as "marginal and inadequate." The relaxed containment levels might well "draw the charge of self-serving tokenism," warned Berg.[32] A group of young biologists in the Boston area—the Boston Area Recombinant DNA Group—went even further, submitting a detailed critique demanding that *E. coli* K12 be phased out as the host organism of choice and that new and safer hosts unrelated to humans be developed.[33]

In response, RAC chair Stetten made an unusual move. Instead of asking the RAC as a whole to reconsider the draft, he appointed a different subcommittee chaired by Elizabeth Kutter to write a new one.[34] Kutter, a critic of the Woods Hole version, had been responsible for stimulating discussion of the draft at the Cold Spring Harbor meeting. Other members of the Kutter subcommittee—Stanley Falkow, Donald Helinski, and Wallace Rowe—were seen by hawks as a predominantly dovish group.[35] Supporters of the Woods Hole draft were irritated, and several—Charles Thomas, David Hogness, Jane Setlow, and Waclaw Szybalski—wrote to NIH officials, decrying the appointment of the new subcommittee. Stetten's reply to Charles Thomas revealed the political calculations behind his decision.

> Personally, I do not believe that any other reaction to the wave of criticism would have been appropriate. There was in addition, as I am certain you are aware, a very distinct implied threat that the Legislature of

the United States would tamper in this business and provide us with legislative sanctions. There was a general agreement that we would be far better off if no laws were written in this area. I have spoken with Jane Setlow and I believe she is in sympathy with the above views.[36]

Allowing the doves to have—and to be seen as having—significant influence within the RAC was deemed necessary by the majority in order to prevent external intervention.

The fourth meeting of the RAC, with all three drafts of the guidelines on the table, was crucial for the formulation of NIH policy: failure to produce controls that were broadly acceptable to the various sectors of the biomedical research community could have jeopardized, in the face of the pressure to get on with genetic engineering experimentation, the entire NIH oversight procedure. As Stanley Falkow wrote to Elizabeth Kutter,

> The La Jolla meeting will definitely be the most pivotal for the future of recombinant molecule research over the next few years. If the Committee fails to provide any guidelines [then] I think things will just disintegrate. If the Committee comes up with guidelines which are considered by too many as being just lip service, there will be an ongoing harassment and polarization that could well lead to federal legislation as crippling as some aspects of the human research bill. Unlike Hans [DeWitt] Stetten I don't despair of reaching a consensus but it will be very difficult to achieve in so short a time span. We presumably will have to lobby before and after the workshop to try to anticipate the problems that will arise so that we have well thought out alternatives to offer.[37]

Kutter also heard from Maxine Singer, who urged haste in completing the guidelines.

> Quite apart from the substance of the document, it is my view that we must finish it in the shortest possible time. It is already very late: many investigators have no doubt already lost patience with the delays. Each investigator who is now working, or begins to work, with his own interpretation of the guidelines, will have a snowball effect on others who are also anxious to start and will be less likely to wait if Joe Smith down the hall is already proceeding. This could result in a very unfortunate situation indeed. Therefore, I urge haste. We should have as our goal guidelines that NIH can adopt immediately after the December meeting.[38]

Thus those committed to the success of the NIH process attempted to maneuver between the Scylla of revolt in the ranks and the Charybdis of external intervention.

In addition, Roy Curtiss was still having problems developing the weakened strain of *E. coli* K12. These were aired at a three-day workshop at Torrey Pines, California (organized by RAC members Curtiss, Helinski, Falkow, and Szybalski), held immediately before the La Jolla meeting. Even after nine months of work with twelve coworkers, Curtiss could not claim a definitely "disarmed" strain. He reported that he was still tinkering with various combinations of mutations. Sydney Brenner, who attended the workshop, later commented: "You see, at the moment my feeling is that EK2 is a jolly well founded rumor and nothing else. I have not seen the pieces of paper with numbers in them. I think they will be made available soon, but I think that one must have that."[39] In fact, many more months passed before Curtiss submitted his data to a RAC subcommittee for review. Yet the pressure on Curtiss to produce a usable *E. coli* strain, as well as on the NIH to certify it, was considerable: the biomedical research community had committed itself to waiting for a disabled strain of *E. coli* before going ahead with any number of projects. The La Jolla minutes recorded the progress on the EK2 host-vector system both evasively and optimistically: "There was a consensus that potential EK2 plasmid and lambda vectors and hosts have been developed which will meet, upon testing, the criteria for EK2 as stipulated in all three versions of the guidelines. . . . additional tests are required to confirm the EK2 properties of these vectors and hosts, but it is anticipated that they will meet the test. Therefore, it is hoped that EK2 vectors and hosts will be available soon."[40]

The La Jolla meeting, like its predecessors, focused on laboratory containment requirements. Eyewitness accounts and even the transcript convey the intense attention given to these requirements and the criteria (related to the purity of DNA and the extent to which it had been characterized) for relaxing them.[41] Beyond the caution of dovish members, two further interests influenced these debates: the strong personal interests of some committee members and their colleagues in performing particular experiments under low-containment conditions, and the universal concern that the RAC avoid outsider criticism.

Much of the intensity of the discussion within the committee may be explained by the fact that assigned containment levels had a direct impact on the ease and expense of doing experiments that were already under way. What appeared to be highly technical debates about, say, the hazards of shotgun experiments were often also conflicts over facilitating a particular experiment with *E. coli* K12 and P2 containment requirements or stalling it by requiring expensive P3 equipment or the yet-to-be-certified "disarmed" EK2 strain. Nicholas Wade reported that complex maneuvering attended containment decisions, even involving ingenious grandfather clauses to allow scientists who had previously used lower containment levels to continue to do so in cases where the RAC decided to raise levels. Some direct pleas were heard on behalf of experiments that scientists known to the committee wished to pursue. Wallace Rowe

Table 4.2   Draft Guidelines Developed by the U.S. NIH Recombinant DNA Advisory
Committee, 4–5 December, 1975: Examples of Containment Requirements

Experiment	Levels
A. Shotgun experiments with *E. coli* (use of recombinants to introduce undefined segments of an organism's genome into *E. coli,* classified by type of organism)	
(i) Eukaryotic DNA recombinants	
Nonembryonic primate	P3EK3 or P4EK2
Embryonic primate	P3EK2
Other mammals	P3EK2
Birds	P3EK2
Cold-blooded vertebrates	P2EK2
Invertebrates and lower plants (ferns to algae)	P2EK1
Higher plants	P2EK2
If cells are taken from embryonic or germline tissue	P2EK1
Higher plants that produce pathogenic or toxic agents	P3EK2
Purification: If a cloned recombinant DNA can be made 99 percent pure on a weight-for-weight basis, the P value of containment may be reduced by one level.	
(ii) Prokaryotic DNA recombinants	
Prokaryotes that naturally exchange genes with *E. coli*	
Class I agents (as classified by the Center for Disease Control), such as enterobacteria	P1EK1
Class 2 agents, such as *Salmonella typhi*	P2EK2
Class 3 and higher	Banned
Prokaryotes that do not naturally exchange genes with *E. coli*	
Nonpathogens	P2EK1
Pathogens	
If of low pathogencity	P3EK2
If of moderate pathogenicity	P3EK3 or P4EK2
B. Use of recombinants to insert genes from viruses, eukaryotic plasmids, and organelles into *E. coli*	
Animal viruses	P4EK2 or P3EK3
Plant viruses	P3EK1 or P2EK2
Eukaryotic plasmids or organelles: As for the shotgun categories, unless the recombinant DNA has been rendered 99 percent pure, in which case eithe the P or the EK value may be reduced by one level	
C. Use of animal virus vectors	
Defective polyoma virus plus class 1 virus or nonpathogens	P3

Table 4.2 (*continued*)

Experiment	Levels
C. Use of animal virus vectors (*continued*)	
Defective polyoma virus plus class 2 viruses	P4
If the polyoma host range has not been changed and the virus segment can be proved harmless	P3
Defective SV40 plus class 1 virus or nonpathogen	P4
Defective SV40 plus nonpathogenic and purified DNA, whether prokaryotic or eukaryotic	P3
Defective SV40 or defective polyoma (lacking late genes) plus prokaryotic or eukaryotic DNA, as long as no virus particles are produced by infected cells.	P3

Source: Nicholas Wade, "Recombinant DNA: NIH Sets Strict Rules," *Science* 190 (19 December 1975): 1175. © AAAS.

Notes: Summary of NIH committee guidelines for containing experiments with recombinant DNA. Each class of experiment has been assigned both a physical level of containment, designated P1 to P4 in increasing order of severity, and a biological level, designated EK1 to EK3.

championed fellow virologist Daniel Nathans's work on the grounds that "things are just moving beautifully for Dan Nathans and he ought to be allowed to [continue]." Rowe's plea was rebuffed by Paul Berg (who attended the meeting as a special consultant): "I think that is totally inappropriate, frankly. I mean, to tell us that this is to keep Dan Nathans in business and keep him moving, I would like to slow him down." Berg's joke, received with laughter for its frank acknowledgment of his own self-interest, momentarily broke through the tension, and at the same time reinforced the claims that the committee *could* act in a disinterested manner.[42]

In the end, compromises and accommodations produced a new draft that was slightly more restrictive in some respects than the general requirements of the Asilomar document. Thus the RAC went part way in responding to doves by requiring high levels of containment for many experiments while mollifying hawks by allowing a fair amount of work to proceed immediately (table 4.2). The most cautious elements of the recommended controls were their use of the highest physical-containment category for experiments with animal virus and primate DNA and their implementation of the sense of the final version of the Asilomar report that there were some experiments and procedures that might pose such serious hazards that they should be deferred. These categories of "prohibited" experiments included cloning of the genes of pathogens; cloning of toxin genes; release into the environment "unless a series of controlled tests leave no reasonable doubt of safety"; transfer of drug-resistance traits to microorganisms not known to acquire them naturally, if this would compromise the use of a drug in medicine or agriculture; and experiments that used more than ten liters of culture, if the DNA used was known to make a harmful product. In the latter case—one of immediate interest to private industry—requests for exceptions might be approved by the RAC.[43]

The intensity of debate at the La Jolla meeting should not obscure the fact that the larger policy framework within which the RAC operated and its discourse were accepted by virtually every member, whether hawk or dove. This policy paradigm can be seen operating in most of the RAC's decisions—and nondecisions—on a host of other issues germane to genetic engineering policy.

Choice of Host Organism

Doubts about the advisability of *E. coli* K12 as the preferred host organism for genetic engineering were raised early in the controversy.[44] In 1975, these doubts were reinforced by documented uncertainties, iterated by Falkow and Curtiss, surrounding the survival of *E. coli* K12 and the transmission of its plasmids and phages to more robust organisms. The Boston Area Recombinant DNA Group took up these questions directly and proposed phasing out *E. coli* as a cloning host and developing other safer systems. "It is erroneous to believe that the nominated 'safe' host-vector systems insure 'safe' experiments," this group wrote.

> Over the many years that these experiments will continue, there may very well be genetic exchange from the constructed "safe" host vectors to wild type vectors. This could occur on lab surfaces, it could occur in someone's body following inhalation or ingestion and preceding death of the crippled original host, or in sewage systems if the crippled host survives to be excreted; it could occur in "safe" cultures and genetic exchange could occur in special natural environmental conditions sometimes encountered by *E. coli,* that we in our ignorance of this bacterium's specialized habitats know nothing about.[45]

The group championed preparing an alternative host organism, one that was "ecologically distant from human beings, and which preferably occupies an unusual ecological niche," as one member of the group, Paul Primakoff, described it.[46] The group also defended delaying research in the interest of ensuring safety: "The social benefits which *may* arise from

DNA recombinant work will be of equal value whether they come in 20 versus 25 years, 50 versus 55 years, 100 versus 105 years. They will endure, if valuable, for many centuries thereafter. For 5 or 10 years *now,* a slow, thoughtful, research-based approach to limiting hazards makes sense."[47]

Primakoff received responses from Rowe, Falkow, and Stetten that sympathized with the Boston group's positions.[48] They agreed that *E. coli* was the wrong choice as a host organism. "I believe that many of your points are well taken—particularly that *E. coli* is too intimate with man to 'trust' as a cloning organism," wrote Falkow. However, all of these respondents went on to indicate that the pressure to proceed quickly made using *E. coli* inevitable. "I still feel strongly that *E. coli* is the wrong host for use with the recombinant DNA molecules," wrote Rowe. "However, I don't think that the people involved in these studies would ever accept a complete moratorium pending development of an alternative host." Rowe was right. Paul Berg, well positioned both to gauge current opinion and to influence it, wrote several months later to the new NIH director Donald Fredrickson, rejecting the delays Primakoff had advocated: "It is pure unadulterated sophistry to argue that a 5 or 10 year delay in achieving the promised advances is acceptable in the long term."[49] In fact, while new host organisms gradually appeared as alternatives to *E. coli* K12, phasing out *E. coli* as a host for cloning experiments was never systematically addressed by the RAC; although the arguments against its use have not been refuted, it remains today a commonly employed host.

Risk Assessment

The Berg committee had written in 1974 that hazards of genetic engineering were unknown, because "there are few available experimental data on the hazards of . . . DNA molecules," and had proposed that "an experimental program to evaluate the potential biological and ecological hazards" of the new field be overseen by the NIH.[50] In due course, the RAC was charged by the NIH with recommending research to investigate the hazards and to guide the formulation of controls (section 4.2). In fact, the RAC paid little attention to sponsoring experiments to assess possible hazards. Risk assessment lagged far behind the drafting of guidelines, and a conference on it with substantial representation from relevant biological fields was not held until roughly a year after the guidelines were released.[51]

The first public discussion of any risk assessment work occurred at the end of the La Jolla meeting, when Sydney Brenner proposed performing a "dangerous" experiment. Brenner suggested inserting polyoma virus (known to cause tumors in mice) into *E. coli* K12 bacteria, exposing the mice through either injection or ingestion of the bacteria, and testing to reveal the effects. The committee debated briefly some issues associated with interpreting the results. No resolution was forthcoming, but the committee unanimously agreed to support such an experiment. Wallace Rowe and his colleague Malcolm Martin at NIAID would take up the challenge.[52] (Rowe's results would not be available until 1979; when they were issued, their interpretation was contested [section 9.6]). Since risk assessment stood in the way of the RAC's permitting research to proceed immediately, the scientists on the committee could appease the community they represented only by reversing the order assumed in the charge to the committee.

Implementation of the NIH Controls

An important set of issues concerned implementing the NIH controls. Who would be responsible for ensuring that the recommended precautions were followed? What checks and balances would be instituted? What kinds of data on conditions in laboratories in which genetic engineering was carried out would be collected and by whom? From the beginning, the NIH and its advisory committee assumed these tasks would be shared by the principal investigator, his or her institution, and the NIH. All procedures for implementing controls were fitted into this framework. Thus, training laboratory personnel, ensuring that medium- and high-containment facilities met NIH specifications, and ensuring that experiments complied with the containment requirements would be assigned respectively to the principal investigator, institutional biohazard committees, and the NIH committee responsible for reviewing the investigator's proposal.[53] Proposals from various scientists for surveillance of the health of laboratory personnel stimulated only a recommendation that principal investigators report serious illnesses among laboratory personnel to the NIH.[54] And monitoring any escape of genetically engineered organisms from the laboratory was simply left as an option for the principal investigator.[55]

Control of Private Industry

One issue raised at the La Jolla meeting by Emmett Redford, who attended the meeting as a nonvoting consultant, was that of controlling genetic engineering activities not funded by the NIH, particularly in the private sector. Redford asked whether the committee should provide safeguards against departure from the NIH standards. Stetten responded that it would be "presumptuous" for the NIH to demand compliance from industry. Another member of the committee persisted, "Industry in the past has been as responsible as they know how to be, and hopefully they could either become part of this body and agree to it or something, but I think we have to start making plans now before someone begins to do plumbing experiments in a pharmaceutical house." Again, however, Stetten demurred: "I think in principle one can support it. I don't quite know what you can do about it."[56] With that rejoinder, the question of industry compliance with the NIH controls was dropped.

These issues were scanted because they conflicted with the overall policy paradigm (that is, enabling research to move forward under voluntary controls) adopted by the NIH and its advisory committee. A comprehensive program to assess risks or to develop alternative host organisms would have required delays of at least a few years—as the Boston critics recognized. Mandating health monitoring or training would have required extensive external institutional machinery to assure uniform compliance. Applying controls to the private sector would have required external mechanisms for inspection and enforcement. In each case, the NIH and the RAC circumvented the problem in its totality, providing a solution that fitted the problem into the original policy framework.

4.4 The Hearing before the Director's Advisory Committee, February 1976

At this point, the controversy surrounding genetic engineering began to ignite debate outside the biomedical research community. One important reason was that at least two prominent scientists, Erwin Chargaff and Robert Sinsheimer, dissented from the Asilomar consensus.[57] Both scientists were well-known members of the National Academy of Sciences, and their achievements had been of fundamental importance for the development of molecular biology. Chargaff, a professor and former chair of the Department of Biochemistry at Columbia University, had demonstrated the equality of the base pairs adenine and thymine and of cytosine and guanine in the DNA molecule.[58] Sinsheimer, chair of the Division of Biological Sciences at the California Institute of Technology and editor of the *Proceedings of the National Academy of Sciences,* was perhaps best known for his determination of the structure of ϕX174, a single-stranded virus.

Both Chargaff and Sinsheimer, impressed by the potential power of genetic engineering and troubled by biologists' ignorance of its possible impacts, challenged not just the details but the most basic assumptions of NIH policy. They saw choosing *E. coli* as the host organism as a fundamental error, and allowing research to proceed under guidelines before the hazards had been investigated as increasing the probability of accidents whose effects, if there were effects, would almost certainly be irreversible. Sinsheimer wrote in a letter to the board of regents of the University of Michigan:

> Is there a potential hazard and if so, is it large? Few people that I know would argue that one could not, by these techniques, design some quite fearsome microorganisms, capable of initiating deadly epidemics. Are such likely to arise, inadvertently? My view is that one cannot honestly calculate the likelihood. Many experiments introduce wholly unknown segments of DNA into these organisms, with patently unpredictable results. We know too little of the intricate ecology of Escherichia coli to state with confidence that, in all circumstances, such organisms are at a disadvantage. The longer-term possible evolutionary consequences of introducing an appreciable genetic intercourse between the prokaryotic and eukaryotic worlds are similarly incalculable.
>
> The potential irreversibility of the process is a further concern. These are self-reproducing organisms. Once released into the biosphere we cannot "recall" them or simply cease their manufacture. In fact we cannot even monitor their presence and number.[59]

Both scientists also held that the advent of genetic engineering raised larger ethical issues about human responsibility for future lives. In a letter to *Science* published in June 1976, Chargaff framed the question in the following terms:

> What seems to have been disregarded completely is that we are dealing here much more with an ethical problem than with one in public health, and that the principal question to be answered is whether we have the right to put an additional fearful load on generations that are not yet born. I use the adjective "additional" in view of the unresolved and equally fearful problem of the disposal of nuclear waste. Our time is cursed with the necessity for feeble men, masquerading as experts, to make enormously far-reaching decisions. Is there anything more far-reaching than the creation of new forms of life?[60]

Sinsheimer similarly asked, in a lecture to the Genetics Society given in 1975: "How far will we want to develop genetic engineering? Do we want to assume the basic responsibility for life on this planet—to develop new living forms for our own purpose? Shall we take into our own hands our own future evolution?"[61]

Both Chargaff and Sinsheimer took the view that genetic engineering, if it was to proceed at all, should initially be confined to a single high-containment facility, and that an organism, not indigenous to human beings and viable only in highly specific environments, be found as the host.[62]

Beyond attacking the substance of the guidelines, Chargaff also unleashed his acerbic pen on the self-interest he perceived in the policy process initiated by the Berg committee's letter and developed at Asilomar: "At this Council of Asilomar there congregated the molecular bishops and church fathers from all over the world, in order to condemn the heresies of which they themselves had been the first and the principal perpetrators. This was probably the first time in history that the incendiaries formed their own fire brigade."[63] Similar charges were made by the public interest organization, Science for the People. Like Chargaff, members of Science for the People pointed out that the policy process had been dominated by those with the most to gain from allowing research to proceed.[64]

By the end of 1975, dissent from NIH policy was growing. The carefully nuanced coverage of NIH guideline development in the scientific press, while often praising the NIH for its leadership on a difficult issue, also revealed the lack of resolution of dissenting views on policy. In addition, Senator Kennedy was known to be tracking NIH decisions with interest and to be planning a conference to air the issues. Concern increased within the biomedical research community that Kennedy might intervene with legislation.[65]

In this context of dissent and criticism, the newly appointed NIH director, Donald Fredrickson, announced a further meeting on the NIH controls—this time a public hearing before the Advisory Committee to the Director, a broadly composed body appointed by the Secretary of Health, Education, and Welfare to advise the NIH director on policy matters. The committee, complemented by consultants invited for the occasion, consisted largely of prominent scientists, physicians, educators, lawyers, and other professionals.

Advice to hold such a hearing no doubt flowed to Fredrickson from several sources. Fredrickson consulted with DeWitt Stetten, Maxine Singer, and David Bazelon, chief justice on the U.S. Court of Appeals for the District of Columbia Circuit, on possible courses of action.[66] In addition, Roy Curtiss had written to Stetten that

> It is apparent to me that we all have the potential of using these techniques and therefore of being accused of having vested interests in the specifics of the guidelines. In order to alleviate these fears and apprehensions, I think it would be most wise if NIH took the ultimate recommendations of the [RAC] and sought approval for these recommendations from another group of individuals who represent not only scientific disciplines but also other segments of our society. Certainly, the scientists who might be present in this committee should not themselves have any interest in performing the types of research for which

the guidelines are proposed. Quite possibly, this committee could be made up of representatives from some of the NIH councils or could be impaneled for this specific purpose alone.[67]

Fredrickson became persuaded that a public hearing was desirable. In April 1976, in informal remarks to the RAC citing the need to "forestall unfortunate and unnecessary and cumbersome legislative approaches to this problem," Fredrickson explained the need for the hearing.

> [Revision of the draft guidelines] is clearly [an event] of very great interest to people who view science from a variety of vantage points, and that includes the legislature. So that we have attempted . . . to make our consideration, and the discussions, the process whereby we come to a conclusion so that research can go on now with guidelines, an open process, a process which is almost quasi-judicial in the sense that it provided for a great deal of input from many points, as much public access as possible to the discussion, and the creation of a record.[68]

At the hearing itself, Fredrickson claimed that it allowed some degree of "participation" by the public in the making of science policy, since it provided an opportunity to air the issues and to allow those who were interested to comment on the NIH process. In addition, the hearing created a "public record . . . that [supplemented] the proceedings of the NIH [Recombinant DNA Advisory] committee."[69] Fredrickson maintained later that the hearing demonstrated that the NIH process had "provided for a great deal of input from many points, as much public access as possible to the discussion, and the creation of a record."[70] In accordance with the goal of permitting "public access," seventeen Washington-based public interest groups were notified of the hearing to be held in February 1976; clearly, the event was meant to include those outside the biomedical research community.

The meeting opened with presentations on technical aspects of the recombinant DNA issue and on the history of NIH policy by those closely involved in developing it—Paul Berg, Maxine Singer, David Hogness, Roy Curtiss, and DeWitt Stetten. A session followed that provided an opportunity for commentary on aspects of NIH policy by others, including critics.

Although prominently featuring those supportive of the fundamental tenets of NIH policy, the meeting also underscored the extensiveness of dissent, not only outside the director's advisory committee but also among its members, who included Robert Sinsheimer, and more generally brought into focus major policy issues concerning the guidelines. Beyond issues relating to the containment requirements of the draft guidelines (which were considered to be too low by a number of scientists in attendance), many questions concerning the overall policy framework that had emerged during the previous year were aired again.[71]

First, Sinsheimer elaborated his concerns about the proliferation of genetic engineering activities in the absence of knowledge concerning the hazards. "What we are doing," Sinsheimer warned the audience, "is almost certainly irreversible. . . . knowing human frailty, these vectors will escape, they will get into the environment, and there is no way to recapture them."[72] Writing to Fredrickson after the meeting, Sinsheimer also questioned the concept of biological containment, claimed by Roy Curtiss and others to mean that the disarmed strains of *E. coli* K12 had an infinitesimally small probability of survival in the environment: "I do not rest easy on Roy Curtiss' 10^{-19} style numbers [for survival] because I expect that other events with which he has not reckoned will come into play at much higher probabilities. The irreversibility of the process—which can only increase over time as more and more laboratories and persons become involved—continues to disturb me."[73]

Second, the question of applying the NIH controls to non-government-sponsored research was raised with particular force by Peter Hutt, a prominent Washington lawyer and a former general counsel to the Food and Drug Administration (FDA). Hutt, who was generally supportive of NIH policy, warned the NIH director that, without uniform controls covering genetic engineering in all sectors, Congress was likely to intervene. He contended later in a letter to Fredrickson, "I believe it extremely unlikely that Congress and the American public will be willing to rely solely on the moral suasion engendered by these guidelines, and the peer pressure that they will carry, to determine that no experimentation will be undertaken in the country in violation of the guidelines. Proliferation of experimentation in violation of the guidelines is, in short, a legitimate public concern." As a model for extending the guidelines' coverage, Hutt recommended the use of section 361 of the Public Health Service Act, a statute that provides powers to the Secretary of Health, Education, and Welfare to prevent the spread of disease.[74]

Finally, minimal public representation in the formation of policy was variously criticized. The issue was raised in connection with NIH compliance with the National Environmental Policy Act of 1969 (NEPA). One of the purposes of this statute was to provide the public with a voice in the decisions on major federal actions "significantly affecting the quality of the human environment." As Richard Andrews, a professor of environmental policy at the University of Michigan, noted in a letter to the NIH director, the NIH decision to release recombinant DNA guidelines seemed to fit the criteria for application of NEPA used by the Council on Environmental Quality particularly well.[75]

Despite the range of views the hearing elicited, it was immediately clear that the NIH leadership did not contemplate disrupting the policy paradigm its scientist clients largely favored. An indication of what was foreseen is provided by DeWitt Stetten's summation, portraying the problem faced by the NIH as one of steering a middle course between the position of those who argued that the guidelines were too permissive and those who considered them to be too restrictive. Stetten advocated compromise—but a compromise that did not deviate from the policy paradigm assumed all along, a course with which Fredrickson was in evident agreement.[76] In subsequent policy documents, Fredrickson depicted the NIH policy in similar terms, as the result of an effort to strike a "reasonable balance" between the concerns of those who advocated "going slow" and of those who wanted research to progress rapidly.[77]

Even if the NIH director had wanted to incorporate into NIH policy the much greater level of caution advocated by Sinsheimer and others, the conditions of international scientific competition virtually blocked such a possibility. Following the hearing, RAC chair Stetten heard from Charles Weissmann, a Swiss molecular biologist who chaired the Advisory Committee on Recombinant DNA of EMBO.[78] Weissmann expressed general support for the La Jolla version of the NIH guidelines but noted that the EMBO committee preferred "not to commit ourselves to all the detailed classifications, at least for the time being." He cautioned, moreover, that EMBO felt that "any further tightening of your Committee's recommendations is unwarranted by the limited evidence that is at hand." (Evidence of expert ignorance could cut both ways.) If the United States were to follow Sinsheimer's recommendations, it could not expect the major European nations involved in genetic engineering to go along.[79] In a report on the evolution of the American controls in April 1976, Nicholas Wade observed that EMBO had won itself "virtual veto power over Fredrickson's decision."[80] It would have been more accurate to say that EMBO was using political leverage that scientists on both sides of the Atlantic could exert by virtue of the existence of scientific competition among laboratories and among nations.

4.5 Promulgating the 1976 NIH Guidelines: Industry and the Public Enter the Policy Debate

Following the hearing before the director's advisory committee, the NIH moved to complete and issue the guidelines and to gain acceptance for them. While the information generated by the hearing, in demonstrating the extent of dissent, was influential in generating a larger public debate about NIH policy, it had little impact on the policy itself, whose basic characteristics remained unchanged. At the last RAC meeting before the guidelines were issued, held on 1–2 April 1976, consideration of a list of changes proposed by the NIH director took up most of the agenda.[81] Virtually all the proposed changes were small modifications of the draft controls considered at the February meeting. Some controversial containment requirements—notably those applying to cold-blooded animals and viruses—were slightly raised. The composition of local biohazard committees was broadened to include one or two nonscientific members drawn from local communities. Perhaps the most significant change was the strengthening of language prohibiting the deliberate release of genetically engineered organisms into the environment. But none of the recommended changes affected the basic policy paradigm, including processes at the national and local levels to implement it and to oversee its future evolution.[82] All of these changes were small. The final version of the guidelines, issued on 23 June 1976, was close in form and content to the La Jolla draft, showing a marked continuity with the principles established at Asilomar (table 4.3).

Before promulgating the guidelines, the NIH director attempted to persuade other sectors to adopt the NIH controls—a step he saw as necessary to head off moves to initiate legislation. As he noted in his informal comments at the April meeting of the RAC, "One of the most important things that we must attempt to advise the government about is the extension of these guidelines to research beyond NIH grantees. And nothing short of wide coverage of the whole national universe of research will do. It is of course our hope and intent that by choosing a wide procedure we can forestall unfortunate and unnecessary and cumbersome legislative approaches to this problem. . . . We know there are lots of anxieties about the extension of this work in industry, cries for prohibition, and so forth."[83]

Meetings were organized with representatives of federal agencies and of private industry to provide information about NIH policy and procedure.[84] The purpose was almost certainly to attempt to establish the guidelines as a standard that would be used, if necessary, by other government agencies. Fredrickson had been warned in a memo from the associate director for program planning and evaluation, Joseph Perpich, that "the issue of regulations . . . will confront you immediately after issuance of the guidelines."[85]

The meeting with representatives of major pharmaceutical and chemical corporations left little doubt about their preferred approach to genetic engineering controls. "The emphasis . . . was voluntary compliance, and if possible, no regulation by the Federal Government," Perpich recorded. Furthermore, industry representatives noted "a number of important areas of interest to industry [that] would have to be taken into account." Prominent among these were the ten-liter limit on culture size,

Table 4.3   The U.S. NIH Guidelines, July 1976: Examples of Containment Requirements

Experiment	Levels
A. Shotgun experiments using *E. coli* as the host	
Nonembryonic primate tissue	P3EK3 or P4EK2
Embryonic primate tissue or germ line cells	P3EK2
Other mammals	P3EK2
Birds	P3EK2
Cold blooded vertebrates, nonembryonic	P2EK2
Embryonic or germ line	P2EK2
If vertebrate produces a toxin	P3EK2
Other cold blooded animals and lower eukaryotes	P2EK1
If Class 2 pathogen,[a] produces a toxin, or carries a pathogen	P3EK2
Plants	P2EK1
Prokaryotes that exchange genes with *E. coli*	
Class 1 agents (nonpathogens)	P1EK1
Low risk pathogens (for example, enterobacteria)	P2EK1
Moderate risk pathogens (for example, *Salmonella typhi*)	P2EK2
Higher risk pathogens	Banned
Prokaryotes that do not exchange genes with *E. coli*	
Class 1 agents	P2EK2 or P3EK1
Class 2 agents (moderate risk pathogens)	P3EK2
Higher pathogens	Banned
In all above cases, if DNA is at least 99 percent pure before cloning and contains no harmful genes, either physical or biological containment levels can be reduced one step.	
B. Cloning plasmid, bacteriophage and other virus genes in *E. coli*[b]	
Animal viruses	P4EK2 or P3EK3
If clones free from harmful regions	P3EK2
Plant viruses	P3EK1 or P2EK2
99 percent pure organelle DNA, Primates	P3EK1 or P2EK2
Other eukaryotes	P2EK1
Impure organelle DNA: shotgun conditions apply.	
Plasmid or phage DNA from hosts that exchange genes with *E. coli*	
If plasmid or phage genome does not contain harmful genes or if DNA segment 99 percent pure and characterised	P1EK1
Otherwise, shotgun conditions apply.	
Plasmids and phage from hosts which do not exchange genes with *E. coli*	
Shotgun conditions apply, unless minimal risk that recombinant will increase pathogenicity or ecological potential of the host, then	P2EK2 or P3EK1
C. Animal virus vectors	
Defective polyoma virus plus DNA from nonpathogen	P3
Defective polyoma virus plus DNA from Class 2 agent	P4
If cloned recombinant contains no harmful genes and host range of polyoma unaltered, reduce to	P3

Table 4.3 *(continued)*

Experiment	Levels
C. Animal virus vectors *(continued)*	
Defective SV40 plus DNA from nonpathogens	P4
If inserted DNA is 99 percent pure segment of pro-karoytic DNA lacking toxigenic genes, or a segment of eukaryotic DNA whose function has been established and which has previously been cloned in a prokaryotic host-vector system, and if infectivity of SV40 in human cells unaltered	P3
Defective SV40 lacking substantial section of the late region plus DNA from nonpathogens, if no helper used and no virus particles produced	P3
Defective SV40 plus DNA from nonpathogen can be used to transform established lines of nonpermissive cells under P3 provided no infectious particles produced. Rescue of SV40 from such cells requires	P4
D. Plant host-vector systems	
P2 conditions can be approximated by insect-free greenhouses, sterilization of plant, pots, soil and runoff water, and use of standard microbiological practice.	
P3 conditions require use of growth chambers under negative pressure and routine fumigation for insect control.	
Otherwise, similar conditions to those prescribed for animal systems apply.	

Source: Colin Norman, "Genetic Manipulation: Guidelines Issued," *Nature* 262 (1 July 1976): 2.

Note: The guidelines define four levels of physical containment, designated in order of increasing stringency, P1 to P4, and three levels of biological containment, EK1 to EK3, and assign experiments to them on the basis of potential risk.

[a] Classes for pathogenic agents as defined by the Center for Disease Control.

[b] cDNAs synthesized in vitro from cellular or viral RNAs are included.

development of new host organisms, and the need to protect sensitive industry information. While the ten-liter limit seemed large for university research, the participants informed Perpich that it was "quite small for the purposes of industrial production." They also noted that non-pathogens other than *E. coli* K12 might be more appropriate for industrial use. Protection of proprietary information was stressed as an "overriding concern." "Companies might be willing to list investigators and the research labs in a voluntary registry, but might be quite reluctant to list the actual research that is ongoing," Perpich observed.[86] Clearly the guidelines would need to conform to these needs if industry was to comply with them.

Correspondence with industry executives following the meeting confirmed industry's independent posture toward the NIH controls—a posture that contrasted with that adopted by NIH-supported scientists.

Cornelius Pettinga, executive vice president of Eli Lilly, informed the NIH director that, while Lilly intended to "adhere to the intent and spirit of the NIH guidelines," it would decide for itself on the operational significance of its "adherence."

> If convinced of the safety of these experiments, we would have no hesitation in conducting work at levels higher than 10 liters. Drug resistance genes are such standard and convenient markers that we would feel these could be used but are confident that we can adequately assess any danger in such experiments. We feel that it is likely that we might develop host-vector systems which we might not wish to submit for approval or accreditation. We would be happy to submit safety data on such systems.[87]

In other words, Lilly, not the NIH, would evaluate the need to follow the letter of the NIH requirements. Nor would it feel obliged to provide information about its activities: minutes of the meetings of a safety committee established by the company would be kept but would "not necessarily be available for public inspection unless there were an unusual reason to make them available." As Pettinga reminded the director, some of Lilly's work would be "proprietary." Quite clearly, private industry would not feel bound by the guidelines, threat of legislative intervention or no.

While NIH officials were making efforts to gain acceptance for the guidelines beyond the biomedical sector, the controversy concerning the validity of the controls began to spread to communities where plans to pursue research were already under way. As we have seen, symptoms of dissatisfaction with the paradigm of NIH oversight emerged at the special hearing before the director's advisory committee in February 1976. Following the hearing, controversy erupted in a number of academic communities—for example, in Ann Arbor, where the debate was largely contained within the University of Michigan, and in Cambridge, Massachusetts, where it drew the attention of the mayor. The mayor's vocal skepticism about the safety of genetic engineering research at Harvard and MIT and his announcement of hearings before the Cambridge City Council made genetic engineering a public issue. The release of the first formal NIH guidelines in June 1976,[88] the outcome of a decision process restricted to the biomedical research community, coincided with the Cambridge City Council's imposition of a three-month moratorium on genetic engineering at Harvard and MIT, symbolizing citizens' insistence on their right to participate in policy decisions that would control the new technology; the city's intervention signaled a growing public awareness of the issue. At this point, challenges to the primacy of pushing rapidly ahead threatened to undermine the social and technical assumptions supporting the policy paradigm that the biomedical research sector had striven to establish.

4.6 The Politics of Genetic Engineering in the United Kingdom

In the United Kingdom as in the United States, the "moratorium" on genetic engineering and the well-publicized outcome of the Asilomar conference aroused various scientific, commercial, and public interests concerned about the promotion, development, and impact of the new field. But the mix of special interests and their expression in political arenas in the United Kingdom differed from the configurations that emerged in the United States.

Following the Asilomar conference, British scientists, especially potential users of genetic engineering, were, like their American counterparts, eager to move ahead. When the Godber committee proposed in May 1975 that genetic engineering hazards be combatted under the aegis of the Dangerous Pathogens Advisory Group responsible to the Department of Health and Social Security, British molecular biologists responded with dismay. Most were anxious to prevent the genetic engineering issue from being subsumed under the larger question of laboratory hazards. In addition, pressure was building to proceed rapidly with research.

A conference at Oxford University in July 1975, supported by the Research Councils and organized by Sydney Brenner, Walter Bodmer (a prominent Oxford geneticist and member of the Royal Society), and others and attended by over one hundred British scientists (mostly potential users of genetic engineering), provided scope for expression of the scientists' interests in the new field. Like Asilomar, the Oxford meeting was a forum for reporting on the prospects for genetic engineering, and it likewise furthered the political purposes of the participants. The ambiance of the meeting was captured in an article in *Nature*.

> Last weekend saw a hundred scientists descending on Oxford to discuss with representatives of the Research Councils and the Department of Health and Social Security the problems of experimentation in the "genetic engineering" field. . . . Several participants were at pains to point out the intellectual pressure that is building up to get moving again in this field. It was wryly conceded that more impatient scientists were already under way and although there was no evidence that they were conducting experiments in an irresponsible way, their action was causing considerable frustration amongst those waiting for more formal guidance on safety. It is, as one participant put it, rather like the vivisection laws—until society has spoken it is left to the vagaries of individual consciences. And there is obvious pressure to go ahead—

participants, on being asked whether they, as individuals, would wish to perform such experiments within the next two years, said "yes" in considerable numbers.[89]

In addition to seeking governmental approval to proceed with research, British scientists by this point were largely aligned behind several related goals. Above all, they wanted to press the government to remove the restrictions on experiments classified at Asilomar as posing a significant hazard. Brenner later recalled: "The Ashby committee had reported, and the question was to get the implementation of [its recommendations]. . . . So [the conference] was done also to try to put pressure on [the government] . . . to do something."[90] In addition, and in opposition to the Godber committee's proposal, British scientists were keen to ensure that oversight of genetic engineering became the responsibility of the Department of Education and Science (the agency responsible for funding scientific research) rather than the Department of Health and Social Security (the department likely to be responsible for controlling use of dangerous pathogens).[91] Such an arrangement not only ensured that the government agency responsible for funding the research would have the lead role in controlling it, but also separated the control of genetic engineering from that of dangerous pathogens. Scientists "did not want to get [genetic engineering] mixed up with dangerous pathogens," Brenner recalled in 1980. Finally, it appears that the consensus emerging at the Oxford meeting was that oversight of genetic engineering should become the responsibility of a small expert committee—a group of peers in the field. This would be a "kind of peer group. At least, that was the discussion," Brenner later stated.[92]

There is little evidence that private industry actively sought to guide the formation of British policy at this stage. Like their counterparts in American corporations, the heads of British pharmaceutical and chemical firms, while actively following and supporting technical developments in genetic engineering (section 2.3), were apparently content to watch from the sidelines. But the fact that the scientists' explicit political goals were also preconditions for the swift exploration of the profit-making opportunities that genetic engineering promised (both for researchers and for corporate investment) suggests that private industry could see that its interests overlapped, for the most part, those of the scientific community.

A third political interest in genetic engineering present in the United Kingdom was that of the trade unions, especially scientific and technical unions such as the ASTMS. In the 1970s, union membership grew rapidly at the same time that workplace health and safety became seen as an important area for negotiation (section 2.2). The ASTMS, whose membership included laboratory technicians and many scientists, hired a full-time national health and safety officer to address the issue. Furthermore, the passage of the Health and Safety at Work Act in 1974 provided unions with a formal arena, the Health and Safety Commission, in which to express their concerns. With one-third of the members of the commission drawn from unions, no procedures had to be invented, nor any industrial action taken, for union concerns about health and safety to become part of the commission's agenda.

Diverse interests around genetic engineering existed within the unions.[93] Unionized scientists, like their nonunion colleagues, were especially sensitive to the competitive dimensions of genetic engineering and keen to see research in the United Kingdom proceed. A memo written in January 1975 and signed by twenty-one ASTMS scientists drawn from Scottish laboratories welcomed the Ashby report and proposed that an advisory committee on the Research Councils, as long as it included union representation, could suitably control research in the field.[94] Briefing notes for the ASTMS written by Robert Williamson, a molecular biologist who was then at the Beatson Institute for Cancer Research in Glasgow, adopted a more nuanced position. Williamson accepted the Ashby report's assessment that the potential benefits of genetic engineering would be great, and appeared as well to think that an advisory committee with trade union representation to oversee all genetic engineering work would be appropriate. However, he noted that the Ashby committee had not addressed the question of establishing statutory control of genetic engineering, and he pointed to the possible relevance of the Health and Safety at Work Act—"an enabling act . . . [that] affects health and safety provisions in laboratories, and includes mandatory provisions for consultations with trades unions." Williamson also proposed that the union insist on a ban on biological warfare research.[95] In notes submitted to ASTMS officials in January 1976, he emphasized the need to move quickly to resume research in genetic engineering and warned that British research was "falling behind . . . with all the usual disadvantages (brain drain, loss of patent rights, demoralization) that follow." While emphasizing the need for safety, he urged that research be allowed to proceed and again proposed an advisory committee to the Research Councils, with union representation, as an appropriate oversight mechanism.[96] Union officials on the other hand, viewed genetic engineering primarily as a model case in health and safety policy under the new Health and Safety at Work Act. They called for statutory controls, with strong union representation, in any central oversight organization.[97] More generally, union members—scientists and officials alike—held that the government should consult with unions no less than with scientists on formulating genetic engineering policy.

Thus the union response was complicated and nuanced, reflecting a commitment to assuring health and safety in laboratories and industry, a general interest in maintaining Britain's competitive edge in new science and technology, an interest in representation in national policymaking and local safety discussions, and the professional interests of scientific members in pursuing a new field. Undoubtedly, differences in perspective and emphasis were discussed and negotiated within the trade unions in this period. By 1976, when their representatives gave evidence to the Williams committee, union interests had solidified behind a specific set of goals, especially statutory controls and representation in policymaking at both the local and national levels (section 4.7).

In contrast to the eruption of public concern about genetic engineering in the United States in 1976, governmental actions on the control of the field evoked little public interest in the United Kingdom, despite increased public awareness of environmental issues in the late 1960s and early 1970s (section 1.4). Only one public interest organization, the British Society for Social Responsibility in Science, appears to have actively followed the issue in the 1970s, and its role was mainly limited to issuing critiques of British policymaking rather than actively challenging the government policy, as did its counterpart, Science for the People, in the United States. As in the United States, possible hazards associated with genetic engineering were covered in the British press, but there were no parallels in the United Kingdom to the controversies that flared in Cambridge and elsewhere in the United States.

Several conditions may have encouraged the general passivity of the British public on this issue. First, the proceedings of the British committees responsible for assessing genetic engineering were not open to the public. Thus questions that fueled the American debate—for example, differences among scientists on the nature of the possible hazards or criticisms of the decision process—were largely hidden from the British public. Indeed, the chair of Britain's first committee entrusted with an assessment of genetic engineering saw one of the functions of this committee as public reassurance (section 3.5).

Second, members of the U.S. Congress could act as critics of government policy in ways not available to members of Parliament. The division of power between the executive and legislative branches of government in the United States meant that Senator Kennedy, chairing a committee with considerable influence on biomedical research policy, could choose to challenge and critique the decisions of the U.S. government and in doing so, bring the issues associated with genetic engineering policy before the American public. No similar powers were possessed by members of the House of Commons. Majority party government in the United Kingdom and the combining of legislative and executive functions limit the role played by the House of Commons. The conventional wisdom that "Parliament 'legitimates' but does not 'legislate'" seems generally accurate.[98] Real power is located in the governing party and in the permanent civil service, which provides—especially through its privileged access to the papers of previous administrations—the essential continuity between governments.[99] Thus members of Parliament in the 1970s had little power and few resources (in terms of staff and support) to challenge government policy. Their influence was largely confined to asking parliamentary questions and—at least for members belonging to the majority party—to informal contacts with cabinet ministers; neither procedure provoked much public awareness.

It is important to recognize, also, that British policymaking was hardly insular. Leaders of science and industry as well as members of the Labour cabinet were keenly aware of policy developments elsewhere, especially in the United States. An enduring concern was that, if the United Kingdom delayed developing controls or developed controls that were significantly more stringent than those adopted elsewhere, the nation would lose its preeminence in the new field. Sydney Brenner's concern, expressed to the Ashby committee in 1974, that "it is important for [Britain] to preserve and generate this technology not only for the future benefits it may bring but also for the power it gives us to understand the complex genetic structures of higher organisms"[100] remained a prominent consideration for the British government.

4.7 The Williams Committee and the Formation of British Policy

In contrast to the early initiation of a process to formulate genetic engineering policy in the United States, no immediate action was undertaken in the United Kingdom, either in response to the Ashby report, issued in January 1975, or in response to the Godber report, issued in May 1975. Ashby, as we have seen, proposed immediate resumption of genetic engineering under appropriate voluntary controls. Godber, in contrast, proposed that the Dangerous Pathogens Advisory Group under the Department of Health and Social Security should oversee research with genetically engineered organisms as well as research with dangerous pathogens, and that policy should have various forms of statutory backing, including that of the Health and Safety at Work Act. However, neither set of recommendations was immediately taken up by the government.

In the spring of 1975, pressures on the Labour government to follow up on the Ashby report and institute a British policy for genetic engineering began to mount. Early in 1975, the ASTMS requested meetings with Reg Prentice, the Minister for Education and Science, as well as with the Health and Safety Commission. A parliamentary delegation representing ASTMS interests also sought out Prentice. It is likely that the union pressed for inclusion in official committees addressing genetic engineering and that it emphasized protecting the safety of its members.[101]

The meeting of scientists at Oxford in July 1975 was clearly designed to convey to representatives of the Research Councils and the Department of Health and Social Security the urgent desire of scientists to separate genetic engineering from dangerous pathogens and to proceed under voluntary controls. A headline in *Nature* ironically expressed the scientists' mounting sense of frustration: "Forever Amber on Manipulating DNA Molecules?"[102] Whether or not government officials also met at this time with representatives of other major interest groups—notably the CBI—is not known. CBI representatives, like their counterparts in the United States, may have been content to track the proceedings without direct involvement.

The government's appointment in late fall of 1975 of a committee chaired by Sir Robert Williams, director of the Public Health Laboratory Service, to develop genetic engineering controls clearly responded to the scientific community's interest in maintaining peer control over the field. Significantly, the Department of Education and Science, rather than the Department of Health and Social Security or the Health and Safety Commission, was chosen as the lead department for future policy on genetic engineering and was given responsibility for organizing and running the Williams committee. In addition, it was decided that the committee would be a "working party of scientists." Williams later recalled that the rationale was that "if it could be kept at the level of scientists, it was less likely to get entangled with what you might call 'politics' in a broad sort of way."[103] Nominations of the committee members came from the Research Councils and the Health and Safety Commission, and Williams was consulted about finalizing the list. The outcome was that all but one of the members were drawn from the scientific establishment (table 4.4). Many had been involved in earlier discussions of genetic engineering: six, including Williams, had attended the Oxford meeting; four had been members of the Godber committee. Williams had sat on the Ashby committee.[104] The interests of the scientific community were clearly to be respected.

In other ways, however, the government's decisions reflected some sensitivity to union interests. Although mandating no direct union representation on the Williams committee, the government later agreed that the unions would be formally consulted. Moreover, the government apparently anticipated that the Health and Safety at Work Act would pro-

Table 4.4   Membership of the U.K. Working Party on the
Practice of Genetic Manipulation

Professor Sir Robert Williams, Director, Public Health Laboratory Service, London
(chair)

S. Brenner, MRC Laboratory of Molecular Biology, Cambridge

J. B. Brooksby, Director, Animal Virus Research Institute, Pirbright, Surrey

J. P. Duguid, Professor, Department of Bacteriology, Ninewells Hospital, Dundee

R. J. C. Harris, Director, Microbiological Research Establishment, Porton Down,
Salisbury

D. A. Hopwood, Department of Genetics, John Innes Institute, Norwich

W. House, Manager of Laboratories, Imperial Cancer Research Fund Laboratories,
London

R. Owen, Deputy Director of Medical Services, Health and Safety Executive, London

D. A. J. Tyrrell, MRC Clinical Research Centre, Harrow

P. M. B. Walker, Professor and Director, MRC Mammalian Genome Unit, Department
of Zoology, University of Edinburgh

Source: United Kingdom, *Report of the Working Party on the Practice of Genetic Manipulation,*
Cmnd. 6600 (August 1976).

vide the statutory backing for the proposals made by the Williams committee. The presence on the committee of Ronald Owen, deputy director of medical services of the Health and Safety Executive, implied an understanding that the Health and Safety at Work Act would cover laboratory work in genetic engineering. How the act might be implemented, however, remained an open question.

The Department of Education and Science charged the Williams committee to "draft a central code of practice and to make recommendations for the establishment of a central advisory service for laboratories using the techniques available for such genetic manipulation, and for the provision of necessary training facilities." It was also asked to consider the relevance of the controls proposed by the Godber working party.[105] Left open were questions concerning the form and composition of the "central advisory service," its relation to the proposals of the Godber committee for subsuming genetic engineering under controls for dangerous pathogens, the precise use of the Health and Safety at Work Act to control genetic engineering research, and of course, the content of safety controls. The Williams committee would deliberate for the next ten months. While the genetic engineering debate heated up in the United States, the British public heard little about the issues addressed by the committee, which met behind closed doors.

Williams and the majority of his committee, in harmony with the strong positions taken at the Oxford conference, appear to have expected that, contrary to the Godber proposals, controls would be entirely separate from those for dangerous pathogens, that they would be largely voluntary (although the general relevance of the Health and Safety at Work

Act was noted), that the central advisory service would consist of prominent scientists with relevant expertise in genetic engineering, safety, and
industrial applications, and that local safety committees would oversee
initial research assessments. These anticipations resembled the voluntary
approach assumed by the Ashby committee and that being developed by
the RAC.

The proposals of the Williams committee differed from these early
expectations of the scientific community not only in introducing an element of regulation but also in broadening participation in the decision
process. Three factors influenced the divergence of the proposed genetic
engineering controls from the initial expectations of the scientific community: trade union interest in pressing for a role in the policymaking
process, the Health and Safety Commission's preexisting mandate to apply the Health and Safety at Work Act to research laboratories, and a
Labour government, which consulted with the trade unions and gave
serious consideration to their proposals.

In the fall of 1975, the effects of union interests on the Williams committee, which neither seated nor formally interacted with them, appeared
to be minimal. But their response to this situation was also an indication
of their ability to exert influence within the British government. By December 1975, the unions were pressing for representation in the policy
process. Their goals are indicated in a letter from Reg Bird, the national
officer for the ASTMS, to union member Robert Williamson, who was
active in genetic engineering. First, the vice president of the ASTMS,
Douglas Hoyle, who was also a Labour MP, would press the Minister
for Education and Science about union representation on the Williams
committee. Second, the Health and Safety Executive would be urged to
solicit evidence from ASTMS on genetic engineering. The letter also
noted that Clive Jenkins, the powerful general secretary of the ASTMS,
would keep the Trades Union Congress (TUC) briefed about developments in genetic engineering policy.[106]

These efforts bore fruit. When Hoyle pressed the Secretary for Education and Science on the floor of the Commons to report on the progress
of the Williams committee's deliberations, specifically asking whether
union representation had been sought or consultation arranged with "the
TUC unions generally, and ASTMS specifically, in respect of the report
they are preparing," the minister, Fred Mulley, no doubt reflecting government sensitivity to a powerful union, announced that "consultation
with other bodies," including the TUC and the ASTMS, was planned
for 1976, although placing trade union representatives on the Williams
committee was not contemplated.[107]

The promised consultations between the Williams committee and the
trade unions occurred early in 1976, when the committee heard from

TUC representatives, the CBI, the Committee of Vice-Chancellors and Principals, the MRC, the Science Research Council, various scientific organizations, and individual scientists.[108] Only the trade unions appear to have taken a position that differed substantially from the original consensus on voluntary controls and peer review. Their oral testimony urged, first, that genetic engineering should be regulated under the Health and Safety at Work Act; second, that the central advisory service should have statutory power as an agency of the Health and Safety Executive; third, that genetic engineering research should be reviewed locally, through the procedures to be mandated by forthcoming legislation requiring safety representatives and safety committees (section 1.3); and finally, that the unions should be represented in decisions at both the local and the national levels.[109]

The response to these proposals of the only nonscientist on the Williams committee, Ronald Owen, underscored that, in addition to the Department of Education and Science, a second government agency had an independent interest in genetic engineering. Owen indicated that many of the union proposals needed to be taken up by the Health and Safety Commission. He observed that any decision to regulate genetic engineering must involve the Health and Safety Commission as well as the Secretary of Employment; further, he served notice that the composition of an advisory committee would be decided after the Williams committee reported. Clearly, Owen anticipated a substantial role for the Health and Safety Commission in future policymaking.[110] A new governmental interest that gave formal recognition to the representation of employees in policymaking was being asserted.

The Williams committee, the Department of Education and Science, and the Health and Safety Commission took many months before releasing a policy proposal. "Why are we waiting?" asked a headline in the journal *Nature* on 22 July 1976. With the NIH guidelines already promulgated, an atmosphere of "frustration and suspicion" was building among British scientists, *Nature* reported. "Inevitably moratorium-breaking has already occurred, and the consequences for those involved have on occasion been unpleasant." The delay was probably occasioned by debate and negotiation on the need for statutory controls among the scientific community, the unions, the Department of Education and Science, and the Health and Safety Executive. In addition, it is possible that government officials wanted to observe the response to the controversial NIH guidelines.[111]

The negotiations culminated in the simultaneous release of two documents in August 1976: the Department of Education and Science issued the long-awaited Williams report with the committee's recommendations, and the Health and Safety Commission issued draft regulations for compulsory notification of experiments involving genetic engineering.[112] These proposals showed the influences of the interest groups on whom they impinged, and reflected the government's concern that regulation of genetic engineering should not cause British research to lag significantly behind that of other countries. A key section of the Williams committee's report recommended categorizing genetic engineering experiments. While differing in detail, the categorization proposals greatly resembled those of the NIH guidelines (table 4.5). Like its U.S. counterpart, the Williams report, although acknowledging the uncertainties plaguing estimates of the hazard of genetic engineering experiments, based its categories of hazard on the phylogenetic distance between humans and the source of DNA to be inserted into a bacterium or other organism. It similarly posited two forms of containment—physical and biological. Despite the claim that the British controls deemphasized the use of "disarmed" strains of host organisms, in fact the British categorization reduced containment when these strains were used. Finally, the Williams report approved *E. coli* K12 as a host organism.

There were also some distinct differences. The first three levels of physical containment prescribed by the Williams committee were somewhat more stringent than the P1, P2, and P3 levels used by the RAC (table 4.3). Biological containment received no precise definition in the Williams report; there was also no restriction, as in the NIH guidelines, to the use of *E. coli* K12 as a host. Nor were there any outright prohibitions or limits on the size of cultures in the Williams proposals. (This difference was not absolute, however: the NIH guidelines made allowance for exceptions to these prohibitions and limits.) On the whole, the British proposals were more general and flexible, leaving details to be decided at a later time, whereas the NIH proposals were quite specific.[113] In particular, the whole question of industrial application of genetic engineering was set aside as a possibility "so remote that the increase in knowledge preceding such development will greatly contribute to their safety."[114]

It has been suggested that the reason for the similarity of the categorizations of experiments recommended by the NIH guidelines and the Williams report was the "common structure of tacit [scientific] knowledge" in the international scientific community.[115] In fact, however, no consensus on the hazards of genetic engineering existed at the time because the field was new and experience with genetically engineered organisms was extremely limited: as correspondence among scientists indicates, scientific speculation about such hazards and how to categorize them varied widely (section 3.4). The second report of the EMBO committee of molecular biologists stated in September 1976: "There are no experimental data on which to base any objective estimates of the conjec-

Table 4.5 The U.K. Code of Practice for Genetic Manipulation, August 1976: Suggested Categorizations for Some Typical Experiments

Source of Nucleic Acid	Specification of Nucleic Acid Sequence	Host-Vector System	Category
Mammals	Random	Phage or plasmid/bacteria, not disabled	IV
	Random	Phage or plasmid/bacteria, disabled	III
	Purified	Phage or plasmid/bacteria, not disabled	III
	Purified	Phage or plasmid/bacteria, disabled	II
Amphibians and reptiles	Random	Phage or plasmid/bacteria, not disabled	III
	Random	Phage or plasmid/bacteria, disabled	II
	Purified	Phage or plasmid/bacteria, not disabled	II
	Purified	Phage or plasmid/bacteria, disabled	I
Plants and invertebrates and lower eukaryotes	Random	Phage or plasmid/bacteria, not disabled	II
	Random	Phage or plasmid/bacteria, disabled	I
	Purified	Phage or plasmid/bacteria, not disabled	I
Mammals, amphibians and reptiles, birds	Random	Virus capable of infecting humans or growing in tissue culture cells	IV
	Purified	Virus capable of infecting humans or growing in tissue culture cells	III
Viruses pathogenic to vertebrates	Random	Phage or plasmid/bacteria, disabled	IV
	Purified	Phage or plasmid/bacteria, disabled	III
Animal viruses, nonpathogenic to humans	Random	Phage or plasmid/bacteria, disabled	II
Bacteria specifying toxins virulent to humans	Random	Phage or plasmid/bacteria, disabled	IV
Plant pathogenic bacteria	Random	Phage or plasmid/bacteria, not disabled	II
Plant viruses	Random	Phage or plasmid/bacteria, not disabled	II
Bacteria or fungi nonpathogenic to humans, animals, or plants	Random	Phage or plasmid/bacteria, not disabled	I

Source: United Kingdom, *Report of the Working Party on the Practice of Genetic Manipulation,* Cmnd. 6600 (August 1976).

[a] *Purified* means fractions with little chance of including any unrecognized extraneous sequences. It is of course possible to have sequences selected because of their pathogenicity, and these would raise the level of containment required.

tured biohazards associated with recombinant DNA research. Schemes
for the classification of experiments according to their conjectured biohazard are, therefore, arbitrary."[116] No scientific work supported the
concept of phylogenetic relatedness. (It was eventually to be abandoned.)
In theory, the Williams committee might well have come up with an
entirely different approach to categorizing hazards. That it did not suggests that the convergence of the British and American systems was primarily the product of political interests, not an international scientific
consensus. The sensitivity of the Williams committee to comparisons
between its recommendations and the NIH guidelines was explicitly acknowledged in its report: "We believe that our own proposals for categorisation of experiments are generally in line with those developing in
the United States."[117] This reassurance, duly noted by the British scientific press,[118] followed from the British need for a system neither significantly more stringent than the American one (which risked stimulating
a revolt among British scientists) nor significantly less stringent (which
risked igniting trade union dissent).

In general, the scientific community could not have been displeased by
the technical content of the British proposals. With respect to their implementation, however, the scientific community was not granted the
voluntary controls with peer review for which it had called. Although it
is impossible, without access to the government records, to assess exactly
why this happened, it is clear that what emerged in the Williams report
and the Health and Safety Commission draft regulations was a compromise between the interests of the scientific community and the trade
unions. The recommended controls were described, in a manner designed to appeal to scientists, as "voluntary in the first instance." A central advisory service, responsible to the Department of Education and
Science, with a secretariat provided by the MRC, would review proposals for genetic engineering research to ensure that experiments were
conducted according to a recommended code of practice. Individual
laboratories would be asked to appoint a biological safety officer and to
establish a safety committee. Experiments would be reviewed both locally and nationally. These mechanisms paralleled those set up in the
United States. But the Health and Safety Commission draft regulations
incorporated a major difference, requiring laboratories to notify the commission and to submit proposals to the central advisory service as well as
to the Health and Safety Commission. Thus, in contrast to the voluntary
NIH controls, British submissions for review would be mandatory.

The influence of the trade unions was also evident in proscriptions for
the various review committees. At the local level, safety committees
were directed to be "properly constituted and representative." This language suggested that such committees would comprise laboratory per-

sonnel as well as research scientists. At the national level, the central advisory service, which was now renamed the Genetic Manipulation Advisory Group (GMAG), would include "not only scientists with knowledge both of the techniques in question and of relevant safety precautions and containment measures but also individuals able to take account of the interests of employees and the general public."[119] Differing significantly from the peer review process generally anticipated by British scientists and used by the NIH in the United States, the British conception of participatory decision making for the assessment and control of genetic engineering was a distinct innovation for science policymaking.

4.8 Forming the Genetic Manipulation Advisory Group

The Williams report laid out a path for future policymaking on genetic manipulation—that is, a regulatory framework defined by the Health and Safety at Work Act; participation in policymaking by trade unions, industry, and those designated to represent the "public interest"; and a broad categorization of potential hazards—but important details remained to be specified. After the completion of the Williams report in June 1976, a series of influential decisions by the British government, especially the Department of Education and Science, spelled out the location, function, and composition of GMAG and the details of the Health and Safety Commission regulations. All of these decisions evoked responses from the government agencies and sectors of British society with immediate interests in genetic engineering.

The Location and Operation of GMAG

The Williams report made recommendations about the function of GMAG but left open the question of how and under whose aegis the committee would operate. Scientists, unions, private industry, and government agencies all had important stakes in the outcome of these decisions. The central issue was controlling GMAG, which involved not only the running and staffing of the committee but also access to the information it generated. If responsibility for it fell under the Department of Education and Science rather than the Health and Safety Executive, the Secretary of State for Education and Science would have the greater influence over the committee's composition and decisions as well as access to the details of genetic engineering work. If, on the other hand, GMAG was defined as a formal advisory committee under the Health and Safety Executive, information it acquired could not, under the terms of the executive's charge, be divulged to other government agencies.

These questions of power and control, hidden from public view, re-

ceived almost no attention in the press, but they were understood by the
principal players in the genetic engineering issue in the United Kingdom.
As the Oxford conference showed, many influential scientists wanted the
central advisory committee to be the responsibility of the Department of
Education and Science. Like their American colleagues, they wanted a
committee housed within the government agency that supported their
research. They particularly resisted locating GMAG within an agency
whose primary mission was health and safety rather than the promotion
of scientific research.[120] Indeed, some British scientists interpreted the
ambiguously phrased Williams report as proposing a system of voluntary
controls administered by the government agencies responsible for research. John Subak-Sharpe and Alan Rowe Williamson, professors of virology and biochemistry, respectively, at the University of Glasgow, and
John Paul, director of the Royal Beatson Memorial Hospital in Glasgow,
wrote approvingly to the Department of Education and Science about the
Williams proposals, which they contrasted with the Health and Safety
Commission's "hasty and microbiologically ill-advised attempt to write
sweeping rules and to assume extensive powers": "[The Williams Committee's] proposal . . . follows the general procedures for the responsible
self-regulation of basic research by the acknowledged experts in the field.
Such a review system, administered like the Grant Review system by the
Research Councils, would have the greatest force and respect from our
scientists, International Science and from industry, without need for the
intervention of any other statutory body."[121] The trade unions, on the
other hand, wanted to see GMAG as a formal committee of the Health
and Safety Executive, and its membership appointed by the Health and
Safety Commission. Clearly in such a case, trade unions, with their
strong representation on the commission, could count on playing an important role on the advisory committee.[122] Industry executives also apparently preferred to house the advisory committee within the Health
and Safety Executive, believing that proprietary information would be
better protected by a government agency experienced in handling trade
secrets. However, the CBI did not press hard for this course of action.

The issue was settled between the completion of the Williams report
in June 1976 and its publication in August. The MRC and the ARC, the
two Department of Education and Science committees with the most
immediate interests in genetic engineering, sought control of GMAG. As
Sir William Henderson, then head of the ARC, recalled in 1980, the possibility that the Health and Safety Executive might control genetic manipulation "raised great anxiety in the breasts of scientists, and so it was
felt that we would have to make sure that this was handled the right
way." The MRC quickly offered to provide a secretariat for the committee, thus ensuring that GMAG became an advisory committee to the

Department of Education and Science. It was, as Henderson recalled, "a deliberate move to prevent it becoming an instrument of HSE." [123] And while Health and Safety Executive officials may have preferred the assignment of GMAG within their agency, they apparently did not resist. By the time the Williams report was issued on 25 August 1976, the decision to make GMAG the responsibility of the Department of Education and Science had been made by the Secretary of State, Fred Mulley.[124]

Tony Vickers, the secretary of GMAG appointed by the MRC, portrayed the Health and Safety at Work Act as a "strong framework" that was being interpreted in a "sensible and flexible way" by the Health and Safety Executive, thus "avoiding putting [genetic engineering] into a straightjacket." But the flexibility of the British approach derived primarily not so much from the act as from its interpretation by a committee run by the MRC. As Vickers also noted, "[I]n the last resort, the best safeguard is that the scientific community should be persuaded of the reasonableness of what is required of them so that their cooperation is willingly and freely given." [125] The negotiations over the form and scope of regulations proposed by the Health and Safety Executive and the establishment of GMAG in the fall of 1976 ultimately provided the required reassurance.

The Form and Scope of the Health and Safety Commission Regulations

The notification regulations proposed by the Health and Safety Commission also provoked vigorous reactions.[126] The draft regulations, which were widely circulated for comment,[127] were meant to codify a requirement that the Health and Safety Executive be informed of all genetic manipulation of microorganisms presenting any hazard to the health and safety of workers or the general public. The form of regulation suggested by the Health and Safety Executive was mild: only notification, rather than approval, of genetic engineering was required. The draft of the key proposal ran, "No person shall carry on any activity intended to alter, or likely to alter, the genetic constitution of any microorganism unless he has given to the Health and Safety Executive notice, in a form approved by the Executive for the purposes of these Regulations, of his intention to carry on that activity." [128]

Criticism of the Health and Safety Executive proposals came from several directions. For example, an internal document of ASTMS suggested that the proposals did not go far enough and that, beyond notification, they should require prior approval of experiments.[129] Strong and mostly negative reactions arose in the scientific community. The immediate object of the scientists' discontent was the definition of *genetic manipulation* contained in the draft proposal. How to define that concept to include hazardous forms and exclude everything else was by no means obvious, and the Health and Safety Executive had erred—whether deliberately or inadvertently is not clear—on the side of including too much. The phrase, "any activity intended to alter, or likely to alter, the genetic constitution of any micro-organism," could be interpreted as including virtually all use of the techniques of modern genetics.

Although the Health and Safety Commission document was only a draft, protest from the British scientific community was explosive. Attacks on the Health and Safety Commission and Health and Safety Executive were delivered in the columns of the *Times* and *Nature*. One scientist claimed that the proposed regulation was "the first outright attack on normal scientific freedom in Britain."[130] Michael Ashburner, a prominent geneticist, suggested that Health and Safety Executive officials must have been inspired "by the mutant monsters so vividly portrayed on television every Saturday evening in *Dr. Who*."[131] The proposals were also the brunt of satirical cartoons in the British popular science magazine *New Scientist*.[132]

Although these attacks mainly targeted the Health and Safety Executive definitions, much more was at issue. Members of the scientific community were worried not only about *what* should be regulated but also about *who* should have the final regulatory authority. This emerged clearly in an exchange between Ashburner and John Locke, director of the Health and Safety Executive, published in *Nature*. Ashburner charged that the proposed regulations would give the Health and Safety Executive (rather than GMAG) the decisive role of screening research proposals, despite the fact that GMAG "alone would include not only technically competent scientists but also individuals able to take account of the interests of employees and the general public." He also noted that the Health and Safety Executive's draft proposal "remove[s] the Advisory Group or any similar body from any central role in the dialogue that must exist between those who do the experiments and those who administer the laws under which they are done. With respect, sir, you do not have the standing in the scientific community required for this job."[133]

Underlying these comments was a threat. If the Health and Safety Executive lost the respect of the scientific community, Ashburner continued, it would jeopardize the goodwill on which it depended for technical advice and warnings about real and potential hazards. If scientists were to find themselves "encumbered with endless red tape," they might be reluctant to speak out in the future. "It would be a tragedy," he concluded, "if certain types of scientific work were driven 'underground' as a consequence of ridiculous regulation."

In response, the Health and Safety Executive backed down. "I really must protest at the suggestion that the Advisory Group would play a subservient role to the Health and Safety Executive," John Locke demurred. "This is simply not the case. It will be for the Advisory Group to lay down the conditions which need to be observed to enable the experiments to proceed safely."[134] The influence over genetic engineering policy wielded by the British scientific community was amply demonstrated by the intensity with which scientists bore down on the Health and Safety Commission draft and the speed with which the Health and Safety Executive yielded. The commission's regulations did not achieve final form until August 1978, but this initial interaction between the agency and the British scientific community established the latter's potency in affecting technical decisions on genetic engineering.

Establishing GMAG

Locating GMAG within the Department of Education and Science meant that the department's head became responsible for key decisions on the committee's charge and composition. To set it up, one civil servant recalled, was "a very time-consuming process . . . [and] taken extremely seriously." Fred Mulley's successor, Shirley Williams, who was appointed in September 1976, took a keen interest in this process.[135]

GMAG's terms of reference, as drawn up by the Department of Education and Science and presented in a press release on 8 December 1976, were the following:

1. In the light of Sections 2 and 3 of, and the Code of Practice in Appendix II to, the report of the Williams Working Party, to advise
   a. those undertaking activities in genetic manipulation, including activities related to animals and plants, as defined in paragraphs 1.3 and 2.2 of the Report, and
   b. others concerned.
2. To undertake a continuing assessment of risks and precautions (and in particular of any new methods of physical and biological containment) and of any newly developed techniques for genetic manipulation and to advise on appropriate action.
3. To maintain appropriate contacts with relevant government departments, the Health and Safety Executive and the Dangerous Pathogens Advisory Group.
4. To maintain records of containment facilities and of the qualifications of Biological Safety Officers.
5. To make available advice on general matters connected with the safety of genetic manipulation, including health monitoring and the training of staff.
6. To submit a report at intervals of not more than a year.[136]

In principle, GMAG's charge appeared to be open-ended. It was, as the undersecretary responsible later described it, "extremely widely drawn" to give GMAG "the greatest possible scope."[137] However, its framework was based on the fundamental assumption that the code of practice proposed by the Williams committee would allow research to move forward. In part, at least, GMAG was clearly expected to implement the Williams recommendations. Whether it addressed other policy issues would depend on how it was run.

For the Secretary of State for Education and Science, Shirley Williams, a major—if not the central—problem encountered in establishing GMAG was how to bring together potentially divergent interest groups without handcuffing policymaking. She anticipated troublesome differences between the trade unions, whose attitude about experiments of unknown hazard tended toward extreme caution, and industry and some of the scientists, who "tended to be extremely cavalier about any controls at all, tended to argue that there were very few risks . . . and that the crucial point was that Britain might lose out in a field of high-level science, and that therefore, the important thing was to have as few regulations as possible to allow the work to go ahead." Selecting the chair and members of the committee was seen as crucial to enable it to transcend these differences and produce policy decisions.[138]

The appointment of the GMAG chair received particular attention. The undersecretary in the Department of Education and Science who originally selected candidates for the committee, Norman Hardyman, searched for a chair who would possess "exceptional sensitivity" to differing perspectives on genetic engineering issues and "exceptional ability to bring conflicting views together and get something out of them." The chair, therefore, could not be strongly identified with any of the main interests that came into play. But Hardyman also held that "we had to have somebody whom the scientists would trust." This requirement meant that "almost certainly . . . he had to be a scientist. But, if other people were to find him acceptable, we judged that he couldn't be a specialist in the field." Thus it was necessary to find someone who "was capable of understanding the scientific [dimensions of the issues] much better than the ordinary lay member . . . because he had to steer the committee. . . . He didn't actually have to understand everything, [but] he had to receive the point of what the scientists were saying."[139]

Sir Gordon Wolstenholme, who was appointed the first chair of GMAG in November 1976, was judged to have precisely the right mix of desiderata. A physician by training, Wolstenholme had served as an officer of numerous medical and scientific organizations and, from 1948 until 1978, as the director of the private CIBA Foundation (supported by the CIBA-Geigy Corporation). Wolstenholme was well known within the scientific establishment (section 3.3). His views on genetic engineering were in harmony with the principal findings of the Ashby committee. He wrote in June 1977: "In regard to prospects for genetic manipulation in the next 10 years, I believe there are high hopes of major successes. If breaches in containment occur, I doubt if the results will be serious—there are second and third line defences in the use of disabled organisms which are most unlikely to survive outside the laboratory and in holding in reserve certain effective antibiotics."[140] Wolstenholme was therefore quite sympathetic to the need for research in genetic engineering to proceed. Although he represented his role as one of "getting this very diverse group of people into a cooperative . . . happily working group," it was also more than this, for the institutional imperative was that GMAG would work most happily when guided by the premise that genetic engineering should advance. Wolstenholme was also sensitive to a need to remain closely aware of developments in the United States. One of his first actions following his appointment was to arrange an early meeting with Donald Fredrickson and other NIH officials.[141]

GMAG's members were also carefully chosen (table 4.6). Department of Education and Science civil servants floated lists of possible names, consulting the chair and others inside and outside the department. But the final selection rested with the minister—and she attended to it carefully. The members of GMAG fell roughly into four types: scientists, employers, trade unionists, and those who were broadly (and vaguely) defined as representing the "public interest."

The criteria and methods of selection varied. In the case of scientists, Department of Education and Science officials looked for expertise in genetic engineering, experience in the handling of pathogens, knowledge of containment techniques, and other relevant scientific experience, and Williams consulted with scientists who were known personally to her.[142] Of the eight scientists eventually seated, three had strong interests in genetic engineering, and the other five were heads of university departments and research institutions that would use the techniques.

Names of members to represent employers were generated by asking the CBI and the Committee of Vice-Chancellors and Principals of British universities for nominations. In this case, presumably, these organizations, not the Department of Education and Science, decided on appropriate criteria for membership on GMAG. The two members appointed in this category were John Gilby, the production director for Beecham Pharmaceuticals Division, a British pharmaceutical firm, and Sir Brian Windeyer, a radiologist and vice-chancellor of the University of London from 1969 to 1972.

Similarly, the TUC was asked to nominate three members to represent employees.[143] Both the number chosen and the unions from which they were drawn provoked extended negotiation among the unions and between the TUC and the Department of Education and Science. The

TUC pressed for four representatives—a proposal eventually accepted
by the Department of Education and Science. But when the TUC applied pressure for a fifth union member, drawn from the Association
of University Teachers, Wolstenholme demurred. The TUC nominees
eventually accepted by the Department of Education and Science were
Ronald Owen, the former Health and Safety Executive representative on
the Williams committee, who had become the TUC's chief medical adviser; molecular geneticist and ASTMS member Robert Williamson;
Donna Haber, a divisional officer of the ASTMS specializing in health
and safety; and Derek Ellwood, a scientist at the Porton Down Microbiological Research Establishment, who represented the Institute of Pro-

Table 4.6   Membership of the U.K. Genetic Manipulation Advisory Group, 1976

Sir Gordon Wolstenholme, Director, CIBA Foundation, London (chair)

Appointed as scientific and medical experts
J. B. Brooksby, Director, Animal Virus Research Institute, Pirbright, Surrey
K. R. Dumbell, Professor, Department of Virology, St. Mary's Hospital Medical School,
  London
H. J. Evans, Director, MRC Clinical and Population Cytogenetics Unit, Edinburgh
R. J. C. Harris, Director, Microbiological Research Establishment, Porton Down,
  Salisbury
B. W. Langley, ICI Corporate Laboratory, Runcorn
K. Mather, Professor, Department of Genetics, University of Birmingham
M. H. Richmond, Professor, Department of Bacteriology, University of Bristol
J. H. Subak-Sharpe, Professor, Institute of Virology, University of Glasgow

Appointed to represent the public interest
M. Jahoda, Professor, Science Policy Research Unit, University of Sussex
J. E. Lawrie, Elizabeth Garrett Anderson Hospital, London
J. Maddox, Director, Nuffield Foundation, London
J. R. Ravetz, Reader in the History and Philosophy of Science, Department of
  Philosophy, University of Leeds

Appointed to represent the interests of employees
D. C. Ellwood, Professor, Microbiological Research Establishment, Porton Down,
  Salisbury (Institute of Professional Civil Servants)
D. Haber, Association of Scientific, Technical, and Managerial Staffs, London
R. Owen, Trades Union Congress, London
R. Williamson, Professor, Department of Biochemistry, St. Mary's Hospital Medical
  School, London (ASTMS)

Appointed to represent the interests of management
J. A. Gilby, Technical Director, Beecham Pharmaceuticals Division (U.K.), London
Sir Brian Windeyer, Chairman, National Radiological Protection Board, London

---

Source: Genetic Manipulation Advisory Group, *First Report of the Genetic Manipulation Advisory Group,* Cmnd. 7215 (May 1978).

fessional Civil Servants. These negotiations were sufficiently lengthy to prevent trade union representatives from attending the first meeting of GMAG in December 1976.[144]

Finally, four GMAG members representing the "public interest" were selected by the Department of Education and Science. These members were qualified more by what they were *not* than what they were: according to Shirley Williams, they were "deliberately chosen because they are not themselves working now in the field, because they are not beholden to either industry or to a research institute."[145] They had to be people who would "inspire sufficient public confidence, and who clearly were not linked to any particular vested interest." Wolstenholme sought to find people "in whom the public could repose faith that they would, in fact, not allow something to happen which could in any way remotely be injurious to the public at large."[146] It was sufficient for them to be able to grasp the general issues raised by genetic engineering without necessarily understanding technical detail.[147] Moreover, Department of Education and Science officials were on the lookout to exclude anybody who was, in the words of one civil servant, "so hostile to genetic engineering as to disrupt [GMAG's] operations."[148] In other words, GMAG had a task to perform, and in order to function smoothly, its public interest members had to accept, as its science and industry representatives did, the basic premises defined by the Ashby and Williams committees. The four representatives of the public interest selected by the Department of Education and Science were: Marie Jahoda, a social psychologist at the University of Sussex; Jean Lawrie, a gynecologist at the Elizabeth Garrett Anderson Hospital; John Maddox, a science journalist and former editor of *Nature* who was then director of the Nuffield Foundation; and Jerome Ravetz, reader in the history of science at the University of Leeds and secretary of the Council for Science and Society.

Department of Education and Science officials insisted that GMAG members sat as individuals rather than as delegates or representatives of specific organizations.[149] In practice, however, most played roles that corresponded to the general interests of the sectors with which they were affiliated. Those drawn from sectors of society with specific interests in genetic engineering—science, industry, and the trade unions—could not (and probably were not really meant to) disentangle a "neutral" viewpoint from their organizational or personal concerns. These latter, moreover, as one observer of GMAG has noted, were sometimes complex; for example, several of the scientists had financial interests in genetic engineering and one—Bernard Langley—was associate director for research at the ICI Corporate Laboratory. Two of the trade union members, Robert Williamson and Derek Ellwood, had strong professional interests in genetic engineering. The foundation directed by public interest member John Maddox funded recombinant DNA research.[150] Although GMAG's formal composition apparently balanced science, industry, trade unions, and general public viewpoints, the crosscutting interests of the actual membership suggested that differences would not overwhelm the committee's common understanding about the need to foster genetic engineering.

Thus the British GMAG was born in the fall of 1976 as a kind of compromise between the interests of the research establishment and those of the unions: the MRC would provide the secretariat; the Health and Safety Commission would regulate; and ambiguously positioned between the two institutions would be the anomalous creature, GMAG, formally existing outside both agencies as a quango. As a broadly constituted committee with representation of the scientific, business, and labor sectors as well as of the vaguely defined "public," it was sometimes referred to as Shirley Williams's "experiment in social democracy." But it was more (or less) than that: a committee whose crosscutting interests would largely support the policy paradigm already adopted; a committee that would advise scientists on the safety of their experiments but would be operated by the MRC; a committee with broadly drawn terms of reference but also one expected to function within the framework defined by the Ashby and Williams committees.

The tensions and ambiguities that characterized GMAG's composition, charge, and institutional framework reflect the conflicting interests that came into play during its formation; in particular, it was supposed to bridge the conflict between fostering a new field of scientific research on the one hand and controlling its potential hazards on the other. At the same time, what GMAG would accomplish was more than simply the outcome of balancing conflicting interests: resolution of their conflict took place within a framework of assumptions and practices formed by earlier phases of policymaking strongly committed to maintaining for the United Kingdom a competitive edge in genetic engineering.

4.9 The American and British Policy Paradigms: Variations on the Asilomar Legacy

As the foregoing analysis shows, American policymaking proceeded along the lines established by the Berg committee. An expert committee drafted voluntary controls for genetic engineering, as watchful leaders of the biomedical research community ensured that the committee struck a careful balance between controls that would be seen as too stringent, thus risking revolt in the ranks, and controls that would be seen as too lax, thus inviting congressional intervention. In the United States, no other interest group had direct influence on development of controls (although

NIH officials were well aware of the watchful presence of multinational corporations): participation on the RAC by members drawn from occupations other than biomedical research was designed—and seen by scientist members of the RAC—not to provide necessary new perspectives but largely to persuade Congress that the public's needs were being attended to. In the United Kingdom, in contrast, an interest group other than the biomedical research community—the trade unions representing technical workers in academia and industry—began to exert some influence on the development of policy, beginning in 1975. Despite the existence of a Labour government that was receptive to union interests, the evidence suggests that the unions had little influence at this point on the conceptual framework for policy established by the Ashby committee: union representatives were not appointed to the Williams committee established to formulate a code of practice, and their evidence to that committee did not affect the technical content of controls. Consequently, at this stage, the basic policy paradigm evolved along similar lines in both countries; moreover, it was reinforced by a series of incremental decisions.

Concentration of Experience of Policy Advisers in the Biological Sciences

Because the recombinant DNA problem was reduced to finding technical fixes to contain research risks, it was also seen as an issue best resolved by technical experts. The Asilomar conference had been organized as a technical discussion of the experimental risks, and the great majority of participants were actual or potential practitioners of genetic engineering. The Ashby committee in the United Kingdom, composed of prominent scientists, similarly foreclosed broader policy issues. When officials at the NIH and the Department of Education and Science established policy committees—the RAC in the United States and the Williams committee in the United Kingdom—these bodies were expected to provide technical advice regarding a solution.

As a result, the expertise and experience of those who assumed the responsibility for advising their governments on solving the genetic engineering problem were concentrated in the biological sciences. Representation of such fields as law, ethics, environmental protection, and industrial safety was limited to the one or two "lay" members eventually appointed to the RAC in 1975–76 and the representative of the Health and Safety Executive appointed to the Williams committee. Consequently, concerns falling outside the paradigm that became dominant within the biomedical community were scanted, while intense attention was focused on the short-term risks of using genetic engineering techniques in research—the most immediate problem confronting scientists. Although other considerations were not entirely ignored, they did not receive sustained examination. Such questions as the deployment of genetic engineering by private industry or the military, if they were broached at all, were set aside.[151]

The Timing and Use of Public Hearings and Other Policy Processes

Other sources of information and public input that could have influenced assessments of genetic engineering in 1975–76 included public hearings, congressional and parliamentary inquiries, and the application of NEPA of 1969.

In the United States, the "sunshine" laws adopted in the 1960s and 1970s—the Freedom of Information Act (1966), the Federal Advisory Committee Act (1972), and the Sunshine Act (1976) (section 1.3)—required the meetings of the RAC (although not the meetings of the Berg committee, the Asilomar conference, or the meetings of agency officials) to be open to the press and the public. As a result, Americans had access to detailed information about the proceedings of the RAC and the debates that occurred within it. Whether information flowed as freely in the opposite direction, however, is questionable.

The only public hearing on NIH policy prior to release of the NIH guidelines was held before the NIH director's advisory committee in February 1976. By that time, the recommendations of the Asilomar conference had been translated into draft guidelines by the RAC. Thus the information generated at the public hearing flowed to the NIH late in the policymaking process, after a consensus on major policy issues had already been established.

Although fundamental questions not only about the validity of the NIH guidelines and techniques of containment but also on the long-term social and environmental impact of the technology and the advisability of promoting its rapid expansion were raised at the hearing, they had virtually no chance of affecting NIH policy. As shown earlier, the RAC's agenda effectively insulated the committee from the concerns voiced at this hearing. These concerns were addressed by the NIH only to the extent that they led to some compromising on details of the guidelines, whose basic form and policy assumptions remained unchanged.[152] In a similar way, the application of NEPA by the NIH, which entailed drafting an environmental impact statement and providing the public with opportunities to comment, while it generated considerable information challenging the basic policy paradigm itself, occurred far too late to carry any weight. Although NIH officials later claimed that "the public [was] consulted no less than the scientific community, through formal hearings, through the publication for comment of guidelines, proposed revisions, and an environmental impact statement,"[153] what was called "public participation" was largely an illusion.

In the United Kingdom, not even an illusion of openness characterized policymaking. The Official Secrets Act closed the doors on all official committee work on the DNA problem. Despite this formal difference, the British process paralleled its American counterpart in accepting input into the policy process almost exclusively from representatives of the scientific community. The Ashby committee consulted no one else; the Williams committee did consult representatives of trade unions, but primarily over implementing policy already laid out. The presentation of the Ashby report to Parliament in January 1975 was expressly intended to stimulate public discussion; however, the minimal media coverage of earlier closed proceedings, the newness of the genetic engineering controversy, and Parliament's limited ability to launch an independent investigation precluded any real public debate. Finally, the intensity of the British scientific community's response to the Health and Safety Executive's draft regulation is an indication of the likely response to any moves to transfer control of policymaking to a government agency that did not have a patron-client relation with the scientific community.

Low Priority for Risk Assessment

One crucial kind of information—one that bore directly on the "technical" side of the issue and that could have been expected to be demanded by policymakers in both countries—was nonetheless bypassed. A thorough assessment of the field required evidence from experiments designed to clarify the nature and level of the risks involved, the uncertainties of which were widely acknowledged. The Berg committee noted the "few available experimental data on the hazards of . . . DNA molecules," acknowledging that "new DNA elements introduced into E. coli might possibly become widely disseminated among human, plant, or animal populations with unpredictable effects."[154] Given these uncertainties, policymaking bodies could have recommended a full-scale program of monitoring and risk assessment *before* deciding on a policy of rapid growth in genetic engineering. (Indeed, the charge to the RAC required guidelines developed on the basis of the results of a research program aimed at clarifying recombinant DNA hazards.)[155] For example, research that carried a risk assessment component might have been selectively cleared; this would have avoided lags between new refinements of the techniques and reliable knowledge of their risks. However, both countries' ventures at risk assessment postdated the drafting of guidelines (in the United States) and a code of practice (in the United Kingdom), despite acknowledgments such as that of the EMBO Advisory Committee on Recombinant DNA that, in the absence of experimental data, schemes for classifying experiments according to possible hazard were arbitrary.[156] Policies to expand research (under controls) were implemented in both countries before risk assessment was initiated. Use of genetic engineering techniques expanded rapidly in 1976, while knowledge about their risks remained as dubious as it had been two years earlier, at the time the Berg committee recommended partially suspending research. Even a consideration of how much research might safely continue in the absence of evidence about possible harmful effects did not appear on the agendas of policymaking bodies in either country in this period.[157]

In conclusion, both in the United Kingdom and in the United States, a single policy paradigm, defining the nature of the problem and the appropriate means to resolve it, which was in place almost from the beginning, governed the development of public genetic engineering policy. This policy was supported by a discourse that recognized that the hazards of the field were still unknown but assumed that they could be contained through the use of physical and biological restrictions that did, however, allow for expansion and spread of the techniques. The series of incremental policy decisions that led to the official promulgation of the American guidelines and the British code of practice articulated this paradigm without giving sustained consideration to alternative policy positions. The resulting restriction of policy options was especially clear in the United States, where the dominant paradigm was actively challenged in 1976. In Schattschneider's terms, a policy of expansion based on a technical solution to research risks was organized *into* the decision-making process, while alternative policies that would have slowed expansion and facilitated a longer look at what genetic engineering might bring about were organized *out*.[158]

The emergence in the United Kingdom of a sector outside the biomedical research community with its own distinct interests in genetic engineering had no immediate impact on key features of this dominant policy paradigm. In many respects, the unions seem to have been prepared to accept it. Nevertheless, the unions' interest in the genetic engineering problem and their access to British policymaking arenas meant that some elements of the policy discourse began to change. For one thing, the unions held that genetic engineering posed important occupational health and safety questions. Defining the issue in this way meant that it was no longer seen as simply "technical" in nature: human behavior in the laboratory and industry was seen as a legitimate problem. This in turn meant that the definition of the appropriate qualifications of policymakers expanded to include those with expertise and experience in occupational health and safety. It also brought the issue firmly under the aegis of the Health and Safety at Work Act, thus requiring the participation of representatives of laboratory employees at the local and national levels. Consequently, British and American policies began to diverge significantly with respect to their implementation. While the British government moved toward enforceable regulation of genetic engineering that encompassed all workplaces, including those in the private sector, the NIH guidelines of 1976 remained voluntary. And while the British government also expanded representation, especially of laboratory workers, in decision making at both the local and the national levels, American policymaking remained primarily in the hands of the scientific community.

This divergence has been explained in terms of the special influence of the trade unions, their access during a Labour government to policymaking arenas, and the willingness of the government to define the genetic engineering issue as involving occupational health and safety, to which a preexisting government statute, the Health and Safety at Work Act, could be applied and for which union representation in policymaking was deemed appropriate. At the same time, trade union influence was by no means unrestricted, and in the last analysis, it did not dominate British genetic engineering policy. Interests in workplace health and safety competed with interests in maintaining the British competitive edge in genetic engineering—and the latter were expressed not only by representatives of science and industry but also within the trade union movement itself. The Labour government, while responsive to union pressures to regulate genetic engineering, nonetheless produced a form of implementation that, while representative, continued to provide the scientific community with privileged channels of communication with the government and considerable political leverage.

Global influence on national recombinant DNA policies was expressed in the sensitivity of the British government to the divergence of the British controls from those in the United States and in its interest in maintaining close contact with American policymakers. The potentially major role of genetic engineering in industry and science meant that all policy decisions in the two countries would be subjected to scrutiny for their impact on global competition. That scientists could hold governments accountable for placing their field at a competitive disadvantage gave them substantial political leverage in policy decisions. (An important instance of use of this political leverage by EMBO was noted in section 4.4). But commercial influence on national policies was potentially even more far-reaching than that of the scientist-clients of national governments. The marked independence with which representatives of the pharmaceutical industry responded to the NIH guidelines was an expression of their power to impose conditions on government policy from the outside. Moreover, they could threaten to move their operations to more inviting regulatory climates, although at this stage, offshore havens were limited by the availability of technical expertise to those nations on the forefront of genetic engineering. Later chapters explore the divergence between the two national systems produced by special interests as well as the limits placed on this divergence by the general conditions of global scientific and industrial competition.

PART THREE
Response to the Recombinant DNA Controversy

FIVE

Defusing the Controversy:
The Politics of Risk Assessment

5.1 The Spread of the Recombinant DNA Controversy

The most immediate challenge to the American genetic engineering policy came from those scientists who questioned the fundamental assumptions of the NIH guidelines and from the attentive public that had followed the sometimes stormy evolution of that policy. By the time the NIH guidelines were promulgated on 23 June 1976, the controversy surrounding these controls and the field whose possible hazards they were meant to mitigate was intensifying, fueled by divisions of scientific opinion. Statements of prominent scientists ranged from dismissal of the issue to serious concern, a spectrum also reflected in the results of a nationwide survey conducted by the *Boston Globe* early in 1977.[1] This divergence was understandable in view of the large areas of ignorance about the behavior of genetically engineered organisms that were acknowledged at this time. As the NIH guidelines themselves stated (accurately conveying the huge set of unresolved issues unearthed by the process of drafting them): "At present, the hazards may be guessed at, speculated about, or voted upon, but they cannot be known absolutely in the absence of firm experimental data."[2]

The *Boston Globe* survey also showed that, while 44 percent of the experimental biologists questioned held that the NIH guidelines were strict enough, an even larger percentage—almost 50 percent—held that they should be even more strict or that a moratorium on research should continue.[3] Questions were raised at several levels. Were the controls technically effective? that is, did they guarantee effective containment? Were they socially effective? that is, would they be followed? But the most fundamental question, striking at the core of the reigning policy paradigm and its supporting discourse, was whether the NIH was the appropriate government agency to develop these controls or whether, as the agency responsible for promoting genetic engineering, the NIH had an irremediable conflict of interest in attempting to control the field.

The controversy over these issues, erupting during the policy process that led up to the 1976 guidelines, spread to some places where genetic engineering was already being conducted: Ann Arbor, Cambridge, San Diego, Princeton, Madison, Bloomington.[4] In Cambridge, the mayor, Alfred Vellucci, held stormy public hearings that coincided with the release of the NIH guidelines. The Cambridge hearings featured a clash of discourses: the mayor, citizens, and some scientists (described as "dissident") argued for the public's right to control a novel technology with unknown consequences, while representatives of the NIH and members of the molecular biology establishment, evoking the discourse established at the Asilomar conference, argued that the problem was being responsibly treated and technically contained. By July 1976, the Cambridge City Council had voted for a three-month moratorium on recombinant DNA research within the boundaries of Cambridge while its own committee—the Cambridge Experimentation Review Board (CERB)—investigated the issue and assessed the adequacy of the NIH controls.[5] The CERB report, issued in January 1977, directly challenged the fundamental assumption of the NIH policy that the biohazard issue was a technical issue, to be addressed by experts: "Knowledge, whether for its own sake or for its potential benefits to humankind, cannot serve as a justification for introducing risks to the public unless an informed citizenry is willing to accept those risks. Decisions regarding the appropriate course between the risks and benefits of potentially dangerous scientific inquiry must not be adjudicated within the inner circles of the scientific establishment."[6]

The CERB proposed a city ordinance requiring uniform compliance with controls in several respects more strict than those of the NIH, and oversight by a biohazards committee appointed by the city.[7] Partly stimulated by the Cambridge developments, moves to develop local controls were also initiated in Princeton, New Jersey (January 1977), New York state (summer, 1976), Maryland (October 1976), Emeryville, California (April 1977), Berkeley, California (September 1977), and Amherst, Massachusetts (January 1978).[8]

As these events unfolded, Edward Kennedy, whose constituency had so vocally aired the issue and claimed its right to enact its own controls, took action that notified the administration that Congress was taking a keen interest in the recombinant DNA issue. On 16 July 1976, Kennedy and the ranking Republican member of the Senate Committee on Labor and Human Resources, Jacob Javits, wrote to President Ford, praising the "great care" the NIH had taken in formulating its guidelines but also expressing "grave concern" that the controls would not apply to all sectors. "Recombinant DNA entails unknown but potentially enormous risks," the two senators warned, and they urged the president to "implement these guidelines immediately wherever possible by executive directive and/or rulemaking, and to explore every possible mechanism to assure compliance with the guidelines in all sectors of the research community, including the private sector and the international community." If legislation was required, the president was urged to send proposals to Congress. The implication was that if the president failed to act to close the loopholes in the guidelines' coverage, Congress might do so.[9] In response, President Ford, with the advice of the Secretary of Health, Education, and Welfare, established an interagency committee consisting of all the federal agencies with responsibilities for either supporting or regulating recombinant DNA activities. The committee was chaired by Donald Fredrickson and charged with reviewing the nature and scope of recombinant DNA activities in the public and private sectors, assessing the applicability of the NIH guidelines, and recommending legislation, if this was deemed necessary.[10] The Interagency Committee on Recombinant DNA Research (IAC) proceeded to meet but took no immediate action. In September, Kennedy held hearings on the NIH policy, challenging, as the CERB had challenged, the Asilomar discourse, especially its assumptions that policy should be made by experts.

> Scientists must tell us what they are capable of doing, but we as members of society must decide how it should be or whether it should be applied. Congress cannot legislate an appropriate answer in this matter. But it can and should take the lead in assuring that these issues are discussed publicly and by as broad a segment of the population as possible. The plain fact is that genetic engineering has the capacity to change our society. How do we want it changed? What uses can we make of this knowledge? What degree of change is desirable and at what rate? . . . These are the questions that must be answered and the development of those answers must involve as many Americans as possible.[11]

The elections in the fall of 1976 and the change in the presidency in January 1977 stalled any immediate government response to the issues indicated by Kennedy. These did not fade from public view, however. The environmental organization Friends of the Earth formed a committee for genetics and began to circulate position papers to its membership, arguing that the NIH and its scientist advisors were enmeshed in deep conflicts of interest in developing controls for research promoted by the agency.[12] In November 1976, two environmental organizations, the Environmental Defense Fund and the Natural Resources Defense Council, petitioned the DHEW to hold extensive public hearings and to develop legally binding regulations to control all recombinant DNA activities in the country, under the authority granted by section 361 of the Public Health Service Act. Noting that the NIH itself had indicated that its guidelines were not a final statement of public policy but rather intended to begin a full public consideration of the relevant issues, the petition observed that "a significant portion of recombinant DNA research and technology is not covered by any mandatory set of safety procedures, leaving the public unprotected from its potential hazards. Furthermore . . . the public did not have an adequate opportunity to participate in the basic policy decisions underlying the NIH guidelines."[13]

The issue grew in salience, with the media covering each new development attentively. Genetic engineering was portrayed as a field of science that frightened "even scientists" (*Atlantic*) or, at the least, a field on which scientists divided (*New York Times*).[14] The issue was projected as embracing a perilous unknown. Headlines asked "New Strains of Life—or Death?" "Progress or Peril?" "Blessing or Curse?"[15] The need for the public to be involved in deciding the outcome of genetic engineering was raised sympathetically. As a *New York Times* editorial stated in October 1976, "[N]on-scientists need to follow, and ultimately join responsibly [the discussion on the effectiveness of safety controls]; for the issues at stake may be as vital for humanity's future as the issues in the debate over the proliferation of nuclear weapons."[16]

Early in 1977, Congress entered the fray. Genetic engineering became a hot congressional issue, with over a dozen bills and resolutions introduced between January and April. The principal bills that emerged by the end of the spring were those sponsored by Edward Kennedy and Paul Rogers, the respective chairs of the Senate and House committees with jurisdiction over the activities and responsibilities of the DHEW. While the House bill was close in its basic provisions to those favored by the administration, essentially codifying the status quo and extending the coverage of the NIH controls, the Senate bill transferred responsibility for regulation of genetic engineering to a freestanding commission, a majority of whom would be nonbiologists. The power to control the issue, previously asserted by the biomedical research establishment, was being broadly challenged. The extent of the biomedical research community's resistance to this challenge would be reflected in its shrill, negative response to the Kennedy bill (chapter 6).

The growing intensity of the issue, and its potential for attracting broad public attention, was evident at a three-day forum in March 1977, sponsored by the National Academy of Sciences.[17] The opening session was disrupted by a citizen's group, the People's Business Commission, led by activist Jeremy Rifkin, whose members, chanting "We Will Not Be Cloned," draped across the platform a banner with the caption, "We Will Create the Perfect Race."[18] Also announced during the meeting was the formation of a new organization of scientists and environmentalists, the Coalition for Responsible Genetic Research, which urged a worldwide ban on genetic engineering until safety issues could be resolved.

Among the organization's sponsors were Nobel prizewinners George Wald and Sir Macfarlane Burnett, author and social critic Lewis Mumford, and Aurelio Peccei, founder of the Club of Rome.[19]

The National Academy of Sciences forum, perhaps the climax of the public controversy over biogenetics, displayed deep and important divisions about virtually every aspect of the hazards and possible benefits of the field. Robert Sinsheimer responded to Paul Berg; Erwin Chargaff to Daniel Nathans, Harvard professor of biology; Ruth Hubbard to the vice president for research at Eli Lilly, Irving Johnson. Claims were challenged, and challenges countered. The NIH policy process itself, elegantly summarized by Donald Fredrickson, was subjected to serious and insightful criticism, its weaknesses deftly pinpointed.[20] By the end of the conference, virtually every essential feature and assumption of that process had been attacked. Even participants who were known to be sympathetic to easing future development of genetic engineering were calling for a reassessment of the social contract between science and society.

Meanwhile, research and development in genetic engineering was expanding. The estimated number of NIH projects involving genetic engineering jumped from a handful in 1975 to over three hundred in 1977 (table 2.1). Research aimed at achieving the expression of the genes of higher organisms in bacteria was moving fast (the first experimental results were announced in the fall of 1977). By 1976, the pharmaceutical industry was developing its own research programs in genetic engineering, and in 1977, the first multimillion-dollar investments in the field were made (table 2.3). This accelerating pace of development provoked scientists in the field to voice the fear that their work would be superseded by research pursued under less strict controls in other countries. Those responsible for national science policy expressed anxiety that the United States could lose its lead in the field.[21]

5.2 The Hazard Problem: A Case Study in the Closure of a Technical Controversy

While pressures to move quickly in the recombinant DNA field intensified, the controversy over its hazards continued. The controversy was not easy to resolve, for the hazards of genetic engineering were multidimensional and complex. First, an enormous variety of microorganisms might potentially be used as hosts for cloning purposes, and an enormous variety of genes might be inserted into these organisms. The action and function of many of these genes was not well understood. In addition, the interaction of genetically engineered organisms with their environment and with other organisms was unknown. In the course of the controversy, dozens of possible risk scenarios were contemplated. A dearth of empirical evidence compounded the uncertainties of any analysis.

Given such complexity and the rapid evolution of genetic engineering techniques, the hazards debate might have continued indefinitely. In fact, it was soon restricted and ultimately closed down. In the period from 1976 to 1978, a distinct shift in the discourse about the hazards of genetic engineering occurred in the United States. Leaders of the American biomedical research community began to argue that their ignorance of the hazards of the field was not as vast as they had earlier supposed, and that important new evidence showed that genetic engineering research was quite safe—at least if it used the officially sanctioned cloning host, *E. coli* K12. Beyond this limited claim for safety, some scientists began to make the far larger claim that the *entire field* was safer than had earlier been supposed. In addition, a less obvious change, not registered at the time, occurred in the discourse associated with NIH efforts to assess the biohazards of genetic engineering. The original goal, implicit in the charge to the RAC, of determining hazards through "worst-case" experimentation and analysis, was displaced and redirected toward the goal of demonstrating that genetic engineering with *E. coli* K12 was safe. Furthermore, these changes happened with remarkable rapidity: a major issue that climaxed in 1977, by 1979 the recombinant DNA controversy had all but faded away.

As a general phenomenon, the restriction of scientific controversy has been addressed in a variety of historical and sociological studies that draw attention to the complexity of scientific evidence, the broad scope often available for development of alternative explanations, and the roles played by nontechnical factors, especially professional interests and institutional interests, in producing closure.[22] Many of these studies limit consideration of the interests acting in a controversy to professional vested interests in particular theories, instruments, and practices. But some have taken the analysis considerably further, showing how such commitments may themselves be shaped by the sociopolitical and cultural contexts in which they are formed.[23] This may be a matter not only of scientists' use of the resources of their society and culture but also of the influence of scientists' social and political commitments and those of the institutions to which they are affiliated.

Despite important observations and programmatic proposals concerning the need to probe these social relations of science, precisely how interests operate in controversies, how some prevail over others to produce closure, and how closure affects later scientific practice have not been addressed in detail.[24] Precisely how political processes result in the "intertwining" of social and technical dimensions of science has not been addressed with any specificity. The purpose of this and subsequent chapters is to show how the constructivist and postpluralist methods described in the introduction can open the study of scientific controversy to an analysis capable of encompassing the operation of sociopolitical interests, at both the macrolevel of corporations and governments and the microlevel of individual scientists, and can also probe the fine structure of the processes that ultimately produce closure.

As discussed in the introduction, postpluralist methods for examining policy decisions may be extended to science policymaking and to decisions conventionally assumed to take place "within" science. These methods are based on the claim that an exclusive focus on decisions taken within any given arena (whether this is a government committee, a scientific meeting, or a laboratory) will miss a great deal, and probably the most important part, of the action. Crucial for the outcome are the institutional environment in which an issue is addressed and the structural bias of that environment—that is, the distribution of access to it and influence within it, especially control of the agenda. Political theorist E. E. Schattschneider warned against the pluralist assumption that interests compete on an even playing field: "The flaw in the pluralist heaven is that the heavenly chorus sings with a strong upper-class accent. Probably about 90 percent of the people cannot get into the pressure system."[25] For the resolution of scientific controversy as well as for the making of science policy generally, it is no less important to examine how the policy arena is structured and who gains access to it. Indeed, the analysis in this case shows that, with respect to the resolution of the hazards controversy (although not necessarily with respect to the shaping of the broader genetic engineering policy), the "heavenly chorus" sang mostly in the language of molecular genetics.

A second dimension follows from the first, for restriction of access to a policy arena means that the issues that receive attention will also be restricted. As decisions in the formal arena incorporate the values and commitments of decision makers, some aspects will be prominent, others obscure, and still others irrelevant. Schattschneider's dictum that "some issues are organized into politics while others are organized out" applies equally to scientific controversy.[26] It is important to ask which aspects of a controversy dominate and which are neglected, and to investigate the reasons behind these choices.

A third dimension is the question of the form in which issues are defined, before they reach the formal policy arena, in the course of decision making, and in the process of dissemination to the public at large. This is the point at which the "linguistic turn"—the attentiveness to discourse and its connection with power—is extremely valuable. The "normalizing gaze" of discursive practice—in this case, the informal rules that evolved to redefine the hazards problem—is an expression of the power relations and interests that gain influence in the policy arena. Furthermore, discursive practices may greatly amplify the effect of initial decisions.

These methods are used here to show how the complex set of perceptions of possible hazards of genetically altered organisms with which the scientific community started were restricted, eventually producing a reduction in the discourse concerning the genetic engineering problem. The initial recognition of the complexity of the problem was largely organized out of consideration, whereas claims that certain types of hazard were minimal were organized in. Each stage of the process simplified the original complexity until what was left was a sharp focus on the properties of one organism, *E. coli* K12, especially a widely accepted claim that the bacterium could not be transformed into an "epidemic pathogen" that could escape the laboratory and run rampant through a population. These methods are also used to show how the risk assessment strategy of investigating a "worst-case" scenario agreed on at the December 1975 RAC meeting was reformulated in a subtle but nevertheless highly significant manner to focus on a demonstration that *E. coli* K12 would reliably contain whatever DNA was introduced into it. Both of these transformations of discursive practice would play highly influential roles in later policymaking.

5.3 The Meetings at Bethesda, Falmouth, and Ascot

Whatever else may be at issue, there is little doubt that three scientific meetings, held in Bethesda, Maryland, in August 1976, Falmouth, Massachusetts, in June 1977, and Ascot, England, in January 1978, were principal sources of the "new information" claimed to support the changed evaluation of recombinant DNA hazards, as well as crucial influences on their interpretation. The first meeting in Bethesda, known as the Enteric Bacteria Meeting, led directly to a decision to organize the second in Falmouth, known as the Workshop on Risk Assessment of Recombinant DNA Experimentation with *Escherichia coli* K12. And the Falmouth meeting strongly influenced the outcome of the third meeting at Ascot, the U.S.-EMBO Workshop to Assess the Containment Requirements for Recombinant DNA Experiments Involving the Genomes of Animal, Plant, and Insect Viruses. The results of these meetings, particularly the second and third, were subsequently cited in testimony, policy documents, and official statements to justify the claim that there was little cause for concern. As we shall see, these claims played a highly influential role in persuading legislators that congressional action was unnecessary and in moves to revise and weaken the NIH controls.

Three main mechanisms at these conferences reduced the complexity of arguments about recombinant DNA hazards: first, the sponsorship and organization of these meetings, which determined the range of scientific and political participation in them; second, informal processes that affected the scope of the proceedings and the reporting of results; and third, the dissemination of the results (addressed in section 5.6).

Sponsorship and Organization

The principal sponsor of the three meetings was the NIH, the main source of government support for biomedical research in the United States. As such, it was the center of a vast research network that connected its leadership closely with large research universities, on the one hand, and with the community of biomedical researchers, on the other. As earlier chapters show, the NIH was also the lead agency for the development of government controls for possible hazards of recombinant DNA technology. Abundant evidence shows that university administrators and biomedical researchers overwhelmingly favored this arrangement.[27] The Bethesda meeting was organized by two NIH virologists, Wallace Rowe and Malcolm Martin; Rowe was also a member of the RAC and of the RAC executive committee, which advised the NIH director on recombinant DNA policy matters. The Falmouth meeting was sponsored and funded by the NIH. The Ascot meeting was jointly funded and sponsored by the NIH and EMBO, a private scientific organization for the support of research in molecular biology supported by several European countries and a major grant from the Volkswagen Foundation. (EMBO was active in monitoring recombinant DNA controls in the United States and Europe and in making policy proposals of its own.) The meeting was jointly chaired by Rowe, Martin, and John Tooze, executive director of EMBO. Thus, scientists close to institutions responsible for the sponsorship of recombinant DNA technology played important roles in deciding on the forms of all three meetings.

The formal purposes of each of these meetings were scientific and technical in nature. The Bethesda meeting, to which infectious-disease specialists were invited, was designed to open up analysis of recombinant DNA hazards to disciplines that had not previously been involved in the recombinant DNA issue. The Falmouth meeting had similar goals but was organized as a larger and more formal event, with solicited papers. The Ascot meeting was designed to involve specialists in animal virus research in a detailed analysis of the hazards of cloning animal virus DNA.

As scientific events, however, the organization of each of these meetings deviated significantly from the classic norm of openness. Each was unannounced, private, and thus known in advance only to a select group of scientists closely associated with the organization of the event. In each case, the wider scientific community and the public learned of the meetings only after the fact. In the case of the Bethesda meeting, even the identities of participants other than the two chairs remain officially unrevealed to this day.[28]

This does not mean that a range of scientific positions was not represented at these meetings. Participation, however, was definitely controlled. In the Bethesda and Ascot meetings, participants were invited by the chairs. All the Ascot participants were virologists with backgrounds either in clinical infectious disease or in molecular biology, and almost all were either actual or potential users of genetic engineering.[29] For the Falmouth meeting, a larger organizing committee was responsible for invitations, but even then only two scientists known to be critical of NIH policy were present. One of those scientists, MIT biologist Jonathan King, later observed that he had to request admission to the meeting from the chair. King also emphasized that the conference "was not announced by the normal procedure for announcing scientific conferences, that is, in the scientific journals, Genetics Society of America, American Society of Microbiology. It was private. It was by invitation of the organizing committee. Many people were rather upset . . . to find out that a risk-assessment conference was taking place and they didn't even know about it until after the fact."[30]

At the Ascot meeting, members of the British GMAG were not invited—an omission that caused many British eyebrows to be raised. One member of GMAG later commented:

> It might be thought a discourtesy to run an international conference on an important policy question without involving the corresponding organization in the host country, particularly when that is the only one in the world to be setting standards that are used internationally. Indeed, it is hard to see why GMAG should have been excluded, except for the strong representation on GMAG of the members representing employees and the public interest. Had GMAG been invited to participate, some of these would certainly have attended, and would have supplied a critical presence.[31]

This evidence suggests that, while technical expertise was well represented, there was much less diversity in representation of political positions on the question of control of recombinant DNA technology. This conclusion is also supported by the fact that, at the more private Enteric Bacteria and Ascot meetings, discussions were characterized by informal understandings about the politics of the recombinant DNA controversy. A strong informal theme of the Bethesda meeting was a shared sense of a pressing need, beyond containing possible hazards of recombinant DNA work, to contain the spread of the controversy as well. These discussions had a siegelike feeling, a shared sense of threat, of polarization, of scientists versus society. Polarized categories—research scientists versus "them," variously described as the "sky-is-falling people," "prophets of doom," those motivated by "political interests"—characterize references to the recombinant DNA issue. The transcript suggests that this group saw the recombinant DNA controversy as symptomatic of a general movement of the nonscientific public to bring biomedical research under external control.[32]

At the Ascot meeting, Wallace Rowe's introductory remarks similarly painted a picture of biomedical research as immensely threatened by external political forces, of the recombinant DNA controversy as a confrontation between the forces of rationality and the forces of antiscience, and of recombinant DNA controls as the beginning of the progressive encroachment of bureaucratic restrictions on biomedical research. "There are very dreadful things on the horizon and it's not restricted to recombinant DNA. [The movement to restrict] recombinant DNA [research] is only the beginning of . . . great dangers to freedom of inquiry."[33] It is unlikely that such statements would have passed unchallenged in open meetings with a wider range of representation.

Informal Processes Affecting the Scope of the Proceedings and Reporting of Results

As Bachrach and Baratz have argued with respect to policymaking on conventional political issues, differential access to a decision arena generally affects the values that shape the proceedings.[34] In this case, the restriction of participation in these meetings meant that the assessment and analysis of recombinant DNA hazards happened in a specific political context characterized by strong informal interests in responding to and containing the recombinant DNA controversy.

THE BETHESDA MEETING, 31 AUGUST 1976. The tone of the Bethesda meeting was set by the chairman, Wallace Rowe, in his opening remarks.

> Part of the agenda today is to get you guys involved and get your voices heard, and maybe if the "Infectious Disease Society of America" comes out and says, "By God, if it's just insertion [of foreign DNA] you are talking about, nobody is worried about this mechanism." That carries a tremendous amount of weight, at least to me, if I could say that to the prophets of doom: "Look, these guys have come out and said there is nothing to worry about here, so let's really start and get on with more serious business." That's what I hope we can accomplish.[35]

With this orientation to the problem, much of the conference was spent brainstorming the hazards of recombinant DNA technology. A wide and complex collection of issues was brought up, the general tenor of which was that unusual and possibly problematic combinations of genes could gradually be transferred into organisms in the environment where they might at some later time be expressed in a way that could cause eruption of novel disease. As one participant stated, "There may be problems of low level endemicity. And, depending on what's created, in special cases, serious endemicity. The *Botulinus* [toxin], the growth-hormone producing *E. coli*. To me, those are frightening" (44).

Consideration of these problems was greatly restricted, however, by the adoption of several assumptions that had the effect of focusing attention on a limited subset of hazards that were generally judged to be of much less concern.

The most important restriction placed on the discussion was the assumption that all recombinant DNA research would be conducted with *E. coli* K12, the strain of the common intestinal organism that had been weakened by many years of use in molecular genetic laboratories. Rowe stated at the beginning of the meeting: "Out of the infinite universe of combinations that DNA recombinant research can involve, the guidelines have narrowed it down, it seems to me, to a very advanced level. Of all the bacteria in the universe, we're only really talking about one particular bacteria with options to find parallel ones that are as laboratory restricted. Okay, so *E. coli* K12 is really the focus of . . . these experiments. No other organism is presently considered as 'licensed' under the guidelines" (3).

Rowe went on to qualify his statement because, even in 1976, there was scope in the NIH guidelines for expansion to the use of other organisms in genetic engineering work. Nevertheless, *E. coli* K12 quickly became the focus of attention.

A second major restriction was the assumption that the assessment would be limited to hazards to communities *outside* the laboratory. Hazards to workers *inside* the laboratory were deemed relatively unimportant. In other words, the group took its concern to be not primary exposure but secondary spread. There appears to have been general agreement on this. As one participant stated, "[The question of epidemic disease] seems to be much more important than infections of laboratory workers." To which someone replied: "I make that as a major condition. I am really not as concerned about laboratory workers as long as the infection is restrained in our midst. Introducing new things in the ecosystem, into populations that are in no way involved in the lab, that is what I worry about. A case in the investigator—[or] the technicians—is bad, but that's not the major question" (15–16).

A third restriction was the assumption that recombinant DNA activities would occur only in technologically advanced countries with adequate public health and sewage treatment facilities. The implication was that epidemics in such environments could not occur under any circumstances. As one speaker put it: "[T]his kind of epidemic just doesn't happen and isn't happening in our society largely because of sanitation. We don't have house rats and we don't [have] house fleas and we don't have lice and we don't eat shit and that's what it comes down to. In our kind of society, this kind of epidemic just doesn't happen" (32).

With these restrictive assumptions, the participants focused their attention on the "epidemic pathogen" scenario, which was generally agreed to be unlikely. Even so, not all concerns were put to rest. The following exchange is characteristic:

> —What I want to know is, living in, say, Washington, can you make an epidemic in Washington? Can you make an organism so virulent that it will make an epidemic in Washington? . . .
>
> —I think the point is that your K12 could be carrying a new product that is quiescent as a genetic entity. It's got virulence, but is not expressing itself. As soon as by some accident of nature it then leaves that environment and gets superimposed. . . .
>
> —Can you arrange this accident? Can you think of a circumstance in which you could make it spread? This is really the heart of the issue. Can it be done?
>
> —I can't answer that. (34)

Another participant, however, cited the occasional large-scale *Salmonella* epidemics, such as an outbreak in Riverside, California, involving twenty thousand people. As someone summarized the concern, "The point we are trying to make is, if you already have an organism that can cause epidemics and if it receives the genes from, say K12, you can get an epidemic of organisms with *those* genes" (35).

Clearly not everyone was persuaded that the new technology posed no new problems. But as the discussion continued, outstanding issues—such as the question of low-level seepage of novel gene combinations into organisms in the environment—tended to be set aside rather than confronted. Instead, the sense that biomedical research was threatened came increasingly into focus. When several people noted that other aspects of biomedical research might pose hazards as serious as those of the new biology, they were warned that scientists must be careful not to stimulate the spread of regulation to other research fields. "Science," someone announced, "is under a very serious attack." "But where is it coming from?" it was asked. "From ourselves," came the answer. "One has to be very careful about the tack one uses and should not say, 'Well, gee, we have been doing much more dangerous experiments for years.' That's murder! You have to use a very positive approach" (40).

In the same vein, someone else (or possibly the same person) warned that "we have a serious political disease. You have to be careful in these arguments that you don't spread it to other people. The big danger about the argument: 'But look! Something else is much more dangerous than what we do already,' is that the 'something else,' all of a sudden, gets in with a big bag of red tape at the very least" (42–43).

Visions of laboratories swathed in red tape dominated the later stages of the morning session. Within the context of concern about the spreading regulation of science, the argument that *E. coli* could be converted into an epidemic pathogen came to be seen, not simply as one consideration among many associated with the problem of defining potential recombinant DNA hazards, but as a leading argument that could be developed specifically for the purpose of defusing the growing controversy. The chair, Wallace Rowe, expressed this sense: "Why I got you here is that I think if somebody acquires data that convinces important people, they'll say, 'It's a bunch of nonsense; you cannot change *E. coli;* you've tried, so and so has done this until he is blue in the face, and I can't see and a thousand other Infectious Disease people can't see any danger in working with a *Salmonella* donor into *E. coli* and a *Drosophila* into *E. coli*'" (43–44).

Exchanges following this statement show that others present accepted this political strategy.

> —Well, who do you have to impress? How does it come to pass that I have to write an application to do a standard genetic cross?
>
> —That's really where it's at. . . . The point, as I understand it, is that the ingredients for infectious disease with *E. coli* K12 are simply not there and the number of unknowns that you have to specify is very large and each probability is very small. You multiply them together and you come out with nothing. You know, numbers that are comparable to 10^{-n}; negative numbers that are comparable to the number of atoms in the universe.
>
> —You are going on the argument that people say you are going to create drastic epidemics, everyone bleeding to death.
>
> —Right, but that's what people are being scared with; that is what the other side is winning with. They are *not* winning with the idea that a few lab technicians or a few scientists are going to get sick. They don't care about that. Nobody cares about that. (44)

When someone at this point attempted to make scientific distinctions about hazards, they were told that the political dimension had to be emphasized. Here is the exchange.

> [We must] really separate these [issues] out and somehow . . . try to get the word going around that informed people are really not

worried about epidemics; that there may be problems of low grade endemicity and, depending on what's created, in special cases, serious endemicity—the *Botulinus,* the growth hormone producing *E. coli.* To me, those are frightening.

—The Mayor of Cambridge doesn't know the difference. What the Mayor of Cambridge is worried about, besides not being re-elected the next time, is the possibility of an epidemic. It's exactly the same issue as the nuclear people have to face. . . . There are serious arguments being made at the level of low levels of contamination, but their popular image is that of explosions. And it's exactly parallel. Serious arguments are about this kind of low level thing, but in terms of the PR you have to hit epidemics, because that is what people are afraid of and if we can make a *strong* argument about epidemics and make it stick, then a lot of the public thing will go away. (44–45)

The select participants at the Enteric Bacteria Meeting, sharing an interest in protecting "free inquiry" in recombinant DNA research, thus carefully concentrated their attention on developing arguments that would convince the public that research hazards were exaggerated. The issue was not *whether* the epidemic pathogen argument was technically acceptable but *how* it should be used politically. Someone summarized the sense of the group at the end of the morning session: "I think [the problem of convincing the public] is what you have to deal with. It may not mean a thing, but that is very easy to do. It's molecular politics, not molecular biology, and I think we have to consider both, because a lot of science is at stake" (45).

THE FALMOUTH MEETING, 20–21 JUNE 1977. Wallace Rowe reported on the Bethesda meeting to the RAC in September 1976. He conveyed the view that, in the opinion of the participants, "enteric epidemics are extremely remote" and that "concepts such as this should be discussed in a public forum."[36] An organizing committee chaired by Sherwood Gorbach, a specialist in enteric disease at Tufts University, was established and the outcome was the two-day workshop held at Falmouth in June 1977.

The agenda of the meeting was limited to the hazards of use of *E. coli* K12 as a cloning host. According to the proceedings, published almost a year later, three basic questions were posed. First, could the addition of "foreign DNA" convert *E. coli* K12 into a pathogenic strain that could either cause disease in an individual or spread through a population? For example, could the addition of DNA enhance the ability of *E. coli* K12 to colonize and survive long enough either to transfer plasmids to other, more robust or more virulent strains or to produce some effect? Second, could DNA inserted into the K12 strain be transferred to other microorganisms or the somatic cells of a host? Third, could "foreign" DNA inserted into *E. coli* K12 encode for harmful products such as toxins, hormones, or proteins capable of inducing an allergic response?[37]

The published proceedings show mixed responses to these questions. Some research appeared to be reassuring, although by no means definitive. For example, efforts to establish the K12 strain in the human intestine showed that the organism generally survived no longer than four or five days.[38] The significance of this result was not altogether clear, however. Feeding experiments had also demonstrated the failure of other, more robust strains of *E. coli* to establish themselves.[39] Further reassurance was claimed for experiments that showed unsuccessful efforts to enhance the pathogenicity of the K12 strain by traditional breeding techniques.[40] However, this work was limited in scope, did not use genetic engineering, and was based on limited knowledge of the location and role of genes controlling pathogenic properties of *E. coli*.[41]

Other research and analysis was clearly inconclusive. For example, troublesome questions were raised about the capacity of *E. coli* K12 to transfer "foreign" DNA to other more robust organisms that could survive more effectively in the environment.[42] Bruce Levin, a population geneticist who attended the meeting, later described the issue.

> There was considerable discussion about the transfer of bacterial plasmids and a general feeling that the rate of infectious transmission in the intestines of healthy mammals would be low. However, in my impression there was absolutely no consensus reached which suggested that the probability of transfer of chimeric DNA by plasmids was sufficiently low to be disregarded. Furthermore, there was very little consideration about transfer via transducing phage or as free DNA.[43]

Further questions were raised about the impact of *E. coli* bacteria that were genetically "reprogrammed" to make novel proteins. At this point in the development of genetic engineering, this major research goal was still unrealized. Jonathan King raised the theoretical possibility that reprogrammed bacteria might generate new forms of autoimmune disease in which secreted gene products caused human or animal hosts to make antibodies against their own proteins.[44] Clearly, Falmouth did not produce a definitive interpretation of recombinant DNA hazards. Indeed, the inconclusiveness of the discussions was underscored by a primary outcome of the meetings: the development of a set of detailed protocols for further risk assessment research.

A sense of the full scope of the Falmouth proceedings was not what reached either the wider scientific community or the general public, however. The public image of the Falmouth results was shaped primarily by a letter sent by Sherwood Gorbach to the NIH director, Donald Fredrickson, immediately after the conference. The epidemic pathogen argument dominated Gorbach's account to the virtual exclusion of other issues. There was, Gorbach emphasized, "unanimous" scientific agreement, backed by "extensive" scientific evidence, "all of which provides reassurance that *E. coli* K12 is inherently enfeebled and not capable of pathogenic transformation by DNA insertion."[45] Other participants were much less sanguine about the behavior of *E. coli*. Population geneticist Bruce Levin and phage biologists Jonathan King and Richard Goldstein, in detailed responses to Gorbach's summary, pointed to the many possible hazard scenarios that Gorbach's conclusions ignored, arguing that the only reasonable conclusion to draw from the conference was not that the available evidence was reassuring but that risk assessment experiments needed to be energetically pursued.[46] "I don't believe that evidence available through existing efforts at risk assessment is sufficient to justify a relaxation of the current NIH guidelines on recombinant DNA research or of the efforts to enforce them," Levin concluded. But these responses were virtually obliterated by the American Society for Microbiology's focus on the Gorbach summary as an element in its campaign to derail legislation aimed at regulating genetic engineering (section 6.3). Consequently, what emerged was a soothing view of the evidence, one in which uncertainties and unresolved issues were obscured by the emphasis on the remoteness of possible hazards.

The new discourse about genetic engineering hazards emerging at the Falmouth meeting was also a distinctly American product, reflecting the political culture in which it was embedded. The fact that technical workers—technicians, maintenance personnel, and graduate students—had no significant political role in the formation of policies affecting the conduct of research meant that, in the United States, the discourse on hazards could be cast exclusively in terms of impacts outside the laboratory. In the United Kingdom and some other European countries, in contrast, the very fact of trade union involvement meant that any discussion of genetic engineering hazards had to address hazards within the laboratory. Thus the epidemic pathogen argument never became a major focus of debate, nor was it used as a rationale for changing policy (chapter 8). Sydney Brenner recalled in 1980: "We were never concerned about creating an epidemic pathogen. In the first instance, our concern was the health and safety of people at work."[47]

The important point here is that how scientists responded to the information developed at the Falmouth meeting depended not on the scientific validity of the epidemic pathogen argument but on whether they saw this result as central to the questions of recombinant DNA hazards. Their judgments on the latter issue were social rather than technical. Scientists who accepted the social assumptions embedded in the epidemic pathogen argument saw Falmouth as a scientific meeting that produced a scientific judgment. Scientists who did not accept those assumptions also saw Falmouth as a scientific meeting—with the difference that they perceived the results as "choreographed" (to use the description of one respondent)—in other words, aimed at emphasizing a preconceived result. One of the scientific members of GMAG expressed this view: "[Falmouth was] a real set-up . . . not a comprehensive scientific debate. . . . [The epidemic pathogen argument] was developed by people who wished to produce a certain conclusion."[48]

THE ASCOT MEETING, 27–29 JANUARY 1978. Gorbach's summary statement of the Falmouth result proved highly influential for the third scientific meeting, at Ascot in January 1978. A major reason for the meeting was growing discontent among virologists in the United States and elsewhere with restrictions on the cloning of animal virus DNA in *E. coli*. These procedures had stimulated some of the original concerns about the potential recombinant DNA hazards and had subsequently been classified as "high risk" in the NIH guidelines. The need for a meeting to reassess these controls was emphasized by Paul Berg in a letter to the NIH director in October 1977 and reinforced by John Tooze, executive director of EMBO, at an NIH hearing in December 1977.[49] Apparently, these concerns were heard sympathetically. The three-day meeting was jointly sponsored by the NIH and EMBO.

As at Falmouth, the Ascot participants addressed only the risks of cloning in *E. coli* K12. Within that constraint, however, the transcript of the meeting shows important differences between the American scientists and some of their European colleagues over where the primary focus of hazard analysis should lie: the primary exposure of the individual worker or secondary spread of infection to the community outside the laboratory. At the outset, some urged that the main focus of analysis should be the possibility of a viral epidemic. Harold Ginsberg, chair of the Department of Microbiology at Columbia University College of Physicians and Surgeons, stated: "The political hue and cry vis-à-vis use of recombinant DNA [focuses on the possibility of] these worldwide epidemics that are going to spread throughout the world, so that one really has to deal with [the question]: Is it likely to produce an epidemic?"[50]

Others stressed the importance of considering the individual laboratory worker. For example, Lennart Phillipson, a Swedish virologist, responded, "I agree completely that we have to look at the epidemiology as the second step but when it comes to the laboratory worker . . . the immunological status . . . of that worker [is very important]" (40). And Helio Pereira of the Animal Virus Research Institute in Pirbright, Surrey, stated, "[T]he safety of the laboratory worker himself is of great concern, for instance, for the trade unionist" (42).

Throughout the conference, however, there were signs of an effort to focus the discussion primarily on hazards *outside* the laboratory. As Edwin Lennette, head of the California State Department of Health Biomedical Laboratories, stated: "It's a universal language we are trying to develop. One that's meaningful to people in all countries. . . . [M]aybe we should think of our language not in terms of . . . laboratory infection for workers but would we want to introduce or take the risk of something introduced into the ecosystem?" (44). Later on, Rowe asserted this view even more strongly.

> I think we ought to make the assumption right off that we are not concerned, our focus is not the laboratory worker. Our focus is the people and animals, or plants, outside of the laboratory. That if the virus itself is a hazard only to the worker, then that's not a hazard that we have to be . . . we know how to protect ourselves. But if it's a perpetuating thing that can move outside of the lab, here's what we have to compare it with. Avian sarcoma virus in you is not going to endanger anyone else. (174)

The tension between a discourse constructed upon the initial exposure of the individual and a discourse constructed upon secondary spread of an organism into a community was never resolved at the Ascot meeting. Eventually, however, the second discourse, validated by the Falmouth meeting, would prove decisive.

In classifying hazards resulting from the use of animal virus DNA in cloning, the meeting explored two major classes of use—cloning of viral DNA in bacteria and use of viruses as vectors to insert foreign DNA into animal cells. In each case, the participants analyzed hazard scenarios originally posited in the Berg committee's letter, exploring in some detail the implications of the type of viral DNA used, how this might be released, and how it might gain access to the cells of a human or animal host (section 3.5). The complexity of this task resulted in part from the variety of types of animal virus (DNA versus RNA, segmented versus nonsegmented; single-stranded versus double-stranded), their mode of replication, and their mode of action in a host organism.

In some cases, there was general agreement that the production of hazards was virtually impossible. For example, it was agreed that, because of the differences between the genetic regulatory machinery of bacteria and higher organisms, particularly the inability of bacteria to splice out the recently discovered intervening sequences in animal virus DNA, bacteria carrying whole viral genomes would be unable to make infectious viral particles and so would be unable to provide a new route of viral infection.

Other possibilities were not so easily eliminated, however. For example, if the gene for a viral coat protein were introduced into bacteria, and if the bacteria made this protein, would human hosts exposed to these bacteria become tolerant to the virus and unable to raise an appropriate immunological response? Or, if an entire DNA copy of an RNA virus such as polio were inserted into a bacterium, could the bacterium produce the intact virus, and thus provide a new route of transmission? Or if a gene known to be responsible for tumor formation were inserted into *E. coli* bacteria and if the bacteria colonized the gut and later died and released this DNA, would this tumorigenic DNA transform exposed cells and cause tumors?

The transcript shows that it was impossible for this group to eliminate such scenarios on theoretical grounds, although the probabilities were generally considered to be low. One participant summarized the conclusions of the group as follows:

> There were certain things that just molecularly seemed [as if] they could not happen, according to our present knowledge of animal viruses. . . . And, so, these were of no concern even if . . . what we know about the safety of K12 and the implausibility of each step for transfer broke down. . . . And there were others where we felt a little bit more uneasy because we could conceive of proteins being expressed or whole viruses being reconstituted, if all the biological safety mechanisms broke down, which is extremely unlikely but not inconceivable. And those I suppose are the sarc genes and similar genes of oncogenic viruses, and I would include whole genomes of positive strand viruses in the same category—things where they can, conceivably, be reconstituted, if all the safety mechanisms that we've built in, biological safety mechanisms, broke down, and if there was full expression, and so on. (1001–2)

The tenor of these discussions also shows that at many points, predictions were speculative. Too little was known about the mechanisms of viral infections and transformation to be able to predict the effects of cloning these genes. One participant remarked: "We do not know that a certain gene product of Marburg or Lassa [virus] is, in fact, highly toxic and is not responsible for the extraordinary . . . pathogenicity of this virus. So, if you had one of these genes making a protein product, I am not sure that I would be willing to say today that it should be reduced to P2. I mean, that is something that we simply do not know" (1005–6). Another participant summed up the essential problem of making these assessments: "You see, the whole discussion has [the feeling of] a sort of Aristotelian academy because we are really just discussing extremely theoretical things and we're deriving models which are based on no experiments whatsoever. I mean, that's why we're talking so much" (284).

A further issue that emerged during the meeting was that, if the high physical-containment levels required by the 1976 NIH controls were lowered, access to the cloning of viruses would be greatly increased and containment barriers would be more likely to be broken. One participant stated, "[I]f there's any concern at all in allowing [the genes of higher organisms or animal viruses] to get into the general environment, you can be sure that if these K12 organisms carrying the clones are generally available in all labs, that they will get out, that they will be mobilized into other strains sooner or later" (1013).[51]

Such thoughts were accompanied by skeptical reflections on the lack of containment provided by P2 conditions. "P2 doesn't seem to be more than closing the door of the lab when you go in, as far as I can see," stated British virologist Robin Weiss (1013). An American, Joe Sambrook, responded, "[I]f you guys take your concerns seriously, then you should be proposing that all of this stuff gets done in EK2" (1017). Sambrook's suggestion was energetically resisted by others. "Everyone I talk to says [EK2] is the god-damndest thing [to work with]," observed American virologist Robert Wagner (1019). Following that observation, little more was heard about using biological containment.

As at the earlier meetings, what eventually neutralized concerns about hazard was a narrowing of the discourse to focus exclusively on the issue of secondary spread outside the laboratory. The shaping of the Ascot assessment is most evident in the final day of the meeting, when the group drafted a summary statement that purported to represent a consensus. Among other things, the draft referred to the conference's recognition that clones of bacteria carrying certain types of viral DNA might "bypass the natural barriers to infection by the virus particle, because it is a conceivable, but extremely remote possibility that all the biological containment barriers might break down" (1035).

This draft was energetically resisted by some who argued that the Falmouth conference had shown that no hazards would materialize under *any* circumstances. Here is the exchange between Wallace Rowe and Harold Ginsberg.

> GINSBERG: I have one concern, if you'll pardon the expression. When this report becomes public, and you talk about the breakdown, the simultaneous breakdown of biological barriers, and . . . when you're before Senator Kennedy's committee and he asks you what does that mean, and then he relates it all the way back to all other recombinant DNA [scenarios] what do you answer? What is this simultaneous breakdown of biological barriers?
>
> . . .
>
> ROWE: The transfers [to other enteric bacteria] that have selective advantage. . . .

GINSBERG: Yes, but you see, the whole Falmouth meeting said that
couldn't occur, and yet you don't make that explicitly clear that this
can't occur. You say it can occur and everything up to this moment
in history has said no. (1050–51)

Some at the meeting challenged Ginsberg's position, attempting to redirect the discussion to hazards to individuals within the laboratory. "If
we ignore [hazards to laboratory workers], we end up looking like a
bunch of virologists with a completely callous and unrealistic approach
to human error," one person commented (1061). Nevertheless, Ginsberg's position, with its emphasis on the Falmouth result, eventually prevailed. One of the European participants later commented, "The trouble
with the Ascot meeting was that the moment one raised a scenario, one
would be shouted down by [those] saying that the Falmouth meeting had
said that the clones were not mobilizable, that they could never get out
of *E. coli* K12 or χ1776, and could not become an epidemic strain."[52]

The final "consensus" statement, which appeared in the *Federal Register* in March 1978, finessed the issue of hazards to laboratory workers and
focused attention on the hazards to the community. The latter, the report
emphasized, were "so small as to be of no practical consequence."[53]

The overwhelming impression produced by the report on the Ascot
conference was one of reassurance. Almost all hazard scenarios were considered "remote," "most unlikely," or "impossible." In general, it was
concluded that the cloning of viral DNA would "pose no more risk than
work with the infectious virus or its nucleic acid and in most, if not all
cases, clearly present less risk."[54] Since the sole risk assessment experiment designed to test the hazards of cloning viral DNA, the Rowe-
Martin polyoma experiment, was a year away from yielding results, these
conclusions were surprisingly emphatic. The scientific community's response to earlier fears about the cloning of viral DNA had come to be
essentially an attempt to tell the public they had nothing to fear.

There is a direct line of development from the Enteric Bacteria Meeting in August 1976 to the new consensus on recombinant DNA hazards
that emerged in 1978. Having discovered that their very success in
achieving a powerful technique for producing novel substances and organisms had led to two distinct threats to their free pursuit of these
methods (costly containment procedures and regulation of research),
molecular biologists closely associated with the NIH and the biomedical
research establishment organized a defense of their interests. They brain-
stormed (under Rowe and Martin) extensively about what *could be perceived* as potential hazards of their techniques; they came up with a way
of approaching the hazards question that could convince both the public
and a fair number of their colleagues that their research was not at all

dangerous; and at the Falmouth and Ascot meetings they succeeded in carrying the day with their approach.

The dominant image of these meetings, as portrayed in press coverage and in official reports, is one of "scientific" meetings with "scientific" agendas. In fact, this analysis has shown that a principal motive for the meetings was the protection of biomedical research from external regulation. At the Enteric Bacteria Meeting, the most private of the three, this motive was made quite clear, and numerous comments—with no explicit dissent—suggest that all of the participants accepted it. The same disparity between image and motive characterized the Ascot meeting. It was "an entirely scientific, analytical process," Rowe later asserted.[55] However, others at the meeting disagreed. "It was very obviously a political meeting," one of the European participants later recalled. "The science was not too bad but I had a strong distaste for the way it was managed. . . . We were being used in the name of being a disinterested group of virologists but it was fairly clear by the end of the meeting that [the organizers] wanted to go back with a result that could be exploited for deregulation."[56]

In the achievement of this consensus, the available scientific data were rarely in question: "the science was not too bad." The analysis was politicized at a different level, namely, through the introduction of restrictive assumptions that allowed a selective and reassuring interpretation of these data. The persistent focus on the question of the conversion of *E. coli* K12 into an epidemic pathogen allowed other considerations to be bracketed.

5.4 Further Sources of "New Evidence"

The Bethesda, Falmouth, and Ascot meetings were crucial events in the development of a new consensus about the nature of genetic engineering hazards. Particularly because they could be claimed to represent the collective position of the scientific community, the conclusions generated gave the emerging consensus legitimacy. These meetings did not occur in isolation, however. The new climate of opinion about the hazard problem also received reinforcement from other statements and publications in this period from scientists who had played prominent roles in this controversy in 1977 and 1978.

The Curtiss Letter

Even before the Falmouth conference, the epidemic pathogen argument began to circulate and gain legitimacy. One important vehicle for this new interpretation of recombinant DNA hazards was a widely circulated letter written by Roy Curtiss to Donald Fredrickson in April 1977. Curtiss had gained recognition in the course of the recombinant DNA debate as a dovish member of the plasmid committee at the Asilomar conference and of the RAC, as the scientist who responded to the challenge of constructing the "disarmed" EK2 strain of *E. coli,* and, not least, as the author of several broadly circulated, extremely long "open" letters on the implications of genetic engineering.

Curtiss's first open letter, written in August 1974, enunciated in detail sweeping concerns about genetic engineering hazards, comparing the dangers to those earlier posed by radiation and urging the Berg committee to broaden its recommendations for a temporary moratorium to encompass a far wider range of experiments.[57] Curtiss's April 1977 letter had a very different tone. Its immediate purpose was to respond to the various legislative proposals emerging in Congress. (The letter was copied not only to more than forty scientists but also to all of the members of three congressional committees with jurisdiction with respect to control of research.) "I am," he wrote, "extremely concerned that, based on fear, ignorance and misinformation, we are about to embark on over-regulation of an area of science and scientific activities." What followed seemed to amount to a recantation of his early caution: "I have gradually come to the realization that the introduction of foreign DNA sequences into EK1 and EK2 host-vectors offers no danger whatsoever to any human being with the exception already mentioned that an extremely careless worker might under unique situations cause harm to him- or herself. The arrival at this conclusion has been somewhat painful and with reluctance since it is contrary to my past 'feelings' about the biohazards of recombinant DNA research."[58]

The reasons for this reversal were detailed in the thirteen pages that followed, which also provided the substance of the paper Curtiss later delivered at the Falmouth conference.[59] Focusing his analysis on the strains required for use by the 1976 NIH guidelines, Curtiss argued that it was "highly improbable" that *E. coli* K12 could be rendered pathogenic, or that it could be transmitted through a population, either through contaminated water, food, or air. Nor did he consider the transmission of foreign DNA inserted into *E. coli* K12 strains to other, more viable organisms to be a problem. While noting that further information was needed to assess this question, Curtiss nevertheless concluded that this possibility was "most unlikely." Nor, in the case that recombinant DNA was released from the cells of *E. coli* K12 bacteria, was transformation (through direct uptake of foreign DNA) of other bacteria likely. Curtiss noted that transformation might be induced by treating *E. coli* cells with a salt and subjecting them to a rapid temperature increase of

42 degrees Centigrade in one minute, but judged that "such conditions were unlikely to be encountered in nature." As Jelsma and Smit have noted, the rhetoric used by Curtiss to address the question of genetic engineering hazards shifted distinctly from his earlier letter of 1974: an emphasis on the uncertainty of the hazards was now replaced by one of certainty; an emphasis on the possibility of hazard by an emphasis on their (very low) probabilities. Evidence on the ability of E. coli K12 to colonize the human gut, described in 1974 as "poorly understood," was now taken to show that the bacterium was "unable to colonize." E. coli itself, described in 1974 as a "ubiquitous organism" and "already one of our largest health problems," becomes in the 1977 letter "a rather harmless inhabitant of the intestines."[60]

Moreover, Curtiss narrowed the hazard question from a concern with the entire range of ecological and health impacts of genetically engineered organisms (in 1974) to an exclusive focus on the hazards posed by the host-vector systems certified to that point under the 1976 NIH guidelines. In other words, the dangers he addressed were those remaining under the putatively effective operation of the 1976 NIH guidelines—a point brought home by Curtiss's acknowledgment that harm might befall "extremely careless" investigators—those who, for example, chose to violate the guidelines by pursuing genetic engineering while their intestinal tracts were disturbed by disease, fasting, or antibiotic therapy, or by ignoring required safety procedures. The emerging American discourse on hazards saw such accidents as simply a matter of individual liability, treating what might happen outside the laboratory as the only pertinent danger.

The Cohen Experiment

Taken alone, Curtiss's letter might not have been influential: there is no evidence that its recipients in Congress immediately began to rethink the need for legislation: on the contrary, in the summer of 1977, the signs were that recombinant DNA legislation in some form would be enacted. The letter was important, however, in reinforcing the message emerging from the Falmouth conference. The fact that it was seen as a "recantation" of his earlier caution also added force to his position.

But Curtiss's arguments were quite limited in scope: like those emerging from the Falmouth conference, they applied only to research using E. coli K12 and assumed that the 1976 NIH guidelines would be followed. Other scientists in this period—notably Stanley Cohen of Stanford University, one of the original inventors of recombinant DNA technology—began to take more sweeping positions, essentially denying that there were, or could be, any special hazards associated with genetic engineering. In an article in *Science* in February 1977, Cohen attacked critics of NIH policy as "a horde of publicists—most poorly informed, some wellmeaning, some selfserving." While dismissing the hazard question generally as "hypothetical," a matter of mere "conjecture" (in contrast to real and substantial benefits), and emphasizing that billions of bacteria containing recombinant DNA had been grown in the previous few years without apparent mishap, Cohen nevertheless held that the 1976 NIH controls were useful "just in case" hazards were proven.[61]

Later that year, Cohen produced what he claimed to be new experimental evidence to support his skepticism.[62] With his assistant, Shing Chang, Cohen attempted to demonstrate that fragments of mouse DNA were naturally taken up by *E. coli* K12 bacteria and joined to pieces of a plasmid inside the bacteria. In other words, "foreign" and bacterial DNA joined naturally in living bacteria (in vivo) as well as artificially in the test tube (in vitro). Cohen concluded in a letter to the NIH director on 6 September 1977:

> These experiments and others have led us to believe that an important biological function (perhaps the major function) of the so-called "restriction" enzymes may be site-specific recombination of DNA, and that eukaryotic DNA fragments formed biologically by restriction enzyme cleavage can link to prokaryotic DNA without *in vitro* recombinant DNA techniques. Our data provide compelling evidence to support the view that recombinant DNA molecules constructed *in vitro* using the *Eco* RI enzyme simply represent selected instances of a process that occurs by natural means.[63]

It should be noted that others would raise important questions about the validity of Cohen's claims (see section 7.3). Most obviously, in order to effect the bacterial take-up of DNA, Cohen and Chang had treated their bacteria with a calcium salt and a temperature rise of 42 degrees Centigrade—the precise conditions that Curtiss had told Fredrickson were "unlikely to be encountered in nature."[64] Curiously, this contradiction was never noted or followed up by the NIH although both letters were used as supporting evidence in policy documents.

5.5 The Politics of Risk Assessment

The epidemic pathogen argument proved to be only the first step toward absolving the entire field of genetic engineering from special hazard. At first, however, the argument supported rather than undermined the legacy of the Asilomar conference. That research was deemed safe if it used an *E. coli* K12 host-vector system left open the possibility that cloning in other systems might be more hazardous. Thus the epidemic pathogen argument reinforced the concept of biological containment as well as the institutions by which the NIH ensured that only weakened systems would be used and that the safety of new ones would be assessed vis-à-vis *E. coli* K12.

Furthermore, the "silence" of the epidemic pathogen argument with respect to the safety of other host organisms left open the possibility of hazard and therefore supported a second legacy of the Asilomar conference—the idea that risk assessment was a desirable and integral part of a policy aimed at ensuring the safety of the field. This need was directly addressed at Falmouth and Bethesda, in the course of which the discourse associated with one key risk assessment experiment changed from worst case testing to demonstration of safety.

The Falmouth conference was expressly aimed at determining the hazards of cloning in *E. coli* K12. Sherwood Gorbach's introduction stated:

> There are potential risks associated with recombinant DNA experimentation. The level of risk and its application to specific experiments remain highly controversial, but all prudent scientists would recognize that certain experiments cannot be exonerated of unexpected, and even serious, misadventures. The second area of commonality is that such risks can be assessed, measured, quantitated, and subjected to the same scientific scrutiny that characterizes the nature of the research endeavor itself.[65]

A brainstorming session later in the meeting produced six protocols for further experimentation. Summarizing them, Gorbach noted that, "in their present crude construction," these were preliminary and would need refinement. But he emphasized the importance of initiating a program of risk assessment: "There must be a beginning, even if it serves to create a focus for disputation; from the cauldron of vigorous scientific debate will finally emerge critical experiments to assess potential hazards in recombinant DNA technology."[66]

The protocols defined the following questions for further investigation:

> Could *E. coli* K12 colonize the human intestinal tract?
> Could *E. coli* K12 transfer its plasmids to other bacteria in the human intestinal tract?[67]
> Did insertion of foreign DNA enhance virulence or the ability to colonize the intestinal tract?
> What were the effects of hormone-producing strains of *E. coli*?
> Did the insertion of DNA shotgun experiments increase the virulence of *E. coli* K12 strains or their ability to colonize the intestinal tract?
> What were the effects of inserting polyoma virus DNA into *E. coli* host-vector systems?

Many of these questions would subsequently be investigated, and by the time of the Falmouth conference, an experiment responding to the

last was already being planned. This experiment, designed to test the
original concerns about genetic engineering, had evolved out of Sydney
Brenner's proposal at the RAC's La Jolla meeting in December 1975 that
the committee should sponsor a "dangerous" experiment to explore possible hazards (section 4.3). Wallace Rowe and Malcolm Martin had taken
on the challenge, designing procedures using polyoma virus as the inserted DNA, *E. coli* K12 as the host, and tests for the ability of the bacteria bearing the viral inserts both to produce polyoma infection and to
induce tumors in test animals.

What was not clearly registered at the Falmouth meeting, or in later
discussions, is that the goal of the Rowe-Martin experiment had been
fundamentally redefined during a freewheeling discussion of risk assessment in the afternoon session of the Bethesda meeting the previous year.
As in the morning session, the exchanges in the afternoon were pervaded
by a sense that what was at issue was not merely the "scientific" purpose
of possible experiments but also their "political" interpretation by the
broader public. Rowe introduced this theme.

> I thought maybe if we could focus on some possibilities here it might
> help to get to the specific types of things [risk assessment suggestions]
> I need to get out of this meeting. That is, if I were to draw up for the
> Recombinant Committee or for NIAID a formal proposal for a pro-
> gram to evaluate risks, whether it's a political or scientific exercise. . . .
> The basic question is, when sampling a lot of insertions of different
> recombinant DNAs into K12 by different vectors, different methods of
> insertion, different topologies . . . anything one can imagine, how
> would one ask the question, "Has it changed the behavior of K12 in any
> unexpected way, particularly in any way that might point it in the di-
> rection of hazard?" This is the nitty gritty that everyone is looking for
> and what might do an awful lot to calm a lot of nerves. It's a huge
> amount of scut work, but I think it is incumbent for NIH to see that
> the scut work gets done, provided that knowledgeable people feel it's
> worth doing, whether for political or scientific purposes.[68]

The discussion that followed addressed the kinds of issues Rowe had
indicated, and the transcript suggests a strong understanding around the
table that "political" and "scientific" purposes meant distinctly different
types of experiments. Twenty minutes or so into the discussion, one person summarized the distinction.

> It seems to me, there are two kinds of experiments that you can think
> about: One kind of experiment involves learning something about
> what's going on and those experiments require the standard sort of
> thing that we are interested in. The second . . . is a class of experiment
> whose aim is to convince people that some of their worst fears don't
> happen very easily. Now, one experiment in the latter case that I would

argue is definitely worthwhile doing is to try every way you can, germ-free or otherwise, to make an epidemic of serious disease with cloned DNA. Just try to do it. I agree that it makes no sense. . . . And the experiment gets published in the *New York Times,* not in the *Journal of Molecular Biology.* That's what ought to be done. . . . A kind of thing that will never live to be a famous experiment in the history of science; it's like all other experiments that end up in the front of the *New York Times.* It's something that captures the imagination of reporters. (52)

Another person agreed: "That's what I'm talking about. Let's get rid of the most outlandish hypotheses that are the cause of the hysteria." Another queried: "Do you think a negative experiment is going to make the front page of the *Times?*" The response was, "I think if it was handled properly, yes" (52). The absence of any direct criticism of such procedures suggests that many around the table accepted the political use of risk assessment to calm the growing public controversy.

Agreement on such use of risk assessment was by no means complete, however. During the meeting, two opposed "philosophies" crystallized with respect to the design of the polyoma experiment. A problem with the design proposed by Rowe and Martin, which used *E. coli* K12 as the host organism, was that it was not a natural mouse pathogen and was eliminated very quickly. As one of the participants observed: "The *E. coli* K12 is probably going to get wiped out. No matter which way you give it, it is going to be wiped out very rapidly" (58). Thus there would be little chance for test animals to be exposed to the polyoma DNA carried by the bacteria. Someone else noted: "That may be the problem with K12. You would have to do too many things to make K12 pathogenic so that it is a slick *New York Times* kind of an experiment" (59). Others agreed with this assessment: "One of the problems with K12 in the mouse is that we may have a very insensitive system for looking at changes in pathogenicity" (63). "My concern is that if you work with K12 and you put a particular factor into it, you may clone something virulent. But you don't detect it because that strain has got how many strikes against it? It doesn't colonize. You could have a biohazard sitting in that K12 and not know it and be led to believe that it is a safe system" (65).

There appeared to be broad agreement about these difficulties with the *E. coli* K12 system. Where the participants divided was on what to do about them. On the one hand, at least one or two participants pressed for the use of a known mouse pathogen, *Salmonella typhi,* as host organism, in order to generate a possibility that the mouse would actually be exposed to the polyoma DNA if in fact it were released.

[W]hy not . . . use a known mouse pathogen; use a *Typhimurium?* . . . Here you have an organism . . . which cuts across the gut barrier. It gets inside the epithelial cells, delivers your polyoma to an epithelial cell. Here is an organism that goes beyond the mucosa and goes into a Peyer's patch. It goes into the liver. It goes into the spleen. It's delivering your "biohazard," and you can see what happens. If you get a negative result with *this,* I think this carries much more weight [than a negative result with the *E. coli* K12 system] against the extremists. (52–53)

Someone else, or possibly the same person, reinforced this view: "I am not saying that everyone should do this experiment. But you have been charged with doing these kinds of experiments. Take the opportunity to do a good experiment."(53) It was also pointed out that an attenuated strain of *S. typhimurium* would persist in the mouse spleen and liver for weeks (in contrast to *E. coli* K12, which was expected to be expelled rapidly). In this way, the possibility of producing a biohazard in the mouse model would be fully explored.

Others objected to the *S. typhi* model. One revealed the leading assumptions of the time: "The point is, the analogy to the human isn't with *Salmonella typhi.* No one is proposing that we clone with *Pasteurella pestis* [plague]" (57). Another (or possibly the same) person noted: "What you are doing is looking at a super germ warfare model. You want to get a bad pathogen and make it worse" (57). The argument being made here was that no one would contemplate deliberately using a known pathogen as a cloning host; therefore the *S. typhi* model did not test accepted practice—the use of *E. coli* K12.

The ground of the argument was subtly shifting. Those who defended the rationale for testing *E. coli* K12 now argued that the purpose of the polyoma experiment was *not* to determine if genetic engineering posed hazards but rather to demonstrate that *E. coli* K12 could contain its DNA inserts. This change of emphasis is made clear in the following argument against the *S. typhi* model:

> I am concerned about . . . changing the host [to *S. typhi*]. I understand that scientifically it makes some sense, but it seems to me that there are two arguments against it. One is that the problem about carrying out the experiment . . . the legalities . . . may not be completely non-trivial. The second thing is that if you tip the balance very close to pathogenicity, then you are also tipping it in a particular direction. . . . I think that one of the things I would like to see documented is that this million-fold business with *E. coli* K12 [the assumption that the probabilities of survival and transmission for *E. coli* K12 were extremely low] is really there. . . . If you find with naked [polyoma] DNA that it works [i.e., infection and tumor induction is measured] and it doesn't work at all with *E. coli,* that difference . . . is the difference we want. That, politically, is going to help if it's true. . . . By using known pathogens, it seems to me we go politically in the wrong direction even though scientifically it does make more sense. (64)

Another person reinforced this view. "I think you are going to be push-ing all your variables to produce a super pathogenic strain [with *S. typhi*] and it's going to get you far away from the main question, which is whether this particular system, which is the one we are all talking about in biology [*E. coli* K12], is a safe system" (65). Others produced practical obstacles: a new system would take time and effort to define. As some-one summed up what appeared to be the sense of the group: "If we want to get these experiments done so we can go about our work quickly, maybe one shouldn't introduce problems of this level" (66).

At that point, the *S. typhi* model was dropped. In the course of the interaction, the purpose of the polyoma experiment had been redefined as a test of containment. What Brenner had conceived as a "dangerous" experiment—one designed to test the possibility of a biohazard—had now been reconceived as an experiment to test the safety of *E. coli* K12. As the Falmouth conference report later described this purpose, the ex-periment would "identify the most important variables that determine whether a nucleic acid molecule can be transferred from an *E. coli* host into eucaryotic cells in vivo."[69] In other words, at the Bethesda meeting the discourse concerning biohazard assessment was reconstituted: con-tainment rather than biohazards would be the object of assessment. Ironi-cally, the results of the polyoma experiment, when they were published and debated in 1979, suggested to some precisely what one of the Be-thesda participants had anticipated—that there might be a "biohazard sitting in that K12 and [you would] not know it and be led to believe that it is a safe system" (65) (section 9.7).

5.6 Dissemination/Legitimation

The proceedings of the Falmouth meeting were not published until May 1978, and the report of the Ascot meeting was never published in a sci-entific journal. (It appeared in the *Federal Register* in March 1978.) Cohen's results were not available in print until November 1977. Long before open communication and assessment within the scientific com-munity could come into play, these new arguments circulated through unusual and nonstandard routes. This dissemination was, as we shall see in chapters 6 and 7, closely connected to deployment of the arguments in Congress, to argue against the need for legislation, and in the NIH policy process, to press for revision and relaxation of the 1976 NIH controls.

Upon their appearance, the hazard arguments were widely circulated in the mainstream scientific and general media in 1977 and 1978. Several national newspapers as well as *Science* covered the Falmouth story and relied on the Gorbach letter to Donald Fredrickson for information. For political reasons examined in chapter 6, scientific criticism of the Gorbach letter received scant attention.[70] The message in almost all of these accounts was the same: researchers at Falmouth had "unanimously concluded that the danger of runaway epidemics [was] virtually nonexistent."[71] An editorial in *Science* in August 1977 that cited the Gorbach summary at length conveyed the view that the risks were now deemed minimal.[72] Many of these accounts also mentioned Curtiss's April letter, portraying it as a reversal of his earlier caution.[73] *Time* magazine, in an article entitled "DNA Research: Not So Dangerous after All?" whose reassuring message contrasted with that of its cover story earlier in the year, quoted Curtiss as saying, without qualification, that "research with the K12 strain poses 'no danger whatsoever.'"[74] In the fall of 1977, Cohen's claims for the results of his experiment were also being mentioned in news stories. A *Washington Post* story quoted Cohen as saying that "it turns out that Mother Nature has been capable all along of doing in cells what scientists can now do."[75] The new slant in the news coverage of the genetic engineering debate was nicely summed up in a headline in the *New York Times* on 24 July 1977: "No Sci-Fi Nightmare after All."

The "new evidence" argument also surfaced at international conferences on the scientific and practical implications of genetic engineering. In these forums, the epidemic pathogen argument tended to be generalized into the much more extensive claim (implicit in the *New York Times* headline above) that recombinant DNA technology posed no significant hazards at all. At one such meeting, held in Milan in March 1978 and sponsored by the World Health Organization and the Fondazione Giovanni Lorenzini, scientists and industrialists involved in recombinant DNA research and development repeatedly assured the audience that the hazards of recombinant DNA technology were no longer significant. Irving Johnson of Eli Lilly claimed that there had been "a steady and persistent decline in concern by informed and participating scientists for any biohazards."[76] Molecular biologist Waclaw Szybalski claimed that the cloning of "practically any DNA fragment in *E. coli* K12" posed "no significant risk."[77] And John Tooze—who undoubtedly was seen as an informed participant from the Ascot meeting—insisted that recombinant DNA was "no more hazardous than many other, now routine, biological techniques whose development—unheralded by well meaning but nevertheless alarmist public statements by those who invented them— rightly excited no concern amongst the general public and entailed no dangers for it."[78]

Claims such as these went virtually unqualified. The Falmouth discussions were cited without reference to the conference's call for further assessment of the hazards of work with *E. coli*. The Ascot results were cited without reference to any need to investigate further any aspect of the cloning of viral DNA or to the fact that the Rowe-Martin experiment had yet to yield results. As a writer for the British science journal *Nature* observed, reporting on the Milan conference and reflecting the skepticism with which many in the United Kingdom were receiving these arguments at this time: "One must now accentuate the positive. The new evidence, however, does not seem substantial: those at Milan witnessed some unseemly clutching at straws."[79]

Finally, the new consensus solidified in 1977 and 1978 as various prominent members of the biomedical research community indicated their approval. Bernard Davis, professor of bacterial physiology at Harvard Medical School, had taken a skeptical position on the laboratory hazards issue from the outset, on the grounds that Darwinian selection would consistently work against the survival and spread of genetically engineered organisms (section 3.4). In the September 1977 issue of the *American Scientist,* Davis reinforced the Falmouth conclusions, arguing that the K12 strain of *E. coli* was so weak that it was "like a hot-house plant thrown to compete among weeds" and that addition of foreign DNA made it even less able to compete and ultimately to survive in nature. Davis concluded that "the risk of laboratory infection by *E. coli* recombinants is much smaller than the risk encountered with known pathogens in medical laboratories." Furthermore, providing reinforcement for the discourse that was now assuming a dominant position in the United States, he asserted, "As long as investigators are not creating undue risk for others, they have the right to take risks for themselves—as they do whenever they work with pathogens."[80] The message here was that the only hazards that needed to be addressed were those to communities outside the laboratory.

In this article and elsewhere, Davis challenged the position developed by Robert Sinsheimer, that genetic engineering might disrupt natural barriers between species, with unforeseeable consequences. To this claim, Davis opposed a position he had first aired in 1974, to the effect that processes similar to those facilitated by recombinant DNA techniques had probably been occurring (albeit at some unknown low frequency) for millions of years in the human intestinal tract with no apparent ill effect. The implication was that we already had evidence that organisms endowed with additional DNA through genetic engineering were ill-adapted to their environment and consistently outcompeted by the "successful" natural strains.[81] While the epidemic pathogen argument presumed that formal controls mandated the K12 strain of *E. coli* as a cloning host, Davis's evolutionary arguments were far broader in scope, foreshadowing the position he would later embrace, that the NIH controls were an unnecessary impedance to biomedical research.

Reinforcing the more radical overtones of Davis's arguments was a stream of articles issuing from the prolific pen of James Watson and appearing in such places as the *New Republic*, the *Washington Post*, *CoEvolution Quarterly*, *Clinical Research*, and the *Bulletin of the Atomic Scientists*. These announced Watson's recantation of the cautious views he had endorsed in signing the Berg committee's letter in 1974 and engaged in strenuous debunking of all concerns about genetic engineering. Watson denounced the Asilomar conference as "an exercise in the theater of the absurd," and the effort to assess and control genetic engineering as "a massive miscalculation in which we cried wolf without having seen or even heard one." Those who advocated caution and federal controls were, according to Watson, "an odd coalition of spaced-out environmental kooks and leftists who see genetics as a tool for further enslaving the masses." Any concern about the hazards of genetic engineering was, in his view, irrational: DNA was almost entirely safe; it was better to "worry about daggers, or dynamite, or dogs, or dieldrin, or dioxin or drunken drivers, than to draw up Rube Goldberg schemes on how our laboratory made DNA will lead to the extinction of the human race."[82]

Watson's harsh denunciations of genetic engineering controls appear extreme, but the same antiregulatory message was transmitted in other, more decorous forms. Rene Dubos, professor emeritus of Rockefeller University, publicly abandoned a previously held "quasi-religious hostility to experiments combining genes from different organisms," bowing instead to the need to proceed with research.[83] Lewis Thomas, president of the Sloan-Kettering Institute for Cancer Research, reversing the arguments of Erwin Chargaff (section 4.4), held it a form of hubris to claim that biological science could easily produce deadly pathogens with genetic engineering: "It takes a long time and a great deal of interliving before a microbe can become a successful pathogen. Pathogenicity is, in a sense, a highly skilled trade, and only a tiny minority of all the numberless tons of microbes on the earth has ever involved itself in it. . . . I do not believe that by simply putting together new combinations of genes one can create creatures as highly skilled and adapted for dependence as a pathogen must be." Beyond the details of the hazard question, Thomas argued that the real issue related to the dangers of allowing the public to become involved in controlling science, because "the easiest course for a committee to take, when confronted by any process that appears to be disturbing people or making them uncomfortable, is to recommend that it be stopped, at least for the time being." That, Thomas averred, was "real hubris" and "carries danger for us all."[84] Similarly, British biologist Sir Peter Medawar argued in the *New York Review of Books* that those who advocated caution were motivated by "fear of the unknown," an indulgence in "imagined dangers." Medawar castigated the public for its "excess of fearfulness," defending the need for the development of controls by scientific peers.[85] All of these articles emphasized the "benefits," in terms of scientific advances and practical applications, that would flow from research. Without engaging the details of the recombinant DNA controversy, the messages of these elder statesmen of science provided the ligase sealing the new arguments that there was little to fear from genetic engineering and that the only real concern was that public intervention would stop the field in its tracks.

SIX

Derailing Legislation, 1977–1978

The dominant discourse about genetic engineering hazards from 1976 to 1978 changed in response to intense public controversy concerning the potential impact of the field and public challenges to the policy developed within the American biomedical research community. The Asilomar conference achieved a first reduction of the discourse surrounding the implications of genetic engineering, limiting it to the technical problem of containing unknown hazards. The Bethesda, Falmouth, and Ascot meetings achieved a further reduction through the claim that new evidence showed that *E. coli* K12 could not be converted into an epidemic pathogen. (And some scientists went even further, claiming that new evidence showed that the entire field posed negligible hazard.) I examine in this chapter how the new discourse was deployed in Congress to counter emerging legislation, and in chapter 7 how it was used within the NIH and its RAC to challenge the real obstacles the NIH guidelines began to pose to the pursuit of certain lines of research and development.[1]

6.1 The Politics of Government Control of Recombinant DNA Technology

The policies of the U.S. government for the promotion and control of science and technology shifted substantially during the early 1970s. At the beginning of the decade, protecting society and the environment from undesirable effects of technological development was widely accepted as a desirable goal, and much environmental legislation, usually mandating public participation, was enacted. By the middle of the decade, however, determined ideological, legal, and political challenges by leaders of science and industry had reordered government priorities for science and technology policy. The Ford administration reinstated the Office of Science and Technology Policy, undertook reviews of the economic impact of regulatory decisions, and increased support for basic and applied research, steps actively followed up during the Carter administration. Despite strains and contradictions, science policy during the Carter years moved toward deregulation, emphasizing economic assessment of the impact of regulation, and toward promoting innovation and stronger linkages between the universities and private industry. Recombinant DNA and other new forms of research-based technology were seen as especially important for development (section 1.4).

The continuity of committees and personnel from Ford's to Carter's prodevelopment recombinant DNA policy was significant. Donald Fredrickson stayed on as director of the NIH. Frank Press, a member of Ford's science policy advisory group, became director of the Office of Science and Technology Policy and Carter's science adviser. In 1977, Press was joined by Gilbert Omenn, a physician at the University of Washington's School of Public Health, who took on responsibility for policy issues associated with the life sciences and health. Other sectors of the Carter administration, particularly the Council on Environmental Quality, favored developing recombinant DNA legislation, seeing it as an essential way to regulate industrial activities over which the NIH had no control. In contrast, Press, Omenn, and other members of the Office of Science and Technology Policy were fundamentally opposed to legislation. Omenn recalled in 1981: "We were confident that the NIH Guidelines would be observed and would prove to be sufficiently stringent. We were determined that a fluid situation not be locked in concrete by legislative rhetoric. We thought the [NIH] Guidelines would be more readily revised, as conditions permitted, than would regulations issued under legislation."[2] Further, officials of the Office of Science and Technology Policy fully accepted the discourse emerging from the Bethesda and Falmouth conferences: "[W]e did not regard the potential hazards of recombinant DNA work to be greater than the known hazards of laboratory work in research institutions and in hospitals with viruses or antibiotic resistant bacteria, or than a variety of chemical exposures." In Omenn's words, the Office of Science and Technology Policy acted as a "broker," representing the interests of science and industry within the government, ensuring that genetic engineering would not be held back by "restrictive" legislation, and persuading other parts of the Carter administration to accept that position.[3]

The federal IAC, established by Ford in September 1976, remained in place. Despite its relative lack of visibility compared to the RAC, the IAC played an important role in American policy for genetic engineering, serving to legitimate the thrust of NIH policy. Two structural characteristics of this group should be noted. First, the committee consisted of all of the federal agencies with responsibilities for either promoting or regulating recombinant DNA activities, comprising a large range of

Table 6.1   Representation of U.S. Government
Departments and Agencies on the Federal Interagency
Committee on Recombinant DNA Research

Constituent agencies of DHEW
  Office of the Assistance Secretary for Health
  Center for Disease Control
  Food and Drug Administration
  National Institutes of Health
Other departments and agencies
  Department of Agriculture
  Department of Commerce
  Department of Defense
  Department of the Interior
  Department of Justice
  Department of Labor
  Department of State
  Department of Transportation
  Council on Environmental Quality
  Energy Research and Development Administration
  Environmental Protection Agency
  National Aeronautics and Space Administration
  National Science Foundation
  Nuclear Regulatory Commission
  Office of Science and Technology Policy
  U.S. Arms Control and Disarmament Agency

Source: Interagency Committee on Recombinant DNA Research, charter, ca. October 1977, reprinted in NIH, *Recombinant DNA Research* 2: 181–83.

federal missions, from defense to the operation of the NEPA (table 6.1). An efficient mechanism for registering the interests not only of the agencies represented but also of their client constituencies in the American public, the IAC advised on executive recombinant DNA policy. Second, the committee's secretariat was provided by the NIH and was chaired by Donald Fredrickson. Much information was solicited from the various agencies represented on the committee, and Fredrickson did not seek to dominate committee decisions; he "tried to hear all viewpoints."[4] However, the NIH did mediate this information, preparing digests of it and developing IAC reports and legislation. Not surprisingly, what emerged was shaped by the needs of the NIH's own client constituency.

Two features of the agencies represented on the IAC also tended to reinforce the central role of the NIH in shaping federal policy. First, neither of the other federal agencies with an immediate regulatory interest in genetic engineering—the Environmental Protection Agency (EPA) and the Occupational Safety and Health Administration (OSHA)—had substantial in-house expertise available to them. Only in 1976 did OSHA begin to emphasize occupational health rather than safety, and there were no biologists or physicians on its staff.[5] The OSHA representative on the IAC was an occupational hygienist. Similarly, the EPA's expertise in molecular biology was limited, and it had not tried to assess the environmental implications of genetic engineering: an ad hoc study group of biologists was appointed to examine these implications only at the beginning of 1977.[6]

Second, regulatory authority for controlling hazardous uses of recombinant DNA was not vested within any single agency. This quickly emerged during the IAC considerations of existing regulatory authority in the fall of 1976.[7] OSHA's authority permitted it to regulate work with genetically engineered organisms, with the important exception of certain state and local government institutions, including universities. Emissions and effluents were regulated in principle by various environmental laws, such as the Toxic Substances Control Act, the Clean Air Act, and the Water Pollution Control Act. In practice, however, application of these statutes to recombinant DNA was complicated because it meant delays to develop standards and because the statutes had been written to address the emission of chemicals, not the release of organisms. To apply section 361 of the Public Health Service Act, which gave the Secretary of Health, Education, and Welfare broad power to control the spread of communicable diseases, was also seen as problematic, since it was directed at problems arising from known pathogens rather than organisms associated with possible hazards.[8]

The complications associated with regulatory authority might not have mattered: after all, the coverage of the British Health and Safety at Work Act did not extend to environmental impacts; the Health and Safety Commission became the regulatory authority nonetheless. An important difference in the United States was that, unlike the Health and Safety Commission, neither the EPA nor OSHA was supported by a constituency with both strength within the majority political party and influence within the government. While environmentalists were politically active and visible, their representation within the government, by officials directly committed to their goals, was weak. Those within the Carter administration committed to protecting human health and the environment effected only superficial changes in U.S. genetic engineering policy. In addition, OSHA and the EPA came under increasing pressure in the late 1970s to relax regulatory decisions on known pollutants and health hazards. Both agencies, beleaguered on other fronts, probably viewed taking on the policy issues associated with a set of hypothetical hazards as unwise.

In the fall of 1976, the IAC deliberated slowly, holding only two meetings between September and the end of the year. The pace of the committee's activities picked up early in 1977, however, following the introduction of the first bill to regulate recombinant DNA, S. 621, by Dale Bumpers in the Senate on 4 February 1977, and its companion bill, H.R. 3191, by Richard Ottinger in the House. This legislation had several features that angered members of the biomedical research community, and certainly spurred the NIH to action. One feature that especially disturbed researchers was a requirement for licenses for all projects. Another was the bill's provision for strict liability, without regard to fault, for injury to people or property caused by the research.[9]

Four IAC meetings were held in February and March 1977. In the same period, several of the agencies represented on the IAC met with constituencies with whom they were closely associated for promotional or regulatory purposes, to elicit views on possible regulation of recombinant DNA activities. The position of large chemical and pharmaceutical corporations was registered with the Department of Commerce at a meeting in November 1976. As they had at their meeting with NIH officials earlier that year (section 4.5), industry representatives took a stand that was notable for its differences with NIH policy, emphasizing that, while they were willing to comply voluntarily, important changes in that policy were needed. They indicated support for a system that would require registration of research but not compliance with details of the NIH guidelines. And they again expressed their desire to keep the nature of their research and the organisms developed secret—in contrast to the requirement for disclosure required by the guidelines.[10]

OSHA representatives met with health officials from the AFL-CIO and, according to the IAC minutes, registered the following concerns: training of laboratory personnel, medical surveillance of laboratory personnel, and the availability of medical records for inspection.[11] The representative from the National Institute of Occupational Safety and Health (NIOSH) also transmitted the view of an AFL-CIO representative that full workplace standards be developed and that compliance with workplace standards be the responsibility of the Secretary of Labor, as well as the less specific views of the pharmaceutical industry that provision should be made for adequate information about possible hazards, medical examinations, record keeping, and monitoring.[12] EPA officials met with representatives of environmental organizations on 15 February. The minutes of this meeting record that it was structured as a question-and-answer session and that the questioners did not press the EPA to adopt particular positions.[13]

Finally, NIH officials met with several prominent biomedical researchers and a larger number of prominent research administrators that included microbiologist Harlyn Halvorson, a professor of biology at Brandeis University and president of the American Society for Microbiology (ASM); Peter Hutt, who at this point was acting as legal counsel for the universities; and one of the younger critics of the NIH guidelines, Richard Goldstein. The group discussed the import of local, state, and national moves to regulate recombinant DNA activities. The IAC minutes record particular concern about the costs of safety controls and of insurance for protection from civil liability, and fear that strict liability for damage from accidents would effectively stop research.[14]

Following the meeting, Donald Fredrickson received detailed letters from several participants. "I am appalled that legislation is being considered to regulate research in any scientific area," wrote Helen Whitley, past president of the ASM. She went on to acknowledge, however, that "the public furor over DNA research has now progressed to the point where some federal regulation is inevitable," and she indicated her support, in this case, for licensing of laboratories and flexible regulations to "permit rapid regulation of new potentially hazardous procedures and relaxation of regulations for procedures or organisms which are proven to be safe."[15] Harlyn Halvorson strongly criticized the strict liability provision of the Bumpers bill, arguing that this would make it impossible for research institutions to obtain insurance. "We are vigorously opposed to the insidious control of scientific inquiry by these punitive means," wrote Halvorson.[16] Both Halvorson and Whitley impressed on Fredrickson that regulatory authority should be vested in the DHEW or in the NIH. (Whitley also proposed uniform regulation of all hazardous organisms—an idea that was never brought up or further addressed by the NIH director.)

At its first meeting in 1977, held on February 25 shortly after the introduction of the Bumpers and Ottinger bills, the IAC was handed a bill of particulars drafted by the NIH legal counsel. This document defined the DHEW as the locus of regulatory authority, regulated the "production or use of recombinant DNA molecules" rather than "research," provided for licensing of institutions rather than projects, preempted local and state regulations, and left inspection and enforcement to the discretion of the Secretary of Health, Education, and Welfare. Under the circumstances, it was a document that surely pleased the NIH's clients.[17]

The records of this meeting and of those on March 10, 14, and 25, where draft legislation and an interim report on the need for it were discussed, show the various agencies pressing for specific provisions and in some cases representing the positions of their client sectors. While the DoD kept a very low profile in public with respect to the agency's interests in genetic engineering, IAC minutes suggest that its representatives were active in attempting to secure DoD control over information concerning future military use of genetic engineering—an indication that the

DoD already anticipated that genetic engineering would have a military impact. At the meeting on 25 February, DoD representatives expressed concern about the vesting of authority for control of genetic engineering in a single agency.[18] In a later meeting, they registered their concerns about the handling of classified material whose disclosure might pose a risk to national security and argued that the mandate of the committee did not extend to such questions, and that these should be addressed by the National Security Council.[19] The representative of the Department of Commerce advocated provisions to protect trade secrets—a position in which she was supported by the representative of the DoD.[20] The representative of OSHA urged provisions for protection of whistle-blowers, for medical examinations, and for inspection of medical records; he also pointed out that OSHA should be given authority to enforce controls in facilities in which the agency had jurisdiction.[21] The NIOSH representative pressed for loss of license as a penalty for violation of controls.[22] The NIH reported that its meeting with biomedical researchers found "general" support for uniform standards (i.e., for preemption of local controls) and opposition to strict liability, which, it was said, would have the effect of stopping research.[23] The minutes record the unanimous agreement of the committee on the location of regulatory authority in the DHEW.[24]

Rounds of drafting by NIH officials and responses from the agencies ensued.[25] The agencies' final comments on a draft bill were submitted to the Office of Management and Budget, where all administration proposals were cleared.[26] At this stage, the Council on Environmental Quality managed to eliminate a provision, drafted by the NIH, that would have exempted the legislation from NEPA provisions.[27] The State Department, with the support of the DoD, submitted a provision for protection of information relevant to national security.[28] And OSHA apparently managed to insert language ensuring that OSHA's role in regulating the workplace was not preempted by the DHEW. As a result, the administration's bill, introduced by Senator Edward Kennedy as S. 1217 on 1 April and by Congressman Paul Rogers as H.R. 6158 on 6 April, included traces of the many influences at work in its construction. Several provisions should have mollified biomedical researchers worried that it would inhibit their work: regulatory authority was vested in the DHEW and decision-making mechanisms were left to the secretary's discretion; state and local laws were preempted; facilities rather than projects were licensed; a "sunset clause" allowed the legislation to expire five years after its enactment; and the strict liability clause was omitted.[29] At the same time, OSHA's role in regulating the workplace and in protecting employees' rights to file complaints was safeguarded. A little-noticed clause allowed the Secretary of Health, Education, and Welfare to exempt certain categories of activities, thus providing a loophole for the DoD to cite national security reasons for protecting information about its genetic engineering activities. Finally, and almost certainly reflecting the influence of Kennedy's constituents, the preemption provision had been removed, and with it went the chances of the bill's being harmoniously supported by prominent biomedical researchers; instead, a clause allowed state or local regulation of recombinant DNA activities if this was at least as stringent as the national controls and could be shown to be "necessary to protect health or the environment and required by compelling local conditions." In the view of one disillusioned NIH supporter, "months' worth of careful negotiation and compromise were washed away . . . and the designation of HEW as the lead agency was obscured."[30]

6.2 Biomedical Research as an "Affected Industry"

The agency maneuvers and drafting of details of the administration's bill were only part of a far larger process of drafting and redrafting of recombinant DNA legislation in 1977 and 1978. All of the intensely contested details became academic by the middle of 1978, however, as the legislative thrust withered away. This section analyzes how an initially hot political issue disappeared in Congress. The analysis focuses on the role of the biomedical research lobby and its allies and the tools used to persuade Congress that legislation requiring certain forms of regulation, and ultimately legislation mandating *any* regulation, was unnecessary.

Only a few weeks after he had submitted the administration's bill, Kennedy revised it, entirely striking out the administration's text. The provisions of this new version of S. 1217 for regulation of genetic engineering were in stark contrast to those of the administration. In a move clearly designed to separate sponsorship of recombinant DNA research from its regulation, regulatory responsibility was vested in a freestanding, eleven-member commission appointed by the president. Of the eleven members, a majority were to be drawn from fields outside the biomedical research sector. In addition, Kennedy's revised bill weakened the preemption criteria, thereby making it easier for local communities and states to develop controls of their own.[31]

Thus by the spring of 1977, the legislation before Congress provided several models for regulating genetic engineering. Kennedy's revision of S. 1217 divided promotion of the field from its control, while the administration's bill, H.R. 6157, vested the DHEW with primary responsibility for policymaking, insuring that the NIH remained at the center of that process. In addition, absolute preemption of local controls (originally envisaged in the NIH draft bill) contrasted with varying degrees of preemption (as in Kennedy's or the administration's bill).

The response of the biomedical research community to the prospect of federal legislation was complex. Many scientists opposed *any* form of legislation. "I am personally opposed to federal legislation . . . and the fact that I think that such legislation will probably pass does not diminish my opposition. Even if we do not express opposition to such legislation openly, I still do not think that it is wise to favor it or support it," Donald Cox, professor of microbiology at the University of Massachusetts, Amherst, and chair of the Public Affairs Committee of the ASM, wrote to Harlyn Halvorson.[32] Helen Whitley, as we have seen, held similar views. "We believe that special enactment of special legislation to regulate recombinant DNA research is unnecessary and unwise," a group of cellular and molecular biologists wrote in an open letter to members of Congress in June 1977.[33] In the spring of 1977, however, many came to believe that some form of legislation would pass and that it was politically expedient to keep such measures moderate rather than to fight a losing battle. Philip Handler stated in his annual report to the National Academy of Sciences, "The principal reasons that many scientists acquiesce to passage of such legislation are (1) to terminate the feckless debate which has offered outlets for anti-intellectualism and opportunity for political misbehavior while making dreadful inroads on the energies of the most productive scientists in the field, and (2) to assure that no state or local government will adopt yet more stringent legislation or, indeed, ban such research entirely."[34]

In this phase of the issue, the ASM and its president, Harlyn Halvorson, emerged to play lead roles. Although neither Halvorson nor other ASM officials had been members of Berg's circle of colleagues, Halvorson had followed the genetic engineering issue closely from its inception, holding a meeting of the ASM's Public Affairs Committee on the issue before the Asilomar conference and appointing, as incoming president, a committee to review the 1976 NIH guidelines.[35] The ASM was becoming a political force to be reckoned with in Washington on biomedical research issues, and under Halvorson's guidance, for the first time in 1976, lobbied hard in Congress for a larger NIH appropriation. The result had been a $30 million increase in the budget of the National Institute of General Medical Sciences and a substantial increase in the budget of the NIAID.[36] Following Halvorson's participation in Fredrickson's meeting with biomedical researchers in February 1977, the ASM launched a lobbying campaign that exceeded any previous political role played by biomedical researchers. Washington politicians were used to dealing with scientists as advisers on funding and as expert witnesses. But on the recombinant DNA issue, biologists behaved, in the words of one congressional aide, like "an affected industry."[37] A close colleague of Halvorson, Frank Young, chairman of the Department of Microbiology at the University of Rochester, wrote to Halvorson that his efforts had "resulted in

Table 6.2   Nine Requirements for Recombinant DNA Legislation Proposed by the
American Society for Microbiology

1. That all responsibility for regulating actions relative to the production or use of recombinant DNA molecules should be vested in the DHEW.
2. That to advise and assist the secretary of health, education, and welfare, an advisory committee should be established whose membership, in addition to laypeople, should include representatives with appropriate technical expertise in this field.
3. That institutions and not individuals should be licensed.
4. That in each institution engaged in recombinant DNA activities, to the maximum extent possible, direct regulatory responsibility should be delegated to the local biohazard committee. These committees should include both members with expertise appropriate to the activities conducted at that institution and representatives of the public.
5. That experiments requiring P1 containment be exempt from these regulations.
6. That license removal is an effective and sufficient deterrent to obtain compliance. Further, ASM is opposed to the bonding of scientists or the establishing of individual strict liability clauses in the conduct of recombinant DNA activities.
7. That ASM goes on record as favoring uniform national standards governing recombinant DNA activities.
8. That the secretary of health, education, and welfare have the flexibility to modify the regulations as further information becomes available. Further, we support the inclusion of a sunset clause in the legislation—that is, that legislation will be reevaluated after a fixed period of time.
9. That ASM expresses its concern that in establishing such important legislation governing research, that this proceed only after due and careful deliberation.

Source: Resolution, Council of the American Society for Microbiology, 77th Annual Meeting, New Orleans, May 1977.

the establishment of the ASM as an important figure in public affairs and hopefully a major constructive force in science policy."[38]

In the spring of 1977, it seemed to Halvorson and his colleagues politically futile to oppose legislation. Instead, an ad hoc task force that included Frank Young, Roy Curtiss, Harry Ginsberg (chair of the Department of Microbiology at Columbia University's College of Physicians and Surgeons), Donald Cox, and Halvorson, drafted a statement of nine requirements supposed to ensure that legislation, if enacted, would be acceptable to the biomedical research community. David Baltimore, Bernard Davis, and Edwin Lennette (president of the Tissue Culture Association) all helped to shape this statement.[39] The requirements included vesting regulation with the Secretary of Health, Education, and Welfare, developing uniform national standards, elimination of strict liability clauses, and fixing a "sunset clause" to force timely reevaluation of the need for controls (table 6.2).

In shaping the ASM's strategy, Halvorson sought the advice of several experienced political consultants, including contacts in Americans for Democratic Action, the American Civil Liberties Union, and the American Bar Association,[40] one of whom recommended that he "devise a plan—decide what it was he wanted from Congress—and stick to it, not changing in mid-course . . . not to be arrogant, and not to embarrass congressmen . . . and to give them an 'escape clause'" if he asked them to change their minds.[41] Halvorson, who followed this advice carefully, began by assembling a coalition of scientific societies to support the nine principles. In May 1977, these were endorsed by the ASM's Council Policy Committee, then by its council at the annual general meeting in New Orleans,[42] to be subsequently published in *Science, Chemical and Engineering News,* and the *ASM Newsletter.*[43] Halvorson then moved to link up with many other organizations, including the Inter-Society Council for Biology and Medicine (ISCBM)—a committee he had helped form in 1976, comprising the executive officers of seven biomedical research organizations.[44] By June 1977, the ASM statement had the backing of the ISCBM as well as that of the huge Federation of American Societies for Experimental Biology (FASEB). Altogether, the coalition could claim to represent half a million scientists as well as to have the support of professional groups like the American Association of University Presidents, the Land Grant Association, and even the American Bar Association.[45]

Other organizations joined the effort to influence the course of legislation. The National Academy of Sciences passed a resolution drafted by thirteen of its members, calling the regulatory commission required by Kennedy's legislation "a wholly new and unfortunate departure" and arguing against legislative approaches that allowed local communities to draft their own controls. Carefully worded, the resolution was endorsed even by Robert Sinsheimer.[46] One hundred and thirty-seven scientists at the Gordon Conference on Nucleic Acids held in New Hampshire in June 1977 endorsed a statement expressing concern that the "benefits of recombinant DNA research will be denied to society by unnecessarily restrictive legislation" and depicting the legislative measures under consideration at the national, state, and local levels as "so unwieldy and unpredictable as to inhibit severely the further development of this field of research." If legislation was deemed necessary, the statement went on, it should be uniform and "carefully framed so as not to impede scientific progress."[47]

Several universities also started lobbying for federal preemption of local controls: Harvard, Stanford, Princeton, and Washington University (St. Louis) hired Nan Nixon to campaign on their behalf; another lobbyist, Donald Moulton, was independently employed by Harvard. Nixon and Moulton kept contact with an informal network of approximately seventy individuals (in over thirty-five universities) that acquired the sobriquet Friends of DNA. This group worked to ensure that any federal legislation would preempt state and local ordinances.[48]

In contrast to the scientific community, large corporations that were closely following the development of genetic engineering preparatory to making large investments in it, kept a low profile. Representatives of the pharmaceutical industry followed developments in Congress and visited members of Congress occasionally.[49] In private, they supported the efforts of the science lobby, indicating that their preference was for no legislation.[50] In contrast to the low-key activities of the large pharmaceutical corporations, the lobbying of at least one of the new genetic engineering firms, Genentech, was far more visible and vocal. (Washington rumors described them as "blanketing the Hill.") In 1977, three Genentech employees—Robert Swanson, the president; Herbert Boyer, its chief scientist, vice president, and cofounder; and Coralee Kuhn, an attorney—formally registered as congressional lobbyists.[51] According to one news report, Genentech opposed any legislation on the grounds that the costs of compliance would put the company out of business.[52]

Generally opposed to the positions of the biomedical research community and the emerging genetic engineering industry, a loose coalition of environmental and public interest groups—Friends of the Earth, the Environmental Defense Fund, the Sierra Club, and the newly formed Coalition for Responsible Genetic Research—looked to legislation as a means to redress the problems they saw with both the substance and the process of NIH policymaking. They saw legislation as a way to extend public involvement in national and local policymaking (and therefore opposed provisions for preemption), to achieve mandated risk assessment for genetic engineering, and to establish a commission to study the ethical, social, and legal implications of the field.[53]

Most efforts to influence Congress were registered in the press coverage of this period. What has not been recognized is the scale and composition of the network associated with the ASM lobby and organized by Harlyn Halvorson and its significance for the politics of biomedical research in this period. As the head of a huge scientific organization, with members in every aspect of biomedical research, Halvorson reached out, by phone and mail, not only to other biomedical research organizations but also to NIH officials, representatives of the Office of Science and Technology Policy, members of the pharmaceutical industry, prominent molecular biologists, university representatives, and scientists in the military sector. Through his contacts and those of ASM's Washington staff, Halvorson was able to map responses to developments in Washington and to mastermind the strategy that fought to delimit the legislation and eventually abrogated it. NIH director Fredrickson was frequently in touch with Halvorson by phone. So was Gilbert Omenn of the Office of Science and Technology Policy. Although Halvorson's telephone notes are often cryptic, they nevertheless show the two in basic agreement, particularly opposing Kennedy's legislation and weak preemption; they shared ideas and strategy, as well as opinions about other actors in this debate, in a frank manner. Fredrickson provided both information about developments within the administration and encouragement, often thanking Halvorson, cheering him "from the sidelines," and urging him on occasion to "hang in there tough." These contacts leave little doubt about the NIH position. There were also conversations with representatives of industry—scientists like Brinton Miller (Merck), Sidney Udenfriend (Roche Institute of Molecular Biology), and Marvin Weinstein (Schering-Plough), and industry executives like Irving Johnson (Lilly) and John Adams (vice president of the Pharmaceutical Manufacturers Association [PMA]). A record of a conference call with Adams and the Lilly legal counsel notes that the participants "reached general agreement on positions." Additionally, Halvorson prodded industry scientists in New Jersey to contact Harrison Williams, chair of the Senate Committee on Human Resources, presumably to oppose the Kennedy legislation. Telephone calls with Johnson revealed that, while representatives of the pharmaceutical industry did not "want to tangle" with the DNA issue because of their interests in legislation in other areas, they were supportive, hoping that Halvorson would "win this fight to help on others to come." Johnson also told Halvorson that he wanted to "see no legislation." Lilly (headquartered in Indianapolis) also helped to activate a meeting with Indiana legislators.

Not all of the scientists in contact with Halvorson agreed with his strategy: several strenuously opposed any legislation, however meliorated. James Watson told Halvorson that he was drafting an editorial for the *New York Times* and *Washington Post* attacking legislation. Waclaw Szybalski, David Baltimore, Norton Zinder, Stanley Cohen, and Joshua Lederberg all appear to have held similar positions and to have used their authority to lobby against legislation at various points. Their disagreements with Halvorson were collegial, however: these contacts were maintained over many months and did not appear to be unduly stressed by differences. Despite differences in strategy, these scientists shared a central concern: to prevent the transfer of policymaking control over biomedical research out of the biomedical research community. Whether and how legislation could be eliminated was a matter of emphasis. Indeed, some congressional aides recalled numerous scientists activated by the ASM network lobbying against legislation rather than for the ASM's nine principles.[54]

6.3 The Rise and Fall of Recombinant DNA Legislation

The interactions among the various DNA lobbies and between these lobbies and members of Congress and their staffs were immensely complex. Congressional politics, the jurisdiction of various committees, the ambitions and commitments of individual legislators, and simply the competition of other legislation played out in tortuous maneuvering on the texts of the DNA legislation as it wound its way through (and eventually out of) the House and Senate. Regardless of the fine structure of these interactions, however, there is little doubt about the larger picture. In the summer of 1977, the biomedical research community, with the support of other interested sectors—private industry, the universities— engaged in one of the largest lobbying efforts with respect to a technical issue ever experienced by members of Congress, and managed to reverse the legislative thrust.[55]

The efforts of Halvorson and the ASM were central to this victory. Using the authority of the biomedical coalition, Halvorson and his colleagues in the ASM campaigned to shape the legislation moving through Congress to their nine requirements. Halvorson's correspondence and phone log in 1977 and 1978 reflect the effort to circulate these widely and to use all possible contacts. In the House, Paul Rogers was seen as responsive to their position and Burke Zimmerman, the chief member of his staff, as someone the organization could work with.[56] Halvorson wrote to Charles Yanofsky in May 1977 that "we have played a major role in rewriting the Rogers bill, which now begins to look quite reasonable."[57] The bill that emerged from the House Subcommittee on Health and the Environment, H.R. 7897, conformed reasonably to the ASM's requirements. In any case, there was also a fallback position, if moderation of the bill was later seen to be necessary. At the suggestion of DeWitt Stetten, it was agreed that Congressman Ray Thornton could request sequential referral of the bill to his Subcommittee on Science, Research and Technology. Thornton was seen as strongly sympathetic to the biomedical research community, recalling later that his earliest concern was "to avoid precipitate action on legislation" and that even entertaining it demanded caution.[58] In the spring and fall of 1977, he held a long series of hearings on the science policy implications of genetic engineering, generating almost thirteen hundred pages of testimony from over fifty individuals and solidifying his committee's claim to review any bill regulating the field.[59] In the Senate, by contrast, the prospects for legislation acceptable to the biomedical research community seemed dim: Kennedy's bill seemed assured of approval; no hurdles appeared to block its passage as it sailed through the Subcommittee on Health and Scientific Research and through its parent committee, Human Resources.

It was at this point that arguments about "new evidence" were deployed in the political arena by the ASM and other scientific lobbies, with notable impact. The conclusions of the Falmouth conference, held on 20–21 June 1977, were written up by the chair, Sherwood Gorbach, in his letter of 14 July 1977 to Donald Fredrickson (section 5.3). Gorbach and Fredrickson communicated them to the ASM, as did Wallace Rowe.[60] Two weeks after Gorbach's letter, Halvorson circulated a general memorandum to all members of Congress, emphasizing that "new information" showed that "the risk involved in recombinant DNA research has been vastly overstated and that recombinant DNA molecules do not increase the potential hazard of any microorganism. The dangers involved in this research are no greater than those encountered when dealing with natural pathogens."[61] Halvorson later acknowledged that, "of course, the whole Falmouth conference was set up just to do that . . . to make that kind of information [on the inability to transform *E. coli* K12 into an epidemic pathogen] available to people who very shortly, within maybe weeks, were going to be voting."[62]

Within a week of the Falmouth conference, Halvorson together with Frank Young, Oliver Smithies (professor of genetics at the University of Wisconsin), Peter Day (chief of the Genetics Department at the Connecticut Agricultural Experiment Station), Tracy Sonneborn (professor of genetics at Indiana University), and Lawrence Bogorad (professor of biology at Harvard) representing the combined forces of the ASM, FASEB, American Society of Biological Chemists, American Association of Pathologists, and American Institute of Biological Sciences, met with Kennedy. According to the FASEB newsletter that month, Halvorson, as spokesman for the group, stressed the significance of the "new information, making many earlier fears groundless and pointing to important benefits," urging upon Kennedy as the best course "legislation [that would] simply . . . extend the applicability of the conservative NIH guidelines to all who use recombinant DNA technology." While indicating some flexibility on the question of the size of penalties for violation and disclosure of proprietary information, Kennedy stood firm on the need for a regulatory commission and on the right of local communities and states to develop their own controls. The FASEB newsletter reported that "there was no suggestion of a disposition to bend [on these issues]."[63] A few days later, on 22 July, the Committee on Human Resources reported the bill.

With Kennedy showing no signs of receptivity to the requirements of the ASM lobby, its members cast around for a committee member who would agree to oppose the Kennedy bill from the floor of the Senate—a search that, according to Halvorson's own account, was not easy, given the influence Kennedy commanded in the Senate. Gaylord Nelson of Wisconsin, with whom Halvorson and others had met on 18 June, was persuaded to take up the cause. Nelson, with the help of the ASM and the NIH director's office, drafted and introduced on 2 August on behalf of himself and Senator Daniel Patrick Moynihan a substitute amendment to S. 1217 (amendment 754)—a piece of legislation entirely in harmony with the ASM requirements. As justification for the amendment, Nelson cited the views of the scientific community that no recombinant DNA hazard had been demonstrated, that the guidelines deliberately erred on the side of caution, that everything that had been learned about recombinant DNA "tended to diminish our estimate of [its] risk," and that "under these circumstances, an unprecedented introduction of prior restraints on scientific inquiry seems unwarranted."[64]

Scientific lobbying on the legislation intensified, with visits to members of Congress, breakfast briefings, and letters and phone calls urging members of Congress to rethink plans to regulate genetic engineering in the light of the new evidence on the hazards. Assessing the credibility of the "new evidence" was not easy for members of Congress beset by many other issues. Senator Jacob Javits was reported to have reacted to the conflicting scientific arguments to which he was exposed by saying, "The problem is I don't know who to believe."[65] Under those circumstances, it was tempting to accept the conclusions of those who spoke most loudly and insistently. Some of the loudest voices claimed that scientists' "freedom" to pursue inquiry would be drastically curtailed by the proposed legislation, with especially shrill criticism aimed at Kennedy's bill.[66] On 26 June 1977, the National Board of Americans for Democratic Action adopted a resolution condemning congressional attempts to "control specific research activities of individual scientists through individual licensing and punitive action" and, drawing a parallel with the "inhuman practices in Nazi Germany [and] the state control of genetic experimentation in Soviet Russia," warning that "strict societal control of the activities of scientists is a primary step in the establishment of totalitarian states."[67]

To marshall support for Nelson's amendment, Halvorson in the last few weeks of August mounted an energetic and comprehensive telephone campaign matching ASM members with senators in each state. The outcome was impressive: by mid-September, Halvorson's phone log recorded encouraging head counts in the Senate and opinions of his Washington office that they had Kennedy "over a barrel." Lawrence Horowitz, Kennedy's staff manager, wanted to find a compromise. Kennedy was reported to have decided to change tactics.[68] On 22 September, a speech given by Senator Adlai Stevenson reinforced the changing sentiment in the Senate on the need for legislation. Stevenson argued that "recent evidence of the decreased risks associated with recombinant DNA research using *E. coli* K12 as the host . . . requires us to weigh carefully the benefits of the proposed regulations against their likely impact on the freedom of inquiry." He urged the Senate to enact interim legislation that allowed great flexibility in accommodating new scientific evidence. Roy Curtiss's letter of 12 April (discussed in section 5.4), correspondence with the director of the Office of Science and Technology Policy, and a report of the Congressional Research Service on changes in perception of recombinant DNA hazards—all of which pointed to "new scientific evidence" as reasons for reassessing the need for controls— were cited by Stevenson as the basis for his views.[69]

On 27 September 1977, Kennedy took the action anticipated by Halvorson and his colleagues. In a speech to the Association of Medical Writers, Kennedy announced that he would introduce new legislation aimed solely at establishing a commission to "study" the recombinant DNA issue, conveying by implication that he was withdrawing support from his earlier bill. Using the "escape hatch" provided by the scientific community—in this case Stanley Cohen's claim to have demonstrated that "by using this technique, scientists can only duplicate what nature can already do"—he argued that "the information before us today differs significantly from the data available when our committee recommended the pending Senate legislation."[70]

Kennedy's reversal was an important moment in the history of recombinant DNA legislation, demonstrating the power of the biomedical research community to retain control over regulating the field and to dictate the terms of technical discourse on hazards. However incomplete and questionable the arguments developed at the Falmouth conference and by individual scientists may have been, they were largely accepted by members of Congress. This, in turn, enhanced their legitimacy.

Initially, legislation in the House aroused less opposition because it was felt that, although some legislation was inevitable, H.R. 7897 could be shaped to conform to the ASM's nine principles. However, other sectors, especially environmental organizations, also had some influence on the House legislation. The bill, as reported by the House Subcommittee on Health and the Environment on 20 June 1977, contained several measures that allowed for some degree of public participation in policymaking. DHEW regulatory authority preempted local controls, as the biomedical research sector wanted, but the secretary's advisory committee was required to have a majority of its members not engaged in genetic engineering. Provisions for licensing of facilities and fines for violations were developed in detail.

As the Senate succumbed to the arguments of the biomedical research lobby, the latter found the House bill—or *any* legislation on genetic engineering—less desirable. Frank Young wrote to Robert Watkins, a staff member in the ASM's Washington office, on 10 August:

> There is considerable support for a more sane approach for recombinant DNA. I believe we are gaining ground and should not, in any way, relax our efforts. If progress is made to the extent that the Senate does wish

to consider more carefully the entire issue, it may be appropriate to raise the concept of a study commission rather than a law regulating recombinant DNA at this time. Therefore, we will have to be very sensitive to the issues as they emerge.[71]

The ASM lobby reevaluated its support for the Rogers bill as more attractive options, like Kennedy's study commission, came on the horizon. Weighed in the balance were both a recognition that ASM would need Rogers's support for other issues, particularly funding of biomedical research,[72] and the ongoing actions in local communities and states, which caused some to hold that federal preemption of local controls was a necessity. This was certainly the view of the Friends of DNA, who pressed throughout this period for strongly preemptive congressional legislation.[73] Halvorson and his associates in the ASM decided to "sit tight" (advice to Halvorson that came from Donald Fredrickson), to continue advocating its nine principles, and carefully to avoid endorsing *any* legislation—a position exemplified by Halvorson's editorial in *Science* on 28 October 1977, which evenhandedly praised the Nelson bill as "reasonable," described the Rogers bill as "accommodat[ing] most of the nine principles," praised Stevenson for his "statesmanlike" reaction to the "new evidence," and conveyed approval for Stevenson's call for interim legislation that permitted flexibility in accommodating to such evidence as it was developed.[74]

At the same time, other individuals and groups were lobbying hard against the Rogers bill. Stanley Cohen urged the chairman of the House Commerce Committee, Harley Staggers, to oppose the bill.[75] Also active were David Baltimore, who argued, with the backing of James Watson and Norton Zinder, that legislation was no longer necessary; RAC member Waclaw Szybalski; and Harvard president Derek Bok, who paid a personal visit to Harley Staggers to argue for the need to preempt local controls.[76] Kennedy, in an unusual move, met with Staggers in mid-October 1977 and reached an agreement that Staggers would block the Rogers bill in the Commerce Committee. Although it was known that Rogers had sufficient support for the bill, Staggers, citing Cohen's view that legislation would harm research, refused to report it to the House floor.[77] Significantly, the ASM did not try to revive the Rogers legislation or to dissuade Staggers from the position he had taken.[78] Staggers's action demonstrated the power of the scientific community to block even legislation that earlier it had deemed acceptable.

This was not the end of the DNA legislation: new bills were introduced in 1978, and Halvorson's log records the further moves of the various lobbies and congressional and administration offices engaged in the issue.[79] The principle bill considered on this second round—H.R.

11192, introduced by Staggers and Rogers in the House with a companion, S. 1217, amendment 1713, introduced by Kennedy in the Senate—showed the hand of the biomedical science lobby writ large: it merely extended the NIH guidelines for two years after enactment, vested authority in the Secretary of Health, Education, and Welfare, and provided protection for trade secrets. To those seeking more extensive public representation in policymaking, it conceded only a study commission appointed by the secretary to examine the long-term implications of genetic engineering. The House and Senate bills differed significantly on a single point: the House bill provided strong preemption of local controls; Kennedy's did not. The House bill made some progress: it was considered and reported by the Commerce Committee, and sequentially referred to the Science, Research and Technology subcommittee (as planned), which reported it on 21 April.

But a distinct change in the climate of opinion, including that of members of Congress, about the genetic engineering "problem" was initiated as scientists and companies engaged in the field began to promote its commercial "benefits." In the fall of 1977, research for Genentech by teams of scientists led by Herbert Boyer and Keiichi Itakura demonstrated the feasibility of expressing genes of higher organisms in *E. coli* bacteria (section 2.4). In a notable departure from standard scientific procedure, the results were announced before publication by Paul Berg and Philip Handler at the hearings held by Adlai Stevenson before the Senate Subcommittee on Science, Technology, and Space in November 1977. Handler, extolling a "scientific triumph of the first order," claimed that somatostatin would prove to have "diverse therapeutic applications." Berg hailed the "extraordinary progress towards the construction of organisms that make therapeutically useful hormones." The announcement of the somatostatin "breakthrough" was orchestrated to highlight politically both the commercial importance of genetic engineering and the need for the field to advance unencumbered by "overly restrictive" controls.[80] That no scientists complained of the departure from the traditional norm of refereed publication in a scientific journal that this event represented suggests the intensity of the scientific campaign to derail the legislative effort.[81]

The argument that "new scientific evidence" revealed genetic engineering hazards to be insignificant also gained ground. At the Stevenson hearing, it was emphasized in virtually all testimony from representatives of the biomedical research—except for critics of the NIH controls. Berg's public renunciation of his earlier caution marked the ascendancy of the new discourse: "More than three years [after publication of the Berg letter], after considerable discussion by experts in this country and abroad and the analyses of past experiences and new findings, I and others have changed our assessment of the risks. I now believe that the possibility that experimental organisms will be hazardous or released is exceedingly small."[82]

In addition, the argument gained the imprimatur of the National Academy of Sciences in a report of the academy's Assembly of Life Sciences, announced by Handler at the hearing: "Taken all together, the panel's conclusion seems entirely justifiable; research with recombinant DNA, performed under the NIH guidelines, offers negligible hazard."[83] Despite rebuttals from scientists such as Robert Sinsheimer and Jonathan King, who insisted that the epidemic pathogen argument and the Cohen evidence hardly absolved the expanding field of hazard, and despite embarrassing testimony that the guidelines had been circumvented, if not violated, by several leading practitioners, the sense emerging from the hearing was that genetic engineering was too important, commercially and scientifically, to be held back by regulation. As Handler asserted: "Cumbersome and punitive legislation is not needed. The financial cost of overly cautious containment and enforcement, the delay in achieving benefits and the penalties incurred by restricting freedom of inquiry are real risks to be considered in setting up regulations."[84]

By the time that the report of the House Subcommittee on Science, Research, and Technology appeared in March 1978, the argument for the need for rapid promotion of the "benefits" of genetic engineering was clearly in ascendancy, as biohazards continued to be played down: "Immediate benefits appear to be more imminent than risks which have been hypothesized although there are so few data available at this time that no absolute conclusions can be derived. . . . types of research currently permitted under the recombinant DNA research guidelines do not seem to pose significantly greater risks than natural diseases routinely confronting the medical community; in many instances they appear to present less risk." The report went on to propose that "the burden of proof should [not] be borne only by proponents of this area of investigation." A subtle change in the discursive practices established at the Asilomar conference was being proposed: the burden of proof was being shifted from scientists, to show that genetic engineering was safe, to the public, to show that it was dangerous.

Obviously sensitive to the preferences of those preparing to make major investments in the field, the report's conclusions reminded readers that "research in this area is going forward rapidly in other nations" and cited the testimony of the PMA vice president, John Adams, that severe restrictions might force relocation of industry elsewhere. "Any significant U.S. lag in genetic technology, for whatever reasons it may occur, could result in diminished power and prestige for this nation on the international scene," it warned. The advent of commercial prospects for genetic engineering moved questions about control of the field firmly into the context of international competition.[85]

By this point, ASM's strategy of shaping federal legislation to its nine principles was being questioned by many close to the organization. Halvorson was told that the principles were out of date, that legislation was unnecessary. Scientists such as Norton Zinder, Arthur Kornberg, and James Watson were reported to be lobbying against the new legislation before the House and Senate. Some in the administration even preferred no bill to the one it had tailored. At the Stevenson hearing, the director of the Office of Science and Technology Policy, Frank Press, was grilled by the chair about the administration's position. The exchange revealed the tentative nature of that support, as well as the expectation that "new evidence" would prove legislation unnecessary.

> STEVENSON: So far I have heard you say that the administration stands behind the administration's bill as if to say that you are not now opposing the administration's bill. You are urging the Congress to support this bill, are you?
> PRESS: As of now, we support this bill.
> STEVENSON: And as of tomorrow?
> PRESS: Well, that's the whole point isn't it, that the data, new data are appearing all the time.[86]

In the end, the bill reported by the two House committees, H.R. 11192, the result of heroic efforts on the part of Rogers's staff to achieve a workable compromise, pleased no one. As concerns about regulation by state and local legislation receded, members of the biomedical research community inside and outside the government increasingly opposed *any* legislation. The study commission, desired by critics of NIH policy, made some scientists nervous that public attention to the issue would continue for several more years. Even those seeking a change in policy saw H.R. 11192 as token legislation doing little to distribute policymaking power since it allotted nearly all of it to the Secretary of Health, Education, and Welfare. Six members of the House Commerce Committee voted against the bill on these grounds.[87]

A final sticking point was the division over preemption, which the House bill incorporated (to the specifications of the biomedical research community), while the Senate bill did not. Whether Kennedy did not wish to engage in a losing legislative debate or whether other reasons were decisive is unclear. Publicly rethinking an issue as volatile as recombinant DNA could not have been politically expedient for a possible presidential candidate. In any event, on 1 May at Kennedy's suggestion, five members of the Senate Committee on Human Resources and Adlai Stevenson decided to take no action on legislation but rather to request the Secretary of Health, Education, and Welfare, Joseph Califano, to regulate genetic engineering under existing statutes. A letter, giving the "new evidence" as a reason, was dispatched to Califano on 1 June 1978. It signaled the disinterest of Kennedy and his colleagues in pursuing legislation and clearly reflected the changed climate of opinion about the need for congressional action. By proposing Califano as the appropriate person to take action, they served notice that the Senate would allow the recombinant DNA legislation that had made its way through the House to die. Not unexpectedly, no action was taken by Califano.[88] After that, congressional interest in the issue all but evaporated. H.R. 11192 disappeared into the House Rules Committee and was never brought to the floor. In January 1980, Stevenson introduced legislation (S. 2234) requiring notification of recombinant DNA activities not funded by the NIH to the Secretary of Health, Education, and Welfare, but with Stevenson's impending retirement in the following November and no support elsewhere in Congress, the bill went nowhere.[89]

6.4 The Political Impact of the Legislative Defeat

The demise of the federal legislation conclusively demonstrated the power of those who joined forces against it to retain control of the direction and management of recombinant DNA technology. The intricate maneuvers and countermaneuvers among members of Congress, the administration, and the various lobbies all played a role in the rise and decline of legislation. But those with a major stake in the development of the field—the biomedical research community, the emerging genetic engineering industry, chemical and pharmaceutical corporations, and the universities—clearly established their ability to dominate the legislative process.

The question of which sectors played decisive roles in the congressional reversal is more difficult to answer. Even though the interests noted above were distinct, they were united in their opposition to any broadening of public control of biomedical research and powerfully reinforced one another in this regard. Harlyn Halvorson's telephone network reveals a great deal about this convergence. The ability of leaders of the biomedical research community to orchestrate a grassroots movement was clearly crucial in persuading Kennedy's colleagues on the Senate Human Resources Committee, and eventually Kennedy himself, to back away from the Senate legislation. But more moderate legislation might still have been adopted. What seems to have produced a turn toward rejection of *any* legislation was the claim in the fall of 1977 that conclusive proof that genetic engineering would provide the basis for a new industry was in. The announcement of the somatostatin "breakthrough" at the Senate hearing in November 1977 confirmed the commercial significance of the field. Berg's reversal of his earlier position of caution at the same hearing provided powerful backing for the claim that the field could be both profitable *and* safe—a claim that echoed through congressional hearings and reports from that point onward.[90] The argument, and implicit threat, from leaders of industry and science that a new field would be held back and its "benefits" squandered if legislation was enacted was influential. No legislator wanted to be accused later on of blocking a potentially valuable source of international trade.

Whether the discourse concerning the hazards of the field, constructed at Bethesda, Falmouth, and Ascot, was instrumental primarily in providing a convenient excuse for backing away from support for legislation or more substantively in persuading legislators that controls were far less urgently needed than the majority at Asilomar had judged is more difficult to assess. Regardless of the precise interpretation of the role of the new discourse in the struggle over legislation, the demonstrated ability of biomedical researchers to control and ultimately to eliminate legislation greatly reinforced the discourse, opening up the possibility that it could be successfully deployed in other arenas without serious concern about further congressional intervention. At the same time, the ability of public advocacy groups and trade unions to argue for regulation was greatly weakened.

PART FOUR
Implementing Controls

SEVEN

Revising the National Institutes of Health Controls, 1977–1978

7.1 The Social and Political Setting

By the fall of 1977, the prospects for congressional legislation were fading, and by the spring of 1978, they were dead, defeated by the determined lobbying efforts of the biomedical research sector with the quiet support of the NIH and the private sector. As we have seen, "new evidence" that claimed to show that genetically manipulated organisms were much less threatening to public health than had earlier been feared was a key element of the arguments against legislation. So too were the claimed rigor of the NIH guidelines and the ability of scientists to regulate themselves.

As congressional interest in recombinant DNA legislation faded, other interests in the field grew in influence. From the fall of 1977 onward, recombinant DNA research began to yield demonstrations of the field's commercial potential. Although many practical problems remained to be overcome at this stage, bacterial synthesis of somatostatin, insulin, growth hormone, interferon, and other products made headlines and began to attract major investments by multinational corporations. As the techniques were further enhanced and as headlines conveying a sense of dizzy progress appeared regularly, interest in the field intensified. For a short period, from 1979 to 1981, the amount of corporate investments and the number of venture capital companies grew exponentially, to reach $800 million and one hundred, respectively. Simultaneously, many of the leading scientists in the field developed ties with private industry as equity owners, executives, advisers, and consultants: Paul Berg (DNAX), David Baltimore (Collaborative Research), Stanley Cohen (Cetus), Herbert Boyer (Genentech), Walter Gilbert (Biogen), Stanley Falkow (Cetus), David Botstein (Collaborative Research), Ronald Davis (Collaborative Research), David Jackson (Genex), and others (chapter 2).

In addition, use of cloning techniques was spreading fast to other nations. By the end of March 1978, the Committee on Genetic Experimentation (COGENE) of the International Council of Scientific Unions estimated that there were at least 367 recombinant DNA projects under way in 155 laboratories in fifteen countries.[1] The accelerated pace of development and the global spread of cloning activities were accompanied by an intensified sense of competition and a fear of falling behind, which worked at all levels of science, from the laboratory to the federal government.

Two conditions exacerbated these fears. For one, national controls for genetic engineering were not altogether uniform. While most countries were following either the British code of practice or the American controls, some controls were noticeably lower by this time. For shotgun experiments with primate DNA, for example, the French controls required physical-containment levels two steps lower than those required in the United States.[2] The Swiss procedures were voluntary.[3] Individual scientists feared that their own work would be outdistanced by scientists working under less strict controls in other countries. "It is painfully obvious," Stanley Falkow wrote to Donald Fredrickson in April 1978,

> that because of the very restrictive nature of the NIH guidelines, as well as the bureaucratic wall that the guidelines have spawned, American biologists can no longer expect to keep pace with either Western Europe or East European science. . . . It is difficult for me to convey to you the intense discomfort many of us felt [at international meetings in April and May, 1978] when we heard the results of reasonable, totally safe experiments being described by Western and East European scientists that are now literally forbidden to U.S. scientists.[4]

And Stanley Cohen wrote to Fredrickson in May 1978: "Work in non–*E. coli* K12 organisms is proceeding at a rapid pace in Eastern Europe and Western Europe; however, unless the wording of the [guidelines] is changed, current work of great medical and social value will halt in the United States."[5] Such statements not only expressed the fears of those engaged in recombinant DNA research that their own work might be rendered obsolete by progress in other countries under less restrictive conditions, but also placed considerable pressure on the U.S. government to weaken the existing controls so that recombinant DNA research with a wide variety of organisms could progress rapidly. Similar pressures were exerted by international organizations such as EMBO (section 7.3).

In the second place, within the United States, the fact that the NIH controls applied only to NIH grantees was creating a double standard between scientists in the universities and those in industry. NIH grantees had to apply to the NIH for approval of genetic engineering experiments, as well as for approval of reductions in containment. Industry scientists needed only to seek the approval of their own biohazard committees. This difference in procedure could mean months of delay for NIH grantees relative to their colleagues in industry.

By 1977, such delays had become immensely significant as teams of researchers competed for patents on specific processes. Bacterial expression of rat proinsulin and human insulin were announced by teams of researchers at Harvard and Genentech in June and September 1978, respectively: the race to achieve commercially viable production processes was on.

Pressures on the NIH to relax controls were sometimes exerted in very concrete ways. Some scientists were rumored to be voting with their feet.[6] By October 1978, at least five research groups, including ones at Harvard and Cold Spring Harbor, were reported to have moved their cloning activities to countries with weaker controls. That academic researchers were going to this trouble suggests the extent of the competition. William Rutter, who moved the cloning experiments of his research group at the University of California, San Francisco, to France, stated bluntly, "The impetus that has resulted in [the transfer of some experiments to Europe] is that there is now an international race and I wouldn't say that we're ahead in many aspects of it."[7]

Far less subtle was the pressure for relaxation exerted by one of the new genetic engineering firms, Genentech. By the fall of 1978, Genentech, having demonstrated the feasibility of bacterial production of insulin and having signed a major contract with Eli Lilly, was eager to initiate a pilot project for large-scale insulin production. On 9 October, Genentech began to threaten privately to disregard the NIH requirements and procedures in order to proceed with plans for a pilot insulin project (section 9.2). The implication was that, if the NIH did not relax its requirements, small firms not bound by them might begin to flaunt them.

The intensified competition among individual laboratories and corporate enterprises also translated into national terms. At this level, concerns focused on the impact of strict controls on commerce and on the U.S. lead in advanced technology. "In all deliberations regarding recombinant DNA research, the nature of its conduct and possible regulations thereof, American authorities must bear in mind that research in this area is going forward rapidly in other countries," warned the report of the House Subcommittee on Science, Research and Technology in March 1978.[8] The report noted that the price of strict controls might be the transfer of private industry abroad. And it warned that the costs of such losses should not be viewed merely in terms of the loss of industry and employment to other countries but also of damage to the nation's balance of trade. The report concluded:

This concern is not merely with industrial relocation, but also with the U.S. balance of trade generally. Further, it is a situation which raises concern beyond industrial research to academic research. Any significant U.S. lag in genetic technology, for whatever reasons it may occur, could result in diminished power and prestige for this nation on the international scene. Such a possibility must not be overlooked as considerations for dealing with the recombinant DNA research issue go forward. (61)

"Falling behind in recombinant DNA technology" had become a major theme of U.S. policymaking.[9] From this point on, it was never far from the considerations of politicians and of policymakers at the NIH, in the White House, and elsewhere in the U.S. government.

The beginning of the commercial development of genetic engineering coincided with the beginning of important changes in government policies for both the promotion and the control of technology. In 1972–76, congressional interest in protective and participatory controls for technology was high, and the wave of legislation aimed at protecting the workplace, the environment, and consumers from the side effects of technology had not yet passed. Although business and science leaders had begun to organize to attack these policies, their efforts had not yet taken effect. In the period that followed, the business sector first formulated and then actively pursued reduction of protective controls and initiation of new forms of government support for technological development in the private sector. As this campaign gained ground, the emphasis of the president and the Congress changed from protecting against the effects of technology to protecting its rapid development. This change in emphasis, which culminated during the Reagan administration, began to take effect during the Carter administration. The strains and contradictions of the latter, with its commitment to environmental and workplace controls on the one hand, and its interest in cutting their costs on the other, meant that a change of policy was only partially pursued and realized. Nevertheless, the fact that recombinant DNA technology promised new products and increases in productivity in the chemical and pharmaceutical industries meant that sectors of the Carter administration as well as the Congress were committed to ensuring that realization of the commercial potential of biotechnology was not delayed by controls that were significantly more stringent than those of other nations.

The tight containment and oversight requirements of the 1976 NIH guidelines—the legacy of Asilomar—continued in force until the end of 1978. But by 1982, a remarkable reversal of the original policy meant that little remained of the original controls and that the philosophy supporting them had been overturned. This process was accomplished in two main phases. In the first phase, completed by December 1978, a relatively mild round of revisions reduced containment precautions and oversight provisions but conserved some of the main features of the initial policy: restriction of cloning to certain certified organisms; exclusion of six types of "prohibited" experiment; oversight at the local level by institutional committees and at the national level by the NIH. In the second phase, from 1979 to 1982, far more drastic changes were effected.

This chapter examines the first phase in which this reversal of policy was initiated, focusing on how this change was made acceptable in the eyes of various attentive sectors (the biomedical research community, private industry, Congress, and various advocacy groups) and how it was represented to the wider public. In shaping policy for recombinant DNA technology, officials at the NIH and DHEW performed a delicate balancing act, serving their primary client, the biomedical research community, while ensuring that policy decisions were palatable to, or at least tolerated by, other interested parties, particularly Congress and the private sector, and managing the decision process so that criticism of policy from advocacy groups could be contained. The biomedical sector functioned in the NIH arena as in Congress as an "affected industry," pressing for major reductions in containment, attempting to weaken the authority of public advocacy groups, and often reducing complex perceptions of hazard to the question of epidemic potential. In this process, arguments claiming a "new consensus" on the hazards of genetic engineering and arguments for the "benefits" to be reaped from the field were influential in justifying changes in policy.

In the continuing controversy over the hazards posed by genetic engineering, a complicating factor was the changing technical ground of the debate. Use of genetic engineering made possible the discovery in 1977 of "introns"—the sequences of bases in the DNA molecules of higher organisms that did not encode proteins and that were removed during the transcription of DNA into RNA. This discovery revealed a natural safety factor, namely, that it was very unlikely that genes of higher organisms or those of viruses infecting higher organisms would be expressed in bacteria by accident. On the other hand, much of genetic engineering research was aimed specifically at overcoming such natural barriers to expression.

A second complicating factor was the continuing lack of firm evidence about the hazards. No harm from the techniques had become evident and no epidemics had swept research laboratories and surrounding communities, but since no records of laboratory infections were being kept, it was impossible to know in detail what the effects thus far might have been.

Furthermore, in the absence of the results of risk assessment experiments, arguments about the existence or nonexistence of recombinant DNA hazards depended mainly on theorizing that could be bent to suit

the predilections of their proponents. The observation of the participant
at the Ascot meeting that arguments about hazards took on the aura of
an "Aristotelian academy" might well have been applied to other features
of the debate.

In addition, the efficacy of controls depended crucially on maintaining
high standards of laboratory practice. Since most laboratories did not fall
under a regulatory code that required inspection, no systematic evidence
on laboratory standards had been collected, although anecdotal evidence
and two documented cases of violation of the NIH controls suggested
that standards were not uniform and in certain cases were slack.[10]

Arguments that the field would yield benefits were similarly highly
uncertain at this stage, if only because major technical problems were
still to be solved before genetic engineering could yield an alternative
technological base for industry (section 2.5). That claims of benefits were
subjected to so little critical questioning by the scientific community was
probably due to at least two factors. First, many scientists at this point
were establishing ties with the emerging genetic engineering industry,
and the industry needed financial backers. Second, since one of the arguments used to justify development of the techniques without stringent
regulation was their very potential to yield benefits, it was difficult to
suggest that these might not materialize. The resulting lack of informed
commentary on the technical problems still to be solved did little to
moderate the media excitement that began to characterize portrayals of
the future of genetic engineering in this period.

7.2 Revisions Proposed, 1977

In 1977, while scientists, scientific organizations, and universities battled
with members of Congress over the recombinant DNA legislation,
pressures to revise the NIH guidelines mounted. While the guidelines
were not strict enough to satisfy their critics (because they allowed expansion and proliferation of recombinant DNA techniques without prior
assessment), they did restrict research considerably. Many experiments,
including those that used human DNA, could be done only in high-
containment facilities not available to most universities; many experiments could be carried out only with the weakened strain of *E. coli* K12,
χ1776, and a limited range of plasmid and phage vectors—and χ1776 was
so weak that some found it frustratingly difficult to work with. Many
experiments could not be done at all. As use of recombinant DNA techniques spread—and this happened rapidly after 1975—the range of desired experimental work also expanded. Cloning human and viral DNA,
developing new host-vector systems such as *Bacillus subtilis, Pseudomonas,*
and *Salmonella,* and inserting foreign genes into whole animals and plants
all attracted scientific and commercial interest. Yet the 1976 guidelines

either banned much of this work or made it difficult and time-consuming to get permission to do it; in addition, the high containment required for some research was prohibitively costly.

Some political pressures worked in the opposite direction, however. As long as Cambridge, Princeton, the state legislatures of New York, Maryland, and California, and the U.S. Congress were making noises about regulating recombinant DNA research, the NIH and its RAC were under substantial political pressure to demonstrate that their recombinant DNA policy was in the public interest. The claimed rigor of the NIH guidelines was used to prove that scientists could best regulate themselves. Furthermore, in the previous years, few defenders of NIH policy had argued that recombinant DNA work might not entail some special hazards. The case for the safety of the research was largely built on the argument that the precautions required by the guidelines—physical containment, use of safe technique, and restriction of work to approved strains of *E. coli* K12—were sufficient to contain possible hazards. The restrictive NIH guidelines, in other words, symbolized the scientific community's willingness and capacity to pursue gene splicing safely.

Politically, therefore, the NIH was subject to two contradictory forces in the early part of 1977: to avert legislation, the guidelines had to appear rigorous; but to enable work to advance and to compete effectively with similar work elsewhere, the rigor of the guidelines had to be reduced and new experiments permitted. The first revisions may be seen as the outcome of these conflicting tendencies. Revisions began to be drafted early in 1977 by a subcommittee of the RAC chaired by John Littlefield. The records of that process show that both the possibility of passage of legislation and perceived needs to press forward with research were principal concerns.

As the RAC began to draft and debate revisions to the 1976 NIH guidelines, the arguments aimed at defusing public concern considered initially at the Enteric Bacteria Meeting in Bethesda in August 1976 began to be deployed to support the position that recombinant DNA hazards were much less than previously supposed. Following the Falmouth conference, a chorus of voices proclaimed that *E. coli* K12 was much weaker and far safer than previously supposed and was incapable of producing epidemic disease. As chapter 6 showed, to oppose legislation, influential sectors of the biomedical research community united in endorsing these arguments. Roy Curtiss's claims, written in April 1977, that *E. coli* K12 posed "no known or expected threat" and consequently that proposed legislation represented a case of "overregulation" of science, were characteristic.[11]

Opponents of recombinant DNA legislation divided, however, on the question of whether the NIH guidelines were themselves a case of "overregulation" of science. Some opposed legislation by arguing that existing guidelines were sufficient to ensure safe conduct of recombinant DNA research. In a letter in March 1977 to John Littlefield, Curtiss urged that the rigor of the guidelines be maintained if not increased. He proposed that revisions include requirements for separating hospital facilities from research laboratories that used recombinant DNA, monitoring of laboratory workers, and isolating of work with approved strains of *E. coli* from work with more robust strains to minimize the possibility of contamination—all precautions that were not addressed in the original 1976 version of the guidelines. His reasons were at once technical and political. In addition to containing the genetically manipulated organisms, rigorous controls would also be more likely to contain the congressional urge for legislation. "In view of the impending legislation," he wrote, "I think it is particularly important that we do the best and most responsible job possible." [12]

Many of the RAC members charged with deciding on guideline revision may have been persuaded initially by this line of reasoning. An early draft of revisions, circulated in May 1977, even slightly increased precautions for some types of experiment. [13] But as the arguments based on the "new evidence" developed at Falmouth gained currency, others argued that it was time to weaken the guidelines. Stanford biochemist and former RAC member David Hogness wrote to the NIH director:

> It is unfortunate that these bills are being considered at a time when there is a gathering unity of scientific opinion that the risks of recombinant DNA research have been grossly overestimated. This unity is based on a considerable body of data, much of it obtained within the last year or so, that has acted to dramatically shrink the range of speculated hazards—to the point where the remaining risks, if they exist at all, are certainly less than those resulting from research on pathogens already in our environment. . . . I hope that decisions to reduce containment and/or control in the light of data and experience indicative of lowered risks are not delayed by some conventional wisdom that one should not tamper with the guidelines while regulatory legislation is under consideration. [14]

The proposed revised guidelines released by the NIH and RAC for public comment in the fall of 1977 may be seen as a compromise between the "conventional wisdom" of Curtiss and the more extreme position advocated by Hogness—a compromise shaped by the general understanding that interest in legislation would surely be reinforced if the guidelines were radically weakened at that moment. [15]

Compared to later changes in NIH policy, the proposed revisions were disarmingly small. Some containment levels—notably those for the cloning of primate and mammalian DNA and of some animal viruses—were reduced by one level; cloning of DNA for species that were known to exchange their genes naturally with *E. coli* was exempted; local safety committees were given authority to approve one-step reductions in containment in cases where the DNA could be shown to be purified and free of harmful genes; and finally, procedures were developed for certifying new organisms as "host-vector systems" for cloning work. Although each of these changes was small in itself, taken as a package the revisions represented a step toward opening up the field to a greater range of experimental work. They made it feasible to work with purified human DNA at the P2EK2 level and to develop new cloning organisms. Work with viruses, on the other hand, was still largely out of bounds, except for laboratories equipped with P3 facilities.

The justification of these revisions followed much the same lines as biomedical researchers were successfully using to scuttle legislation: "new information" showed that recombinant DNA hazards were much less than the scientists at the Asilomar conference had feared. According to the introduction to the proposed changes, "everything we have learned tends to diminish our estimate of the risk associated with recombinant DNA in *E. coli* K12."[16] A document circulated by the NIH in the fall of 1977 showed that what had been "learned" was primarily the information about *E. coli* generated by the Falmouth conference and Cohen's and Chang's claimed demonstration of "natural" gene splicing (section 5.4).[17]

These claims about the significance of the "new evidence" were optimistic and sweeping. The real state of scientific opinion on the hazards of cloning was much less settled: most of the "new evidence" was still unpublished. A small circle of scientists closely associated with NIH policy had already concluded that it provided justification for accommodating the individuals and institutions who were pressing for reduction or elimination of governmental oversight. But the normal process of achieving consensus, through publication, discussion, and debate, and further experimental and theoretical work had barely begun.

7.3 The Director's Advisory Committee
Meeting, December 1977

Following the procedures he had established for earlier decisions on the NIH controls, in December 1977, the NIH director held a hearing before his twenty-two-member advisory committee, seventeen of whom participated.[18] Also attending the hearing were six members of the RAC and twenty-four witnesses—some invited and some not—who were allowed to make five-minute commentaries on aspects of the proposed revisions. Of these participants, approximately half were biomedical researchers and others whose generally supportive positions with respect to the NIH controls were well known to the NIH. Approximately twelve were known to hold positions critical of NIH policy. In addition to a range of biomedical sciences, members of the advisory committee were drawn from law, ethics, and public policy. Their occupations included an environmentalist, a technician, and a specialist in occupational safety.

In his introductory remarks, Fredrickson described the hearing as a means to "preserve due process and provide for full public participation" (212). The generation of information at this hearing did indeed appear to be participatory in the sense of representing a broad spectrum of positions. The control and use of this information was much less participatory, however. The agenda and timing of agenda items were controlled by the organizers. Speakers not designated to present the NIH proposals were limited to five-minute presentations and abruptly cut off. Committee members were given no time to deliberate among themselves. The "new" data used to support the revisions were still largely unpublished and unavailable. Even Peter Hutt, known as a warm supporter of NIH policy, complained that the proposed revisions had been brought forward with "undue, unnecessary, and unseemly haste" (476). Inevitably, the hearing produced little more than a debate. Decisions on how to use the information generated were taken by the NIH director, without collective advice from his advisory committee.

Much of the discussion turned on familiar issues. The position of critics of NIH policy was essentially unchanged, not only because the new information was largely unavailable for scrutiny but also because, even if accepted, it had in their view little relevance to the main problems they saw in NIH policy. They continued to emphasize the uncertainties in the guidelines' basic assumptions. Robert Sinsheimer, for example, charged that the guidelines were anthropocentric (in focusing on risks to human health rather than to the environment as a whole), parochial (in assuming the advanced sanitation and sewage treatment available only to technologically advanced nations), and based dangerously on an "extraordinary confidence in the completeness of our knowledge in microbiology and in our ability to predict the consequences of scaling things up by factors of 10^{10}" (480). Critics argued again that oversight that was voluntary and did not require uniform training and monitoring of workers was woefully inadequate. They continued to press for comprehensive assessment of the hazards of the technology as well as for fundamental changes in NIH policymaking structures, particularly the need for representation at every level of policymaking by the sectors affected by recombinant DNA technology. They persisted, in other words, in a vision of a policy and of mechanisms for policymaking for a new technology that differed radically from the course set by the NIH, and even more radically from the new course contemplated.[19]

Supporters of NIH policy, on the other hand, portrayed the proposed revisions as cautious, responsible, and scientifically supported. The "new evidence" generated by the Falmouth meeting and by the Cohen and Chang experiment was used repeatedly to justify the claim that the hazards were much less than previously assumed. Bernard Davis pointing to this evidence, charged that the risks had been "enormously exaggerated," that the danger of epidemic spread of disease was "virtually nil," and that the dangers of infection of laboratory personnel were "very much less than the danger of working with everyday pathogens." The experiments of Cohen and Chang, Davis claimed, showed that "a million DNAs" had "entered bacteria many times in the past" (224–26).

Since information on the Falmouth meeting was unavailable, discussants had no way of evaluating evidence claimed to have been considered there. On the other hand, some debate was possible on the implications of the Cohen-Chang results, which had been published in the *Proceedings of the National Academy of Sciences* in November 1977 and circulated in advance of the hearing.[20]

Cohen and Chang claimed to have shown that pieces of "foreign" DNA and of bacterial DNA joined naturally in living bacteria (in vivo) as well as artificially in the test tube (in vitro). Cohen's procedures raised some questions: for example, he had to treat the bacteria with calcium salt and heat shock. Was that a "natural" process? Was the frequency of the event he had claimed to demonstrate anything like the natural frequency? A most critical question was whether Cohen had actually done what he had claimed to do, namely to demonstrate in vivo recombination of "foreign" and bacterial DNA. The following exchange between Sinsheimer and RAC member Donald Helinski did much to reveal the dubious experimental foundation of Cohen and Chang's result:

> SINSHEIMER: [What Cohen and Chang did] was to take a plasmid containing a piece of mouse DNA, cut that plasmid with Eco RI, so that you had some Eco RI fragments, take some other Eco RI fragments from another plasmid, mix those, provide them to a bacterium which had been treated with calcium and heat shock, and had in it Eco RI, and find that lo and behold! you have got a plasmid put back together from those two pieces inside that microorganism. . . .
>
> HELINSKI: That is not entirely accurate as a description of the experiment. This was an *in vivo* cleavage and ligation cutting and joining of DNA and you described it as something that you did outside the cell and had the cells take up.
>
> SINSHEIMER: I don't think you are right. . . . There were *in vivo* recombinations, but the ones with eukaryotic DNA were *in vitro* mixtures added to the cell. (328)

It was an illuminating moment. Significantly, references to the Cohen-Chang experiment—the "crucial" experiment providing an escape clause for Kennedy—were dropped from all later NIH policy documents.

On the basis of much the same information, the Senate Subcommittee on Science, Technology, and Space concluded in August 1978 that, while the 1976 guidelines adequately provided for physical and biological containment, they were inadequate with regard to provisions for oversight of work and deficient in not covering activities in the private sector.[21] At the director's advisory committee hearing, in contrast, a much less cautious view of the issue, one that supported laissez-faire development of the technology, appeared to be gaining ground.

Bernard Davis, for example, argued that scientists had been "victims . . . of a gross misunderstanding" of the potential impacts of recombinant DNA technology, and that the revisions, which he considered modest, should be followed by further relaxation (224). RAC member Waclaw Szybalski urged that all experiments with *E. coli* K12 host-vector systems should be excluded from the NIH controls (379). Stanford biologist Allan Campbell, also a RAC member, insisted that there was no "valid theoretical reason . . . for worrying about the danger of recombinant DNA" (444). The climax came when James Watson appeared to apologize to society for having signed the Berg committee's letter. The letter, Watson charged, was an artifact of liberal guilt and fear of cancer: there was no evidence that recombinant DNA could do anyone any harm; the guidelines had been an "enormous diversion" from the main mission of the NIH; and it was time to give them up completely (438–42). MIT molecular biologist Alexander Rich commented later that Watson was "speaking for a large group of scientists."[22]

Symptomatic of this shift in the center of the controversy was a marked change of response to the question of who should bear the burden of proof on hazard. No one at the first hearing before the director's advisory committee in 1976, and few before that except for Watson, had argued against the need for controls on the grounds that the hazards were unknown. Now, Watson energetically defended this position, as shown in the following exchange with environmentalist Karim Ahmed:

> WATSON: Will anyone in this audience stand up and show one piece of evidence, one item from the literature, in which recombinant DNA is dangerous? . . .
>
> AHMED: The same question could have been posed 30 years ago on asbestos. We didn't have evidence then.
>
> WATSON: But until someone gets sick, you can't fight against it. You have to have a reason. There are so many things one can get worried about, we have to limit ourselves to those things for which there is evidence. (442–43)

In contrast to the reception of Watson's views at Asilomar, policymakers and colleagues alike were now listening—and becoming persuaded. Support for weakening the NIH controls was also reinforced by the presence at the hearing of Sir John Kendrew and John Tooze, director general and executive secretary, respectively, of EMBO. One main point of their testimony was that several European countries were pursuing recombinant DNA work under less restrictive controls than those in the United States. Tooze, for example, maintained (inaccurately, in this case) that this was so in the United Kingdom.

> HUTT: Now is there a significant amount of experimentation that is so characterized under our Guidelines—not just the current ones, but the proposed new ones—that requires less stringent EK or P containment abroad, and therefore that can be done abroad at this time, that can't be done here? . . .
>
> TOOZE: In the United Kingdom primate and other mammalian shotgun experiments are category three, physical containment.
>
> HUTT: And they are [physical containment level] four here?
>
> TOOZE: Yes. . . . You can argue whether category three is greater than the American P3, which it is, but it is certainly not as stringent as American P4, which is extremely difficult to achieve.
>
> HUTT: Now, is research going on in that area in England?
>
> TOOZE: To be sure, yes.
>
> HUTT: Well, that answers that.
>
> TOOZE: Very much so, yes. I mean, someone [in the United Kingdom] is busily cloning human X chromosome DNA, absolutely. Why not? That is what we wish to do. (349–50)

The message was clear. It was repeated several times during the hearing and relayed the next day to the general public by the *New York Times:*[23] if the 1976 guidelines were maintained, the United States would fall behind in a field of research that promised important practical applications. In the policy decisions that followed, the claimed need to compete with the rest of the world was never far from the center of attention.

7.4 The Position of Private Industry, December 1977

News coverage of the second hearing before the NIH director's advisory committee focused mainly on the still lively debate about the NIH controls and on the case for revision and the procedures being used by the NIH for that purpose. The positions of Sinsheimer and Watson were used as exemplars of critics at either end of the spectrum, and it is probably indicative of the changing climate of opinion that it was now Watson's position that stole headlines: "Discoverer Would Lift Curbs on DNA," the *Washington Post* announced.[24]

Little attention was paid to representatives of private industry attending the hearing, perhaps because it was the NIH policy, not the industry position, that was the continuing focus of controversy. Before, during, and after the meeting, however, industry representatives pressed for important changes in the NIH controls, and their position would prove influential.

From the outset, as we have seen, industry representatives had taken the position that they would prefer to see voluntary rather than mandatory controls. But they had also emphasized their reluctance to provide details of their procedures and host-vector systems and their need to be assured that "proprietary information" would be protected. This need was regularly aired by industry representatives at congressional hearings and meetings with NIH officials.

By the time of the second hearing, a new factor had entered the picture. Industry officials were interested in developing new cloning hosts. As Ronald Cape, president of Cetus, stated at the hearing, "[W]e do have a specific interest in these . . . [since] they may [provide a] favorable . . . means of getting some practical use out of recombinant DNA techniques and there also may be safety factors." [25] They could also foresee that they would want to use cultures of genetically engineered organisms in volumes much higher than the ten-liter limit permitted by the 1976 NIH guidelines. [26] As Cape insisted at the hearing, "[The NIH provides] no review process regarding our plans, our protocols, our containment, and so forth. . . . we can't . . . officially interact with or enlist the assistance of our government in adhering, as we are committed to do, to the NIH Guidelines. What we want is diplomatic recognition." [27]

However, as John Adams, a vice president of the PMA, demonstrated later in the hearing, industry reluctance to reveal trade secrets to the RAC was as strong as ever: "[Host-vector systems] could be considered and probably will be considered industrial property rights." Adams made clear that scrutiny of those systems by the RAC was something private industry preferred to avoid. Under questioning from Harold Ginsberg, Adams conceded that the use of two or three consultants would be acceptable if they were subject to criminal penalties for revealing trade secrets. [28]

A few days later, the industry goals were developed more fully at a meeting held at the headquarters of the PMA, attended by John Adams, representatives of three leading pharmaceutical houses (Lilly, Upjohn, and Hoffmann–La Roche), Ronald Cape, and representatives of the Department of Commerce, the NIH, and the Office of Science and Technology Policy. The meeting had been called to respond to questions issued by Senator Adlai Stevenson to the Department of Commerce concerning the details of voluntary compliance with the NIH controls by private firms.

A memorandum on the meeting written by one of the Department of Commerce participants recorded two issues of preeminent concern. In the first place, as always, was the issue of protection of proprietary information "either as trade secrets or as the basis for later patent applications." The second concern was "the development of a monitoring system capable of maintaining close contact with the private laboratories performing RDNA research, in order to collect sufficient data regarding protocols, containment systems, and research results to provide a basis for surveillance decisions, while at the same time guaranteeing maximum protection of confidential technical data."[29] Further, it was agreed that, "if properly constituted, [institutional biohazard] committees can probably be used as a substantial part of the monitoring operation as well as provide a means of communication between a company and the community where it is located."[30] The example of "an apparently successful biohazards committee" at the Upjohn Company in Kalamazoo, Michigan, composed of six Upjohn executives and three prominent members of the local community—a physician, a clergyman, and a biology professor—was cited. The memorandum went on to observe that

> it was pointed out that picking of [the members] of a committee must be discreet and selective, but must assure the presence of the highest type of persons, who will make sure that the public interest is properly served. Any implication of scientific elitism must be avoided. Selection of a biohazards committee by a company . . . is likely to bring accusations that the regulated is picking his own regulator. The existence of biohazards committees poses also the problem of safeguarding . . . proprietary information; there are no legal restrictions [on] the release of this information by irresponsible committee members. Upjohn has attempted to solve the problem by judicious selection of members and by setting up bylaws stipulating the protection of confidential information.[31]

Having agreed on the desirable qualities of a "successful" biohazards committee, the group assembled at the PMA headquarters went on to consider the nature of sanctions needed to ensure effective compliance. They recognized that penalties would be difficult to enforce. However, they agreed that "exposure of wrong-doing and consequent adverse publicity" as well as "possible exposure of confidential information" would be a "sufficient deterrent for most potential non-compliers." Clearly, the participants in this meeting believed that increased reliance on local biohazard committees should be a crucial element in the development of a voluntary compliance scheme. Their views proved to be influential in subsequent revision of the guidelines.

7.5 Cloning Viral DNA: The Original Problem Reassessed

Indicative of the pressures to relax restrictions on recombinant DNA research was criticism directed at the proposals for revising the 1976 guidelines for not permitting or facilitating a wide range of cloning experiments involving viral DNA. Paul Berg's plan to clone SV40 tumor virus DNA in *E. coli* had prompted some of the original concerns about genetic manipulation and had led to the Berg committee's letter, the moratorium, the Asilomar conference, and ultimately to the NIH guidelines. The latter required high containment (P4, and in cases where the hazard was considered less extreme, P3) for two main uses of animal viruses: the cloning of animal virus DNA in *E. coli* and the use of animal viruses as vectors to carry "foreign" DNA into animal cells. The revisions of the 1976 guidelines proposed in September 1977 reduced these requirements by at most one step.

Toward the end of 1977, these restrictions were proving increasingly frustrating to those who wanted to clone animal virus DNA. Paul Berg, for example, in the fall of 1977, questioned the necessity for P3 or P4 for the cloning of animal DNA in *E. coli*. Animal DNA was known to contain cryptic viruses, and for shotgun experiments with mammalian DNA, the proposed revisions required P2EK2. Why P3 or P4 for viral DNA, Berg asked, if the DNA of a cryptic virus might be cloned at P2?[32] Berg was supported by Ann Skalka of the Roche Institute of Molecular Biology and by David Baltimore (now a major shareholder in Collaborative Research), both of whom challenged the need for high containment for work they wanted to pursue in the cloning of DNA from retroviruses, a type of virus known to cause cancer in animals. This work was being "seriously hampered," Baltimore complained. "The availability of clones containing fragments of retroviral genomes would have advanced the field enormously already and are still badly needed." Both Baltimore and Skalka acknowledged that the use of these viruses was not free from hazard because of the known connection with cancer. "I of course recognize that these clones containing retrovirus DNA may present special concerns and I am not suggesting that they should be made at will," wrote Baltimore.[33] But to enable the work to go forward at all, he proposed that it be allowed at P3EK2. Skalka went further. "Why the special concern?" she asked. Using the argument developed at the Enteric Bacteria Meeting and promulgated at Falmouth, she argued that the hazard, if there was a hazard, would affect only laboratory workers, not "the public."

> As far as I can tell, [the special concern about this category of experiment] relates to a scenario in which DNA of an animal tumor is presented to a cell (presumably via *E. coli* K12 infection of the gut) in a form for which it has no natural resistance. Since it already seems clear that the *E. coli* K12 carrying a DNA insert cannot become an epidemic pathogen (especially if it is an EK2 strain) the only potential hazard is to the *investigator* and not to the *public*. Therefore, the basis for concern and conservatism in this versus other types of viral research seems to have disappeared.[34]

The proposed containment levels for work using animal viruses as vectors to transfer foreign DNA into animal cells were also criticized. Berg claimed that there was no difference between the use of a phage as a vector to transfer foreign DNA into *E. coli* and the use of SV40 to insert foreign DNA into monkey cells. Political, rather than scientific, considerations, he suggested, had restricted this work.

Berg concluded with a proposal that the "entire question" of appropriate containment for the cloning of viral DNA be reexamined by "an appropriate group of experts," assembled to "make recommendation on this matter for the next cycle of revisions in the guidelines." [35] Additional weight was added to Berg's argument by John Tooze, who told the director's advisory committee in December 1977 that the Europeans considered the American containment requirement for the cloning of viral DNA "hopelessly, excessively stringent" and that "the Europeans do not go along with this naive overstringency." The Europeans, according to Tooze, were going to go their own way. And he pointed to the example of France, where the cloning of viral DNA was already under way. He also proposed a joint European-American effort to reconsider the containment levels for viral DNA. [36]

Tooze's proposal that a meeting to reconsider virus containment levels should be a joint American and European venture was evidently accepted. Just over a month after the director's advisory committee meeting, the U.S.-EMBO-sponsored workshop, chaired by Malcolm Martin, Wallace Rowe, and Tooze, was held in Ascot, England.

The express purpose of this meeting was to reassess the hazards of cloning viral DNA. Martin introduced the proceedings: "I want to stress that you are here as competent and expert virologists representing nobody except yourself, and what we're trying to come up with here is . . . the best scientific judgement, what we can say about some of the potential risks of using viral DNA inserts in various categories, various types of recombinant DNA experiments." [37] As one of the participants put it later, the meeting was to be a "virological Falmouth." [38]

As chapter 5 relates, despite the immense complexity of the issues addressed, the fact that the main risk assessment experiments in this area had not yet produced results, and unresolved differences among the participants about what should count as the central issue (hazards to individuals versus hazards to the community), the participants, under the strong influence of the Falmouth results, concluded that hazards to surrounding communities, at least, would be insignificant (section 5.3).

The assessment of hazard performed at the conference was intertwined with a more immediate political goal, namely, the desire of the organizers to take back to the United States recommendations for revised containment requirements. Although this purpose was not stated explicitly,

the organization of the second part of the meeting around the development of a statement making containment recommendations meant that
it was implicit in the meeting's structure. One participant later recalled:
"It was fairly clear by the end of the meeting that Rowe, Martin, and
Ginsberg wanted to go back with a result that could be exploited for
deregulation. On the other hand, they were probably scientifically right.
Nevertheless, one felt that one was under pressure to consent."[39]

The recommendation eventually produced at the Ascot meeting, as
reported in the *Federal Register,* was that the cloning of viral DNA should
use P2 as a minimum level of physical containment with an EK1 host-
vector system. Higher levels of containment should be used where these
were required for handling the virus under question. The report also
emphasized the need for training in safe laboratory practice.[40] However,
whether these proposals represented the consensus of all the participants
at the Ascot meeting, and even the meaning of *consensus* in this case, is
debatable. In the first place, approximately half of the twenty-five scientists left before the end of the meeting.[41] It is doubtful in any case that
the European scientists saw themselves as responsible for recommendations for research requirements in another country. Second, calls for
some level of caution concerning such matters as the production of viral
gene products or the cloning of whole viruses seem to have been largely
set aside. For example, proposals for the use of physical containment one
level higher than that used to handle the virus itself and for the use of the
weakened EK2 strains of *E. coli* did not survive the discussion.[42] Finally,
the stringency of these recommendations depended critically on how
they would be implemented. In countries like the United Kingdom and
Sweden, where laboratory work was regulated by comprehensive occupational health laws, such recommendations would carry the force of
law, and laboratory workers would participate in their implementation.
In countries like the United States, where laboratory work was unregulated, the same recommendations would have the status only of advice
to research institutions, and the hierarchical structure of the laboratory
meant that training and emphasis on good technique would be a function
of the attitudes and values of laboratory directors. Nevertheless, these
proposals were represented as the outcome of a new expert consensus
and taken back to the United States (as planned) as evidence for reducing
restrictions on the cloning of viral DNA.

At this point, the Ascot recommendations underwent a further muta-
tion. A Virus Working Group of twenty virologists, chaired by Harold
Ginsberg, was convened to review the Ascot report and to translate its
findings into recommendations to the RAC. This group recommended
that the Ascot recommendations (which were already lenient compared
to the requirements in the United Kingdom and the United States) be

decreased even further. Many of the recommended containment levels for insertion of virus DNA into *E. coli* were changed from P2 to a modified P1 that included a ban on mouth pipetting.[43]

The Virus Working Group's proposals, which enabled much virus cloning to be conducted on the open bench, were defended to the RAC by participant Wallace Rowe.

> I feel that it is now time for the RAC and NIH to make it very clear that quality microbiological technique is indeed expected and demanded of grantees working with recombinant DNA, and that it is the responsibility of the institution itself to ensure that such practices are followed. NIH cannot oversee this itself; it must rely on the institution to assume this responsibility. Further, we as a committee cannot be in a position of requiring all researchers to use unnecessary high levels of physical containment because we feel that a certain proportion of researchers may be incorrigible. *We must not wind up penalizing the careful and trained workers.* We only lose credibility and the trust of the bench scientist.[44]

The containment-level reductions that Rowe and others were seeking were justified with reference to the Ascot recommendations as reasonable rewards to a scientific community demonstrably capable of policing itself and maintaining respectably safe procedures. The primary goal, as one of the leading proponents of relaxation of controls saw it, was to avoid "penalizing" properly trained workers and properly run laboratories because of incompetence or irresponsible behavior elsewhere. Protection of laboratory workers, whether in properly or improperly run laboratories, was not seen by Rowe as the principal responsibility of the NIH.

7.6 Making the Changes: Initiating a Policy Reversal

By April 1978, the prospects for the enactment of legislation were dim, and the new consensus on genetic engineering hazards was rapidly gaining acceptance. International conferences such as that held in Milan in March 1978 reinforced and transmitted this new consensus (section 5.6). At Milan, calls for caution, influential only a year earlier, were replaced with calls for lifting the fetters from research. The *real* disease, insisted chair Luigi Cavalli-Sforza, was an "epidemic of fear" of the unknown that had spread widely among scientists. The *real* danger, insisted Waclaw Szybalski, was that recombinant DNA research would be stifled by repressive regulation.[45] Concern for potential hazards of research and industrial application was replaced by visions of practical "benefits" in the form of industrially important enzymes, vaccines, and cheap, pure human insulin.[46]

This change in the climate of opinion was also reflected in the report of the House Subcommittee on Science, Research and Technology in

March 1978 (sections 7.1 and 6.3). Pointing to the economic importance of the new technologies likely to be based on recombinant DNA techniques and to the rapid progress of research in other countries, the subcommittee concluded that "the burden of proof of safety should not be borne exclusively by proponents of recombinant DNA research." "Opponents," the report affirmed, "must now assume a corresponding burden."[47]

In this new climate, so different from that of only a year earlier, the RAC met in Bethesda at the end of April 1978 to advise on the final form of the revised guidelines. The tone of the meeting was set by the NIH director, who introduced proposals for further revision of the guidelines. In his opening remarks, Fredrickson compared the meeting's importance to that of Asilomar, but his argument indicated that a new consensus had developed. Emphasizing the conclusions of the Falmouth, Ascot, and other meetings convened to readdress the hazard problem, he asserted that "no evidence has come to light that the thousands of DNA experiments or application of DNA techniques over the last five years have yielded any product harmful to man or his environment. . . . No new scientific evidence not considered in the promulgation of the guidelines has emerged to support the fears that the techniques will create a harmful product." Falmouth had shown that *E. coli* K12 could not be converted to an epidemic pathogen, and Ascot that the hazards of cloning viral DNA were not greater and would usually be much less than the hazards of handling the virus itself. Echoing the conclusion of the House Subcommittee on Science, Research and Technology, Fredrickson put forward the major change of principle that would influence all future decisions on recombinant DNA controls: "We are approaching a point where the burden of proof is shifting more and more to those who would restrict such activities" on the grounds of potential hazard. The RAC, it was clear, was now expected to reverse, not continue, the policies born of Asilomar.[48]

In a move not detailed in the proposed revisions to the 1976 guidelines and not addressed at the 1977 director's advisory committee hearing, but certainly responsive to the needs of private industry, Fredrickson asserted that it was also "the consensus that the locus of responsibility for use must now shift toward the institution conducting this research." He went on to explain that "the present requirement for approval by the NIH before an experiment can proceed has caused considerable delay and more importantly not delays that are justified by any proof that safety has been enhanced as a result of this practice." Implying—at least partly inaccurately, as we shall see—that procedures and containment requirements were less stringent elsewhere in the world, Fredrickson insisted that there was "no factual basis on which to defend the greater stringency of the United States controls."

Fredrickson left the RAC with no doubt as to what he wanted to see. In addition to the revisions already proposed at the director's advisory committee hearing, he placed particular emphasis on several far-reaching changes. He urged, first, that the guidelines be redefined to exempt certain classes of experiment involving organisms judged to exchange DNA naturally; second, that authority for approval of experiments be transferred from the NIH to the institutions involved in recombinant DNA work (that is, to the local biosafety committees); third, that containment levels for experiments involving virus DNA be reduced; fourth, that containment levels for experiments involving plant pathogens be reduced; and fifth, that the committee contemplate provision of a formal mechanism for demonstrating voluntary compliance with the guidelines on the part of the private sector.

Fredrickson claimed that the proposed changes had been "clearly mandated" by the director's advisory committee, although the committee as a group had not addressed some of the changes he proposed and had not made any collective recommendations. The committee heard critics who called for continued caution and increased representation of nonusers of recombinant DNA technology in decision making, and users of the techniques, who largely called for weakening containment requirements and oversight. Fredrickson's proposals largely spoke for the latter.

Throughout, a sense of urgency informed these directives to the RAC. Fredrickson warned that other countries were doing experiments that were not permitted in the United States and that, unless the guidelines were revised, they were likely to be cast into "rigid standards" by the passage of legislation. Noting the possibility of approval of legislation in the House, Fredrickson urged the committee to act on his recommendations in order to avoid the conversion of the 1976 guidelines into "an undesirably stringent and . . . much too complex set of regulations." As the committee's chair later put it, "speed is of the essence because there seems to be an advantage in getting guidelines into the *Federal Register* before passage of legislation."[49]

The RAC proceeded to approve, with little resistance and with generally minor modifications, the proposals Fredrickson had brought to it. Permissive criteria were adopted for exempting classes of experiment from the guidelines on the grounds that they involved only exchanges of DNA that also took place naturally. Even critics of NIH policy conceded that some classes of research could be exempted—for example, the combining of DNA from a single species. Other criteria for deciding what took place "naturally" proved to be much more contentious. If Stanley Cohen's claim to have shown that all recombinant processes were "natural" had been widely accepted, for example, *all* recombinant processes would have deserved exemption, but the majority of RAC members did not go that far.[50]

With minor changes, the proposals of the Ginsberg-Martin working group on containment levels for animal viruses were accepted. Thus containment levels for cloning viral DNA were largely set at P1 and P2. A major consequence of the RAC's decision was to lift the ban imposed by the 1976 guidelines on oncogenic viruses classified by the National Cancer Institute as "moderate risk"—a decision defended on the grounds that the National Cancer Institute list was "outdated" and "conservative."[51] Requirements for recombinant DNA activities using plant and insect pathogens were similarly relaxed.

As a consequence of its decisions on the cloning of viral DNA, the RAC readdressed the question of containment requirements for the shotgun experiments with primate DNA. Containment requirements for these experiments had previously been set by the 1976 guidelines at P4EK2 or P3EK3, and the proposed revisions contemplated P4EK1 or P3EK3. Having decided that the cloning of viruses posed little danger, however, the committee concluded that the inadvertent cloning of animal viruses associated with experiments with primate DNA also posed little danger. It voted unanimously to reduce containment to P2EK2.[52]

The RAC also approved two important procedural changes proposed by Fredrickson, thus making notable concessions to the users of recombinant DNA technology while overriding the criticisms and apprehensions of those who drew attention to the demonstrated weaknesses of the NIH procedures. First, the committee approved Fredrickson's proposal to transfer authority for approval of experiments to institutional biohazard committees—now renamed "institutional biosafety committees" (IBCs). The group discussed the need for community representatives on these committees, but it finally recommended by a slim 7–5 majority that one or more public members and one or more technicians should be *suggested* rather than mandated participants.[53]

Second, the desire of the private sector for a "voluntary compliance" arrangement with the NIH controls (to which, of course, it was not formally subject) was accommodated. The RAC agreed to review requests from the private sector for such matters as certification of new host-vector systems and exceptions from the guidelines and to protect confidential information. This was a significant step, involving the RAC members in criminal liability for release of such information. It was not taken without a certain airing of reluctance on the part of some who noted that the private sector wanted NIH approval but was unwilling to make available information needed to assess hazards. This proposal was a preliminary move by the NIH toward an arrangement that responded to the needs of both the NIH and the private sector: it simultaneously provided a bulwark against legally mandated controls and reduction of the risk of liability suits, should some hazard be blamed for causing injury.[54] No immediate action was taken. As it turned out, the private sector wanted even more effective protection of commercially sensitive information (section 7.7).

Only days after the RAC meeting, any concerns that Congress might act on legislation were finally relieved. On 2 May, Harlyn Halvorson's telephone network—including ASM members, officials at the White House, and officials at the NIH—was buzzing with the news of the meeting among aides to Kennedy, other members of the Senate Health and Scientific Research Subcommittee, and Senator Stevenson at which it had been decided to toss the issue back to the administration, at least symbolically. For now, the NIH associate director for planning and evaluation, Joseph Perpich, advised Halvorson that action on legislation should be put "on ice."[55] Confirmation of this change of mind soon followed. On 2 May, Kennedy wrote to Fredrickson, addressing only the concerns that the 1976 guidelines might be holding back "scientifically and medically important work" and that the United States might be "falling behind" other nations in the recombinant DNA field.[56]

Meanwhile, the NIH revisions moved forward. Justified in the terms in which Fredrickson had proposed them to the RAC, a new version of the proposed revised guidelines based on the changes approved by the RAC at its April meeting was published in the *Federal Register* for a further round of comment in July 1978.[57] The proposals represented an important turning point in NIH policy for containing recombinant DNA hazards. Reductions of containment requirements opened up research to a much larger circle of users. Lowering the containment required for cloning mammalian DNA made it much easier and cheaper to pursue work with the genes that coded for human growth hormone, insulin, interferon, and other substances deemed to be of primary commercial, medical, or scientific interest. Transfer of authority for initiation and oversight of experiments from the NIH to the institutions doing the research removed the delays inherent in the requirement for prior review and approval and accelerated the pace of development. Procedures for NIH approval of new host-vector systems prepared the way for dramatically expanding the scope of genetic engineering. Whether the proposed changes were justified or not, they responded largely to the changed perceptions and interests of users of recombinant DNA technology.

In addition, the proposed revisions paid some attention to the interests of the private sector by indicating a scheme providing for voluntary registration and certification of projects that allowed private industry to demonstrate its compliance with the guidelines.[58] The language of this section gave little reassurance that trade secrets would be fully protected, however. "Proprietary information," the section stated, might be released to the extent "necessary in the judgment of the Secretary of

DHEW to protect the public or the environment from unreasonable risk of injury to health or the environment" and under several other conditions. Companies were advised to consider applying for patents before submitting information to DHEW, a procedure that would have protected the NIH but was certainly seen as unsatisfactory by the private sector, since many of the NIH procedures involved the entire RAC.

In contrast, the proposed revisions responded minimally to the substantive and procedural recommendations of critics of NIH policy. A requirement that one member of the IBC should not be affiliated with the institution was added. But proposals to maintain strict controls until firm evidence against possible hazards was achieved and to mandate oversight of laboratory standards, training of employees and investigators, health monitoring, and representation of workers on IBCs were ignored.

7.7 Revisions Released, December 1978

When the proposed revisions were published, Secretary of Health, Education, and Welfare Joseph Califano announced, in a somewhat surprising move, that he would hold one further round of public hearings before a panel composed of DHEW general counsel Peter Libassi (chair), NIH director Donald Fredrickson, and two assistant secretaries of health, education, and welfare. The panel also held two private meetings, with representatives of the PMA and of environmental and other public interest groups.

By this point, the prospects for legislation were becoming ever more remote. That the NIH guidelines would remain the only form of public control of genetic engineering was ensured, moreover, when Califano closed the door on the possibility, proposed earlier in the summer by Kennedy and his colleagues in the Senate, that the department should apply existing law to control use of recombinant DNA technology in the private sector. In a response to the senators on 12 September, Califano cited "new scientific information on the safety of *E. coli* K12" as the reason for the lack of need to act and returned the regulatory ball to the senators' court. "The department does not intend to invoke existing statutory authorities to regulate DNA activities at this time and continues to support legislation if it embodies the moderate approach of [the House Bill]," he stated. Further, he strongly implied that regulation of any kind was not his first choice: "We are pleased with the progress made in the absence of legislation and believe that invocation of existing authorities, however appropriate, would not contribute materially to our objectives."[59] Thus the NIH guidelines were the focus of all interested parties. Califano's schedule of meetings was clearly intended to hear from all sides.

The September hearing and the correspondence associated with it exhibited a by now familiar clash of positions. In general, leaders of biomedical research welcomed the revisions but pressed for further ones. Harlyn Halvorson, for example, spoke approvingly of the generation of "essential, valuable" information by the expert groups that met at Falmouth, Ascot, and elsewhere, of the increase in the responsibilities of the IBCs (which the ASM had proposed as one of its nine principles), and of the exceptions made to the guidelines. On the other hand, he also noted that research with organisms classified as "class 3" by the Center for Disease Control was still prohibited and implied it would be desirable if research with these organisms could be initiated.[60] Paul Berg, now a consultant for DNAX Corporation, welcomed the revisions in general but recommended a more streamlined decision process with increased use of expert groups and decreased opportunity for public notice and comment: "I believe the process is already public enough in as much as the Director's Advisory Committee and the RAC have public members and meet in full public view [and] changes in the guidelines are made in close consultation with these groups. . . . Must every technical decision be the subject of the same interminable review process now in progress?"[61] Others, however, held that the proposed revisions did not go nearly far enough. David Baltimore, now a consultant for Collaborative Research, expressed his hope that "early abandonment of the guidelines or replacement with simple, good sense, safe microbiological procedure will be a matter for discussion in the near future."[62] Ronald Davis, also a consultant for Collaborative Research, claimed that it was unclear that the proposed revisions would alleviate the stifling of creativity imposed by the 1976 guidelines.[63] With this stand being increasingly taken by others, the position of James Watson had become mainstream.

Critics of the NIH policy continued to press for an alternative that was anticipatory and protective rather than reactive in thrust.[64] A principal defense of the guideline revisions was the argument that no hazards had emerged over the preceding six years of gene splicing and that this experience suggested that the technology was low risk. Critics claimed that this argument was logically flawed. According to Jonathan King, the argument confused the safety of handling organisms containing foreign genes under strict containment conditions with the safety of the organisms themselves. It was, claimed King, "formally akin to arguing that since iodine[125] can be handled safely, there is little risk associated with iodine[125]; therefore we can relax the procedures for handling iodine[125]."[65]

Critics also attacked both the quality of the evidence used by the NIH to justify revisions and the procedures used to collect this evidence. The Natural Resources Defense Council pointed out that the published proceedings of the Falmouth conference appeared only shortly before the revisions were issued, that the full proceedings of the Ascot meeting had not been released, and that the Rowe-Martin results, cited in the director's decision document, were incomplete and unpublished.[66] King faulted the NIH for its "heavy reliance on unrefereed, unpublished, unexamined reports of small appointed committees, meeting in private, in complete contradiction to the canons of the scientific process and the governmental process."[67] Representatives of the Natural Resources Defense Council insisted that, since the NIH still had not instituted a comprehensive risk assessment program, it was hardly surprising that it had found no evidence of hazard. "The failure to find evidence of harm primarily reflects the NIH's failure to look," they concluded.[68]

In addition to questioning the quality of the evidence used to support revision, critics asserted that this evidence was highly selective. "The most relevant observations which form the basis for the need for containment in the first place [are absent]," wrote King. He cited the NIH's claims for the safety of *E. coli* K12 without a corresponding discussion of the pathogenic properties of related wild strains, and its citation of the Falmouth conclusion on the inability of *E. coli* K12 to produce an epidemic without a corresponding discussion of its potential to transfer plasmids to wild strains. "The proposed revisions," King concluded, "are premature. They are not based on data or positions gathered under the normal canons of scientific investigation, modeling, and verification."[69]

Equally flawed, in the critics' view, were the institutional structures and procedures by which the NIH made and implemented its recombinant DNA policy. Critics returned repeatedly to what had been, all along, a major theme of their position, namely, that participation in policymaking was largely restricted to the sector most interested in promoting the technology. According to Philip Bereano, a lawyer and professor in the University of Washington's Social Management of Technology Program, NIH decision making via the RAC was dominated by an "old-boy network"—the people most often consulted on policy issues were precisely those scientific colleagues of theirs most heavily involved in pursuing recombinant DNA research.[70] Critics noted that growing industrial interests in recombinant DNA meant that the RAC would be called on increasingly to make policy recommendations with far-reaching social implications. Fredrickson's proposal to "broaden [the committee] modestly as needed for expertise" and "to include, perhaps, a dissenter from NIH policies" was seen as an inadequate response to those responsibilities. "The statement provides the appearance of liberality without its substance," Bereano commented.

The composition of the IBCs bore out this criticism. Pointing to preliminary results of a Stanford University study that showed that the proportions of workers, students, and members without ties to the institutions doing the research averaged less than 2 percent of the total and that almost 70 percent of the membership was drawn from scientists and administrators closely connected to fields using recombinant DNA techniques, Bereano called the IBCs "hopelessly skewed, demographically."

Consequently, critics claimed that the proposed revisions, in shifting further responsibilities from the NIH to IBCs, would perpetuate the existing bias. Since documented violations of the guidelines had occurred during the previous two years at Harvard Medical School and the University of California, San Francisco, they suggested that it was inappropriate to enhance the responsibilities of local committees. "NIH proposes a form of enforcement which institutionalizes conflict of interest and provides even less accountability for the Director of NIH and the IBCs than the previous guidelines," wrote Marcia Cleveland and Louis Slesin of the Natural Resources Defense Council.[71]

At a substantive level, opponents of the revisions called for retaining the containment levels required by the 1976 guidelines until recombinant DNA hazards were more precisely known; they also proposed that revision be tied to the results of a comprehensive risk assessment program, and called for strengthening of the 1976 controls through mandatory monitoring of the health of laboratory workers, mandatory training programs, and NIH oversight of safety standards.[72]

Procedurally, they proposed that the NIH give those affected by the technology a voice in decisions. They proposed, for example, that the RAC include representation of labor unions, environmental organizations, and other public advocacy groups, that IBCs include representation of laboratory technicians and of citizens from local communities, and that, at each level, measures be taken to prevent the control of decision making by those with interests in promoting the technology.[73]

That these proposals and criticisms would sway NIH policy, in the face of the death of the recombinant DNA legislation and the fading of interest in developing community controls, seemed unlikely. The NIH, bolstered by strenuous support for the revisions and pressure for even further reduction of controls from the biomedical research community, had little need to compromise on its proposals. By October 1978, an article in the *Boston Globe* predicted that the changes proposed in July would soon be put into effect.[74] However, critics hoped that they might still be able to alter the procedures by which policy was made, if not the content of policy. They were supported by Senator Stevenson, who sent a strongly worded response to Califano's letter of 12 September, supporting many of the critics' proposals and noting that he would consider introduction of further legislation in the new congressional session.[75]

Further airing of positions on the proposed revisions occurred a few weeks after the September hearing, at meetings called by the DHEW general counsel Peter Libassi with representatives of the PMA and of public interest organizations. By the time of the PMA meeting on 13 October, corporate interest in moving rapidly with large-scale cloning projects was heating up. Three days after this meeting, NIH officials learned that Genentech intended to proceed with its pilot insulin project under P1EK1 containment (an action that informally violated the NIH controls) later that month.[76] Genentech's maneuvers were of course intimately connected with the issues of the ten-liter limit and the desire of corporations for a "voluntary compliance" scheme that both demonstrated their compliance and protected their proprietary information. The existence of the ten-liter limit placed some pressure on the NIH to oversee commercial applications; otherwise it was vulnerable to the charge that it had no authority to ensure that industry complied with its requirements.

An important obstacle to resolution of the problem, which was recognized in different ways by the NIH and the PMA, was the prospect that information submitted to the NIH for review would be subject to the Freedom of Information Act, which provides for release of information held in government records to the public.[77] Although the act exempted trade secrets from disclosure, the criteria for determining precisely what might be exempted and what released were vague. Much discretion therefore rested with the NIH in determining how much industry information, submitted on a voluntary basis, could be released. Clearly, the scope of this discretion could become the subject of lengthy and expensive litigation. At the October meeting, PMA representatives emphasized that the imposition of the ten-liter limit was a "timely issue" because "production of insulin is imminent," and that they doubted that, under the arrangements proposed for the revised guidelines, the industry would "go to full disclosure."[78] The NIH director, for his part, was less than eager to be burdened with the responsibility of interpreting the Freedom of Information Act. Fredrickson later wrote to PMA president Joseph Stetler: "NIH does not now have the powers to protect completely proprietary rights and trade secrets. Such protection would be necessary for whatever agency performs certain of the registration and consultative functions required for full voluntary compliance with the Guidelines by researchers not funded by the Federal Government."[79]

For the time being, it was agreed that the issue would be shelved and that the question of voluntary compliance would be addressed after publication of the revised guidelines. Meanwhile, the PMA president was urged to propose mechanisms for protection of trade secrets and handling of proprietary data. Apparently neither side was anxious for the issue to be aired in public at this point: both preferred to address the question of protection of proprietary information after the revisions were published. Possibly neither side wanted a plan so essential to protecting themselves from various forms of litigation to be debated in a public hearing by individuals and organizations capable of eventually bringing suit.

The meeting with environmentalists on 18 October addressed only procedural matters—a feature that suggests that public interest groups believed that conserving the substantive requirements of the 1976 guidelines was a lost cause. Perhaps mindful of remarks made by Califano in December 1977 at a University of Michigan commencement address, that science was "too important to be left to scientists" and that policy decisions needed to be made "democratically, through wide consultation, not by special elites,"[80] those attending the meeting proposed increasing representation of the public on the RAC, retaining oversight of recombinant DNA experiments by the NIH, and increasing representation of community members and laboratory employees on the IBCs. The minutes of the meeting with the environmentalists suggest that DHEW officials were thinking in terms of negotiating a trade-off between the interests of users of recombinant DNA in accelerated decision making and those of their critics in expanded participation in policymaking. As the minutes record, Libassi wondered "how one would balance the delegation of authority to the IBCs with requirements for their composition, procedures, structure, and function."[81]

The negotiation of that trade-off took two further months, until the end of December 1978. In that period, the question of representation of sectors other than biomedical research in policymaking became a focal issue. The argument that representation on the RAC needed expansion in order for the RAC to address major policy issues apparently carried some weight, especially with Joseph Califano. But such expansion raised the contentious question of who should make the new appointments. Selection of RAC members had previously been determined by the NIH officials. However, as the final version of the revised guidelines was being renegotiated, the office of the Secretary of Health, Education, and Welfare asserted its authority—a move strongly resisted but ultimately agreed to by scientists at the NIH.[82]

Amid rumors of considerable tension between the leadership of the DHEW and of the NIH as to the final composition of the new RAC, Califano issued the new guidelines at the end of December 1978.[83] The reductions in containment requirements and the delegation of authority to research institutions proposed in July 1978 were granted, almost intact. Califano's reasons for approving the changes echoed the NIH director's justification of the revisions and the new consensus among users of the technology: "Since the likelihood of harm now appears more remote than was once anticipated, the scientific community has now concluded that this down-grading is appropriate."[84]

In addition to the substantive changes in the guidelines, Califano also announced several other policy changes. First, he issued a directive to the commissioner of the FDA and a request to the head of the EPA to take regulatory measures within their agencies to ensure that research in the private sector complied with the NIH guidelines.[85] This move was intended, Califano claimed, to ensure that "virtually all recombinant DNA research in this country would be brought under the requirements of the revised guidelines."[86] It was also a move likely to ensure that members of Congress would take little interest in any new legislation.

Second, Califano directed the NIH to design and pursue a program of assessment of the hazards of recombinant DNA. "While our knowledge about the risks of recombinant DNA has increased dramatically, much remains unknown. . . . In my view, the more risk assessment experiments the NIH conducts or supports, the better we can judge whether the guidelines—and actions taken under them—afford appropriate protection for health and the environment," he stated.[87]

Third, public representation on the RAC and the IBCs was expanded. The revised guidelines specified that at least 20 percent of the members of local biosafety committees should represent the public and have no connection with the institutions involved. The same percentage of the membership of the RAC had to be drawn from "persons knowledgeable in such matters as applicable law, standards of professional conduct and practice, public and occupational health, and environmental safety."[88] To represent these additional fields, the membership of the RAC would be expanded from eleven to twenty-five.

Fourth, new procedures for future policymaking were instituted, and these promised to speed up the process of revision. Decisions defined as "major," such as further revisions, granting exceptions to the prohibitions and exemptions from the guidelines, and certifying new host-vector systems, would be made by the NIH with the advice of the RAC after a public comment period of thirty days. Lesser changes would require only announcement.[89] Califano thus accommodated the interests of scientific and industrial users of recombinant DNA technology while attempting to defuse the criticisms of NIH policy that had been voiced in Congress and elsewhere. The reductions in containment and delegation of authority to the IBCs made it possible for research to move ahead rapidly and for research institutions to act autonomously on containment requirements covered by the guidelines. At the same time, criticism of the nonuniformity of the guidelines, of continuing lack of knowledge of the hazards, and of conflicts of interest within the RAC was assuaged by Califano's requests to the FDA and the EPA to regulate industrial users, by the call for comprehensive risk assessment, and by the expansion of membership of the RAC and the IBCs.

In conclusion, the changes in decision procedures accommodated the needs of both researchers and the government and established less cumbersome—as well as less visible—ways of changing the guidelines in the future. The new mode of decision making meant that the expanded RAC would combine the roles it had shared with the director's advisory committee. Just as important, although they escaped notice at the time, were two other implications. First, differences of perspective were now brought inside the RAC, in contrast to the previous polarization of the positions between the RAC and its critics. As the NIH director later put it, the public policy role would now be performed "in the midst of experts."[90] Second, revision of the guidelines could now proceed through small, incremental changes, taken as frequently as the RAC met, rather than in large quantum leaps.

For the time being, Califano's directives successfully contained criticism of NIH and DHEW policy. Congressional initiatives were defused, the critics' energies were diverted to action inside the RAC rather than outside it, and the new procedures would help to shield further changes in policy from the glare of polarization and publicity.

EIGHT

Operating the Genetic Manipulation Advisory Group, 1977–1978

8.1 The Social and Political Setting

In contrast to the somewhat raucous scenes in the United States, the new British committee for overseeing genetic engineering, GMAG, began its operations at the beginning of 1977 in a comparatively peaceful political environment. As an experiment in science policymaking, even with all of its ambiguities, it was apparently accepted by the British public. While the U.S. Congress flexed its political muscles, calling officials in the DHEW and the NIH to account, Parliament and the British public appeared willing to accept calmly the regulatory policy that emerged from the Williams committee and Health and Safety Executive proposals. The public remained largely inert.[1] The policy process, in other words, was widely perceived to be working, and British officials congratulated themselves on having worked out a low-key solution to a potentially volatile issue.

The decision to appoint GMAG was reinforced by developments abroad, which were carefully monitored by British officials. At the beginning of 1977, there was little reason to believe that the American approach was superior to the British one. Although GMAG was certainly a deviation from the committees of technical experts with which both British and American officials were most comfortable, the political upheavals that surrounded the NIH guidelines in 1976 and 1977 suggested that the British approach, with uniform control of genetic engineering by a representative policymaking body, had advantages in terms of assuaging public concern. Referring to likely legislation in the United States, GMAG chair Sir Gordon Wolstenholme wrote to the NIH director, Donald Fredrickson, in August 1977 that "we are watching developments in the USA with anxiety and sympathy."[2] Moreover, a few months before it was implemented, the British approach received important reinforcement from abroad when the American and British approaches were reviewed by two European scientific organizations, EMBO and the European Science Foundation (ESF). EMBO's standing advisory committee on recombinant DNA, meeting for the second time in September 1976, produced a detailed comparison of the newly produced NIH guidelines and the Williams proposals, arguing that, since "there are no experimental data on which to base any objective estimates of the conjectural biohazards associated with recombinant DNA research," the classification of hazards was arbitrary. Endorsing neither the British nor the American approaches, the EMBO committee recommended that European countries should adopt one or other of these approaches but not some mixture of the two; however, it supported the minimal British category I containment on the grounds that it offered greater safeguards than those provided by the American P1 and P2 levels. The report also called for training, health monitoring, and hazard assessment.[3] An ad hoc committee of the ESF, composed of lawyers and physicians as well as biologists, explicitly endorsed the British approach in preference to the American one. Meeting in September 1976, the ESF committee emphasized the need for uniform control of both public and private laboratories and the advantage of a legal requirement to ensure the effectiveness of a central advisory committee. The committee also preferred the greater emphasis in the British system on physical containment, since in its view biological procedures were still debatable.[4] Thus commended, GMAG held its first meeting in December 1976. The scientific and industrial communities appeared to be persuaded that the British policy was reasonable and seemed ready to cooperate. Throughout 1977, as the genetic engineering controversy climaxed in the United States and legislation became the object of a fierce political struggle, GMAG proceeded to implement the Williams proposals in relative calm, with little or no public controversy and with the general support of science and industry.

By the beginning of 1978, however, support for the British controls from these two sectors began to decrease for several reasons. In the first place, by 1978, interest in commercial applications of genetic engineering was growing, and after a slow start, British companies were launching genetic engineering projects. Two companies, Searle and ICI, were operating category III facilities.[5] Scientists were exploring possibilities for industrial applications: a meeting organized by the National Research and Development Corporation (NRDC)—the British government agency responsible for patenting the results of government-supported research—was reported to be "lively and well-attended."[6] As such interest grew, GMAG's restrictions began to be depicted as "oppressive."[7] Industrial scientists attending the 1978 meeting of the Biochemical Society were said to be "unanimous in their lukewarm regard for GMAG," some seeing it as a "foolish innovation, a misguided experiment in participatory politics," others proposing that GMAG be replaced by a full-time inspectorate.[8] And John Maddox, one of GMAG's "public interest" representatives, warned in the *Financial Times* that GMAG's controls for industrial activities (described below) would "undoubtedly be an encumbrance for many companies."[9]

At the same time, the prospects for controls elsewhere that were significantly weaker than those in the United Kingdom were increasing. In the United States, as congressional interest in legislation waned, it grew more likely that the American controls would be substantially reduced. Other countries, notably France, were also contemplating weaker controls. Furthermore, the international scientific community was hearing a stream of claims that recombinant DNA hazards were much less significant than originally anticipated. The epidemic pathogen argument—which emerged at the Bethesda meeting in August 1976, was legitimated at Falmouth in 1977, and applied at the Ascot meeting in January 1978—was being transformed into a claim that most genetic engineering hazards were trivial.

The response of the British scientific community to the diffusion of these arguments was complex. As in the United States, many scientists were receptive to them. Writing in *Nature* in June 1978, Robert Pritchard of the University of Leicester claimed to speak for many of his colleagues in depicting "the recombinant DNA debate as the longest running and most expensive farce in town."[10] GMAG, bowing to the same pressures, stated in its first report: "If the present trend continues, of increasing reassurance and evidence that the conjectured hazards have been overstated, more work will be judged appropriate to the lower categories of containment. We very much hope this will come about; we are anxious to avoid unnecessary—and thus uneconomic and wasteful—restrictions."[11] By the fall of 1978, Sydney Brenner was sardonically referring to "six months in category IV" and asking why one branch of science should be "singled out for social punishment."[12]

The unions representing scientific and technical workers provided a countervailing force to the pressure from researchers and industrialists to weaken and eliminate controls (see section 3.2). At an ASTMS conference in October 1978, union commitment to the regulation of genetic engineering and to the representation of technical workers in policymaking at the local and national levels was forcefully asserted before an audience that included leading scientists, representatives of private industry, and the Secretary of Education and Science, Shirley Williams. As ASTMS general secretary Clive Jenkins made clear, the union wanted more, not less, control of genetic manipulation.[13]

The unions' claim that laboratory standards and procedures were not effective enough was particularly pointed, since only months before, ASTMS member Janet Parker, a photographer at the University of Birmingham Medical School, had died from smallpox contracted in a laboratory that experimented with the virus.[14] The event gave substantial force to the union's opposition to weakening of containment provisions and oversight. The smallpox event also provided a counterargument to proposals that containment requirements for genetic engineering should be no more stringent that those required for the source of foreign DNA used in genetic manipulation. ASTMS health and safety officer Sheila McKechnie asserted that the fact that controls for genetic manipulation were more stringent than for work with pathogens might indicate not undue tightness in GMAG's requirements but inadequacy in the regulation of work with pathogens.[15]

As noted earlier, a consequence of union engagement with the genetic engineering issue in the United Kingdom was an emphasis on controls that protected laboratory workers. The epidemic pathogen assessment, which became so widespread in the United States, helping to sidetrack regulatory legislation and weaken the NIH guidelines, did not achieve comparable influence in the United Kingdom or within GMAG. Members of GMAG and its subcommittees interviewed in 1979 and 1980 saw the argument as interesting, possibly valid (although this was challenged by at least one GMAG member), but irrelevant to what they saw as the central issue of worker protection. According to Sydney Brenner, "We were never concerned about creating an epidemic pathogen. In the first instance, our concern was the health and safety of the people at work."[16] A survey of ten GMAG members conducted in 1980 indicated that this was a majority view. Six members indicated that they did not believe that the epidemic pathogen argument justified reducing controls, two indicated the opposite, and two had no opinion. In addition, six felt that the argument was not influential, and of those who recalled the argument being used on GMAG, two indicated that it was influential only with certain people.[17]

The belittling of recombinant DNA hazards that the Falmouth and Ascot meetings promoted circulated quickly in the United Kingdom. Such claims were met with skepticism, partly for scientific reasons and perhaps because GMAG had not been officially invited to participate. One GMAG member described Falmouth as "a real set-up . . . not a comprehensive scientific debate" and the epidemic pathogen argument as a position developed "by people who wished to produce a certain conclusion."[18] Similar things were said about the Ascot meeting. Both were seen as directed for political reasons toward preconceived conclusions. Wolstenholme later expressed the views of many GMAG members when he recalled being "very suspicious [of the results of these meetings because] the experimental evidence really wasn't there."[19] While the "new evidence" was being used energetically on the RAC in April 1978 to justify reductions in controls (section 7.6), GMAG's first report, issued a month later, was distinctly more circumspect about the significance of the new hazard arguments. "If those findings stand up to careful scrutiny, it will appear that many of the more extravagant predictions of risk can probably be disregarded," the report stated. But it also emphasized that "at the present time there is little evidence about the hazards of this work, whether to the worker or to the general public," and warned that, "however strong an individual worker's intuitive feeling may be that the hazards have been greatly overestimated, we—and the public generally—have a right to expect of him, until we know more of the reality of the situation, that he should do the work either with scrupulous care or not at all."[20]

In general, GMAG was not at first inclined to accept the arguments against recombinant DNA hazards that were sounded so emphatically on the other side of the Atlantic. But to criticize those arguments or to dismiss them as mere political expediency did not dissolve the dilemma that any reduction of American recombinant DNA controls presented. Researchers immediately understood that if GMAG did not downgrade its containment requirements and the RAC did, the United Kingdom could be placed at some scientific and economic disadvantage. A report in *Nature* in January 1978 expressed this dilemma:

> The main concern [in Britain about the proposed U.S. revised controls] is not so much their scientific content, but the fact that, if adopted, serious discrepancy between British and American laboratory practice could result, with potentially embarrassing consequences.
>
> It could, for example, be cheaper to send a research worker to the U.S. to carry out a set of experiments requiring little more than a conventionally equipped laboratory bench, than to install—or even hire—facilities providing the higher physical containment levels required to carry out the same experiment in the U.K. Such embarrassment is likely to be reinforced if other European countries, many of which have so far followed the British guidelines as suggested by the Williams committee in 1976, decide to break ranks and, in line with the U.S., introduce significantly lower containment levels (as the French are, indeed, now proposing to do). The members of GMAG are conscious that if they impose too harsh a set of restrictions on industrial research programmes as compared with other countries, companies will merely transfer their research programmes—and their revenue-earning potential—elsewhere; "and we don't want a repeat of the penicillin story," according to one GMAG member.[21]

Controls that were stricter than those elsewhere could handicap British science and industry in competition for leadership in the recombinant DNA field, and no one wanted the United Kingdom to be left behind.

8.2 The Politics of GMAG

Four intertwined interests—those of science, employers, employees, and the general public—were represented on GMAG, although its members were deemed independent of any particular constituency (table 4.6). The scientist members, like their colleagues elsewhere, were keen to proceed with research, at least as rapidly as their peers in other countries; they were most likely to object when British controls looked significantly more strict then those adopted elsewhere. The two representatives of the employers—one representing universities and the other private industry—shared this concern, and the latter insisted on the particular needs of private enterprise. The four representatives of employees had been nominated by the TUC to ensure the protection of workers, from maintenance personnel to professors. They were also interested in promoting the goal of expanded participation in science policymaking, which Clive Jenkins, the outspoken general secretary of the ASTMS, was widely known to advocate.[22] According to GMAG member Robert Williamson, one of the two ASTMS members on GMAG, the trade unions' larger objective was "to establish the principle that a broadly based and representative committee has the ability and right to advise on questions of safety policy and national policy in a scientific area." Rather than being the exclusive preserve of scientists and research councils, safety policy is "something which the whole community, through representative committees, has a right to participate in at the highest level." The unions saw GMAG as an important model for participatory policymaking—one that might be used elsewhere.[23] In contrast, those who did not share this goal tended to see the unions as simply self-interested.

The representatives of science, industry, and labor tended to take uniform positions on many issues and could weigh into discussion with the full muscle of their constituencies. At the same time, influential linkages between interest groups, particularly science and industry, were embodied by individual members, permitting crosscutting alliances, especially with respect to questions of scientific and industrial development. In contrast, those said to represent the public had no specific constituency behind whose agenda they united. Moreover, they generally lacked the technical expertise required by some issues addressed by GMAG. Although the public interest representatives brought to the proceedings novel disciplinary perspectives on the making of science policy, which might have been used to the same extent as expertise in a science, one member speculated that the main expectation of those responsible for managing the committee was that the public interest representatives would act as "buffers" in cases of clashes of interests among the groups represented by other members.[24] Wolstenholme later explained, "You needed someone else in the sandwich to represent, if you like, the very informed, neutral opinion that could understand the union's concern for their own members, but at the same time could understand the national need for research to be carried forward."[25]

A recurring question in the early history of GMAG involved the precise role of committee members. The Department of Education and Science position was that they were appointed as individuals to represent certain interests in genetic engineering but not as delegates of particular constituencies. In contrast, the trade unionists saw themselves as mandated TUC delegates with the right and obligation to consult their constituency. The issue was eventually taken to Shirley Williams, who insisted that the trade unionists act as individuals. GMAG, in other words, would not be used as a negotiating forum. Nevertheless, the tensions associated with the roles of the trade union members surfaced in connection with certain questions, especially the question of the confidentiality of industry applications, as we shall see (section 8.3).[26]

Much of the basic work of GMAG was conducted, not by the full committee, but by subcommittees established to examine specific areas of policy. Five such committees were formed during GMAG's first year of operation, each chaired by a GMAG member drawn from either science or industry. All were open to other GMAG members to attend. Most of the members of these subcommittees were coopted (forty-five out of fifty-six). With the exception of a committee on medical monitoring, which included ten physicians and a health professional, coopted members were drawn exclusively from science and industry.[27] Thus while GMAG's composition was broad, that of most of the subcommittees was narrow, resembling the composition of the RAC.

A major difference between GMAG and the RAC was that GMAG's proceedings were conducted entirely behind closed doors and shrouded by the Official Secrets Act. Whereas American officials had to be alert to the possibility of reactions from Congress and the public to the RAC's proceedings, GMAG was insulated from immediate public scrutiny. As a result, GMAG's institutional environment—primarily the Department of Education and Science and the MRC—and the important constituencies some of its members represented shaped, almost exclusively, the committee's agenda.

GMAG's charge, which mentioned "a continuing assessment of risks and precautions," provided ample scope for a broad consideration of the risks of genetic engineering and a reassessment of the Williams categorization of hazards. However, implementing the Williams proposals was the principal business put on GMAG's agendas during its first year by the MRC secretariat and the chair.[28] These agendas were rarely extended to address other policy issues. Questions spun off from the meetings of the committee and its subcommittees also focused on the implementation of the Williams report. Like those of the RAC, GMAG's agendas were overflowing, producing reluctance to add new issues for consideration. The chair later recalled that the secretariat "did not circulate [a memo asking] 'have you anything for next time's meeting?' because we *always* had more than we could cope with. We always had an enormous task to get through the agenda."[29]

Broader questions such as risk assessment received little attention, and efforts to put such matters before the committee failed for the most part because committee members were content with the agendas developed by the secretariat. Interviews conducted with ten members of GMAG in 1980 show that most believed that the committee had addressed the right questions and that it was run in a satisfactory manner. Only two of the ten interviewed—both public interest representatives—expressed reservations about the running of the committee and held that important policy issues had been ignored. Both expressed frustration at their lack of influence, one recalling that "I did not need to be taken seriously on the things the committee considered its prime concerns."[30]

In conclusion, GMAG's composition and operation were designed to ensure that it would function within the policy paradigm established by the Ashby committee and at the Asilomar conference, avoiding the possibility of being mired in the difficult long-term issues of hazard analysis and social application that challenged that paradigm. The group's existence would reassure the public that the genetic engineering problem was being addressed, by demonstrating that those pursuing genetic engineering would abide by the rules and at the same time allow research to advance.[31]

8.3 Implementing the Williams Proposals, 1977

Implementing the Williams proposals took two main forms. First, from the second meeting of GMAG onward, the committee reviewed and categorized proposals for genetic engineering experiments submitted by British scientists. In the words of one member, processing the applications was "the core of every session."[32] Those responsible for GMAG saw this as an imperative. The chair recalled in 1980, "They always had to be done, regardless of anything else."[33] Clearly the desire of researchers to move ahead with genetic engineering was behind most of the committee's efforts. To process research proposals was "certainly necessary

Table 8.1   Issues Addressed by the U.K. Genetic Manipulation
Advisory Group, 1976–77

Issue	Index
Validation of safe vectors	0
Medical monitoring	0
Confidentiality of proposals	2
Genetic manipulation of plants	0
Definition of genetic manipulation	0
Coverage of the use of the products of genetic manipulation	0
Composition of safety committees	1
Scientific merit as a criterion for evaluation	2
Siting of laboratories	1
Empirical risk assessment	—
Training	0
Physical containment	0

Sources: Genetic Manipulation Advisory Group, *First Report of the Genetic Manipulation Advisory Group,* Cmnd. 7215 (May 1978); interviews with members of the first GMAG, 1980.
   Notes: 2 = widely seen as highly controversial; 1 = widely seen as somewhat controversial; 0 = not mentioned as controversial; — = not actively addressed.

in order for GMAG to survive at all and retain the good will of the scientific community."[34]

Second, the committee took up specific issues raised by the Williams recommendations and by the Health and Safety Commission consultative document: the definition of *genetic manipulation,* training and monitoring of laboratory workers, the development and testing of safe vectors, the risks associated with an increase in the scale of genetic engineering, the special questions posed by the genetic manipulation of plants, problems related to the confidentiality of submissions from private industry, and hazard assessment (table 8.1). As noted, subcommittees to address such matters were regularly formed and conducted a significant part of GMAG's business.

In many of these cases, issues were defined largely as technical problems—an approach reinforced by the fact that most of the coopted members were drawn from science and industry. According to Bennett, Glasner, and Travis, "Scientific and medical expertise [and] technical elements in policy decisions were the basic currency of the Group's deliberations."[35] This is an important point of similarity between the operation of GMAG and of the RAC. Nevertheless, the broad composition of GMAG sometimes interfered with the sweeping reduction of policy issues to their technical components that characterized the RAC's proceedings.

Compared to the controversy over the RAC's application of the NIH guidelines in the United States, GMAG generally handled its business quietly, debate rarely spilling out beyond the committee's meeting room at the CIBA Foundation. In addition, the controversies that emerged had a different character from those in the United States, since they arose primarily from interactions within the committee rather than from interactions between the committee and the larger public. The nature and extent of the controls themselves, points of heated controversy in America, provoked little debate in the United Kingdom, perhaps primarily because GMAG was insulated from external scrutiny that might have challenged the overall basis of its activity.

Despite the eruption within the British scientific community over the Health and Science Executive's definition of *genetic manipulation,* subsequent discussions in GMAG about the definition appear to have proceeded smoothly. Health and Safety Executive officials consulted with the committee, reaching a consensus that appears at the time to have been entirely uncontroversial. The definition eventually used by the Health and Safety Executive was "the formation of new combinations of heritable material by the insertion of nucleic acid molecules, produced by whatever means outside the cell, into any virus, bacterial plasmid, or other vector system so as to allow their incorporation into a host organism in which they do not naturally occur but in which they are capable of continued propagation."[36]

Apparently only toward the end of GMAG's first year were questions raised about this definition's coverage of certain aspects of recombinant DNA research and development, particularly the use of products of genetic manipulation. GMAG's first report noted that there would be a "brisk traffic, from laboratory to laboratory, in the intermediate products of genetic manipulation."[37] But the group's elaboration of the definition of genetic manipulation left certain gray areas with respect to regulation of this "traffic." A guidance note issued in 1977 stated that the definition covered the construction of hybrid DNA and its subsequent insertion and replication in an organism.[38] This definition did not cover the activities of laboratories or industries that imported genetically engineered organisms from other sources. (In contrast, the 1976 NIH guidelines encompassed the construction of recombinant DNA molecules, their replication in organisms, and their subsequent use.)[39] Despite the concerns of the trade unions,[40] GMAG made little effort to revise the definition, apparently accepting the official reason for proceeding with the more limited definition, that addressing use would require a new round of consultation and further delays in issuing the notification regulation. Moreover, this gray area in the control of genetic manipulation would be useful later, when certain committee members argued for relaxing controls.

GMAG's response in its first year to at least two issues—medical monitoring and the composition of safety committees—differed strikingly from that of the RAC. The different balance of political interests within the two committees offers an explanation for this contrast. Both the NIH guidelines of 1976 and the Williams report made reference to the desirability of monitoring the health of laboratory workers. The guidance note issued by GMAG during its first year required health monitoring for all research in categories II–IV (roughly 70 percent of the research screened between January 1977 and mid-February 1978). The requirements for health monitoring included health checks, taking serum samples before research was initiated, flagging the health records of all workers through the registry of the National Health Service (with specific notations on deaths and cancers) to facilitate long-term studies, and reporting unusual illnesses.[41] In contrast, the American arrangements for health monitoring were left entirely in the hands of the head of a project and were voluntary and vague.[42] While the presence of a monitoring scheme appears to have been taken for granted in the United Kingdom, its absence was almost equally taken for granted in the United States, except by critics of the NIH guidelines. This suggests, first, that the monitoring was not a priority for the NIH and its advisory committee; second, that critics of the NIH had no power or influence on this issue; and third, that MRC officials (who drafted the monitoring note) recognized that either some members of GMAG or the Health and Safety Executive were committed to monitoring and would insist on it. Clearly the existence of a large union representing laboratory workers in the United Kingdom explains the contrasting behaviors of the two governments: questions of employee health could not be set aside in a policy arena in which unions were represented, nor could they be successfully addressed in a policy arena where employer interests were overwhelmingly ascendant.

GMAG's decisions on the composition of local safety committees provide a further contrast between British and American implementation of recombinant DNA policies. The NIH guidelines called for local safety committees appointed by the institutions in which research was pursued, with membership specified only in terms of competence.[43] In practice, this meant that such committees were managed by research scientists in the institutions pursuing genetic engineering and that there was little scope for the expression of other interests. In the United Kingdom, in contrast, it was anticipated that regulations under the Health and Safety at Work Act would require representative safety committees in all workplaces, including research institutions.[44] Thus those responsible for GMAG operated within an uncontested framework of assumptions that recognized differing interests in laboratory safety and the desirability of their representation on safety committees. GMAG's guidance note on the composition of safety committees, requiring that trade unions, wherever they existed at the local level, were to be consulted about the establishment of safety committees and that these committees would be composed of "at least as many employees as managers," was accepted by the committee without significant controversy. Furthermore, GMAG required the establishment of a safety committee and a review of its composition before any review of proposals from an institution.[45] What would have been considered absurdly radical by the NIH and most of the members of its advisory committee was accepted as the norm by GMAG and the MRC.

Other issues addressed by GMAG proved to be more contentious, evoking strongly differing political positions within the committee.[46] For example, the question of examining the scientific desirability of genetic engineering experiments at the local level was debated by GMAG at length. This question was never entertained by the RAC. According to both the 1976 guidelines and the revised 1978 guidelines, the roles of the principal investigator and the local biohazard committee were largely confined to the technical function of assigning containment levels; the question of desirability, if it emerged at all, was assumed to be primarily the responsibility of the NIH peer review panels and, in the case of an appeal, the RAC itself. In the United Kingdom, in contrast, the members of the Godber and Williams committees had expected laboratories to address the merits as well as the hazards of experiments. However, such expectations predated the restructuring of laboratory safety committees required by the Health and Safety Commission in 1978.[47] By the time of GMAG's consideration of the question of "scientific merit" in 1977, traditional hierarchical structures were being replaced by participatory ones featuring equal representation of laboratory employees.

The issue of scientific merit arose directly for GMAG when the committee developed a guidance note laying out the information needed from scientists and laboratories proposing genetic engineering experiments.[48] The debate within GMAG defined the issue in part as pitting participatory democracy with union representation against peer review, and in part as a question of competence. The trade union representatives held that local safety committees and GMAG should address the safety and the desirability of experiments. Committee member Robert Williamson recalled in 1980: "The trade union representatives argued that it was impossible to look at safety exclusively in terms of categorization and that GMAG had a wider remit to advise the government and to advise laboratories . . . not just on the narrow question of safety but on somewhat wider issues as well. . . . We didn't want the safety committees to become purely a rubber stamp for looking at the [categorization] of experiments."[49] That view provoked resistance within GMAG, especially from some of Williamson's scientific colleagues who saw union interest in evaluating scientific merit as a move toward extending "worker power"[50] and who held that local committees were not the appropriate bodies to consider scientific merit. In a letter to the secretary of GMAG, John Subak-Sharpe insisted that the function of the safety committee was to address safety, not questions that he believed were the special province of scientific peers.[51]

When the guidance note was revised in May 1979, the original text, indicating that safety committees should provide "a statement . . . of the desirability or otherwise of carrying out the work proposed" was replaced by the observation that "the safety committee may wish to consider the scientific content of each proposal and to formulate provisional conclusions about the desirability of conducting the work as proposed." Despite this weakening of the language of the GMAG note, a role in judging the merit of experiments, denied to laboratory employees in the United States, was conserved—a reflection of union influence within GMAG.[52]

Perhaps the most intense debate—the only issue that led GMAG to a vote—was the question of the confidentiality of industrial proposals. Two diametrically opposed positions arose. On the one hand, industry representatives, like their counterparts in the United States, pressed hard for the protection of trade secrets. Under British patent law, products could be patented only if there had been no prior disclosure. But showing proposals to GMAG under conditions of confidentiality did not, in itself, constitute prior disclosure.[53] Since the British controls required review of all genetic engineering activities, industry representatives were concerned not only about possible denial of patent coverage if confidentiality were not assured but also, regardless of the stringency of formal requirements, about the possibility of leaks of information concerning specific experiments, general intentions, and progress toward products. On the other hand, what concerned GMAG's trade union members was the possibility of restrictions on their freedom to consult with local safety committees and with union officials about the details of all genetic engineering work, including that proposed by private industry. The scientist and public interest members do not appear to have been prominently involved in this dispute, although their presence on GMAG might have aggravated industry fears of jeopardizing trade secrets.

The intense conflict that resulted reveals the ability (and the limitations) of the trade unions to oppose private industry interests on GMAG and the strong interest of private industry in defending proprietary rights.[54] In the summer of 1977, a GMAG subcommittee consisting of a majority of industry representatives recommended limiting scrutiny of industry applications to a subcommittee of four members with no personal interest in the work under review. At the July 1977 meeting of GMAG, this met with flat rejection from the trade unionists on the grounds that it was unnecessarily restrictive. After prolonged negotiations in the fall of 1977, the chair issued a warning (and veiled threat) that failure to work out a solution might make it desirable for the government to find an alternative to GMAG. With that possibility in the air, a compromise was reached, giving all members who did not have a conflict of interest and who were willing to sign a confidentiality agreement not to divulge the details of industry proposals the right to review them. It was agreed further that trade union representatives might discuss industry experiments and processes with the particular local safety committees in the industry under review, and were free to divulge other aspects of applications—for example, the composition of safety committees.

The compromise was difficult. Industry representatives remained uneasy and dissatisfied with the outcome. As one of them stated in 1980, "People are almost bound to twitter . . . and by the composition of GMAG, the sort of people who will know are the very sort of people who will go and accidentally maybe, do [a similar experiment] at their own lab or drop the hint if they are consulting with somebody else."[55] In other words, private industry wanted stricter guarantees of confidentiality than they believed GMAG could ever provide. Their concerns about this issue figured prominently in testimony to Parliament's Select Committee on Science and Technology a year later, when industrial organizations aired their discontent with GMAG's procedure and pressed to have control of genetic engineering transferred to a specialized inspectorate of the Health and Safety Executive.[56] For the trade unions, this was the only issue on which they seriously divided: the TUC representative, Ronald Owen, refused to sign the confidentiality agreement.

GMAG rarely looked beyond the task of implementing the recommendations of the Williams report during its first two years of operation.[57] As the British Society for Social Responsibility in Science noted in a response to the group's first report, GMAG did not address broad social and ethical issues posed by genetic engineering—for example, the goals of a science and technology policy for genetic engineering, social and ethical issues posed by potential medical applications, or the danger of possible biological warfare applications.[58]

Nor did the group pay more than cursory attention to the problem of assessing recombinant DNA hazards, despite its formal charge to undertake "a continuing assessment of risks and precautions." The need for risk assessment had been reinforced by the reports of international organizations such as EMBO. Moreover, the largely arbitrary nature of the Williams categorization of hazards was understood by most members of the committee. One member later described the scheme as "obviously nonsense."[59] Another recalled that "one had this relatively objective, expert discussion of categorization which I eventually realized was all relative to these largely arbitrary classes."[60] Yet a proposal in July 1977 that the group take a fresh look at the analysis and classification of hazards was resisted by the majority of GMAG members on the grounds that airing possible hazards might alarm the public and that, in any case, the group had more pressing matters to attend to.[61]

Furthermore, experimental investigation of hazards was not a high priority. "I think, when it did come up, the whole committee was wholly in favor of risk assessment experiments. The question was, who was going to do them, and who was going to pay for them," Wolstenholme recalled in 1980.[62] A member of GMAG confirmed that view: "GMAG as a whole . . . always tended to say that risk experiments would be worthwhile. But it was difficult to see how they could be done, and no one wanted to do them."[63] The problem was that risk assessment conflicted with pursuing genetic engineering itself, which was generally seen as a much higher priority. Several GMAG members felt that most scientists were not interested in hazard assessment. One GMAG member recalled that, when the possibility of an MRC hazard assessment program was raised, it was "rapidly pushed on one side" by scientists who were not keen to see research support diverted to addressing such questions.[64] Those members who held that these experiments were essential to carrying out GMAG's charge perceived little support within the committee for their position.

GMAG's response to the results of the two hazard assessment experiments that were conducted during its first two years further reflects the committee's disinterest in such research. As it turned out, these results were not entirely reassuring. An experiment in 1978 demonstrated that the American "disarmed" strain of E. coli K12, χ1776, appeared to lose its sensitivity to the salts of the intestinal tract, suggesting that it might survive quite well in humans. The experimenters concluded that χ1776's bile salt sensitivity was "phenotypically variable"—suggesting that this property could not guarantee "biological containment."[65] Other experiments, conducted by scientists at the Central Public Health Laboratory in London, showed that, in one out of four subjects who ingested cultures of E. coli K12, the bacterium became the major component of the gut for at least 116 hours after ingestion.[66] One GMAG member later recalled "the shock that went around the technical panel when they saw the data."[67] This surprising result was published only in GMAG's third annual report, and no further experimental investigation was undertaken.

In general, the pressure to review research projects—GMAG's predominant agenda item—prevented the committee from focusing on designing an experimental reassessment of genetic engineering hazards, and competition with other research priorities prevented many experiments from being supported or pursued. However, recognition toward the end of 1977 that American requirements for experiments would be lowered eventually produced a reversal of GMAG's earlier resistance to a *theoretical* reassessment of the Williams scheme, as the group's response to a proposal from Sydney Brenner would show.

8.4 Developing the Brenner Scheme, 1977–1978

As American intentions to weaken the 1976 NIH guidelines became clear, technical and economic pressures to revise the Williams categorization of hazard increased. The Falmouth and Ascot meetings, attended by several members of GMAG and its subcommittees, served notice that the American containment levels were not likely to remain at the 1976 levels for long. Moreover, the vociferous opposition to *any* special containment requirements mounted by some prominent American scientists promised further reductions in the future. But many GMAG members, especially those representing unions, would not countenance abandoning the Williams scheme simply to compete with the United States. Since the epidemic pathogen argument would not sway GMAG, something else had to be found.

One way out of this impasse was proposed by Sydney Brenner to the GMAG's Committee on Validation of Safe Vectors early in 1978.[68] Brenner had played a central role in the formation of British policy—testifying to the Ashby committee and sitting on the Williams committee and the Safe Vectors Committee. He was well versed in the complexities of the hazards issue and sympathetic to the point of agreeing, privately, with many who advocated caution in the use of cloning technology. On the other hand, as a leading British molecular biologist, Brenner was also committed to maintaining Britain's competitive position in the field and in its industrial applications. He was, as a sympathetic observer put it, "the archetype of the chap in the middle who on the one hand thinks the public exaggerates the dangers and on the other thinks that the scientists could make a mistake if they advocated the abolition of controls."[69]

Under pressure from some members of GMAG, Brenner proposed that the entire hazard question be reanalyzed. Instead of classifying hazards on a phylogenetic scale (the basic assumption of the NIH guidelines and the Williams categorization), Brenner focused on how they might develop. In accordance with British—as opposed to U.S.—policy, his analysis addressed the possibility of harming laboratory workers rather than the possibility of an epidemic. From an initial exposure of an individual to an organism carrying foreign DNA, the scheme attempted to trace all the pathways by which this DNA might be transferred, be expressed, and produce some harmful effect, and then to calculate the probability of such a sequence. Brenner described this method in a later report.

> This is a tree of branching paths with some transition probability at each node. The nodes involve natural genetic mechanisms of transfer, recombination, and rearrangement, the intrinsic frequencies of which are known or could be determined. The paths can therefore be ranked in order of likelihood. Furthermore, the trees cannot ramify infinitely and must terminate at limiting points which are just as likely to be reached by a similar concatenation of natural events but originating from a natural source.[70]

In principle, this approach encompassed any organism, any piece of DNA, and any interaction between the organism and its exposed human host. For example, the dangerousness of an organism carrying a foreign gene that coded for a hormone might be estimated by examining the probability that the organism would either survive or transfer the foreign DNA to some other organism that was able to survive in the human host, the probability that the gene would be expressed and that the hormone would be secreted, and the probability that the hormone would damage the host.

The scheme also encompassed the understanding that biological systems, unlike toxic substances, might replicate and amplify the hazard they posed. Such amplification would be selective since organisms carrying recombinant DNA would compete "in a world densely populated with similar elements which are already well adapted for survival."[71] In addition, the scheme could accommodate new information such as the discovery of introns in the DNA of higher organisms.

In this way, many pathways to harm, including infection of the urinary and respiratory tracts, and the generation of immunological effects, rarely discussed on the RAC, were acknowledged as possibilities. For such scenarios, Brenner emphasized the uncertainties of prediction and the need for further experimental assessment and evidence.

> In considering the effects of proteins and polypeptides produced by organisms, such as *E. coli,* which inhabit the large intestine, the question arises as to whether the products will be taken up and presented to targets on cells internal to the organism. Many proteins are, of course, destroyed by the abundant proteolytic enzymes present in the gut, but there is good evidence that large molecules can be taken up by mucous membranes including the lining of the intestinal tract. *E. coli* can establish itself in the urinary tract and there are reports of colonization by *E. coli* of the nasopharynx. . . . Immunization by way of the intestinal

tract is well established and the gut and other mucous membranes have a specialized immune system producing the secreted IgA antibodies. In addition, food allergies show that proteins and other determinants can also sensitize and excite an IgE response by way of the gut. The respiratory tract is also notoriously susceptible. . . . [The question of the dose required to produce an immunological effect] needs to be further explored by experiment.[72]

Brenner's original conception of the scheme involved an important ambiguity: although presented as a *program* for assessing hazards (and it soon became clear that an actual empirical assessment would take many years to accomplish), it was also conceived as an *alternative* to the Williams categorization—that is, as a classification of hazards that could be applied to assess proposals immediately. This ambiguity persisted throughout the evolution of the scheme, evaporating only after it was transformed into an immediately usable form and implemented.

Serious problems arose in the effort to translate the Brenner scheme into a workable system of hazard classification. In most cases, numerical values for the probabilities of events in Brenner's hazard generation trees did not exist. Furthermore, estimates of hazard did not in themselves indicate appropriate levels of containment, and little was known about the relative efficiencies of the latter. The analytic advantages of Brenner's approach were clouded by these multiple layers of uncertainty. Moreover, GMAG showed virtually no interest in launching an experimental program to measure the variables assumed in the Brenner scheme. The development of a theory of hazard assessment was disconnected from an experimental program.

GMAG's reception of the scheme in the early months of 1978 was mixed. By this point, some British scientists wanted either to follow the American lead or to have no controls at all. GMAG member Derek Ellwood later said, "Put it this way: if the scientists had had their way, the Brenner scheme wouldn't have been adopted; there wouldn't have been *any* scheme."[73] The difficulty of translating the scheme into an operable system of hazard classification was also seen as a major drawback. According to Ellwood, "Brenner's scheme got a very cool reception on GMAG, partly because it was questioned whether numbers could be attached to hazards. [This was] a fairly general perception." Molecular biologist Peter Walker commented, "Sydney [Brenner] muddied the waters by thinking of outcomes that other people are not intelligent enough to think about."[74] Historian of science Jerome Ravetz, a strong supporter of hazard assessment, recognized that, unless GMAG also initiated an experimental program to measure the probabilities of each event in the chains assumed by Brenner, the scheme would generate vast uncertainties. Ravetz wrote to Wolstenholme in January 1978 that

a serious difficulty [with the Brenner approach] is that by making so
many things more explicit, the analysis would create many new areas
for debate. This could occur on GMAG and more seriously it could
occur in subsequent negotiations between GMAG and a dissatisfied
scientist. . . . It is also possible that there will be so many identified
uncertainties at critical points that they will frequently swamp the estimated values of hazards in an assessment. We would then be in the
position of legislating on the basis of publicly admitted ignorance.[75]

Brenner himself apparently had doubts about whether implementing the
scheme was the right course for GMAG. He was later characterized by a
GMAG member as "unsure whether to put his energies behind his own
scheme or to argue for the NIH guidelines."[76]

In the first few months of 1978, discussion of the scheme remained
confined to the Safe Vectors Committee, where Subak-Sharpe was not
inclined to act quickly. By the spring of that year, however, the momentum in the United States for a major weakening of the guidelines was
apparent. Reports in *Nature* relayed news of the RAC's progressive revisions and undoubtedly the news filtered through informal channels as
well.[77] In May, the ESF liaison committee endorsed the conclusions of
the Ascot workshop, implicitly supporting the American drive to reduce
containment levels for experiments with viral DNA.[78] British sensitivity
to the American developments was growing. One member of the Safe
Vectors Committee later recalled: "There was obviously a lot of discussion as to how we were going to proceed vis-à-vis what was happening
in the States. If the American guidelines revision had gone through and
nothing had been done to alter the GMAG guidelines, Britain could have
surrendered its interests in molecular biology. We could have closed
down the labs and given up."[79] Later in the year in a television interview,
GMAG chair Wolstenholme recalled that the committee knew "that the
new revised American guidelines were coming into force in the very near
future. [We knew] that we really did need to provide a viable alternative
because we're not pleased with the new NIH guidelines, nor the logic
behind them."[80] Clearly the pressure to act was growing. Jerome Ravetz
later summarized the politics of hazard classification in the summer of
1978: "[By that time] there was a growing sense that something ought to
be done, but almost as much to protect ourselves from NIH [as to develop a valid system of risk assessment.] NIH was the danger, not the
plasmids."[81]

Thus in the spring of 1978, implementing the Brenner scheme became
an urgent consideration, and two meetings of the Safe Vectors Committee were held in June to address this question. Despite the scheme's complexity and uncertainty, agreement emerged that, failing precise values
for probabilities of harm, it could still be used to calculate rough, order-

of-magnitude values, which could yield a qualitative ranking of hazards. The new strategy was worked out by molecular biologist Peter Rigby, a member of the Safe Vectors Committee, who applied it to calculate probabilities of harmful outcomes for a series of examples—designated "canonical experiments"—whose outcomes were deemed to be known with high probability. For example, the probability of harm from inserting the gene for cholera toxin into wild-type *E. coli* so that the gene would be expressed and the gene product secreted was assumed to be high (close to one), and the experiment was assigned to category IV containment. On the other hand, the probability of harm from inserting the tryptophan operon (the regulatory elements and the structural genes involved in the synthesis of tryptophan, an amino acid) from *E. coli* into *E. coli* K12 so that the gene would be expressed was considered to be very low (close to zero), and the experiment was assigned to category I containment.[82] Considered at a meeting of the subcommittee in John Subak-Sharpe's laboratory in Glasgow on 30 June 1978, these examples helped to produce a consensus that the scheme could be made workable.[83]

A revised paper with notes by Jerome Ravetz and John Maddox was presented at GMAG's meeting in July 1978. The group welcomed the report, although some members expressed reservations about specifying numerical values for key variables in the scheme and about winning acceptance from scientists and the general public. At the same time, the acknowledged likelihood of the NIH guideline revisions pressed the committee on. Wolstenholme appointed a working party composed of John Maddox, who chaired the group; GMAG scientists Mark Richmond and Keith Dumbell; and Robin Weiss, a virologist from the Safe Vectors Committee, to develop a revised version suitable for promulgation.[84] One member, Mark Richmond, had no illusions about the way the committee needed to proceed: "The first thing we did on the Maddox committee was to convince ourselves that Brenner was not workable." To Richmond, the committee's goal was to convert the scheme into an operational system.[85] Ravetz, who was a keen supporter of the Brenner scheme and one who would have wanted to use the scheme to explore the uncertainties of hazard analysis, was not invited to join the Maddox group.

After several drafts, the Maddox working party produced a new version of the Brenner scheme, which was discussed by GMAG at its October meeting; approved, with revisions, at its November meeting; published in *Nature* on 9 November 1978; and presented to a group of scientists, representatives of safety committees, and GMAG members on 22 December 1978. This new version greatly simplified Brenner's hazard-generation tree, essentially reducing the analysis in terms of three composite factors.

Access (A)—the chance that an escaped organism would survive in the human body and gain access to susceptible cells

Expression (E)—the chance that a foreign gene carried by an organism would be expressed and that its gene product would be processed and secreted by the cell

Damage (D)—the chance that the released gene product could cause damage by contact with susceptible cells [86]

Like the original version, the published version of the Brenner scheme was flexible and able to accommodate new information (and seen as advantageous precisely on those grounds). For example, it was possible to distinguish between the hazards posed by experiments that aimed at the expression of genes and by those that did not, and between hazards posed by processes in which a gene product was secreted outside the cell and those in which the gene product remained inside the cell. In the absence of results from empirical studies, the uncertainties associated with Brenner's original scheme remained.

First, estimates of the probability of harm had to be matched with appropriate containment categories—an exercise that necessitated arbitrary assumptions. An internal GMAG memorandum observed that this had advantages: "The analysis is flexible because it provides a scale of relative risk; and the index of risk does not itself determine the containment category, which can then take policy into account as well as science." [87] In other words, GMAG could apply the scheme as cautiously or as freely as external (extrascientific) considerations permitted.

Second, experimental evidence supporting values assigned to the factors A, E, and D used to calculate hazard probabilities was still lacking. Members of the working party were aware that the numbers in most cases were guesses. One member described the procedures and assumptions of the group: "The first thing we did [on the Maddox committee] was to convince ourselves that the Brenner scheme was not workable as formulated. But by arbitrarily imposing certain plausible constants, which are only 'guesstimates' of very unequal certainty, one can come up with numerical values you can use." [88] GMAG's announcement of the new scheme in *Nature* acknowledged this problem: "Given the present incompleteness of biological understanding and the lack of data bearing on the numerical determination of these factors, it is clear that, for the time being decisions will continue to depend on judgment rather than calculation. . . . Refinement of present knowledge of the relative efficacy of the different categories of containment may be though an urgent need."

This version of GMAG's report also placed the primary burden of providing the necessary data on the individual investigator: "In the first instance, it would be for applicants to provide such empirical data as they can on which the technical panel's consideration of the three factors would be based. It is hoped, however, that the technical panel would be a powerful stimulus of the research needed to make risk assessment more objective and widely applicable."[89]

At least one member of GMAG, Ravetz, saw the absence of hard data on hazards as well as the relegation of responsibility for collecting new data to individual investigators as major problems. In a letter to GMAG's secretary in October 1978, Ravetz urged the committee to recommend, as "an early and urgent task," the establishment of an experimental program of risk assessment.[90] Ravetz's plea had little effect within GMAG, however. The scientists and industrialists were not interested; the union representatives had other priorities. The chair's dubbing Ravetz at the November meeting a "professor of hazardology" struck some as an indication of official disinterest. One GMAG member later commented, "Unfortunately, GMAG doesn't see risk assessment as a part of its role."[91] Despite a surprising announcement on British television from Secretary of Education and Science Shirley Williams that "tens of millions of pounds" would be available for assessment of genetic engineering hazards, no such program ever materialized.[92] Substantial interest within GMAG in experimental determination of hazards disappeared at the end of 1978 when Ravetz was not reappointed. As a trade union member later observed, "It would [have] been difficult to argue [about such factors as access, damage, and expression] without mounting a major [risk assessment] program. And I couldn't see anyone doing that."[93] The MRC and research scientists alike had no enthusiasm for pursuing an expensive and time-consuming program designed to measure the variables of the risk assessment scheme. The prospects for genetic engineering were too tantalizing, and no one particularly wanted to tie up Britain's limited facilities with "unproductive" research.

As disseminated to the scientific community and to the wider public, the new risk assessment scheme's programmatic and operational ambiguity persisted. On the one hand, the *Nature* article described it as a "new approach" that would be used "when there are enough data to employ it."[94] Other press accounts described it as a set of "new proposals for assessment of risks from individual experiments," as "a new basis for determining the conditions under which experiments in genetic manipulation should be made," and as "a system for assessing more rationally the potential hazards of genetic engineering experiments."[95] At the December meeting of GMAG with scientists and safety representatives, Peter Rigby noted that any calculation of hazard "would be in vain if the precise containment effectiveness of the four containment categories" were unknown and that this "must be measured."[96] On the other hand, the scheme was expected to be pressed into service immediately: the

Nature paper proposed to operate it in parallel with the Williams system "for as long as may be necessary." Although noting that relatively few experiments might immediately be reclassified according to the new scheme, the article anticipated "a steady transfer of experiments and classes of experiments" from the old system to the new. Despite the uncertain parameters it assumed and the reluctance of British scientists and funding agencies to pursue research to define them, there were many indications that GMAG would in any case proceed.

The scheme was also promoted as an effective competitor with the revised NIH guidelines. Shortly before the publication of the *Nature* article, Wolstenholme stated on television: "I think that many countries will discover that whilst the [revised NIH guidelines] look important . . . the British plan . . . will be so full of sense in the end that it will be the one which will predominate."[97] But this hope would fade quickly. The revised NIH guidelines were gaining support in other countries. Even at the time of the announcement of the new scheme, there were strong indications that Canada and several European countries would follow the American lead.[98] A further blow to British hopes was dealt by EMBO's standing advisory committee, which met in December at London's Heathrow Airport. The committee came out in support of the extreme position, exemplified in the United States by James Watson, that "the overall probability of a hazardous event [from the use of recombinant DNA technology] was negligibly low," endorsing the American proposal to shift the burden of proof to those who advocated special precautions. The goal of GMAG's scheme, to develop a quantitative assessment of hazard, was challenged: "The Committee not only agrees with GMAG that risk analysis with inadequate numerology can lead to a spurious sense of precision but also believes that in many cases it is difficult to use necessarily incomplete risk analyses for assessing the *relative* magnitude of *conjectural* hazards." Specific judgments of hazard and containment assignments—for example, those for experiments with conceivable immunological or oncogenic consequences—were dismissed. And ridiculing the plan's claims to universal appeal, the committee concluded that "experiments involving *E. coli* K12 are of parochial concern and there is no scientific reason for attempting to achieve international uniformity in this matter."[99] The prospects for broad European support for the scheme, particularly if it was used to argue for controls that were more restrictive than those being adopted elsewhere, appeared dim.

PART FIVE
Dismantling Controls

NINE

Dismantling the National Institutes of Health Controls: From Prevention to Crisis Intervention, 1979

In the final stages of American policymaking examined in this study, the NIH, with the advice of its restructured RAC, reversed the course on which it embarked in 1975. In the process, NIH oversight of recombinant DNA activities was virtually abandoned (with the exception of a few limited classes of experiment), containment requirements were radically reduced, and most important of all, the entire spectrum of living things was opened up to genetic manipulation at the same time that industrial and military applications were initiated.

These policy changes were accompanied by a marked reversal in the discourse on hazards. By the beginning of 1979, the American biomedical research community's derailment of congressional legislation was assured. This and the achievement of the first major revision and relaxation of the NIH controls had demonstrated the power of the biomedical research community to control American regulatory policy for genetic engineering. At Falmouth and Ascot, the "new evidence" and "epidemic pathogen" arguments generated a highly influential discourse to support the political position that regulation was unnecessary and the guidelines overrestrictive. Having established their power to promote this position, American biomedical researchers could now change the discourse about hazards once again in order to justify dismantling controls.

Michel Foucault has observed that power and discourse are intertwined: the truths produced by a specific discourse are "linked in a circular relation with systems of power which produce and sustain it, and to effects of power which it induces and which extend it."[1] Once the power to turn back congressional interest in regulation had been demonstrated, it could be deployed further to reconstitute the discourse on recombinant DNA hazards. The first indication of this possibility had come the previous year, when Donald Fredrickson, anticipating the demise of congressional legislation, announced to the RAC the imminent shift in the burden of proof from practitioners, to show that their work was safe, to the public, to show that it was hazardous (section 7.6). Genetic engineering would be innocent until proven guilty. Although Fredrickson did not immediately press the point, his prediction also implied that controls might be dismantled unless they were demonstrated to be necessary—a position that Fredrickson elaborated at the first meeting of the newly reconstituted RAC in February 1979 (section 9.5). In Foucault's terms, the new discourse would "induce" a new "effect" of power in the form of the dismantling of controls. The position that unknown hazards posed by recombinant DNA activities should be contained by using special precautions—the Asilomar legacy—gave way to the position that such activities posed no extraordinary hazard and that therefore containment need not exceed that employed for nongenetically manipulated organisms.

This transformation of policy and discourse was actively supported by the corporations that were developing genetic engineering for industrial purposes, especially the pharmaceutical corporations. Throughout earlier stages of the formation of NIH policy, representatives of industrial interests in the new field had kept a low profile. In general, the predominant interest shaping NIH policy had been that of the biomedical research community, with the quiet support of the private sector. As industrial applications of genetic engineering expanded, commercial ties with academic scientists and laboratories grew in number and strength, and interests in the formation of NIH policy became more complex: increasingly these interests were not simply in protecting biomedical research but also in enabling the infant genetic engineering industry to compete in the international market. As the genetic engineering industry emerged, American policymaking in this period began to exhibit more visibly the power of industry's interest. The scientist clients of the NIH continued to shape its policy; at the same time, the concerned private sector wielded its influence directly and through scientists, with considerable effect.

9.1 The Social and Political Setting

By the end of the 1970s, technical developments in genetic engineering—especially the bacterial synthesis of proteins like insulin, previously made only by higher organisms—meant that commercial applications were imminent. The pharmaceutical industry anticipated initiating production of new drugs and vaccines. The chemical and mining industries expected to develop bacteria to make new products as well as to clean up old and problematic ones. In agriculture, genetic engineering promised plants with new genes coding for nitrogen fixation, resistance to pests, and resistance to pesticides. Animals were also targeted for genetic alteration, and it was conceivable that ways would be found to inject new genetic information into human cells as well. Headlines confidently announced the "promise of gene-splicing"—a prospect reinforced by a major report of the congressional Office of Technology Assessment in 1981.[2] As the *New York Times* concluded in January 1980, "The pace of recombinant DNA research is quickening. Small new companies are leading the effort but big drug firms and even a few oil and chemical giants are entering the field. . . . No doubt there will be exaggerated claims of breakthroughs as small companies vie for contract work and venture capital. But recombinant DNA technology seems poised at the threshold of advances as important as antibiotics or electronic semiconductors."[3]

The commercial development of genetic engineering and other biotechnologies received strong support from the tax, patent, and science budget allocation policies of the Carter and Reagan administrations. From 1979 onward, these policies encouraged joint university-industry research programs in biogenetic technologies, provided strong incentives for private investment in basic research, and permitted universities to patent the results of research funded by the federal government. In the same period, the Carter administration initiated and the Reagan administration zealously pursued deregulatory policies aimed at reducing the costs to private industry of environmental, workplace, and other regulatory controls (section 1.4).

Propelled by the support and incentives provided by government agencies, corporations, and venture capital firms, the race to develop recombinant DNA techniques expanded far beyond the academic laboratories in which they originated. Multinational corporations, venture capital firms, private investors, universities, and individual states within the United States, depending on their needs, sought high returns on investments, support for research, and development of new, high-technology industries that promised to alleviate the weakness and unprofitability of older ones. A transformed recombinant DNA field saw its sponsors and its practitioners assume new roles resulting from this commercial development. As developers of patents on basic techniques, as recipients of corporate grants for research, and as cosponsors of new biotechnology companies, leading research universities established important interests in the industrial development of the field. So too did research scientists, as they became equity holders, advisers, consultants, and executives for new genetic engineering companies and multinational corporations (sections 2.5 and 2.6).

After 1978, competition for these new roles intensified as investment in the field accelerated: multinational corporations competed for "windows" on the new technology; genetic engineering firms and multinationals alike competed for products; universities competed to attract corporate support for research; nations and individual states within the United States competed to attract new biotechnology plants and laboratories; and scientists increasingly engaged in all of these developments. Perceptions of the need to pursue rapid development of the recombinant DNA field intensified and multiplied.

Much less visible but also initiated in the early 1980s were military applications of genetic engineering. Until the last year of Carter's presidency, the DoD had insisted that it was not sponsoring and did not plan to pursue genetic engineering. Such declarations did not signify a lack of interest in the field, however. DoD representatives sought in 1977 to protect future classified information in this area from the requirements of the Freedom of Information Act (section 6.1). As early as 1978, the army commissioned a (presumably theoretical) investigation of the biological weapons implications of genetic engineering.[4] And in 1980, the DoD, citing the possibility that genetic engineering could be used to "implant virulence factors or toxin-producing genetic information into common, easily transmitted bacteria such as *E. coli*," argued that use of the new technology posed new biological warfare threats. The same report proposed new research, the goal of which would be to "provide an essential base of scientific information to counteract these possibilities and to provide better understanding of the disease mechanisms of bacterial and rickettsial organisms that pose a potential BW threat, with or without genetic manipulation."[5] By September 1980, the department had apparently reversed its earlier position, since it was advertising contracts to support research and development using genetic engineering and other biogenetic techniques. At about the same time, the Defense Advance Research Projects Agency began to make known its interest in supporting applications of biotechnology for diagnostic and detection purposes.[6]

In the climate of superpower rivalry and the deterioration of East-West dialogue that characterized President Reagan's first term, military development of genetic engineering expanded rapidly (although constrained by the 1972 Biological Weapons Convention).[7] In the early years of the Reagan administration, however, the Pentagon maintained a low profile with respect to biological defense activities and genetic engineering applications. Had the DoD argued that genetic engineering and other biogenetic technologies would make biological warfare easier, cheaper, and far more effective, as it would later in the decade, it would have risked undermining the claim that genetic engineering posed no special hazard and reigniting biohazard controversy. But the "new biowarfare" discourse did not emerge publicly until the mid-1980s, after deregulation had been achieved.

9.2 Industry, Academe, and the Politics of the NIH Controls

The revised NIH guidelines adopted in December 1978, while weaker than the original 1976 controls, still placed significant containment and oversight restrictions on recombinant DNA activities, especially for cloning DNA from higher organisms and for large-scale cultures. For example, although shotgun experiments with human DNA in *E. coli* K12 could be performed in P2 physical containment, it was necessary to use a debilitated (EK2) strain of *E. coli*—and these strains were gaining reputations as finicky and fragile. Such requirements meant that large-scale work—if it could be done at all—would be both costly and difficult. Whether laboratory containment requirements as high as P3 and P4 could be applied to industrial-scale fermentation machinery, and at what cost, was unclear. Keeping a huge fermenter under negative pressure would clearly not be cheap, and perhaps not even feasible. In addition, large-scale processes, falling under one of the six prohibitions to the NIH guidelines, had to be granted exceptions by the RAC. Proposals for new host-vector systems—an emerging interest for the new genetic engineering industry—also were subject to the RAC's review. These procedures blocked fast development of the technology. Furthermore, gaining the RAC's assent meant divulging full details of sensitive corporate information to the committee.

As industrial applications became imminent, the private sector's interest in weakening these controls intensified. Publicly, firms small and large united to plea for the need to maintain their lead in the emerging genetic engineering industry in the face of strong competition from abroad. At a Senate hearing held in May 1980, the issue was framed as a race between American companies and competitors abroad. As Peter Farley, president of Cetus Corporation, put it, "We are in a race with groups in Europe and Japan and it is a race in which a period as short as a few weeks can be critically important."[8] The ten-liter limit and the requirements for approval of host-vector systems were repeatedly cited as obstacles to industrial development. According to Farley,

> In a competitive world, the [1978 NIH] guidelines as they presently exist still represent a major handicap. . . . We are handicapped by the fact that only one organism has been approved as "safe." There are quite a number of other bacilli we would like to be working with that, from a commercial point of view, would be much better utilized, and yet, it would literally take 6 to 12 months and the time of the very scientists we have working on the project to develop the data to prove that it is safe, et cetera. One might say, "So what?" Well, the "so what" is "compared to the European situation." Our competitors in Europe are under no such restriction. . . . The Europeans are very much in [the race to develop the commercial potential of genetic engineering.]

> Everything I have just said [concerning approval of new host-vector
> systems] holds in spades for the 10-liter scaleup.[9]

Farley was followed by Ralph Hardy of Du Pont, Irving Johnson of Lilly,
and Nelson Schneider of the brokerage firm E. F. Hutton, who reinforced the theme, comparing the American controls with those in Europe and linking arguments for relaxation with the "new consensus" that
hazards had been exaggerated.

> HARDY: . . . we should not let future regulations or lack of financial
> support put U.S. industry behind foreign companies in achieving the
> goals of this research. . . .
> JOHNSON: . . . in most of Western Europe, new regulation is not cur-
> rently under consideration. . . . There appears to be a developing
> consensus that the potential hazards associated with this technology
> were initially overestimated and continuing changes in the Guidelines
> are desirable. It seems logical that the other countries will also de-
> crease unnecessary restrictions on recombinant DNA research as the
> U.S. has. It is fairly obvious that under these circumstances, all other
> factors being equal, that research and development will tend to mi-
> grate to those political environments where the progress of research
> is least impeded.[10]

Privately, corporations large and small began to pressure the NIH to
be more responsive to their needs. At their meeting with the NIH director in the fall of 1978, PMA representatives had drawn attention to the
difficulties posed, in their view, by the ten-liter limit and had made it
clear that they were unlikely to comply unless further measures were
developed to protect sensitive corporate information (section 7.7). At a
further meeting with DHEW and NIH officials in May 1979, their requirements for modification of the NIH controls were set forth in detail:
first, transfer of decision-making authority to institutional biohazard
committees, especially approval of new host-vector systems; second,
industry representation on the RAC; and third, expedited approval of
large-scale industrial processes.[11] A steady lobbying effort pursued these
goals.

The primary need of the new genetic engineering firms pushing the
frontiers of the technology was to move quickly in order to secure patents, contracts with larger firms, and ultimately markets. Michael Ross,
Genentech's director for protein biochemistry, told the *Wall Street Journal*
in 1981: "If you come out a year after someone else, forget it. You'll have
10% of the market, and he'll have 90%."[12] Tension between the NIH and
the emerging genetic engineering industry erupted in the fall of 1978
when Genentech, eager to test a pilot plant for production of insulin,
began to chafe under the ten-liter limit. Without explicit permission for
large-scale production from the NIH, however, the process could not be

said to comply with the 1976 guidelines. Apparently undaunted, Genentech's president, Robert Swanson, announced in a letter to the NIH director on 9 October 1978 that the company's biosafety committee had approved pilot plant processes for production of insulin at the P1EK1 level, and that the company planned to proceed on 17 October. Swanson closed with the transparently contradictory statement that Genentech would "continue its activities in voluntary compliance with the guidelines." [13] Following a delay of several months, William Gartland, director of the NIH Office of Recombinant DNA Activities, replied, informing Swanson that, under the 1976 guidelines, "it would appear that P2 + EK1 conditions, rather than P1 + EK1, are appropriate for the propagation of the clones mentioned in your letter." Gartland's letter also noted that the revised guidelines, which came into effect on 28 December 1978, required prior approval by the RAC. [14]

Genentech's action undoubtedly placed the NIH and the DHEW in a difficult position. A public challenge to Genentech's behavior was tantamount to an acknowledgment of the weakness of the NIH controls. On the other hand, if the NIH took no action, the gap between the stringency of controls for researchers in universities and in industry might increase. On 14 March 1979, Swanson called Gartland to inform the NIH that "due to problems of proprietary information [a reference to the continuing lack of procedures in the revised guidelines for protection of trade secrets] Genentech would make most of the decisions assigned by the revised Guidelines to the Director, NIH." [15] Two weeks later, *Nature* reported that Genentech intended to exceed the ten-liter limit without seeking formal NIH approval. At the DHEW meeting with representatives of the PMA, Genentech's flaunting the NIH controls was noted by DHEW counsel Peter Libassi. [16] By June 1979, the *New York Times* reported—and Genentech confirmed—that Genentech had proceeded along this path. Genentech justified its action on the grounds that to go to the NIH would be to "risk divulging information to Genentech rivals who might force it from the Government under the Freedom of Information Act" and on the grounds that "the risks involved in this are minuscule." [17]

Following the DHEW meeting in May 1979, both the NIH and Genentech found technical reasons to argue that the company had been in compliance all along. But such technical adjustments are less significant than the memoranda and letters between Genentech and the NIH, revealing a company threatening to make policy decisions by itself and a government agency taking no action to prevent it from doing so. The incident elucidates the political pressures on the NIH following release of the 1978 revised guidelines. It shows the extent to which the NIH was being pressed behind the scenes both to lower containment levels for large-scale production and to protect trade secrets. To avoid congressional intervention, the NIH left such backstage arm-twisting unpublicized, and companies like Genentech took the opportunity to press their agendas with impunity.

The Genentech affair also brings into focus the difficult double standard in NIH relations with scientists in academia and scientists in industry. The NIH had no legal authority to require companies to comply with the guidelines: Genentech was flaunting only a moral requirement, not a legal one. In contrast, scientists funded by the NIH or working in universities receiving NIH support had no such license. Their violation of the controls could result in long and time-consuming inquiries and, if they were found at fault, in revocation of funding. While their counterparts in commercial laboratories could move quickly with experiments, university researchers were often delayed by requirements that they seek NIH permission for reduced containment and other procedures not explicated in the 1978 guidelines. University scientists had to divulge, prior to publication, full details of their research—again a discriminatory requirement in an increasingly competitive field. By 1978, this double standard—a voluntary track for industry, a mandatory one for the academic community—was already being perceived as discriminatory (section 7.1). Robert Swanson noted in a press interview in January 1979: "The guidelines haven't been a problem. What has been a problem, at least to the people in the academic community, is all the government red tape involved in getting the approvals for the research." [18]

In contrast to the influence that the scientific community and the private sector could exert to shape the NIH controls to respond to their needs, the influence of employees in affected industries was weak. One reason was that workers in the American pharmaceutical industry often were not unionized. Nevertheless, unions representing the interests of employees did follow the development of the NIH controls, generally opposing NIH efforts to accommodate industry interests through a voluntary compliance scheme and pressing for mandatory controls for the private sector. [19]

Union officials saw OSHA as the appropriate agency to control industrial processes and urged it to intervene. In 1979, the heads of OSHA and its associated research arm NIOSH, Eula Bingham and Anthony Robbins, were in principle receptive. Certainly neither administrator was ready to dismiss the issue. Bingham, a biologist who had studied deep fungus infections of the lung in the workplace, later recalled her skepticism that activities using genetically engineered organisms could pose no problems: "I watched day after day workplaces, still out there, that are *pits:* workers being taken advantage of, not being told what they're working with. . . . I just don't think we know the ecology . . . of the organisms being used. And the whole field is not based on one or two organisms. . . . So we have to ask: 'What do we know about the natural ecology of these organisms?' And gene splicers are not apt to have that knowledge. That just isn't in their training."[20]

But OSHA and NIOSH were already overwhelmed with pressing occupational health issues. Recombinant DNA was "a minor concern in the overall NIOSH world. . . . We tried to . . . stay ahead of [the issue], but in terms of overall time and energy invested, it was very small," Robbins later recalled.[21] Furthermore, in the later years of the Carter administration, OSHA was coming under considerable pressure to justify the introduction of new regulations (section 1.4). Finally, OSHA and NIOSH had little expertise in genetic engineering and were therefore dependent on NIH information. While the heads of both agencies held that OSHA, not the NIH, should regulate any industrial use of genetic engineering, it was difficult to take steps toward that end independently of the NIH. Furthermore, Reagan's attack on regulation after the 1980 election meant that unions needed to devote all their energies to more pressing occupational safety and health issues and that the interest of OSHA and NIOSH evaporated. Support for protection of occupational safety and health in the emerging genetic engineering industry, always weak in the United States, virtually disappeared.

The following examination of NIH policymaking shows how these various interests acted in the policy arena. Industrial and academic interests were similar but not identical. The strongest desire of academic scientists was in not being restrained more than colleagues in industry or abroad. Commercial firms certainly shared this, and thus interest in both sectors converged on the weakening of the NIH requirements. But there were also important differences: industry was far more strongly committed to maintaining secrecy of research and production processes; moreover, most corporations felt the NIH controls were a useful shield against litigation. Consequently, and unlike some of their academic colleagues, industry representatives were unlikely to press for abolishing the guidelines.

9.3 The Status of the Hazards Debate

Despite many hundreds of pages and hours of time devoted to debating the hazards of genetic engineering, empirical evidence about the behavior of genetically altered organisms when the NIH guidelines underwent their first formal revision (December 1978) was scant. The only point of strong scientific consensus was the "epidemic pathogen" argument, the outcome of scientists' efforts to assuage public anxieties about genetic engineering in the 1970s. As earlier chapters have shown, this argument represented a radical reduction of the range of possible hazards and lacked any relevance to what the GMAG defined as the primary issue, namely, hazards to individual workers.

Thus as the new RAC assumed its duties in 1979, uncertainties regarding recombinant DNA hazards persisted. Few risk assessment experiments had been completed, and advances in genetic engineering were opening up several hazards raised previously only as theoretical possibilities (section 3.4). What impacts could be expected from engineered bacteria making insulin, or growth hormone, toxins, or enkephalins in large quantities? Might bacteria whose products were not quite identical to human proteins trigger the immune system in a human host to mistakenly turn its defenses against the body's own tissues? The novelty of the techniques precluded swift answers to such questions. Not surprisingly, there was little consensus. The tendency of some scientists to dismiss these concerns was responded to in the following terms by Robert Williamson, one of the trade union scientists on GMAG: "I'm astounded that anyone could [categorically dismiss such possibilities]. My colleagues admit that they don't know much about the way bacterial infections with antigenic determinants are involved in autoimmune disease. . . . Who are the people who are saying that they understand what goes on in the infectious potentiation stage of rheumatoid arthritis [an autoimmune disease]?"[22]

Moreover, organisms other than *E. coli* K12 were actively surveyed for genetic alteration, and little was known about these new hosts' survival in nature or ability to transmit their plasmids to other organisms. Furthermore, the large-scale fermentation processes using genetically engineered organisms that had reached the drawing boards in the pharmaceutical industry also posed novel problems. A report to the Commission of the European Communities issued in 1979 cited possible new hazards resulting, not from the novelty of the organisms used in the processes, but from "the very large scale on which some micro-organisms or their products are about to be grown for the first time, the very large number of people who will encounter these newly available microorganisms or their products, and the possibility that the environment might be altered by their use or inadvertent release." The report cautioned that workers might be subjected to dense aerosols of micro-organisms at various stages of fermentation and processing, with unpredictable results. Inadequate techniques for monitoring the organism in a fermenter might miss changes in its characteristics or contamination of the culture that could make virulent a large volume of organisms thought to be harmless. Organisms released into the air, water, or sewer system outside a factory could transfer harmful characteristics to other more robust organisms or facilitate the colonization of new niches. Expansion of the fermentation industry could swamp sewer systems or demand the widespread use of hazardous chemicals with special handling and waste-disposal problems.[23]

The establishment of a risk assessment program as a part of the new procedures announced in December 1978 acknowledged that only further experimental evidence could accurately assess the evolving technology. A memorandum from Secretary of Health, Education, and Welfare Joseph Califano to the Assistant Secretary for Health and the NIH director issued at that time seemed to confirm this view.

> With the issuance today of revised Guidelines for recombinant DNA research, the responsibility of the National Institutes of Health to conduct and support experiments designed to determine the risks of recombinant DNA research becomes even more important than it has been in the past. The revised Guidelines now require a finding by the Director of the NIH that each proposed action under the Guidelines presents no significant risk to health or to the environment. It is critical that these judgments, to the maximum extent possible, be based on the firm foundation of documented research that is subject to peer review.
>
> Experience and knowledge gained from the broad range of recombinant DNA research already underway will provide much information for assessing risks. But in many areas special research and careful attention will be needed. To discharge our responsibility to assess risk before certain research is conducted on a wide-spread basis, NIH should formulate a plan for carrying out a balanced program of more such risk-assessment experiments. . . . The first such plan should be ready for publication and submission to the Advisory Committee [RAC] by March 30, 1979.[24]

The duly-initiated program did produce new experimental data, whose interpretation consequently provoked further technical controversies (section 9.6). Generally, however, debate over genetic engineering hazards did not much impede changes in policy, and the results of risk assessment efforts usually lagged far behind the adoption of the deregulatory proposals whose safety such efforts were intended to appraise.

9.4 The Wye Meeting

International meetings continued to catalyze development of a shared discourse about the recombinant DNA problem as well as of a political solution among scientists, industrialists, and government administrators on both sides of the Atlantic. In the period examined in this chapter, perhaps the most influential meeting was held in April 1979 in the United Kingdom, at Wye College in Kent, shortly before the second meeting of the newly constituted RAC. The event was sponsored by the British Royal Society and COGENE, a private, nongovernmental scientific organization. Members of both organizations had addressed the issue from the beginning; some of the most influential, present at the Wye meeting, strongly favored dismantling controls.

The organizers presented the meeting as an expert group whose purpose was to review the current status of the field and to reassess the need for controls.[25] Indeed, the design of the meeting appeared wide-ranging both in content and in representation. The 150 participants were drawn from all parts of the industrialized world; there were also a few participants from nonindustrialized countries. The topics addressed included recent advances in genetic engineering, applications, risk assessment, and government controls. However, something other than the free exchange of ideas in an open arena was perhaps anticipated. Participation was limited and expensive, and press coverage was inhibited. An editorial in *Nature* later observed, "First the meeting was open; then it was closed to journalists. Then it was open to three journalists picked by the Association of British Science Writers, but the proceedings were to be 'off the record.' Then at the last minute all restrictions were lifted."[26] The removal of restrictions was too late for further members of the press to attend, however. The result was a mainly private meeting at which strong views on the recombinant DNA issue, particularly views critical of caution, could be aired without risking banner headlines in the British press.

Furthermore, on the matters it planned to reassess, representation of views was noticeably skewed. The positions of scientists from countries other than the United States and the United Kingdom probably varied, but the more vocal scientists and officials, drawn mainly from the latter countries, displayed a firm commitment to dismantle controls. Only a few scientists present—and none from the United States—favored continuing the existing controls.

The tone of the meeting (as well as a position on the more restrictive British controls) was set in the introductory remarks of British biologist Michael Stoker, foreign secretary of the Royal Society and member of COGENE. He lambasted the British safety legislation and genetic engineering controls, depicting the British as "hoist with [their] own petard" and claiming that those responsible for developing the controls were now "trying to devise a means of escape from the safety net or should I say prison bars, which prevent us from reacting rapidly to changing ideas evident in the rest of the world" (xx).

Throughout the meeting, this tone and attitude toward safety controls persisted. The loudest voices, especially those of American scientists, reiterated the arguments of the new discourse. Former RAC member Waclaw Szybalski, long hawkish on the NIH controls and now a zealous proponent of deregulation, opened a session on government controls by claiming that "many novel experiments are currently blocked by regulations" and that the problem could only be solved "by abolishing [the guidelines] or replacing them by a single sentence" (147). In a paper delivered in absentia, NIH director Donald Fredrickson described the revised NIH guidelines as a "new set of rules painfully formulated during this unprecedented curtailment of experimentation in biology" (151), and predicted "their eventual elimination when the need passes" (156). Unbecoming slides of critics of NIH policy shown during this presentation left no doubt about Fredrickson's views. British historian of science Edward Yoxen later observed, "It was a most extraordinary performance by a supposedly neutral public official."[27] The new RAC chair, Jane Setlow, expressed the same cynicism about the recombinant DNA issue and the involvement of the wider public in policy, commenting that the RAC "would not be allowed to die as a committee until we have vomited enough."[28] Robert Pritchard, a professor of biochemistry at the University of Leicester and a vocal critic of the British GMAG, urged the signatories of the Berg committee's letter to retract their earlier position (227). And John Tooze, executive secretary of EMBO, dismissed the need for risk assessment experiments (a reversal of EMBO's earlier position) and advocated—as leading proponents of deregulation would also do at the forthcoming meeting of the new RAC—that all work with *E. coli* K12 should be exempted from control. The only worthwhile way to do risk assessment was, Tooze insisted, to "encourage its use in unfettered scientific inquiry" (173–74).

Participants at the Wye meeting also witnessed remarkable recantations by several of those closely associated with the initial recombinant DNA "moratorium." Maxine Singer called her activity at the Gordon conference in 1973 "one of the worst things that ever happened to me" (185). James Watson portrayed the Berg committee's letter as a product of "liberal guilt . . . a silly emotional response which we as scientists should be ashamed of" (191). He offered to atone by going to jail. "I was," he now insisted, "a jack-ass" (236). When MIT science historian Charles Weiner questioned this revision and recalled the original reasons for concern, he was coldly received (281–87). The dominant discourse surrounding genetic engineering hazards was being made over, and those promoting the change did not want to be reminded that only a few years earlier they had conceptualized the implications of genetic engineering very differently.

In this atmosphere, more cautious attitudes toward the issue of recombinant DNA controls began to appear radical, even anachronistic. A paper presented by former GMAG chair Sir Gordon Wolstenholme, defending the broad composition of GMAG and its value for generating "mutual understanding, compromise, and public reassurance" (167–70), got a cool reception. The only recorded response to it came from molecular biologist and Biogen scientist Charles Weissmann, who observed that in Switzerland responsibility for safety was vested solely in the principal scientists and heads of research laboratories pursuing recombinant DNA work, a remark that drew loud applause (167–70). A proposal by international lawyer Edith Brown Weiss that scientists should take steps to ensure that the Biological Weapons Convention's ban on possession of biological and toxin weapons applied to genetic engineering was transformed by Watson into an irrelevant discussion of a specific biological warfare scenario (251–57). A pointed request by ASTMS representative and GMAG member Donna Haber that scientists debate their second thoughts about genetic engineering hazards in public rather than out of sight of the public in private meetings was ignored (235–37). Scattered comments from those who did not share the rosy revisionism of the majority, who still insisted on the rights of affected sectors of society to participate in policy decisions, who were not fully persuaded by "new evidence" on hazards, or who raised the question of possible abuses of the technology were generally dismissed, sometimes crudely.[29]

The participants briefly reviewed experiments in the NIH- and EMBO-sponsored risk assessment programs. Malcolm Martin presented some recently published results of the Rowe-Martin tumor virus experiment as well as further unpublished results; preliminary findings from a parallel European study were also presented (195–203). Ann Skalka, a molecular biologist at the Roche Institute of Molecular Biology and chair of a COGENE working group on risk assessment, cited work by Sagik, Freter, Levy, and Chatigny on the survival of strains of E. coli K12 in various environments and the monitoring of laboratory workers for exposure to E. coli K12 strains by Mark Richmond at the University of Bristol (211–14).

These new studies produced a complex array of results, some surprising, others contested, and many begging further questions (section 9.6 and appendix B). In general, however, the cursory discussion at Wye overlooked complexities and emphasized the most reassuring features of these new results. Martin, for example, concluded that "taken together these results indicate that it is far safer working with E. coli containing recombinant viral DNA than intact virions and are highly reassuring with respect to the safety of cloning in such systems. . . . We feel that the results of this experiment can be extrapolated to many other groups of viruses" (198, 209). Skalka, while acknowledging the incompleteness of the NIH studies, held that "no special risk inherent to recombinant DNA research has been defined" and that, "[w]ith respect to experiments involving E. coli K12 systems, it appears that little if any grounds for concern remain. Regardless of the nature of the cloned fragment or even its ability to be expressed, accepted microbiological practice should be sufficient to prevent escape and survival." (She qualified this optimism by noting that the COGENE working group did "not consider it justified to extrapolate from experience with E. coli to all possible experiments in the future" [213]) In general, the broad claim for safety being promoted met little scrutiny. A *Nature* editorial following the Wye meeting asked: "What really are the arguments that the risks are negligible? Although there were some interesting talks which were relevant to the question, there was no scientific *debate* of the matter. Where were the critics— and there are scientific critics, with genuine arguments—to bring matters to a head? To the outside observer the view that risks are negligible went through on the nod." [30] The Brenner risk assessment scheme, which GMAG had initiated on a trial basis only months earlier, was not addressed. When asked why not, one participant's response was that the meeting's organizers did not like the scheme. [31]

What the absence of a balanced technical analysis and debate suggested to one of the British participants was that Wye "was another political meeting like Ascot. There was no real forum for dissent. From the chairman's introductory remarks onwards, it was clear that the meeting was being held to say that 'recombinant DNA is not dangerous' and not to discuss whether we have new knowledge. I think that they're mainly right [about the absence of hazard] but [that idea] was being sold as a scientific discussion and it was more than that." [32]

Perhaps the most important political message of the Wye meeting concerned American intentions. "Reentry" (into the traditional system in which scientists controlled safety issues) and "disengagement" (from what was defined by the majority as the pernicious intrusion of the public into science) were dominant themes (as though genetic biology's venture into politics was an aberration). Although the organizers failed to get a call for dismantling controls, they succeeded in conveying a sense that the American controls were on their way out, with the support of the NIH and leading scientists. The message to the United Kingdom and other nations was that maintaining existing controls would shackle those racing to develop the recombinant DNA field. A British participant stated a few months after the meeting: "European countries know that America is powerful, Britain is weak. Whatever the virtues of the British system, they'll go for the American one—with the powerful backing of EMBO, which has no time at all for the British scheme." [33] Once the new discourse of "unfettered science" and "negligible hazard" became established in the United States, the powerful geopolitical position of that country virtually assured the diffusion of the discourse elsewhere.

9.5 The New Recombinant DNA Advisory Committee

In the United States, the immediate instrument of recombinant DNA policy was the reconstituted RAC—the product of the policy changes made in December 1978, which modified the mechanisms through which NIH policy was made, along with the revisions of the NIH controls (section 7.7). The composition of the RAC was broadened to include representation of fields other than science, such as law, public policy, environmental safety, and ethics. The lengthy procedure for changes in the guidelines (RAC recommendations followed by public hearings and review by the director's advisory committee) was telescoped, with both technical and policy review responsibilities vested in the RAC. Significantly, however, the RAC remained a "technical committee, established to look at a specific problem." The principal change in the charge to the committee, reflecting the new discourse that hazards had been overrated, was that developing controls was no longer tied to producing evidence about the hazards.[34]

The procedural change meant that revision of the guidelines could now occur incrementally, at each of the quarterly meetings of the RAC, rather than in quantum jumps. NIH policymakers perceived it as less cumbersome and more efficient than the earlier procedure. As Fredrickson noted in a retrospective assessment of the policy process in 1982, the new mechanism avoided the "intolerable delays and confusion" that marked the first revisions of the 1976 controls.[35] It also avoided the full glare of publicity accorded by the media to public hearings before the director's advisory committee. A single focal event could no longer fix public attention on moves to dismantle the NIH controls. The RAC thus became the crucial instrument for realizing and legitimating NIH policy. Interestingly, its charge remained virtually unchanged: according to its charter, it remained a "technical committee, established to look at a specific problem."[36]

Appointment of the New RAC

In December 1978, the composition of the expanded RAC, which included representation of such fields as public policy, law, and ethics, became the object of extended negotiation. These struggles over the appointment of the new members made partially visible what is often obscure: the importance attached to the composition of a high-level advisory committee by the government agency to which it will give advice.

Until 1978, the practical responsibility for selecting RAC members—although formally vested in the DHEW—had fallen to the NIH. Its recommended slates had been automatically approved. The controversy over the 1978 revisions and determined lobbying by public interest organizations in the fall of 1978 propelled the department into an active role in deciding the composition of the new RAC. In the last few weeks of 1978, the precise complexion of the committee became the object of prolonged and sometimes tense negotiation between the NIH leadership and the office of the Secretary of Health, Education, and Welfare, Joseph Califano.

The process started off in the usual manner: a slate of new members to add to the nine whose appointments carried over into 1979 was proposed by the NIH director to DHEW Secretary Califano. To generate this list, Fredrickson solicited nominations from private industry and the biomedical research community, but the industry nominations were not finally accepted.[37] According to PMA president John Adams, NIH officials justified this exclusion on the grounds that several RAC members drawn from academia also served as industry consultants.[38]

DHEW officials modified the list by applying several criteria. An effort was made to assure balance in terms of gender and ethnic and regional background—a standard applied to all government committees. To help ensure that broad policy issues might be addressed within the RAC, representation of fields in addition to the biomedical sciences was sought. And an informal criterion was also used by the DHEW officials: candidates for the committee were screened in terms of their position on NIH policy: "for" or "against" as one DHEW official later recalled.[39] In the words of another official, "we looked for a balance [of positions] to the extent that there were two extremes."[40] Responding to pressures from environmental organizations, trade unions, and other groups for a cautious, preemptive policy for recombinant DNA technology, DHEW officials nominated several people who had strongly questioned previous NIH policy decisions.

Few details of the negotiations between the DHEW and the NIH that followed are part of the public record, but it was later acknowledged that these were difficult. One DHEW official recollects a "strong negotiating process."[41] Some of the difficulties surfaced in an angry editorial by Maxine Singer in *Science* in January 1979, which complained that DHEW's handling of the revision process was "rapidly eroding" the scientific community's confidence in the wisdom of official efforts to deal with the recombinant DNA issue. As evidence, she cited the transfer of authority to appoint RAC members from the NIH to the DHEW and subsequent action in which, "contradicting its own definition of good process, HEW considered new members without adequately consulting the scientific public and in disregard of much of the NIH's advice." Singer went on to add, however, that "in response to the NIH urging and intervention by an alerted community, the most misguided inclinations were finally corrected." And she noted that these "inclinations" included "questioning the qualifications of an eminent molecular biologist, himself one of the first to call for caution [this was Nobel laureate David Baltimore, who was also chair of the board of scientific advisers and a major shareholder in the genetic engineering company Collaborative Research], questioning the independence of NIH scientists serving as RAC members; and consideration of individuals known to be intractably opposed to the research."[42] Singer charged that the DHEW's "insensitivity to vital issues" was still apparent in the final composition of the new committee. But her complaints reveal the NIH and its constituency as by far the most powerful influence at work: only the NIH was in a position to insist that its candidates be appointed, and the NIH alone was repeatedly consulted until the composition of the new RAC was settled.[43]

The DHEW–NIH negotiations produced a RAC which, while its composition was diverse in terms of gender, ethnic and regional background, field of interest, and position on NIH policy, nevertheless included at least ten members who could be expected, on the basis of their previous statements or role within the NIH or affiliations with the genetic engineering industry to support further relaxation of controls (table 9.1).[44] In contrast, only three members (nominated from outside the biomedical research community) had established positions critical of NIH policy. A further member, also nominated independently, could be expected to focus on the interests of employees in the new genetic engineering industry. The remaining ten members had few views on public record. Five were research biologists, working in institutions that were engaged in recombinant DNA activities. Three from other professional fields had had little exposure to the genetic engineering issue. And finally, two were dovish members of the previous RAC. Thus, although no clear majority on the 1979 RAC openly supported further relaxation of the NIH controls, its largest and most influential block could be expected to do so, and only a small minority had embraced more cautious positions. Furthermore, a solid majority of the members had primary professional ties to institutions engaged in biomedical research, including private industry.

A new chair was also appointed: Jane Setlow, a research biologist at Brookhaven National Laboratory who had served on the committee from its inception. As one of its most outspoken and hawkish members, she left observers in no doubt about her commitment to dismantling the NIH controls. Her often cynical views about the issues posed by genetic engineering, her conviction that these issues were best handled by research scientists, and her resistance to the involvement of other sectors in decision making were well known and occasionally expressed vociferously. But among the nonscientist members of the RAC, only those who held other views found bias in her handling of committee business, perceiving her as impatient, intolerant, and unreceptive to points of view that differed from her own. This was all but invisible to those who shared the chair's commitments.[45] Setlow chaired the committee until June 1980, when she was succeeded by Ray Thornton, a former chair of the House Subcommittee on Science, Research and Technology. From his handling of the genetic engineering issue in that committee, Thornton was known to NIH officials as a sympathetic supporter of genetic engineering research, committed to maintaining the U.S. leadership in

Table 9.1   Membership of the U.S. Recombinant DNA
Advisory Committee, February 1979

Jane K. Setlow, Biologist, Brookhaven National Laboratory, Upton New York (chair)
(continuing)

Karim A. Ahmed, Senior Staff Scientist, Natural Resources Defense Council, New York
(new*)

David Baltimore, Professor of Biology, Massachusetts Institute of Technology,
Cambridge (chair, Board of Scientific Advisers, Collaborative Research) (new)

Francis E. Broadbent, Professor of Soil Biology, Department of Land, Air, and Water
Resources, University of California, Davis (new)

Allan M. Campbell, Professor, Department of Biology, Stanford University, Stanford,
California (continuing)

Zelma Cason, Supervisor of Cytopathology Laboratory, University of Mississippi
Medical Center, Jackson (new)

Peter R. Day, Chief, Division of Genetics, Connecticut Agricultural Experiment Station,
New Haven (continuing)

Richard Goldstein, Assistant Professor of Microbiology and Molecular Genetics, Harvard
Medical School, Cambridge (new*)

Susan K. Gottesman, Senior Investigator, Laboratory of Molecular Biology, National
Cancer Institute, National Institutes of Health, Bethesda, Maryland (continuing)

Richard B. Hornick, Chairman, Department of Medicine, University of Rochester School
of Medicine, Rochester, New York (new)

Patricia A. King, Professor of Law, Georgetown University Law Center, Washington,
D.C. (new)

Sheldon Krimsky, Acting Director, Program in Urgan, Social, and Environmental Policy,
Tufts University, Medford, Massachusetts (new*)

Elizabeth M. Kutter, Member of the Faculty in Biophysics, Evergreen State College,
Olympia, Washington (continuing)

Richard P. Novick, Chairman of Plasmid Biology, Public Health Research Institute, New
York (new)

David K. Parkinson, Associate Professor of Occupational Health, University of
Pittsburgh, Pittsburgh (new*)

Ramon Pinon, Assistant Professor of Biology, University of California, San Diego (new)

Samuel D. Proctor, Professor of Education, Rutgers University, New Brunswick, New
Jersey (new)

Emmette S. Redford, Ashbel Smith Professor of Government and Public Affairs,
University of Texas, Austin (continuing)

Wallace P. Rowe, Chief, Laboratory of Viral Diseases, National Institute of Allergy and
Infectious Diseases, National Institutes of Health, Bethesda, Maryland (continuing)

John Spizizen, Member and Chairman, Department of Microbiology, Scrips Clinic and
Research Foundation, LaJolla (continuing)

Ray H. Thornton, Executive Director, Joint Educational Consortium, Henderson State
University, Ouachita Baptist University, Arkadelphia, Arkansas (new)

LeRoy Walters, Director, Center for Bioethics, Kennedy Institute, Georgetown
University, Washington, D.C. (continuing)

Luther S. Williams, Associate Professor of Biology, and Assistant Provost, Department of
Biological Sciences, Purdue University, West Lafayette, Indiana (new)

Frank E. Young, Dean, School of Medicine and Dentistry, University of Rochester,
Rochester, New York (contract with Miles Laboratories) (new)

Milton Zaitlin, Professor, Department of Plant Physiology, Cornell University, Ithaca,
New York (consultant, Plant Genetics) (continuing)

Source: Recombinant DNA Advisory Committee, minutes, 15–16 February 1979; Sheldon
Krimsky, data on links with pharmaceutical and genetic engineering firms.
*Members nominated by public interest organizations and trade unions.

the field. He was also an experienced parliamentarian and universally seen as a skilled chair, handling issues in an impartial manner.

What could be predicted from the composition of the new RAC—and was soon apparent in its proceedings—was a characteristic opposition between two main points of view: the majority position, forcefully articulated by influential spokesmen for biomedical research such as David Baltimore, Wallace Rowe, and Allan Campbell; and a minority view espoused by the few public interest and trade union nominees. The majority was characterized by a strong belief in the wisdom of self-regulation in research (reflected in frequent invocation of the need to protect "scientific freedom" and harsh criticism of NIH control over recombinant DNA activities) and increasingly, as time went on, by the new discourse promoted at the Wye meeting, that recombinant DNA hazards were negligible in almost all cases. The minority, in contrast, persisted in using and applying the fundamental assumptions and rules of the older discourse that was being displaced. These members framed the primary goal of NIH policy in terms of the need to anticipate emergent hazards to workers and to the environment of a new form of technology. They called for maintaining cautious controls and for pursuing risk assessment research, worker training programs, and health and environmental monitoring. Pointing to emergent problems with previous forms of technology—toxic wastes, the greenhouse effect, and the depletion of the ozone layer—they insisted on asking the question increasingly ignored by the majority: could similarly unanticipated harm emerge in applications of genetic engineering? Virtually all of the major proposals to weaken the NIH controls were seen by these members as premature.

Expression of this minority position on the RAC plainly annoyed and frustrated those RAC members who wanted to move quickly to dismantle the NIH controls. In addition, this position, carrying the possibility of stimulating further congressional action, may have been threatening. At the Wye meeting, Setlow and Fredrickson obviously wanted it dismissed. In fact, the minority view, however often aired on the RAC, rarely presented any significant challenge to the course of dismantling the NIH controls. With one or two exceptions, minority positions were voted down. Moreover, the lingering of the displaced discourse had an important function. At the hearings of the Senate Subcommittee on Science, Technology, and Space in May 1980, David Baltimore portrayed the RAC in the following way:

> The RAC is a remarkable organization because about 50 percent of its membership consists of public members who are responsible to a variety of constituencies other than the scientific community. The RAC also contains people who have been committed to the notion of extraordinary hazard in recombinant DNA work, as well as people who have

been committed in the opposite direction. The RAC deliberations thus take into account a variety of perspectives. . . . it is hard to imagine a more appropriate body to [review industrial proposals]. Having both public and scientific members, and being open to public scrutiny, the RAC's activities involve multiple perspectives.[46]

A contained public voice within the RAC legitimated the policy role of the committee while constituting little more than an annoyance to those whose goal was to eliminate most controls.

Dynamics of RAC Decision Making

RAC procedures for setting and addressing the agenda made it likely that issues associated with a preventive approach to policymaking would remain in the background. The RAC agenda was largely set by the NIH director and an executive committee (often referred to as the "kitchen RAC") of NIH officials and scientists.[47] These agendas were extremely large, reflecting four principal functions of the committee: (1) implementing the 1978 guidelines, especially reviewing the large flow of requests for NIH approval of an ever wider array of hosts, vectors, and source of DNA for cloning purposes; (2) revising the 1978 guidelines; (3) developing procedures for responding to requests from the private sector; and (4) implementing the risk assessment program initiated in December 1978 as a part of the new NIH policy.

In principle, new items to the RAC agenda could be added either by RAC members or by the public. However, various formal and informal procedures ensured that the committee rarely strayed far from the agendas accepted by the NIH director and his "kitchen RAC."

The tone of many of the meetings of the new RAC, especially those that addressed important changes in policy, was set personally by Donald Fredrickson. At its February 1979 inaugural meeting, Fredrickson depicted the committee as an important "experiment" in policymaking. But he did not hesitate to project its outcome, emphasizing that delays would be "completely intolerable to all the scientists . . . who are dependent on the decision making capacity represented around this table . . . and disastrous not only for [the immediate problem at hand] but also for attempts to broaden the [application of genetic engineering in science and industry]." Introducing the theme that he would develop for the Wye meeting, Fredrickson characterized the committee as an "instrument which will permit continuing revision of the guidelines to go on, proceeding to where they must take us, certainly when we understand enough, to the elimination of this set of rules." Although he confirmed that Califano's request for risk assessment would be given "high priority," he expected such experiments merely to reinforce the prevailing consensus: "I believe we will come to an understanding of these techniques and their potential for creating both good and harm to a degree where it will not be necessary for us to have guidelines for their use. I am convinced of that. But I don't know when that will occur. That will occur when we understand enough to all agree that it should take place." Significantly, no member chose to examine the contradiction involved in launching an experimental program and predicting its outcome.[48]

Fredrickson's directive to the committee not to hamper genetic engineering was reinforced by the organization of RAC meetings. To a certain extent, the running of these meetings might be seen as a function of the personality of the new chair, Jane Setlow, who frequently registered her enthusiasm for releasing the field from what she plainly perceived as an unwarranted bureaucratic intrusion. Beyond the question of the individual bias of a particular chair, however, organizational dynamics also inhibited significant changes in committee agendas. Consistently heavy, their supporting material often ran to hundreds of pages. Furthermore, the rapidity with which genetic engineering advanced and the RAC's confrontation with state-of-the-art experiments and processes meant that even the scientist members had difficulty comprehending technical and policy issues posed. For members outside the field, the problem was compounded. "Much of the discussion excluded non-technical people . . . and was sometimes used as a way of pushing things through. There was too much willingness to accept data on hearsay," observed a scientist member who often voted with the majority.[49]

A crucial limitation on altering the agenda was revealed at the first meeting. Briefed in general on recombinant DNA experimentation, the committee immediately took up the tightly packed agenda, for which roughly half the supporting documents had been received only days earlier. ("This is really two RAC's worth of meetings," an NIH official commented at the time.)[50] On some issues, discussion revealed differences in expert opinion and that few members had had the time to examine primary sources. How the size and nature of the agenda gave structural leverage to the RAC's majority came into sharp focus when one of the new members, David Parkinson, suggested that votes on further requests for approval be postponed and that the second day of the proceedings be spent coming to an understanding of the issues raised by these proposals. Parkinson's move triggered a minor explosion. The chair insisted that the "gigantic agenda" required that the committee keep voting. David Baltimore warned that "to do anything that blocks approval [of agenda items] is going to lead to rebellion of the scientists." Parkinson's motion was voted down, and the meeting proceeded.[51]

Later votes on other procedures that would have helped to elucidate the policy issues embedded in requests for approval of experiments reinforced this initial decision. A motion by Elizabeth Kutter at the May 1979 meeting that NIH officials provide summaries in lay terms of each agenda item met a countermotion from David Baltimore to table; the tied vote on the latter was broken by the chair in favor of tabling.[52] Such contests are important not only in showing how the committee's decisions were shaped but more generally as indicators of the location of power within the committee: they confirm that control of its agenda rested with the NIH and the RAC's biomedical research majority.[53] "Participation" by those members chosen to represent other sectors was mainly limited to questioning positions accepted by the majority without displacing these positions. The few exceptions to this generalization are revealing. They occurred when the NIH-appointed scientist members of the RAC were seriously divided (section 10.2).

In summary, the independently nominated public interest members of the new RAC played a role that differed in some respects from that played by the public interest members of GMAG. Within GMAG, a committee that proceeded mainly by achieving consensus, public interest members were expected, when necessary, to mediate between the opposed positions of labor and management, or of labor and science. Within the new RAC, a committee that proceeded by voting, the interests of the public interest members were largely contained. But despite the differences in the political interactions of each committee, the public interest members of both had similar legitimating functions: in each case, the responsible government agency could claim that the public interest was represented.

9.6 The Rowe–Campbell Proposal: The First Move toward Dismantling the NIH Controls

Following its first meeting in February 1979, the new RAC embarked on an extraordinarily complex series of decisions that eventually reversed the legacy of the Asilomar conference (the assumption that unknown hazards of genetic engineering should be contained biologically and physically), permitted an almost unlimited expansion of cloning hosts and vectors, removed most of the six original prohibitions, and virtually dismantled procedures for review and oversight (table 9.2).

Many of these changes were accomplished through a series of major proposals that were addressed by the RAC in 1979–82 (table 9.3). The first of these was a sweeping proposal from NIH virologist Wallace Rowe and Stanford professor of biological sciences Allan Campbell to exempt from the NIH controls all nonprohibited experiments that used the approved *E. coli* K12 host-vector systems.[54] Because this proposal represented the first major change of policy to be addressed by the new RAC, it was seen by its supporters and critics alike as setting a crucial precedent

Table 9.2 Revisions of the U.S. NIH Guidelines, 1978–82

Date	Containment Requirements (selected experiments)	Oversight for Government-Funded Work	Oversight for Work in the Private Sector	Prohibitions
December 1978	Containment levels for work with *E. coli* K12 ranges from P1EK1 to P3EK2	Prior review and approval by IBCs; registration with NIH	Voluntary registration	Six classes; cloning of DNA from class 3, 4, or 5 from P1 to P3 pathogens or from oncogenic viruses classified as moderate risk Cloning of genes encoding toxins. Creation of plant pathogens likely to have enhanced virulence or host range. Deliberate release into the environment. Transfer of drug resistance trait not known to occur naturally Large-scale experiments (greater than 10 liters)
January 1980 (response to Rowe-Campbell proposal)	Experiments using *E. coli* K12 that are not prohibited and not exempt: P1	*E. coli* K12 experiments that are not prohibited and not exempt: registration with IBCs; no prior review, except for experiments involving expression of genes Other experiments: prior review and approval by IBCs; registration with NIH	Voluntary compliance scheme for private sector, with procedures for protection of trade secrets and commercial and financial information	Six classes retained
November 1980 (response to Singer proposal)	Experiments using *E. coli* K12 and *S. cerevisiae* systems that are not prohib-	Elimination of NIH oversight for all experiments assigned containment levels in	RAC responsible only for setting containment requirements for L–S processes	Five classes of prohibition; prohibition of the creation of plant pathogens with increased host range and virulence deleted

ited and not exempt: P1		the guidelines; IBCs responsible for reviewing research for compliance with the guidelines	Prohibition on the formation of rDNA containing the genes for biosynthesis of toxins limited to a few lethal toxins; other four classes retained
July 1981 (response to IBC chairs' proposal)	Experiments using approved E. coli, S. cerevisiae, and B. subtilis systems that are not prohibited: exempt; P1 recommended	Nonexempt experiments assigned containment levels in guidelines; prior review and approval by IBCs	Prohibition lifted on large-scale processes using approved E. coli, S. cerevisiae, and B. subtilis systems; such processes require only IBC prior review and approval
October 1981 (response to Lilly proposal)			
April 1982 (response to Baltimore-Campbell proposal)	For nonpathogens: P1.	Elimination of NIH oversight for all experiments and processes except for three classes of previously prohibited experiments; IBCs and/or principal investigators responsible for all other experiments and processes	All prohibitions removed; cloning of genes encoding highly lethal toxins, deliverate release into the environment, and transfer of drug resistance trait not known to occur naturally require RAC review and NIH and IBC approval before initiation.

Sources: Department of Health, Education, and Welfare, National Institutes of Health, "Guidelines for Research Involving Recombinant DNA Molecules," *Federal Register* 43 (22 December 1978): 60108–31; Department of Health, Education, and Welfare, National Institutes of Health, "Guidelines for Research Involving Recombinant DNA Molecules," *Federal Register* 45 (29 January 1980): 6724–49; Department of Health, Education, and Welfare, National Institutes of Health, "Recombinant DNA Research: Actions under Guidelines," *Federal Register* 45 (21 November 1980): 77372–81; Department of Health and Human Services, National Institutes of Health, "Guidelines for Research Involving Recombinant DNA Molecules," *Federal Register* 46 (1 July 1981): 34462–87; Department of Health and Human Services, National Institutes of Health, "Recombinant DNA Research: Actions under Guidelines," *Federal Register* 46 (30 October 1981): 53980–85; Department of Health and Human Services, National Institutes of Health, "Guidelines for Research Involving Recombinant DNA Molecules," *Federal Register* 47 (21 April 1982): 17180–98.

Table 9.3   Major Decisions by the U.S. NIH Recombinant DNA Advisory Committee

Rowe-Campbell proposal (4/13/1979)

Proposal: To exempt from the NIH guidelines nonprohibited experiments with *E. coli* K12 host-vector systems.

NIH decision (1/29/1980): Experiments with *E. coli* K12 host-vector systems require P1 containment and notification to IBC prior to initiation; IBC prior review and approval required for experiments involving expression; no registration with NIH. Such experiments are assigned to category III-0.

Singer proposal (8/21/1980)

Proposal: To eliminate NIH review, registration, and approval for all experiments assigned containment levels in the guidelines.

NIH decision (11/21/1980): Approved with the specification that experiments be registered with IBCs and requirement for prior review and approval for experiments not in the III-0 category.

Proposal of the IBC chairs (RAC minutes, 8–9 January 1981; 3/20/1981)

Proposal: To exempt experiments in category III-0 from the guidelines. Three options proposed: review by an institutional official, elimination of registration, and exemption.

NIH decision (7/1/1981): Approved exemption of experiments in category III-0 with exception of experiments using DNA from Center for Disease Control class 3 organisms.

Baltimore-Campbell proposal (3/20/1981)

Proposal: To convert the guidelines into a voluntary code of practice, with containment requirements at least as high as the Center for Disease Control requirements for the laboratory use of the organisms used as hosts, vectors, or sources of DNA. The prohibitions would still apply.

Modified proposal developed by the RAC (12/4/1981): To convert the Guidelines into a voluntary code of practice; the function of the RAC would be to modify the guidelines not to implement them; IBC oversight would be optional.

Gottesman proposal (12/7/1981): To retain the guidelines but to relax oversight and restrictions as follows: Eliminate the prohibitions; three classes of previously prohibited experiment (cloning of genes encoding highly lethal toxins, deliberate release into the environment, transfer of drug-resistance traits not known to occur naturally) would require review and approval by the RAC.

Assign all other experiments to three procedures: prior review and approval by IBCs (e.g., experiments using Center for Disease Control class 2–5 organisms as hosts or vectors; experiments using whole animals and plants; large-scale processes); IBC notification only (e.g., experiments involving nonpathogenic prokaryotes or nonpathogenic lower eukaryotic host-vector systems; experiments involving formation of recombinant DNA molecules containing no more than two-thirds of the genome of any eukaryotic virus); and exempt experiments (e.g., all experiments using approved host-vector systems not assigned to other categories).

NIH decision (4/21/1982): Accept the Gottesman proposal with minor changes.

Note: Dates in parentheses are for the *Federal Register*.

Table 9.4   Evolution of the Rowe-Campbell Proposal: Meetings and Reports

1. Proposal: National Institutes of Health, "Recombinant DNA Research: Proposed Actions under Guidelines," 44 (13 April 1979): 22315.
2. First consideration by the RAC: RAC minutes, 21–23 May 1979; Wallace Rowe, statement to the RAC, May 1979.
3. Consideration by the RAC working group on *E. coli* K12 host-vector systems: Background Document for Proposed Exemption of *E. coli* K12 Host-Vector Systems from the Recombinant DNA Guidelines, n.d., RAC doc. 708, distributed to the RAC on 6–7 September 1979 with supplement.
4. NIAID Ad Hoc Working Group on Risk Assessment, 30 August 1979: transcript of meeting (NIAID); memo John Nutter, Chief, Office of Specialized Research and Facilities, NIAID, to the Record, 4 September 1979.
5. Second consideration by the RAC: RAC minutes, 6–7 September 1979.
6. Proposed decision of the NIH director: National Institutes of Health, "Proposed Guidelines for Research Involving Recombinant DNA Molecules," *Federal Register* 44 (30 November 1979): 69210–34; "Recombinant DNA Research: Proposed Actions under Guidelines," *Federal Register* 44 (30 November 1979): 69234–51.
7. Revised guidelines incorporating final decision on the Rowe-Campbell proposal: National Institutes of Health, "Guidelines for Research Involving Recombinant DNA Molecules," *Federal Register* 45 (29 January 1980): 6724–49.

for further decisions. Consequently, it was intensely debated at each stage of the NIH's new decision process (table 9.4).

The proposal came before the committee at its second meeting in May 1979. A vote following lengthy debate established a working group to analyze the state of evidence concerning the hazards associated with *E. coli* K12 host-vector systems and to report its findings to the committee, along with proposals for alternatives to the guidelines for ensuring that safety standards would be met. The working group duly met and produced a brief report on the evidence on hazards. In addition, NIAID, the agency overseeing the NIH risk assessment program, established the Ad Hoc NIAID Working Group on Risk Assessment to review assessment experiments on the survival of *E. coli* K12 and on the transmission of its plasmids to other bacteria. At a one-day meeting held on 30 August 1979, this group recommended against continuing these experiments and endorsed the Rowe-Campbell proposal. Following further extended debate at the RAC's meeting in September 1979, a slightly modified proposal was approved. This recommendation went to the NIH director, who, rather than exempting *E. coli* K12 activities, proposed assigning such experiments to a special category (designated III-0) that required only the submission of a registration document to an institutional biosafety committee—a decision that went as far as possible toward exemption without having that formal designation. An accompanying document justified the NIH director's decision. Following a further comment period, revised guidelines were issued on 29 January 1980.

The Rowe-Martin Experiment and Other New Evidence:
Surprises, Gaps, and Contested Meanings

The claim that new information supported a radical weakening of controls provided a fundamental justification for the Rowe-Campbell proposal. So it is important to examine the nature of this claim and its bearing on the safety of the research. The new information flowed from a tangle of evidence and arguments—some published, some not—related to possible biohazards (table 9.5 and appendix B).[55] Several experiments expressly designed to assess hazards produced results in this period—most prominently, the long-awaited results of the Rowe-Martin experiment testing the possibility that *E. coli* bacteria bearing tumor virus DNA would provide a novel route of exposure (table 9.5, I). The experiment, originally cast as a "dangerous" or "worst-case" test by Sydney Brenner at the RAC's La Jolla meeting in December 1975 (section 4.3), had been recast at the Bethesda meeting in August 1976 as a test of the containment provided by *E. coli* K12 (section 5.5).

Four experiments supported as part of the NIH risk assessment program tested the survival of strains of *E. coli* K12. Two of these responded to the first two risk assessment protocols proposed at the Falmouth conference: the colonization and transmission of plasmids by *E. coli* K12 in the human gastrointestinal tract, and tests for survival and transmission of plasmids in germ-free mice (table 9.5, II and III). The other experiments examined the survival of weakened strains of *E. coli* in sewage and on laboratory surfaces (table 9.5, IV and V). The results of this work were published in the *Recombinant DNA Technical Bulletin* in the summer of 1979 and were reviewed by the Ad Hoc NIAID Working Group on Risk Assessment on 30 August 1979.[56] In addition, evidence on the survival of *E. coli* K12 in laboratories also came from British microbiologist and GMAG member Mark Richmond, who monitored laboratory workers for exposure to the strain (table 9.5, VI).[57]

Further evidence of a preliminary nature on the survival of *E. coli* was contained in brief, unpublished memoranda submitted to the RAC by several laboratories. An Eli Lilly scientist tested the survival of strains of *E. coli* K12 carrying insulin sequences in the guts of rats and mice; Donald Brown of the Carnegie Institution of Washington examined competition between strains of *E. coli* and the same organism containing a variety of recombinant DNA molecules; David Botstein of MIT, together with Wallace Rowe, Malcolm Martin, and Hardy Chan, tested *E. coli* K12 bacteria bearing fragments of yeast DNA for enhancement of virulence (table 9.5, VII–IX).[58]

Finally, some informal theoretical analysis of the scenario raised at the

Table 9.5 Use of and Response to New Evidence and Arguments Related to the Rowe–Campbell Proposal

Experiment/Analysis	Published?	Citation				Further Investigation Advocated by Investigators?	Only χ1776 Tested?	Significance of Results Debated?
		Rowe	NIAID	RAC	NIH			
I. Rowe–Martin	Yes	Yes	No	Yes	Yes	Yes (qualified)	Yes	Yes
II. Levy	Yes	No specific discussion	Yes	Yes	No	Yes	Yes	Yes
III. Freter	Yes	No	Yes	Yes	No	Yes	No	No
IV. Sagik	Yes	No	Yes	No	No	Yes	Yes	Yes
V. Chatigny	Yes	No	No	No	No	No	No	No
VI. Richmond	No	Yes	No	Yes	Yes	No	No	No
VII. Burnett	No	No	No	Yes	Yes	No	No	No
VIII. Brown	No	Yes	No	Yes	Yes	No	No	No
IX. Botstein	No	No	No	Yes	Yes	No	No	No
X. Beckwith	No	Yes	No	No	Yes	Yes	N.A.	Range of opinion

Sources: See appendix B.
Note: N.A. = not applicable.

Falmouth conference, that bacteria making proteins similar but not identical to human proteins might trigger autoimmune reactions in individuals exposed to them, was carried out. Harvard molecular biologist Jonathan Beckwith had drawn attention to the unresolved problem in October 1978.[59] In response, Rowe solicited comments from seven immunologists.[60] Their analyses of the problem were purely theoretical: no experimental tests of bacteria designed to produce proteins similar to those of humans had been carried out (table 9.5, X).

In many respects, this new evidence posed more problems than it resolved. First, many in the scientific community, including several of those responsible for the risk assessment experiments, saw some of the results as surprising and therefore as raising new questions about hazards. Stuart Levy, before analyzing the survival of the weakened strain of *E. coli* K12 in the human intestinal tract, had predicted entirely negative results: "Believe me," Levy had told a meeting of NIH contractors for risk assessment experiments in the summer of 1977, "we all know the answer's going to be zero." The results Levy actually obtained were not nearly so unambiguous. When the weakened strains were fed to human volunteers, only the strain without a plasmid died out relatively quickly. The strain with a plasmid, designated $\chi 2236$, was recovered from the feces of the subjects for an average time of eighty-two hours after ingestion—a result the researchers described as "surprising" and "unexpected."[61] In this case, genetic alteration had enhanced the weakened strain.

Other surprises included Rolf Freter's observation that, while $\chi 1776$ did not persist long in the guts of mice (a trait that he saw as "impressive evidence of its safety for experimental work"), mutations of the strain resisted bile in the gut. The bile-resistant mutations persisted and could even colonize germ-free animals.[62] Bernard Sagik and C. A. Sorber showed that various weakened (EK2) strains of *E. coli* K12 passed through a model sewage treatment plant in viable condition, as did a lambda phage vector. The data on rates of survival suggested that one of the weakened strains (called DP50supF) might even have multiplied instead of progressively dying out.

A second complication was that these results were quite widely seen as incomplete and requiring further experimentation. The surprising implication that the weakened strains of *E. coli* might persist in various environments particularly raised the question of their ability to transmit their novel DNA to other microbes. Such studies had become, in Levy's view, "more critically important."[63] Roy Curtiss, the scientist responsible for constructing the weakened strain of *E. coli,* agreed.[64] Furthermore, many risk assessment studies had used the weakened EK2 systems; less evidence pertained to K12 strains.[65] The Levy-Marshall experiments and the

Rowe-Martin experiments had employed only χ1776. To extrapolate from the results on the weakened strain to the behavior of the K12 strain was not justifiable.

In addition, no studies had examined *E. coli* altered to produce biologically active polypeptides. To the NIH, Jonathan Beckwith argued that autoimmune effects warranted consideration: many applications of genetic engineering aimed to enable bacteria to express novel proteins on their external surfaces and a modified human protein on the outside of bacterial cells seemed more likely to reach sites in the body where an autoimmune response could be stimulated.[66] The comments solicited by Rowe mainly conveyed indifference about the autoimmune scenario, but this was by no means uniform. At least one indicated that, if certain pathways to triggering an autoimmune reaction were considered significant, they could be tested.[67] And the most detailed response, from William Paul and Richard Asofsky, the respective heads of the Laboratory of Immunology and the Laboratory of Microbial Immunity at NIAID, indicated reasons for continuing concern, especially if controls were weakened and laboratories or industries used robust strains of *E. coli* or other organisms.[68] While these scientists saw little potential risk of generating an autoimmune disease through the use of weakened strains of *E. coli,* they cautioned against using those capable of growing outside the laboratory or of transferring genes to wild-type *E. coli.* Some theoretical possibility of hazard remained. The paucity of experimental evidence drove the RAC's Risk Assessment Subcommittee to recommend a conference on hormone-producing strains of *E. coli* that would address possible adverse effects.[69] The conference took place in April 1980, but in 1979, there was no focused discussion of the issue, let alone experimental investigation.

A third problem was that the interpretation of many results reported in 1979 was contested. The results on the survival of strains of *E. coli* confirmed for some the essential safety of *E. coli* K12 host-vector systems; for others, they disconfirmed it. Paul Berg summarized the former view.

> When [the Berg committee] expressed . . . concern about potential risks associated with recombinant DNA research, we had in mind the possibility that such experimentation might employ a wide variety of *E. coli* strains and transmissible vectors as host-vector systems. We imagined that some of these would be able to establish themselves in nature or in the intestinal tracts of man and animals. Had we known then that *E. coli* strain K12 and nontransmissible plasmids or phages would be widely adopted as the preferred host-vector systems, and that this system would be as secure as current expert opinion and all the risk-

assessment experiments have shown, I doubt that we would have raised the issue in the manner we did. I am persuaded by the evidence I have seen that molecular cloning of any DNA segments in *E. coli* K12 using the array of present day cloning vectors is no longer of any real concern, certainly not enough to warrant the containment requirements or the bureaucratic, confusing, and time consuming reporting procedures demanded by the present Guidelines.[70]

Others argued that the surprising survival of certain strains made the need to investigate transmission even more important. As the weakened *E. coli*'s creator, Roy Curtiss, wrote to Donald Fredrickson, "Essentially all data obtained since 1977 by NIH contractors conducting risk assessment experiments to evaluate the safety of the *E. coli* K12 systems have indicated that host strains and bacteriophage lambda vectors survive better in many environments than previously believed. . . . As a consequence, transmission of recombinant DNA from K12 to other microbes is a more probable event than was previously believed."[71]

Above all, the results of the Rowe-Martin experiment, contested almost as soon as they were publicly announced, dramatized the difficulties and uncertainties of producing a quick assessment of the hazards. The decision at the Bethesda meeting in August 1976 to use this experiment to test the containment provided by *E. coli* K12, not a worst-case scenario, was approved by the RAC at its meeting in January 1977 and subsequently by participants at the Falmouth conference in June. But the NIH director denied a request for an exception to the 1976 guidelines to allow Rowe and Martin to run trials with the K12 strain carrying the DNA of polyoma virus, a virus that infects and causes tumors in rodents. As a result, the Rowe-Martin experiments were carried out only with the weakened strain of *E. coli,* χ1776, which was sensitive to bile salts and unable to colonize mice.[72]

The strategy adopted in these experiments was to introduce polyoma virus DNA into plasmid and phage vectors and to insert these into the weakened *E. coli* bacteria. The phage, plasmids, and bacteria bearing the polyoma virus DNA were then tested for their capacity to produce polyoma infections in mice and tumors in newborn hamsters. Two forms of polyoma DNA were used in these experiments: single copies, or monomers; and double copies, or dimers, in which two copies of the polyoma genome were joined end to end. In accordance with the design of the experiment, the frequencies of infections and tumors in each case were compared with the frequencies of infections caused by free polyoma DNA and by the polyoma virus.

Positive results from these experiments had various implications. Polyoma infection meant that the inserted DNA had been excised from the plasmid, phage, or bacterial cell and had entered intact into the cells of

Table 9.6   Results of the Rowe-Martin Experiment: Infection, Plasmid System

Form of DNA	Feeding[a]	Injection[b]	Page Reference
Polyoma DNA	—	Yes (9/10)	Table 2, 885
Plasmids with 1 copy of polyoma DNA	—	No (0/5)	Table 2, 885
Plasmids with 1 copy of polyoma DNA, cleaved with restriction enzyme	—	Yes (5/5)	Table 2, 885
Plasmids with 2 copies of polyoma DNA[c]	—	—	886–87
E. coli with 1 copy of polyoma DNA inserted in plasmid	—	No (0/45)	Table 3, 885

Source: Mark A. Israel et al., "Molecular Cloning of Polyoma Virus DNA in *Escherichia coli*: Plasmid Vector System," *Science* 203 (2 March 1979): 883–87.

[a] Results of feeding experiments were not reported. The investigators commented that "since the parenterally inoculated bacteria did not induce PY infection, it seems highly unlikely that oral administration of the PY plasmid in strains of E. *coli* capable of surviving in the intestinal tract would have given positive results."

[b] Numbers in parentheses are number infected/number inoculated.

[c] Unable to obtain plasmids containing two copies of polyoma DNA (886–87).

the test animal. Tumor induction required the activity of only a single gene and did not depend on the transfer of the intact viral genome into the cells of the host animal.

In 1979, Rowe and Martin reported the results of three sets of experiments. The first set tested plasmids carrying polyoma DNA and E. *coli* χ1776 bacteria carrying the recombinant plasmids. Mice fed or injected with these were examined for polyoma infection. The second set used phage lambda as the vector. A third set tested for tumor induction by the E. *coli* plasmids and phage systems in newborn hamsters.

The results are summarized in tables 9.6–9.8. Reporting in *Science* in March 1979, the investigators observed no cases of infection in the mice when a single copy of the polyoma DNA was used. When dimer DNA was used, however, infections were found in two of the five mice injected with phage DNA and in six of the eleven injected with the E. *coli* DNA containing the recombinant phage. Such levels of infection were comparable to that caused by free polyoma DNA and roughly 10^4-fold less infectious than polyoma virus itself. (For reasons that are unclear, the investigators were not able to obtain plasmids containing two copies of polyoma DNA.)[73]

Moreover, the investigators demonstrated that polyoma DNA retained its potency when spliced into a plasmid or a phage. When these latter were cleaved with the same restriction enzyme used in their construction, the freed polyoma DNA could still cause infection, thus confirming a route by which the test animals had acquired the virus.

In the third set of experiments, reported in *Science* in September 1979, tumor formation was observed in three out of sixteen hamsters injected

Table 9.7   Results of the Rowe-Martin Experiment: Infection, Phage System

Form of DNA	Feeding	Injection[a]	Page Reference
Polyoma DNA	Yes		Table 3, 891
		Yes (5/5)	Tables 2, 3, 891
Phage with 1 copy of polyoma DNA		No (0/5)	Table 3, 891
Phage with 2 copies of polyoma DNA		Yes (2/5)	Table 3, 891
			Table 2, 891
DNA from phage with 2 copies of polyoma DNA		Yes	Table 2, 891
			Table 3, 891
E. coli with 1 copy of polyoma DNA		No (0/5)	Table 2, 89
			Table 3, 891
E. coli with 2 copies of polyoma DNA[b]		No (0/23)	Table 2, 891
			Table 3, 891
DNA from E. coli with 2 copies of polyoma DNA		Yes (2/5)	Table 2, 891
			Table 3, 891

Source: Hardy W. Chan et al., "Molecular Cloning of Polyoma Virus DNA in Escherichia coli: Lambda Phage Vector System," Science 203 (2 March 1979): 887–92.

[a] Numbers in parentheses are number infected/number inoculated.

[b] The investigators expected mice injected with live E. coli carrying the phage with polyoma DNA in the dimer form to be infective, but they obtained a negative result. They hypothesized that phage replication was probably inhibited by the subcutaneous environment, which was anaerobic: optimal phage reproduction was known to require aerobic conditions.

with lambda phage containing the double copy of polyoma DNA (a level comparable to the level observed with free polyoma DNA) and two out of twenty-seven hamsters injected with plasmids containing the monomer. No tumor formation was observed with E. coli bacteria containing plasmids; however, tumor induction produced by E. coli bacteria bearing double copies of polyoma virus DNA carried by plasmid vectors and by E. coli bacteria bearing single or double copies of polyoma virus DNA was not tested.[74]

A further result, reported by Rowe and his colleagues in the Proceedings of the National Academy of Sciences in August 1979, had the campus of the NIH buzzing.[75] Cleaving the polyoma DNA with a variety of restriction enzymes had markedly enhanced its tumor-inducing potential. In one case, 100 percent of the animals tested with the fragments of polyoma DNA developed tumors.[76]

Debate about the significance of the Rowe-Martin experiments erupted with publication of the first results and continued both inside and outside the RAC for much of 1979. The papers on infection in mice together with a preprint of the paper on induction of tumors in newborn hamsters were discussed at the May 1979 meeting of the RAC. The September meeting took up the whole series of results, for they had bearing not only on the capacity of E. coli to transmit tumor virus genes to host

Table 9.8   Results of the Rowe-Martin Experiment: Tumor Induction

| | Result of Injection[a] | | | |
| | Uncleaved | Cleaved | | |
Form of DNA		Eco RI	Bam HI	Page Reference
Polyoma DNA	Yes (4/73)	Yes (29/64)	Yes (11/35)	Table 2, 1141
Plasmid with 1 copy polyoma DNA	Yes (1/11) (1/16)	Yes (6/9)	Yes (3/8)	Table 2, 1141
E. coli with plasmid and 1 copy polyoma DNA	No (0/23; 0/9)			Table 1, 1142
E. coli with plasmid and 2 copies polyoma DNA	Not tested			No explanation.
Phage with 1 copy polyoma DNA	No (0/20)	Not tested		Table 2, 1141
Phage with 2 copies polyoma DNA	Yes (3/16)	Not tested		Table 2, 1141
E. coli with phage and 1 copy polyoma DNA	Not tested			No explanation.
E. coli with phage and 2 copies polyoma DNA	Not tested			1142

Source: Mark A. Israel et al., "Molecular Cloning of Polyoma Virus DNA in *Escherichia coli:* Oncogenicity Testing in Hamsters," *Science* 205 (14 September 1979): 1140–42.

[a] Numbers in parentheses are number with tumors/number tested.

cells but also on the RAC's impending decision on deregulation of *E. coli* K12 experiments and, indeed, on deregulation of recombinant DNA activities in general.

Throughout these discussions, Rowe and other supporters of the Rowe-Campbell proposal heavily emphasized the negative results of the polyoma experiments. The experiments measuring polyoma infection had found none produced by plasmids or phage containing single copies of polyoma DNA, or by *E. coli* bacteria containing either single or double copies of polyoma DNA. Rowe stated in the conclusions to his second paper: "The most striking feature of our results, then, is the extremely low or absent infectivity of the recombinant molecules. In no instance was a recombinant molecule with a monomeric insert infectious, and in no instance did oral or parenteral administration of massive doses of live recombinant-containing *E. coli* induce polyoma infection."[77] As for the positive results showing infection with phage carrying double copies of the polyoma DNA, Rowe and his collaborators emphasized the *reduction* in infectivity, by about seven orders of magnitude, for polyoma DNA in recombinant form.[78] They also claimed that production of the double copies of polyoma DNA was, in any case, "highly unlikely."[79]

The tumor induction results were likewise selectively emphasized. Rowe and his colleagues concluded:

> Since polyoma virus is so highly oncogenic in hamsters, and since non-infectious subgenomic segments of polyoma DNA are oncogenic in this host, these experiments constitute a "worst case" analysis; that is, the experimental design maximizes the chances for obtaining positive results. Thus, it is all the more striking that neither the live χ1776 harboring the recombinant plasmids nor the phage preparations containing the monomeric polyoma DNA insert induced any tumors.[80]

The investigators said nothing about the significance of tumors produced by plasmids containing monomer polyoma DNA. And in explaining those produced using phage containing dimers of polyoma DNA, they emphasized that the oncogenic activity was "4–5 orders of magnitude less than the activity of the polyoma virus itself." The only comment on the fact that *E. coli* bacteria containing phage with dimer inserts were not tested in this experiment was that it was "quite likely" that "the same reduction in activity would be seen."[81]

To the scientific community and to the general public these results were announced as convincing evidence of the safety of cloning viral DNA. In their published papers, Rowe, Martin, and their collaborators called them "highly reassuring with respect to the safety of cloning viral genomes in *E. coli*," and further indications of "the safety of cloning other postulated oncogenic gene segments."[82] At a press conference on 1 March, Rowe and Martin asserted that "this form of research will be perfectly safe" and that "there is nothing you could cut out of a smallpox virus" that, inserted in *E. coli*, would be dangerous to work with on an open laboratory bench. The claim applied to all the DNA tumor viruses and even to lethal Lassa fever viruses.[83] Martin went on to say that the cancer scenario was "at the top of the list of proposed scenarios" for the dangers of gene splicing and that the experiment had shown the scenario was unrealistic. Not only was the cloning of tumor virus DNA in the weakened strain of *E. coli* pronounced safe, but, as a *New York Times* report put it, "the risks are considerably less than had been feared."[84] A casual reader might have inferred that the entire field of genetic engineering had been cleared of hazard.

A diametrically opposed response came from other scientists, however. Their interpretation, developed in papers and in testimony at RAC meetings, emphasized the importance of the *positive* results obtained in the infection and tumor-induction experiments. For example, in a letter sent to *Science* on 11 April 1979, a brief version of which was published in the *New York Times* on 3 May, MIT biologists Jonathan King and Ethan Signer; Stuart Newman, a developmental biologist at the State University of New York; and David Ozonoff, a professor of environmental health at Boston University, argued that the infectiousness of phage containing a double copy of polyoma DNA should be cause for serious concern.

> A phage lambda polyoma hybrid when introduced parenterally into mice at routine laboratory concentrations (2×10^8/ml) induced polyoma infections in 6/11 mice. Clearly cloning of polyoma into lambda has converted one isolate to a source of potential polyoma infection, as proposed in the original risk scenarios. Prokaryotic viruses have very different routes of propagation through the ecosystem than do polyoma or any other animal viruses.[85]

Similarly, Barbara Rosenberg, a biochemist at the Sloan-Kettering Research Institute, and Lee Simon, a microbiologist at Rutgers University, in an article published in *Nature* in December 1979 maintained that the positive results from the phage-dimer recombinants showed that polyoma DNA inserted into phage was not recognized as "foreign" by animal cells and, further, that the polyoma retained its activity after cloning.[86] The general conclusion reached by these scientists was that under suitable circumstances (e.g., use of a more robust strain of *E. coli* and the occurrence of a route of exposure in which *E. coli* bacteria survived and multiplied), the *E. coli*–phage-polyoma system could bring polyoma virus into new ecological niches that the virus alone was unable to reach. The Rowe-Martin results, far from demonstrating the safety of recombinant DNA technology, had raised the possibility of generating a "new route or reservoir of infection through splicing viral genes into microorganisms."[87]

Critics of the Rowe-Martin interpretation argued further that these positive results, establishing that phage carrying the dimeric polyoma DNA entered mouse cells, also called into question the significance of the negative results of the polyoma experiments obtained with the lambda-monomer and plasmid-monomer recombinants. Possibly these genomes *also* entered mouse cells in noninfectious form, and might subsequently induce tumors; Rowe and Martin had already shown that preparations of polyoma DNA, cleaved with restriction enzymes, were tumorigenic without being infectious.[88]

Finally, critics of the Rowe-Martin interpretation argued that the basic design of the experiments excluded several important possibilities bearing on the safety of cloning viral DNA. The investigation had been restricted to the weakened strain of E. coli, χ1776, and a lambda vector manipulated so that it could not be maintained in its bacterial host in a stable fashion over long periods of time.[89] Martin later acknowledged that K12 strains of E. coli disappeared so rapidly from the mouse gut that "exposure to potentially infectious polyoma recombinant DNA was extremely limited."[90] More robust strains of E. coli might multiply (hence delivering larger doses of tumor virus to the cells of the host organism over longer periods of time). In addition, the only virus tested—a DNA virus—needed to be excised from its bacterial host in order to generate some effect. Significantly, critics of the Rowe-Martin results arrived by a different route at the understanding that had pervaded the discussion of risk assessment at the Bethesda meeting in 1976: the Rowe-Martin experiment was by no means a "worst-case" experiment but aimed at the far more limited goal of showing that E. coli K12 was safe. In summary, the polyoma results gave little reassurance to those who advocated caution with respect to genetic engineering. As RAC member Sheldon Krimsky wryly commented in a letter to Donald Fredrickson, "If I were a mouse, with intellect, I would surely be concerned about the possibility of spreading polyoma DNA in E. coli."[91]

9.7 A Turn in Discourse and Policy

The new data and their associated surprises, uncertainties, gaps, and contested questions were discussed by the RAC in 1979 at its May and September meetings, as fundamental to the Rowe-Campbell proposal. At this stage, the discourse about genetic engineering hazards was shifting. The old discourse—"containing unknown hazards" (with the associated operational assumption that containment should be cautious) was now subtly giving way to a new discourse—"negligible hazard" (with the associated operational assumption that containment and oversight precautions could be largely removed). The RAC's approval of the Rowe-Campbell proposal may be seen as the critical move that initiated the transition.

The proposal was launched by Rowe in a strongly worded memorandum put before the RAC in May.[92] Rowe's argument was cast in the old discourse. Focusing on the hazards of using the biologically contained strains of E. coli, he claimed that important new technical developments were dispelling the naïveté that had prompted the original concerns. Summarizing both the epidemic pathogen argument generated by the Falmouth meeting as well as the results of his own and a few other more recent experimental results, Rowe concluded in sweeping terms: "The basic message from every one of these experiments is that there is no cause for concern about recombinant DNA research with K12. I do not know of a single piece of new data that has indicated that K12 recombinant DNA research could generate a biohazard." While prejudging gaps in relevant evidence, asserting a clear-cut interpretation where others saw gray areas, and ignoring controversy concerning the results of his own and other's experimental work, Rowe nevertheless employed the old discourse of hazard prevention to justify the Rowe-Campbell proposal: the controls for work with *E. coli* K12 could go because the risks had been shown to be negligible.

At the same time, the statement angrily denounced the NIH guidelines as "wasteful, expensive, inefficient, inflexible, and inhibitory." Although the Rowe-Campbell proposal affected only genetic engineering with *E. coli* K12, Rowe anticipated more extensive deregulation: if the guidelines were not needed, as he believed was the case for research with the K12 bacteria, then they were "an intolerable, capricious, irrational, offensive intrusion." The old discourse of hazard prevention was being used to open up the possibility of a transition to a new discourse based on the claim that virtually all hazards associated with genetic engineering were "negligible."

The transitional quality of the arguments about controlling genetic engineering hazards is also evident in the RAC's discussion of the Rowe-Martin polyoma results. Supporters of the Rowe-Campbell proposal focused on the K12 strain, claiming that, since it had been proven safe, it was time to loosen controls. Having perceived that the discourse that supported caution as an obvious policy depended on assuming the *possibility* of hazard, they wanted to use the polyoma-insertion experiments to counter that point, at least for one class of experiments. Opponents of the Rowe-Campbell proposal on the other hand saw the polyoma-insertion results as confirming the potentiality of hazard and consequently resisted any intrusion of the "negligible hazard" thinking, seeing this as an attack on the foundations of the Asilomar discourse. The clash of discourses is exemplified by an exchange at the May RAC meeting between Jonathan King and his colleague at MIT, David Baltimore. Like Rowe, Baltimore argued that the mere decrease in infectivity observed for the lambda-dimer recombinants over the infectivity of polyoma itself demonstrated the safety of cloning the virus in *E. coli* K12.

> BALTIMORE: [Lambda with the dimer insert] causes less infection than polyoma DNA; it causes less infection than polyoma by many orders of magnitude; but it does cause a little infection. It is however a defective virus; it cannot grow on its own. And there's no way that I can see that it can become propagated in the environment.

KING: Only if it is contained.

BALTIMORE: But it is contained.

KING: Only if you guys vote that it's a hazard. If you vote that it is not a hazard, it doesn't have to be contained.

BALTIMORE: We are not voting whether it is or is not a hazard. That phage which Wally constructed . . . is carefully designed to be unable to grow in the environment. . . . The guidelines have been carefully drawn, the experiments have been carefully done, to avoid all of the hazards that you keep harping on. It's a waste of our time to have to go through that form of argumentation. . . . The experiments that were done [were not designed to find out] whether polyoma was a hazard. [They asked] if a DNA molecule was inserted into a plasmid or a phage, could it get out and produce a problem?[93]

Baltimore's emphasis was firmly on the biologically contained system, although he tacitly extrapolated to the more robust K12 strain. King, in contrast, insisting that biohazard scenarios had not been disproved, only circumscribed, tried to forestall the consequences of presuming that genetic engineering involved no new danger.

To carry the Rowe-Campbell proposal at this point, Baltimore had to deny that it was the opening move toward deregulation. King pinned him to the containment-oriented discourse to prevent its being swept aside. That the Rowe-Martin experiments provided support for the Rowe-Campbell proposal depended on their being evaluated in terms of the Asilomar discourse; that the proposal was a reasonable first step toward deregulation depended on the presumption of "negligible hazard." (Only two years later, with the new discourse ascendant, Baltimore would propose full deregulation on the grounds that "the hazards appear to be non-existent.")[94]

Intriguingly, Baltimore's position reflected precisely the agreement on the purpose of the polyoma experiment that emerged from the Bethesda meeting in August 1976: the experiment would *not* be designed to test the hazards of cloning polyoma virus in general. Rather, it would serve the "political" function of demonstrating that the use of the biologically contained strains was safe (section 5.5).

While the discussion at the May meeting indicated an emerging consensus that deregulation of experiments using *E. coli* K12 was desirable, any immediate decision on the Rowe-Campbell proposal would have seemed premature. Most of the supporting evidence cited by Rowe was still unpublished. Forming a committee to examine current research on the hazards of cloning in the bacterium was called for. At the same time, and reflecting the antiguideline sentiment of many on the RAC, the special committee was also directed to investigate how the guidelines could be replaced without jeopardizing safety standards.[95]

The Meeting of the Ad Hoc NIAID Working Group on Risk Assessment, 30 August 1979

Between the May and September meetings of the RAC in 1979, support mounted for the Rowe-Campbell proposal. The NIH received many letters from biomedical researchers advocating it, including one signed by 183 scientists at the Gordon Conference on Nucleic Acids and on Biological Regulatory Mechanisms. This lobbying marshalled two arguments: first, no doubt reflecting the discrimination academic researchers felt vis-à-vis their colleagues in industry, the guidelines were condemned as an unnecessary, wasteful, bureaucratic procedure; second, the evidence that experiments with *E. coli* K12 as a host did not pose an extraordinary risk was claimed to be strong.[96]

Reinforcement was also generated by an ad hoc group of consultants assembled on 30 August 1979, at the behest of the director of NIAID, the institute responsible for the NIH risk assessment plan, to appraise the accumulated experimental evidence on the capacity of *E. coli* to survive and to transfer plasmids to other bacteria. It attended specifically to the significance of results reported by NIAID contractors Stuart Levy, Rolf Freter, and Bernard Sagik in the *Recombinant DNA Technical Bulletin* that summer. Unannounced, except to invited participants, the meeting comprised NIAID director Richard Krause, Wallace Rowe, plasmid biologist Stanley Falkow, who chaired it, and several microbiologists with experience in enteric diseases. Absent were both the scientists whose results were being reviewed as well as any biologists known to oppose the Rowe-Campbell proposal.

Rowe's introduction defined the fundamental question before the group as the problem of transmission: could plasmids in *E. coli* K12 move into indigenous *E. coli*—a test that had never been done. However, he questioned whether this issue was not now "obsolete," asking, "Is there really still a need to check out these small plasmids or do we know enough about them now that the experiments are just not worth the trouble and the difficulties?"[97] In addition, Rowe explicitly related these considerations to the movement to exempt genetic engineering with *E. coli* K12 as host.

> The second reason why this group was brought together . . . grew out of the last meeting with the RAC. I presented a proposal to the committee that we exempt K12 research with nonconjugative plasmids and lambdoid phages; that these be exempted from the guidelines for most classes of DNA inserts. In general, the committee was very receptive to this and very close to being able to accept it, but two scientist members of the RAC felt that they could only go along with this proposal once they had been satisfied about the nonconjugative plasmids. In effect,

they were saying, "Let's see the results of the Falmouth protocol; then we will go along with it." So that is the importance of this meeting. The RAC will meet next week and will reconsider a modified version of this amendment and the feelings of this expert group will be a very great part as to how this motion is resolved. (2)

From the outset, then, an interesting and important inversion of the decision making process took place. Instead of risk assessment guiding practice into safe channels, the anticipated deregulation of K12 cloning provided the framework for addressing risk assessment.

Indeed, the desired outcome pervaded the discussion that followed. The transcript records a long discussion of the new results on the survival capacity of *E. coli* K12. In general, the participants held to theoretical assumptions that set these results in an optimistic light. Expectations were reiterated that the probabilities for survival and transfer of plasmids in *E. coli* K12 would under normal circumstances be low, that the addition of DNA would handicap the recipient organisms, and that adding a single gene for a pathogenic property would not, by itself, turn *E. coli* into a pathogen. Consequently, while the researchers had found their surprising observations provocative, the NIAID group tended to treat these and other surprising results as immaterial. The subject in a British experiment who persisted in excreting *E. coli* K12 for a full thirteen days after ingestion was dismissed as an "ornery cuss" (4). Sagik's and Sorber's data on the survival of *E. coli* K12 in sewage was used only to stress that complete treatment destroyed most organisms; that a significant percentage survived intermediate stages and remained in sewage sludge was never mentioned (4). That an antibiotic-rich environment would raise the probability of transfer of plasmids was no more than noted. That laboratory workers' carelessness might increase exposure risks was acknowledged but set aside. A prominent geneticist's flaunting the NIH guidelines by refusing to decontaminate cultures before pouring them down the drain was noted with disapproval, but the implications of this behavior were not pursued (33–34).

The *transfer* of plasmids from *E. coli* K12 to other organisms was acknowledged, however, to require further investigation, as Sagik and Levy had emphasized, and discussion concerned not only how this should be done but how such experiments might affect the Rowe-Campbell exemption proposal. Since *E. coli* K12 did not generally establish itself in the human intestinal tract, a very large number of human volunteers would be needed to detect transfer from the K12 strain to other organisms. As a consequence, the group agreed that transfer events would be more easily detected if a more robust strain of *E. coli* were used, since it would colonize the gut more effectively than K12, thus increasing the probability of observation of DNA transfer in a manageably small number of human subjects.[98]

But discussion returned frequently to the concern that a recommendation for further experiments on plasmid transfer or other risk assessment work would have the effect of delaying the RAC's decisions on the Rowe-Campbell proposal. One person asked the group: "[Is the RAC] going to say, we had better keep [the research] under guidelines if it turns out that you have a lot of transfer from HS [a robust strain]? Or, are you going to say, okay, let's drop the guidelines for K12 since it didn't transfer?" The transcript suggests that the response to this question was not uniform. One person responded: "You know if I were there and it came to me and there was a lot of transfer, I would go back and say I think we had better go and run it again with K12." But others dismissed this cautious position:

—Okay, so now we wait two years!
—But, is saying that you are going to get a lot of transfer again saying, if miracles occur?
—I think it is going to be hard put to find transfer. I think they are asking for miracles. (52)

Several participants insisted that, whatever the detailed results of transfer experiments, the probability of transfer would in any case be low, and that therefore the RAC could be reassured beforehand that experiments with *E. coli* K12 were not especially hazardous. The development of this line of reasoning went as follows:

—I am trying to come up with the words where I can reassure the RAC that they should not wait forever to make decisions, pending every last detail but some people will recommend they do. I don't want to bring in the politics but I have to.
—Well, I can't think of the language. I don't know how many years I have been associated with the RAC and I can't think of any language that is going to make a number of people feel better no matter what we say. I can't think of even negative experiments that satisfied them.
—I am trying to see that a reasonable group of lay people and the few scientists on the committee come to real terms with themselves and say we can get rid of the K12 guidelines but that really doesn't stop us from doing further risk assessment experiments. (53)

This rationale garnered acceptance sufficient to confer informal assent to a proposal framed by the meeting's chair, Stanley Falkow, that approved the exemption proposal with the additional requirement that all research with *E. coli* K12 use P1 containment. The summary report recorded that "the group expressed their solid support for the proposed exemption . . . providing P1 laboratory practices were employed uniformly in conducting this and other categories of exempted research." The group also concluded that it "could not foresee how recombinant DNA technology as currently practiced could confer pathogenic properties on recipient *E. coli*." [99]

Thus the NIAID working group not only boosted the Rowe-Campbell proposal but also subtly reinforced the inversion of previous procedure, by contemplating weakening controls in advance of risk assessment results. The RAC's own working group, which met over the summer of 1979, appeared more ambivalent. While allowing the proposal to be published in the *Federal Register,* the working group's report to the RAC refrained from taking a position, merely listing relevant evidence and observing noncommittally that "the task at hand is a determination of whether the recently-generated experimental evidence *is both relevant to the original concerns* as reflected in the Guidelines and *is sufficient to justify the exemption.*" [100]

Approval of the Rowe-Campbell Proposal

At the RAC's September meeting, with both reports before the committee, the Rowe-Campbell proposal came to a vote, after more lengthy discussion, which was eminently a *debate* rather than a collective investigation of evidence. As in some unruly courtroom, RAC members who supported or opposed the Rowe-Campbell proposal played adversarial roles, unencumbered by enforcement of rules of evidence by an impartial judge. Supporters of the proposal used available evidence selectively, with little attention to its status—whether published or not, whether experiments had generated surprises that required further investigation or not, whether evidence was substantial or radically incomplete. The virtual absence of evidence of bacteria designed to make biologically active polypeptides was described as posing "one or two unanswered questions." [101] Rowe and Baltimore now insisted that the Rowe-Martin experiments were "sensitive" tests that "magnified the chance of delivering the polyoma DNA," ignoring other interpretations of the design of these experiments, including the view presented originally at the Bethesda meeting in August 1975 that "K12 in the mouse [is] . . . a very insensitive system for looking at changes in pathogenicity." [102] Opponents could argue that traditional scientific standards compelled delaying a decision on the proposal until more substantial evidence became available.

But the old discourse of hazard prevention was giving way to the new discourse of crisis intervention. The minutes of the meeting contain many examples of this change. "One cannot spend one's life worrying about hypothetical hazards. Rather, we are bound to deal with hazards as they surface," one member put it.[103] Baltimore commented on the possible impact of bacteria producing novel proteins, "One [would] want to have a positive reason to worry about the polypeptide problem."[104]

In the end, the RAC approved the Rowe-Campbell proposal with the provision that the exempt experiments would be registered with local biohazard committees. The vote was ten in favor, four opposed, and one abstention, with eight members absent and not voting. Three of those nominated by public interest organizations—Karim Ahmed, Richard Goldstein, and Sheldon Krimsky—opposed the motion; the fourth—David Parkinson—was absent.[105]

The approved proposal was duly forwarded to the NIH director's office, whose final decision was subject to further lobbying. With some important exceptions, biomedical researchers largely aligned behind approval, while environmentalists and labor leaders pleaded for rejection on the grounds that the decision was "premature." Letters from certain members of Congress showed that the question of genetic engineering controls had not entirely faded from view. Senator Adlai Stevenson, writing to the new Secretary of Health, Education, and Welfare, Patricia Harris, questioned the Rowe-Campbell decision in much the same terms as its other opponents.

> First, is it wise to proceed with so significant a relaxation of the Guidelines before completion of the risk assessment studies, many of them relating specifically to K12 . . . ? Secondly, does the brief experience under the revised Guidelines justify confidence in the willingness and ability of local institutional review committees to maintain laboratory standards and practices which the RAC believes should still be followed? Finally, if the Director's recommendation differs from the RAC's proposal, should there be another opportunity for public comment and review?[106]

Stevenson also revealed that he planned to introduce legislation requiring notification of recombinant DNA activities conducted by institutions not subject to the NIH controls. But such announcements no longer pressured the NIH: Stevenson needed Kennedy's backing to enact recombinant DNA legislation, and there were no signs of interest from that quarter. The recombinant DNA battle in Congress would not be refought.

The decision of the NIH director was issued for a last round of public comment on 30 November 1979: while not formally exempt from NIH purview, all nonprohibited experiments with *E. coli* K12, henceforth assigned to category III-O, could be conducted under P1 containment and required only registration with a local biohazard committee. NIH oversight was removed. In effect, these measures provided an exemption in all but name.[107] The proposed policy change was accompanied by a final official justification, lengthy but no more complete than its predecessors. The justification was selective: results supporting the "negligible hazard" position was cited; those suggesting ambiguities and a need for further investigation were not. The status of results—whether published or unpublished, generally accepted or contested—was hardly addressed. The highly contested Rowe-Martin results were blandly described as providing "valuable data."[108]

The NIH's acceptance of the essence of the Rowe-Campbell proposal represented a critical turn in NIH policy. In practical terms, roughly 80 percent of all genetic engineering activities under way in 1979 could be carried out under P1EK1 conditions and no longer came under the purview of the NIH. The political salience of the decision is clear: the change reduced containment requirements for industrial processes at a moment when this was being demanded by the private sector. Furthermore, it eliminated the additional oversight requirements for academic research, which academic scientists had perceived as unjust and discriminatory. Equally if not more significant in the long term, approval of the Rowe-Campbell proposal was a major step toward overturning the original policy aimed at prevention of exposure to unknown hazards and replacing it with a policy of crisis intervention. For a large part of genetic engineering activities, the burden of proof previously borne by those who wished to pursue these activities shifted to those who wished to control them. As RAC member Richard Novick stated: "[The RAC's approval of the Rowe-Campbell proposal] is not based on hard data. It's based on a change of philosophy. There's no point in trying to exercise prior restraint in the face of imaginary hazards."[109] The achievement of this transformation would virtually eliminate public oversight of the field.

Press coverage of this major decision was mixed. Some accounts registered the spread of positions on the RAC and the split vote of the committee (*Nature, Science, Chronicle of Higher Education, New York Times*), while others suggested that the vote represented a change in "the consensus of knowledgeable scientists" (*Washington Post, Chemical and Engineering News*). Only one (*New Scientist*) indicated a possibility that the decision might have been a response to political pressure.[110] All published accounts missed the political significance of the decision on the Rowe-Campbell proposal: the turn of discourse within a committee that could be claimed to incorporate and represent the public interest and its implication for future policymaking.

TEN
Dismantling the National Institutes of Health Controls but Preserving Quasi-regulation, 1980–1982

The approval of the Rowe-Campbell proposal not only marked a major turning point in NIH policy; it also signified a decisive defeat for those inside and outside the RAC who still supported a policy aimed at prevention of genetic engineering hazards. Given this defeat, it would be far more difficult to insist on caution as interest surged in genetic engineering new hosts and vectors, in using genes (such as those encoding toxins) that were seen in 1975 as too hazardous to be used, and in increasing the size of cultures to industrial volumes.

Furthermore, congressional interest in the issue, which had previously provided some leverage for advocates of caution, disappeared in the early 1980s. Adlai Stevenson, who as chair of the Senate Subcommittee on Science, Technology, and Space had long contemplated legislation establishing mandatory compliance with the NIH controls, decided not to seek reelection in 1980. Ronald Reagan's landslide victory and the election of a Republican majority in the Senate in November 1980 meant strong administration and congressional support for deregulatory policies.

At the same time, the lens of the press refocused: the hazard question blurred into obscurity while the promise of genetic engineering, shaped by the stream of announcements of advances in bacterial production of commercially important substances from genetic engineering firms, came into sharp relief (section 2.5). As headlines announced "the promise of gene-splicing," "the genetic payoff," and the onset of the "cloning gold rush," controlling possible hazards disappeared as a salient issue.[1]

The development of American policy in the early 1980s is especially interesting because at that point the interests of the biomedical research sector and private industry, which had previously converged on opposition to regulation, diverged to an extent. From 1976 to 1979, industry leaders played a supportive though low-key, low-visibility role, backing the biomedical research community's efforts to block congressional intervention and to liberalize the NIH controls. They quickly adopted the new hazard discourse as it evolved, echoing arguments that hazards had been exaggerated. In some important respects, however, industry interests diverged from those of the biomedical research sector. Industry leaders favored the maintenance of weak controls by the NIH, which would enable them to argue against the need for local regulation and at the same time provide strong protection for trade secrets (section 9.2). Academic researchers, in contrast, favored dismantling the NIH controls entirely and opposed the transformation of the NIH into a surrogate regulatory agency whose committees were legally bound, under penalty of criminal charges, to protect industrial secrets. Nor were they anxious to see their influence over the NIH policymaking system diminished through direct industrial representation on advisory committees. An industry proposal in June 1980 for industrial representation on the RAC was voted down by a clear majority on the grounds that the RAC should represent expertise rather than interest groups.[2] This chapter examines how these diverging interests found expression in NIH policymaking in the early 1980s as the controls were weakened.

10.1 Dismantling Controls

The Rowe-Campbell decision, codified as revised guidelines,[3] addressed only one element of the NIH controls: experiments conducted with strains of *E. coli* K12. Other elements of the original guidelines were retained, especially the principle that experiments should employ approved host-vector systems and the principle that some experiments might be—potentially—sufficiently hazardous to be prohibited and to require special exceptions before being attempted. Indeed, the official justification reinforced those principles by insisting that experiments in the K12 strains could be given preferential treatment precisely because of their safety. Rowe stated in defense of his proposal in May 1979:

> It must be emphasized that this is in actuality only a tiny fraction of all the possible combinations of donors and hosts that fall under the scope of the Guidelines. In no sense does this proposed exemption constitute "getting rid of the Guidelines." It does not affect cloning in wild type *E. coli, B. subtilis,* yeast, mice, *Drosophila, Salmonella,* corn, *Agrobacter,* or any of the thousands of other possible hosts; it does not affect the prohibited experiments with K12—they remain prohibited and subject to the control of the RAC; the exemption does not affect the 10-liter limit, which is one of the prohibitions.[4]

Similarly, when Adlai Stevenson held a further round of hearings on industrial applications of genetic engineering in May 1980, David Baltimore argued that new host-vector systems still needed to be scrutinized on a case-by-case basis.

> I am not arguing that our vigilance relative to recombinant DNA research should be abolished. Especially in the use of the new host-vector systems, the potential for widespread dissemination of recombinant DNA molecules will always exist and should be carefully considered. This is, to my mind, the present function of the RAC. The RAC has, to a great extent, deregulated the use of *E. coli* K12 as a host organism for recombinant DNA molecules. It may well do the same for yeast in the near future and there are probably other systems that either naturally or by genetic manipulation can be made so completely safe that minimal regulation is necessary. But where recombinant DNA research is to be performed in wild type organisms that can easily disseminate through the environment, the function of the RAC as a screening body for such proposals remains important and will remain important into the foreseeable future.[5]

Yet even before the new guidelines were issued, influential members of the biomedical research community anticipated further deregulation, viewing the decision only as a first step in the dismantling of controls. "This is really a rather conservative recommendation, but perhaps it is the best that would be acceptable at this time," wrote Harold Ginsberg, chair of the Department of Microbiology at Columbia University's College of Physicians and Surgeons, to Donald Fredrickson in December 1979.[6] Bernard Davis wrote:

> As you know, I have believed from the start that the hypothetical dangers from recombinant DNA were greatly exaggerated, and I feared that the expensive and time-consuming procedures introduced for regulating this research might stay with us, like some "temporary" buildings in Washington, for many years to come. . . . I recognize that you may be responding realistically to conflicting pressures. I would only emphasize the value of proceeding further as rapidly as possible in eliminating restrictions that are seen by the bulk of the concerned and knowledgeable scientific community as useless and burdensome.[7]

Both scientists recognized that the Rowe-Campbell decision represented the limit of political feasibility, however. Under the Carter administration and with Stevenson's persisting indications of interest in regulation, it was still necessary to demonstrate that the controls were not being irresponsibly dismantled.

But pressures from biomedical researchers mounted as genetic engineering expanded. A veritable avalanche of requests for the use of new hosts, new vectors, previously prohibited sources of DNA, and previously prohibited large-scale procedures packed the RAC agendas. Non-human vertebrates, higher plants, yeast, *Bacillus subtilis,* and *Neurospora crassa* received approval as cloning systems in 1979. Many organisms were exempted from the controls on the grounds that they naturally exchanged genes; use of viruses as vectors was expanded.[8]

At the same time, the RAC began to weaken the original six prohibitions. In June 1980, the prohibition banning pathogens as sources of genetic material was reduced to cover only the two most lethal classes. Cloning of genes from pathogens deemed less harmful—including those causing anthrax, plague, tularemia, brucellosis, and yellow fever—was allowed, and the third prohibition, on plant pathogens, was deleted. From September 1979 onward, under procedures that were heavily debated, the committee reviewed requests for and granted exceptions to the sixth prohibition on the use of large volumes of genetically manipulated organisms (section 10.2).

These decisions signified not only the willingness of the committee to permit an expanding range of cloning procedures but also the increasing legitimation of the argument that the techniques were safe *in general*. In accordance with the reversal of the burden of proof established by the Rowe-Campbell decision, the majority of RAC members countenanced desired practice as the measure of safety.

The RAC's slide toward deregulation cannot be attributed to definitive evidence negating the hazards of genetic engineering. A workshop organized in Pasadena, California, by NIAID in April 1980 as part of the NIH risk assessment program evaluated the hazards of producing biologically active polypeptides with *E. coli* K12.[9] The transcript of the final plenary session records a discussion encompassing a myriad unresolved possibilities.[10] It was also apparent that, as at earlier risk assessment meetings, the focus on *E. coli* K12 had a soothing effect, allowing the larger issues of cloning in more robust organisms to be set aside. While the consensus among this group of invited participants appeared to be that the risks of producing disease either through exposure to an active substance released by bacteria or through stimulation of an autoimmune response would be low, this sense was by no means unqualified. Although bacterial production of hormones such as insulin and growth hormone did not appear to pose significant risks, it was unclear that the same could be claimed for bacterial production of more potent proteins. Those present agreed that further empirical information on the effects of synthetic polypeptides was desirable.[11]

Consideration of autoimmune scenarios similarly produced no compelling consensus. A vote following a long discussion recorded strong support for animal studies of possible autoimmune effects of recombinant bacteria, although the majority, no doubt aware of the political implications of approving further risk assessment studies, voted against doing these studies for risk assessment purposes.[12] But despite the persistent uncertainties surrounding the possible effects of bacterial production of proteins, the outcome of this meeting would be cited in support of continued weakening of controls on both sides of the Atlantic.[13]

In the same period, a preliminary report on the performance of nineteen biosafety committees in California conducted at Stanford Medical School indicated considerable variation in how these committees interpreted their responsibilities and handled research proposals. Training in safe procedure varied, being left in some cases entirely to the principal investigator. The majority of committees surveyed offered no formal training in safe technique. And not all institutions with P3 facilities appointed safety officers even though required by the NIH controls to do so. Representation of the public also varied: in seven cases, the "public" representative turned out to be an individual engaged in recombinant DNA research in another institution. None of the committees publicized their meetings, and in some cases decisions were made almost entirely over the phone or in writing. While these observations did not totally impugn the IBCs, they did raise doubts about their operation—doubts that a later study of safety procedures in several new genetic engineering firms showed were justified.[14]

Nevertheless, in August 1980 Maxine Singer proposed that the RAC transfer responsibility for all experiments listed in the NIH guidelines from the NIH to IBCs. Despite the doubts raised by the Stanford report, the RAC approved the Singer proposal by a large majority, with only three of the members originally nominated by public interest organizations voting against it.

Paul Berg, for one, believed that these changes did not go nearly far enough. On 2 July 1980, Berg wrote to David Baltimore that

> I find it increasingly difficult to accept the maintenance of a non-trivial government bureaucracy to contend with the mythical possibility that recombinant DNA experimentation is hazardous. . . . RAC has elected a strategy of evolutionary change in The Guidelines, but in my view, it has not been without paying a substantial price: the time and energy of ORDA, RAC members, institutional committee members and the scientists doing the research, as well as the expenditures to maintain ORDA, RAC, and the new layers of bureaucracy spawned in the universities. I do believe it is time to seriously consider whether the ORDA-RAC format should continue. . . . I would favor its dissolution and replacement by something like the advisories put out by the CDC.[15]

To Singer, he wrote that "frankly, as you know, I favor even more substantial eradications, but perhaps it is best to begin as you have suggested."[16]

The Republican victories of November 1980, which removed the threat of legislation and provided general reinforcement for antiregula-

tory sentiment, no doubt accelerated the downward plunge of the NIH
controls. Decisions that responded to the interests of leaders of biomedical research like Berg came quickly in 1981 and 1982. An early move was
made at a meeting of the chairs of IBCs organized by the NIH in late
November 1980. While some participants commended good laboratory
practice, monitoring of equipment, and surveillance for illness as essential protections against potential laboratory hazards, a majority favored
exempting the large class of experiments classified as III-0 (which now
encompassed not only experiments with *E. coli* K12 but also with yeast
host-vector systems) and opposed systematic investigation of the operation of the IBCs.[17] The big vote for exemption in particular echoed antiregulatory sentiment, justified as opposition to "paperwork."[18]

By the time of the RAC meeting in April 1981, the thrust toward deregulation was evident throughout the agenda. By a vote of 13-7-0, the
committee endorsed a modified form of the IBC chairs' proposal and
exempted experiments in category III-0. It went on to sweep away prohibitions against cloning toxin genes, leaving only those encoding the
three most lethal (botulinum, tetanus, and *Shigella* neurotoxin) in the
prohibited category. Recombinant DNA work with all toxins, including
many of interest to the military, was now permitted at varying levels of
containment and supervised only by IBCs. The cloning of anthrax toxin,
ricin, and certain neurotoxins present in snake venom were consequently
assigned to category III-0 and were thus exempt from oversight. That
the RAC had little interest in debating deregulation affected its reaction
to a letter received from a former safety officer at Fort Detrick that decried danger in cloning genes controlling bacterial toxins.

> Based on my past experience, I would suggest a cautious approach to
> this relaxation. In the mid 1960s, staphylococcus enterotoxin was being
> investigated as a potential incapacitating agent in the biological warfare
> program. . . . we found that many of our laboratory technicians devel-
> oped an allergic state when we decreased some of our containment cri-
> teria. In some of the people the allergic reaction involved the eyes and
> in others the reaction was a severe respiratory distress. . . . Staphylo-
> coccus enterotoxin is not classified as a lethal toxin but its reactivity as
> a hyperallergen created conditions in some individuals that were nearly
> lethal.[19]

Whether these observations indicated that cloning the staphylococcus en-
terotoxin did not deserve exemption from NIH control was probably
impossible to judge immediately. But that the RAC simply chose to ignore the question is an indication of its deregulatory momentum at this
stage.

The key agenda item in April 1981 was a proposal framed by David
Baltimore and Allan Campbell. Effectively, the proposal converted the

guidelines into a recommended code of practice, eliminating penalties for violations, institutional biosafety committees, and registration, and sought containment reductions for many experiments and processes (many to the P1 level) by assigning containment levels on the basis of those in general use for the host or vector, not those for the source of the inserted DNA.[20] Little was left of the original controls except for the section specifying the prohibitions, and here, also, Baltimore and Campbell anticipated further relaxation. Furthermore, the proposal's implicit elimination of the concept of biological containment was a crucial policy reversal: the fact that organisms were genetically manipulated was now seen as irrelevant to their safety.

The supporting discourse made no reference to the special safety of biologically contained strains of bacteria: instead, it asserted the innocuousness of the entire field. Thus Baltimore and Campbell in their justification stated:

> Since 1976, neither experimental evidence nor solid theoretical arguments have been advanced to support the position that recombinant DNA research poses any danger to human health or to the integrity of the natural environment. At this point, we doubt that the beneficial side effects of continued regulation justify the expenditure of time and money required to maintain a regulatory apparatus that has been developed to protect society from hazards that appear to be non-existent.[21]

The proposal received prominent support from some who had been closely associated with the Berg committee's letter calling for a moratorium. The early concerns were described as "unsupportable" (Maxine Singer); the NIH controls were said to have "outlived any usefulness they may have had" (Norton Zinder); and Paul Berg, citing his letter to Baltimore in July 1980, claimed that "there is more to fear from the intrusions of government in the conduct of scientific research than from recombinant DNA experiments themselves." All urged the dissolution of the controls.[22]

The RAC's response in April 1981 was to again assemble a working group to evaluate the need for the NIH controls and the accumulated evidence bearing on this need and to report back to the RAC at its next meeting. Among those coopted for this purpose were Zinder (a former member of the Berg committee) and former RAC member Edward Adelberg, both vigorous supporters of deregulation.[23] Two positions crystallized during deliberations of the group, chaired by Susan Gottesman. The first retained mandatory controls but considerably reduced oversight requirements and containment levels. The second followed the Baltimore-Campbell lead by converting the guidelines to mere recommendations. That the first position was now seen as moderate demonstrated how far and how quickly the center of the debate on controls had swung toward deregulation since the beginning of 1979. Both positions abandoned biological containment, sanctioning virtually any organism as a host or a vector for cloning.

The abolitionist position dominated the ensuing debate at the September meeting of the RAC. Baltimore and Zinder claimed that the only justification for controls was political, not scientific, and that the changed political climate would now eliminate even the political reason. Advocates of the more moderate position were put on the defensive, forced to argue that the remaining controls would not obstruct science.[24] Those few who attempted to challenge the new discourse of "negligible hazard" and the radical shift toward deregulation represented by *both* positions were indeed voices crying in the wilderness. The abolitionists' influence was reflected in a vote of 16-3-1 for a proposal that outdid the Baltimore-Campbell one, striking virtually all requirements except two prohibitions (those on transfer of drug resistance and the cloning of genes for the biosynthesis of the four most lethal toxins) and the provisions for protecting proprietary information.[25] Registration of IBCs with the NIH and penalties for noncompliance were abandoned. Even RAC chair Ray Thornton, who had opposed mandatory controls and had supported weakening of the guidelines, found this too radical. Following the vote, Thornton reminded the committee of Maxine Singer's testimony in March 1977 to his congressional committee, reassuring her audience that "scientists today recognize their responsibility to the public that supports scientific work in the expectation that the results will have a significant positive impact. . . . Scientists also accept the need to restrict certain laboratory practices in order to protect the safety and health of laboratory workers and the public." Thornton rejected Zinder's claim that the guidelines were no longer needed because the political interest in them had evaporated: "Human experience has shown that any tool powerful enough to produce good results of sufficient importance to shake Wall Street and offer hope of treating diabetes is also powerful enough, wrongly used, to produce bad results of equal importance. . . . I believe that the recommendation that the nature of the guidelines be downgraded to a simple statement of good laboratory practice raises issues as serious—though opposite from—the ones considered by Congress several years ago."[26]

The NIH proceeded to publish the revised Baltimore-Campbell proposal for public comment.[27] At the same time, it published an alternative proposal by Susan Gottesman, which retained the guidelines as controls for government-funded research and the institutions—the RAC and the IBCs—to oversee them.[28] Like the majority position, however, the Gottesman proposal made radical reductions in containment and oversight, essentially requiring containment sufficient for the most pathogenic host or vector. It also eliminated the prohibitions, leaving a required RAC and NIH review for only three types of experiment (the cloning of genes encoding highly lethal toxins, deliberate release of genetically engineered organisms into the environment, and the transfer of drug resistance traits not known to occur naturally). The Gottesman alternative stopped just short of the final coup de grace that the modified Baltimore-Campbell proposal would have delivered to the NIH controls. The end product was almost the same: for most types of genetic engineering, the Gottesman proposal provided little more than paper controls. But on paper, if not in substance, the Gottesman proposal provided NIH oversight of experiments and local biohazard committees. The RAC's final consideration of the two proposals in February 1982 would reveal the power of another highly influential "community"—private industry.

As controls were removed, the risk assessment program mandated by Califano in December 1979 faded into insignificance. Despite the recommendation of the participants at the 1980 Pasadena meeting that the effects of bacteria engineered to produce novel proteins should be investigated, few experiments were ever initiated. One proposal to NIAID to investigate an autoimmune scenario, submitted by a group that included an immunologist, a plasmid biologist, and a developmental biologist, while praised by its reviewers, was rejected by the agency.[29] Calls by Australian biologists for investigation of the hazards of cloning oncogenes (genes encoding proteins that are assumed to play important roles in transforming normal cells into cancerous ones) some years later were ignored.[30] After 1982, the NIH ceased publication of reports on the progress of its risk assessment program.[31] Thus, in a period in which the NIH policy facilitated virtually unimpeded cloning without restriction to specially designed host-vector systems, a program that might have provided an early warning signal for hazardous processes was phased out of existence.

10.2 Evolution of the NIH Industrial Policy

While the interests of industry leaders and biomedical researchers powerfully reinforced each other on the drive for weaker containment and review requirements, it is important to recognize that they also differed significantly. While many scientists wanted to entirely dismantle the NIH controls, industry leaders had made plain to the NIH and DHEW leadership over the years that they wanted strong protection of trade secrets, and they preferred the central, quasi-regulatory role of the NIH to the possibility of local regulations that might spring into the void if the NIH controls were to disappear entirely.

Since in contrast to GMAG the private sector was not directly represented on the RAC, business interests in NIH policy were generally mediated and emerged indirectly in the proceedings of the committee. A growing industry presence at RAC meetings from 1979 onward made it clear that decisions were being watched attentively. Their representatives at RAC meetings averaged 7 in 1979, 16 in 1980, and 18.6 in 1981, and dropped off only in 1982, after most of the NIH oversight functions for private corporations had been eliminated.[32] In this process, the NIH director, with the support of the Secretary of Health, Education, and Welfare and the RAC chair, worked to structure committees and agendas so that the strong *shared* interests of industry and biomedical research could be used to persuade the RAC to respond to industry needs and to restrain countervailing interests, especially those of employees.

Establishing the Voluntary Compliance Scheme

In the NIH policy framework, controls for genetic engineering activities in the private sector had always been problematic. The only effective sanction behind the NIH controls—the withdrawal of funding for research—had no force against private business. Joseph Califano, in promulgating the revised NIH guidelines in December 1978, implicitly recognized this gap by requesting the FDA commissioner and the administrator of the EPA to consider how their regulatory authorities might be used to cover genetic engineering activities in the private sector (section 7.7). Only the FDA responded, with an announcement of an "intent to propose regulations," which was quickly challenged by the PMA and other organizations in the private sector and was soon dropped. The EPA took no immediate action.[33]

The gap in controls began to pose more immediate problems as Genentech and Eli Lilly designed pilot plants for human insulin production and rumors circulated that Genentech intended to proceed without government approval.[34] At that point, Donald Fredrickson, with the support of the FDA commissioner, Donald Kennedy, laid the ground for NIH oversight by drafting the detailed procedures for voluntary compliance by private institutions that had been promised to representatives of the PMA when they met with Fredrickson and DHEW representatives in October 1978 (section 7.7).[35]

A draft proposal for voluntary compliance was floated to the RAC at its meeting on 21–23 May 1979, and to representatives of the PMA and of labor unions and environmental organizations on 25 May. The scheme enabled private companies to register their biosafety committees with the NIH and to have their projects and host-vector systems reviewed by the RAC while providing broad protection for sensitive corporate information. Companies could designate information as confidential and consult with the NIH before release of any corporate information under the Freedom of Information Act; their submissions would be reviewed in closed sessions of the RAC; they would be given fifteen days' notice before release of information under the Freedom of Information Act unless earlier release was considered "necessary" to protect public health and the environment; and unauthorized release of information would be subject to criminal penalties under U.S. Code 1905. The plan was fully responsive to the PMA's requirements.[36]

Just one element of the draft supplement—the protection of proprietary information—was brought before the RAC at its May meeting. Members were informed of their responsibilities if they reviewed proprietary data and of the criminal penalties they might incur for unauthorized release of such information. The RAC, however, was uneasy. The members asked whether there was any guarantee that voluntary compliance would work. They were told by NIH officials that there was no guarantee that companies would register with NIH. The only aspect of the scheme that seemed entirely clear was that RAC members, rather than corporations, would be liable for criminal penalties for violations of its provisions. RAC member Sheldon Krimsky claimed that the proposal would produce a "bizarre situation in the history of science and technology in which large-scale work can be done" without mandatory controls.[37] The outcome was a vote of 9-6-6 to recommend *mandatory* compliance with the NIH guidelines for institutions that were not supported by NIH. The vote is significant in demonstrating that the RAC's response was hardly sympathetic to industry's goal of voluntary compliance with strong protection of sensitive industry information. The stiff penalties for divulging trade secrets brought home to committee members that at least one aspect of the committee's interaction with industry would be regulatory in nature, and many questioned taking on that role.

The response to the NIH proposal from the PMA and from labor unions and environmental organizations was more predictable. The PMA, having resisted mandatory compliance all along, plainly favored the scheme, only registering concern over the handling of trade secrets and the delays entailed by RAC reviews of corporate submissions.[38] The unions and environmental organizations questioned the effectiveness of voluntary compliance and proposed instead that agencies with statutory authority, such as OSHA, should develop regulations.

The decision on the government's response to the RAC's resistance to the voluntary compliance scheme and to the conflicting views of labor and environmental organizations on the one hand and the PMA on the other appears to have been jointly formulated by Califano and Donald Fredrickson, who met on 8 June. Califano, Fredrickson later reported, "was not inclined to seek legislation" requiring mandatory compliance with the NIH controls for the private sector.[39] Instead, he reactivated the federal IAC, the last meeting of which had been held in October 1978, requesting its advice on the voluntary approach and on the NIH draft supplement.[40]

The meeting of the IAC on 17 July 1979 was chaired by Fredrickson. An official account in the *Federal Register* stated that it met to "consider the draft supplement and alternative approaches to extend the revised guidelines to the private sector."[41] Its minutes show, however, that discussion concentrated on the voluntary compliance scheme. Fredrickson set the agenda by noting that Califano "had requested the advice of the Interagency Committee on the voluntary approach and the NIH draft supplement." According to some participants at the meeting, only two options were seriously entertained: either the voluntary compliance scheme would be developed or there would be no scheme at all.[42]

Several regulatory officials expressed reservations about the scheme. The representative of the FDA proposed that at least registration of projects should be mandated. (Fredrickson responded by arguing that even mandatory registration would require new legislation and that there was general agreement that legislation was not warranted.) The representative from the Department of Energy wondered whether NIH had the expertise necessary to handle industrial submissions. Representatives of agencies funding recombinant DNA activities, on the other hand, generally favored the voluntary scheme. While the IAC reached no consensus on the effectiveness of voluntary compliance, they did agree that the NIH should publish for public comment a revised draft supplement to the guidelines, and include expanded and strengthened procedures for protection of proprietary information along the lines indicated by the PMA. The supplement duly appeared in the *Federal Register* in August 1979.[43]

When the RAC met in September 1979, the voluntary compliance scheme was on the agenda for final action. The NIH director again personally intervened, appearing before the RAC to press for adoption of the scheme and warning the committee that not to adopt it would undermine the future of American science.

> If the fruits of recombinant DNA are something that is going to be achieved then clearly this is something that is inescapable and we will have to go through this, at least for a while. To create a second RAC to handle propriety information would be undesirable, counterproductive and clearly quite impossible and destructive of the whole system that we're trying to allow you with all of your effort to create. Because you're serving more than American science but a whole universe of people trying to use safely and effectively and efficiently this new potential set of miracles in biology.[44]

Fredrickson also impressed on the committee that there was no alternative to voluntary compliance. Asked how the NIH and the DHEW would respond to the RAC's previous vote to endorse mandatory compliance, Fredrickson stated that Califano was "not inclined to seek new legislation at this time." He also relayed the "unanimous opinion" of the IAC that the NIH should provide "this opportunity for voluntary compliance."

On the initiative of RAC chair Jane Setlow, the committee also heard from John Adams, president of the PMA. Adams conveyed a strong sense that PMA firms were eager to comply with the guidelines and work closely with the NIH—although he also noted that the PMA would like to see even tighter control of proprietary information. The image of responsible corporate behavior was reinforced by Irving Johnson, vice president for research of Eli Lilly, who insisted that the compliance of his company would be absolute. The scheme had strong proponents within the RAC: Ray Thornton and LeRoy Walters argued that, on the strength of the assurances from private industry, the scheme should be run for a test period; David Baltimore pointed out that "this takes the universities off the hook as far as having to deal with legislation [which was] one of the most threatening things that has ever happened to the freedom of science." The PMA, Baltimore averred, would be "extremely responsible." Sheldon Krimsky's observation that Genentech was already using large-scale cultures without the committee's permission drew no discussion. [45]

The RAC proceedings were thus shaped so as to preempt resistance to the voluntary compliance scheme and to generate a climate of opinion that was favorable to it. The issue was framed as one of either accepting the scheme or having no controls at all, and of either supporting American science or blocking its future progress. Aware that requests from Eli Lilly and Genentech that the ten-liter limit be waived for their insulin projects were slated for a closed session later in the meeting (a further signal of what the NIH leadership wanted and expected), the committee voted to approve the supplement. The RAC thereby became the principal (and effectively the only) arena for the formulation of industry controls.

Implementing the Voluntary Compliance Scheme

The development of industrial policy now took several forms. First, the committee immediately began to review, in closed sessions, proposals from the private sector for use of genetically engineered organisms on a commercial scale. (In all, some thirty-nine proposals were reviewed and approved by the RAC between 1979 and 1981.) [46] Second, the committee developed containment requirements as standards for the review of large-scale applications. Finally, the committee confronted requests from the private sector to weaken both oversight and containment provisions for large-scale work.

But while policymaking proceeded, unease concerning the effectiveness and advisability of the NIH's assumption of responsibility for controlling industrial applications persisted. The issue found the RAC seriously divided. Doubts about procedures repeatedly surfaced at RAC meetings in December 1979, March 1980, and June 1980. Many members felt that the committee was technically ill-prepared to rule on the safety of processes that used microorganisms on an industrial scale. It became clear that ensuring the safety of fermenters in which the organisms would be grown and processed, controlling the disposal of wastes and spills, and controlling the organisms and the cultures used in fermentation processes were beyond the RAC's expertise. Because technology was at an early stage of development, fermentation systems were not standardized and could be evaluated only through a site visit. Some RAC members dismissed these problems, but others were bothered. Plasmid biologist Richard Novick admitted at the meeting in September 1979: "I'm already technically over my head. It's a universe I have no connection with. Do we have any business getting into it?"[47] Microbiologist Frank Young said in March 1980: "We have tried to translate from laboratory safety into a fermenter situation. But this is not an area we are qualified in or experienced in at the same level that I think we can judge the biological aspects of the technology."[48]

The RAC's proposed standards for industrial processes—up before the RAC for a vote at its March 1980 meeting—were also criticized for neglecting training, medical surveillance, and plans for dealing with accidents and contamination of personnel. Moreover, its standard-setting process omitted the industrial workers most likely to be exposed to industrial hazards. RAC member Sheldon Krimsky stated at the March 1980 meeting:

> In the regulation of new technologies, there are many unfortunate instances where the subpopulations at greatest risk had little or no input into the decision process. Decision makers and risk takers are often separated. Those at risk have no power to direct acquisition of information on health data and have no control over the technology. Those determining standards have a moral responsibility to seek input from representatives of those subpopulations potentially at high risk. . . . It is irresponsible of this committee to make regulations without having representatives from unions to speak to these issues.[49]

Efforts to provide labor representatives access to the debate met with strong resistance, however. A motion by David Parkinson at the RAC meeting in September 1979 to convene a public meeting on worker safety in the fermentation industry produced a tied vote (7-7, with 4 abstentions); Setlow cast the chair's deciding vote against the motion. Similarly, NIH officials resisted and Setlow rejected Krimsky's efforts to involve labor representatives in the formulation of standards for industrial recombinant DNA work. The discomfort of certain RAC members at this exclusion continued, Krimsky stating in March that "certifying proposals [for industrial use] without input from representatives of the workers in fermentation industries involves a strong derogation of fairness."[50]

Additionally, the NIH's lack of legal authority to regulate private use of recombinant DNA technology remained troublesome. RAC member and lawyer Patricia King stated at the June 1980 meeting that "voluntary compliance is the worst of all possible worlds. . . . You achieve none of the objectives of regulation and none of the benefits of being regulated. All you're saying is 'I give a stamp of approval to what I see here before me' without any authority to do anything."[51]

But powerful voices continued to champion voluntary compliance. Representatives of the pharmaceutical industry insisted on their commitment to comply with the NIH standards, claiming that companies needed to maintain public images that charges of violations could sully (to which Krimsky, citing the case of dioxin, responded that such embarrassment had not always been a sufficient sanction). RAC members sympathetic to the idea of voluntary compliance argued that the possibility of civil suits if accidents occurred was sufficient to ensure compliance and that voluntary controls were better than nothing.[52] Very strong pressure to maintain the voluntary compliance scheme came from David Baltimore. As a Nobel laureate, Baltimore had great stature on the committee; as a major equity owner in a new genetic engineering firm, Collaborative Research, he was probably eager to maintain the scheme. He wielded once more the arguments that had successfully deterred Congress from passing legislation and had persuaded the RAC to lower controls on research. Fundamentally, he argued, the RAC was assigned to catalyze development of recombinant DNA in both its scientific and industrial applications: "If we don't do the job, industry either has to go outside the United states or without regulations, or we have to get ourselves in the very complex, nasty situation of either developing legislation or precise regulations, which would mean that industry cannot do anything for a minimum of, I would guess, three years. That would probably put American industry at a significant disadvantage with respect to the rest of the world."[53]

Many RAC members remained unpersuaded, however. At the March 1980 meeting, an impasse seemed imminent when some who had supported majority positions expressed discomfort with the voluntary compliance arrangement, and several walked out before the closed session for review of industrial applications. The balance of forces on this issue is illustrated by the committee's response to Krimsky's motion that the RAC should limit its review of large-scale projects to those involving no proprietary information, and refer all other proposals, with recommendations, to OSHA. Baltimore's subsequent motion to table passed, but only by eight to six, with four abstentions.[54]

Noises were again heard on Capitol Hill. Adlai Stevenson wrote to the new Secretary of Health, Education, and Welfare, Patricia Harris, criticizing the voluntary compliance scheme. Noting that "voluntary registration provides no assurance . . . that all firms will register their research nor that any single firm will register all of its work," Stevenson concluded that the scheme was "seriously flawed."[55] In January 1980, he introduced a forlorn bill, S. 2234 (the last bill aimed at regulation of recombinant DNA research and industrial production), requiring registration of recombinant DNA activities in the private sector with the DHEW.

Moreover, OSHA, the agency formally responsible for occupational safety and health, was stirring. In December 1979, Eula Bingham, the administrator of OSHA, wrote to the NIH director noting that OSHA supported the voluntary compliance scheme but nevertheless felt that "more formalized enforcement of procedures and rules within the regulated private sector is warranted." Bingham proposed that an expert panel investigate exposure to genetically engineered organisms in the workplace, and indicated that registering users would facilitate evaluation. In addition, NIOSH director Anthony Robbins contemplated on-site surveys of industries involved in genetic engineering. Bingham anticipated that these studies would underlie further action by OSHA—establishing a registry of users and possibly developing enforceable standards.[56]

Fredrickson's reaction was twofold. He set up an Industrial Practices Subcommittee of the IAC, chaired by a biomedical research scientist and physician, Gilbert Omenn, of the Office of Science and Technology Policy. The committee proceeded to hold several meetings (in parallel with those of the RAC) that were of an information-sharing nature. Omenn had earlier opposed recombinant DNA legislation and held that estimates of recombinant DNA hazards had been exaggerated. His view was that informing industries of possible hazards provided "significant power, short of regulation."[57] The group did not take any action; after the November 1980 election, it dissolved. But the Industrial Practices Subcommittee did have one useful function. When questions arose on the RAC concerning gaps in the NIH controls for genetic engineering in the private sector, such matters could be conveniently deflected to the subcommittee.[58]

At the June 1980 meeting of the RAC, Fredrickson once again checked the incipient rebellion against the committee's quasi-regulatory role with respect to private industry. When the voluntary compliance scheme came up for review, Fredrickson addressed the committee. Noting the "malaise of the RAC," he acknowledged the need to relieve it and to "move NIH at least one step back from the brink of regulatory involvement." But he insisted that controlling recombinant DNA technology without passage of legislation was essential for the "maintenance of parity among the nations for safe use of this technology and access to its benefits." He warned the committee that "it would not be appropriate for it to decide in exasperation that it would just like to avoid handling *any* proprietary data." Claiming that no agency other than the NIH had "anything like the in-house competence to perform for industrial applications all of the present tasks of the RAC," he warned that the creation of "one more committee" (presumably advisory to more than one government agency) would be inefficient, destroying the "unity of information sharing and decision making" achieved to that point for recombinant DNA technology.[59]

Fredrickson offered the committee two options: it might continue with the existing arrangements for large-scale work, or it might restrict its responsibilities to setting physical-containment levels for the organisms being used and give up ensuring that fermentation facilities met its requirements. With Fredrickson's alternatives as the basis for discussion, the RAC eventually opted for the second. The large gap in control this decision permitted was unlikely to provoke Congress, already preparing for the November election, to action.

Dismantling NIH Oversight of Industrial Processes

As biomedical researchers pressed for and progressively obtained the downgrading of containment and oversight requirements for recombinant DNA research, industry representatives could, with increasing persuasiveness, deny any need for rigorous controls for genetic engineering in the private sector. The discourse justifying the Rowe-Campbell proposal in 1979 was taken up by industry officials (and by scientists closely associated with industry) to argue for the safety of *E. coli* K12 when used in bulk. Referring to the Falmouth, Ascot, and Pasadena workshops and to the Rowe-Martin experiments, Lilly's Irving Johnson assured the Industrial Practices Subcommittee of the IAC in June 1980 that "the overwhelming conclusion from these data is the unquestionable safety of *E. coli* K12 hosts utilizing non-conjugative plasmids with carefully and fully characterized DNA inserts." Johnson also argued that the large-scale cultures would be completely contained in closed, stainless steel vessels.[60] To the RAC earlier the same month, Johnson used a similar rationale to argue for excluding large-scale work with approved *E. coli* K12 host-vector systems from the list of prohibited experiments and processes. Such systems were, he claimed, "inherently safe." The effect would have been to transfer decision-making authority for the approval of large-scale processes using *E. coli* K12 entirely to IBCs. But this argument, made during the committee's intense debate on overseeing private sector activities, proved to be an idea whose time had not quite arrived, and the RAC voted it down. Even Allan Campbell, one of the architects of the reasoning behind weakening the NIH guidelines, doubted whether there was no increased risk with large volumes. One should not, Campbell asserted, "transfer the logic behind the [Rowe-Campbell] proposal" to large-scale activities.[61]

As deregulation accelerated and the discourse justifying the safety of genetic engineering expanded to include the entire universe of possibilities for cloning, the RAC became more receptive to the industry arguments (table 10.1). Industry representatives also used actions on the other side of the Atlantic, whether real or contemplated, for additional leverage. The RAC and GMAG moved into a synergistic relation, each closely watching the other's moves and using these to argue for further relaxation (see also section 11.5).

After Baltimore and Campbell proposed converting the guidelines into a voluntary code of practice in the spring of 1981, Johnson again called on the RAC to exempt large-scale processes with approved host-vector systems.[62] Alerted by Johnson that GMAG was about to consider a similar proposal and accustomed now to limiting its oversight, the RAC responded favorably, reversing its decision of the previous year.[63] Furthermore, both modifications of the Baltimore-Campbell proposal published in the *Federal Register* in December 1981 removed NIH oversight for large-scale processes. Ironically, Campbell's reservations about their hazards and his last-minute plea for continuing NIH oversight, articulated in a lengthy letter to the NIH director in January 1982, received no attention at the RAC's next meeting.[64] The momentum behind laissez-faire cloning was too strong.

Private industry wanted minimal oversight and control of genetic engineering, but it had no desire to see the NIH controls entirely dismantled—an important divergence from the interests of many scientists. When faced with a choice between the revised Baltimore-Campbell proposal and the somewhat more moderate Gottesman proposal that left in place certain symbols of control—notably the voluntary compliance scheme and the IBCs—industry opted for the latter. The NIH guidelines were seen as the best form of protection against maverick local communities who might impose their own regulations. As the president of the newly formed Industrial Biotechnology Association, a trade association

Table 10.1 Evolution of U.S. NIH Procedures
for Reviewing Industrial Processes

Date	Source of Policy	Requirements
6/23/1976	1976 NIH guidelines	Prohibited. Exceptions granted by NIH if process of "direct social benefit" and after review by RAC.
12/22/1976	1978 NIH guidelines	Prohibited. Exceptions granted by NIH if approved by NIH and after review by RAC.
10/30/1981	1981 NIH guidelines; NIH approval of request from Eli Lilly and Company to allow IBCs to review large-scale processes with organisms otherwise exempted from the NIH controls (10/31/1981)	Responsibility delegated to IBCs to review large-scale procedures involving *E. coli* K12 EK1, *S. cerevisiae*, and *B. subtilis* HV1 host-vector systems.
8/27/1982	1982 NIH guidelines	Responsibility delegated to IBCs to review all large-scale procedures covered by the guidelines, with the exception of those experiments specifically requiring RAC review and approval (cloning of genes encoding highly lethal toxins, introduction of antibiotic-resistance genes that do not occur in nature; deliberate release into the environment).

Sources: Department of Health, Education, and Welfare, National Institutes of Health, "Guidelines for Research Involving Recombinant DNA Molecules," *Federal Register* 41 (7 July 1976): 27911–43; Department of Health, Education, and Welfare, National Institutes of Health, "Guidelines for Research Involving Recombinant DNA Molecules," *Federal Register* 43 (22 December 1978): 60108–31; Department of Health and Human Services, *Federal Register* 46 (1 July 1981): 34462–87; Department of Health and Human Services, National Institutes of Health, "Guidelines for Research Involving Recombinant DNA Molecules," *Federal Register* 47 (21 April 1982): 17180–98.

representing fourteen companies engaged in genetic engineering, stated in a letter to the NIH in February 1982, "We regard the [formal institutional structure of the RAC and the IBCs] as currently more of an aid for well managed technical growth than an impediment. . . . One significant reason for adherence to a uniform system of federal guidance and overseeing is our belief that such an approach is more compatible with commercial development and the benefits it brings to society than would be a system of varying local requirements."[65] Michael Ross, a research director at Genentech, expressed a similar view: "The alternative to the attentiveness of the NIH could well be a patchwork of local and state regulations which might impede our national progress and do nothing to assure additional safety."[66]

Industry wanted a predictable regulatory climate. And while they did not say as much, it seems likely that they also saw the NIH controls as protection against future litigation. Some academic institutions, especially those in areas whose local governments had enacted their own controls based on the NIH guidelines, agreed or even went further, arguing that both proposals were premature. William Lipscomb, chair of the Harvard Biosafety Committee, warned that, "should these changes be adopted and local regulations not be altered, recombinant DNA research and development activities in Massachusetts would be conducted at a definite disadvantage compared to other regions."[67]

In contrast, most academic scientists pressed for the complete dismantling of the NIH controls. Out of roughly twenty-nine letters to the NIH from academic scientists, twenty-two favored converting the NIH guidelines into a recommended code of practice and eliminating NIH, RAC, and IBC oversight. Paul Berg's letter to the NIH in January 1982 conveyed the majority view.

> I believe that the Guidelines for Recombinant DNA research are now dispensable. Based on the substantial amount of experience and experimentation with the recombinant DNA methodology during the last six years, there is widespread agreement that the risks that were once thought to be so plausible are actually remote or possibly nonexistent. If that judgment is indeed correct, and I know of no evidence to indicate otherwise, then it seems wasteful of effort and money, even counterproductive, to maintain the elaborate procedures and organizations that were set up to guard against hypothetical threats.[68]

It was an interesting moment. Scientists wanted to abandon the controls. Private industry and some institutions wanted to keep them. After many years of responsiveness to the former, the RAC, at its meeting in February 1982, proved most receptive to the latter. This time, David Baltimore's defense of unfettered science fell on deaf ears. In debating the Baltimore-Campbell and Gottesman proposals, RAC members echoed the industry argument that to abandon the guidelines invited local intervention. That the State of California had recently held hearings on recombinant DNA technology provided a reminder of this possibility. Repeatedly, RAC members warned that without the guidelines the recombinant DNA issue could once again become, in the words of one member, "a political football."[69] As Reagan appointee and California lawyer Robert Mitchell summarized the concern, if the guidelines were abandoned, "should Congress ever again consider national legislation, scientists could no longer argue that they were following a policy of self-regulation."[70]

But it was mainly industry, rather than science, that was most vulnerable to local intervention at this point. The scientific community had demonstrated to its own satisfaction that it could persuade the public that there was nothing to fear from *research,* and scientists felt legally invulnerable. Industrial processes, on the other hand, might still engender local controls or even lawsuits.

At the February meeting, a strong majority of the RAC backed away from the committee's original position favoring complete dismantling of the NIH controls and voted in favor of the Gottesman proposal. In the end, the committee proved more sensitive to industry and institutional arguments for containing public concern than to the arguments of prominent scientists for complete freedom of inquiry. The RAC's vote for the Gottesman proposal, while primarily a vote in favor of dramatically weakening the NIH controls, also represented a positive response to the industry need for a predictable regulatory climate.[71] The vote also made clear how effectively industry needs had been transmitted to the committee. By February 1982, the principal goals specified by representatives of the major pharmaceutical corporations at their meeting with DHEW officials in May 1979 had been achieved: strict protection of proprietary information that provided scope for legal intervention prior to release of such information and transfer of decision-making authority for approval of large-scale processes and the use of new host-vector systems to the IBCs.

10.3 The Politics of the RAC: Industry, Science, and the Public

The history of the dismantling of the 1976 NIH guidelines and the overturning of the policy of prevention that supported those controls suggests that the new policy emerging in 1982 resulted from the interplay of two influential interests—the scientist-clients of the NIH and private industry—operating in the absence of any significant countervailing force. These interests, while initially distinct, converged on the goal of reducing the NIH controls. In the political and economic climate of the late

1970s and the early 1980s, neither sector would countenance significant delays resulting from oversight or costly containment. Furthermore, these interests overlapped as a result of government policies that forged links between academic research and the emerging genetic engineering industry. That is, to the extent that scientists became industry consultants, members of scientific boards, and equity owners for new genetic engineering firms or for the pharmaceutical industry, academic and industrial interests became the same interest.

Where these interests converged, their effects on the decisions of the RAC were powerful: the pressure exerted by industry and many academic scientists to reverse the Asilomar legacy and to replace both the original policy of prevention with a new policy of crisis intervention and the "novel biohazard" discourse with the "no extraordinary hazard" discourse was irresistible, and the small minority of RAC members who remained committed to the original policy and discourse could not withstand it.

Where academic and industry interests only overlapped but did not converge, the effects within the RAC were less powerful and provided the minority members nominated by public interest groups with their only significant leverage in the committee's decisions. The key issue here was that of voluntary compliance, the approach favored by the PMA and the new genetic engineering firms but not necessarily by academic researchers without ties to private industry. As the committee records show, protection of proprietary information under the voluntary scheme—the *only* aspect of the NIH controls prescribing criminal penalties for violation and a provision whose basic philosophy ran counter to the traditional scientific value of "communal science"—divided the committee, producing at the May 1979 meeting a vote in favor of mandatory compliance.

The committee's resistance to the scheme revealed the extensive responsiveness of the NIH director, backed by the Carter administration, to industry interests in having the arrangement. Skillfully extracting a vote seeming to favor voluntary compliance from the IAC, NIH director Fredrickson persuaded a reluctant RAC to begin to implement the scheme in the fall of 1979. When the committee continued to resist, the NIH director overcame its "malaise" (Fredrickson's word) by orchestrating further meetings with the IAC to demonstrate the interest of the federal government in industry compliance, and by personally exhorting the RAC to continue to review industry proposals in closed sessions despite its desires to the contrary. The lengths to which the NIH director was willing to go to achieve that goal revealed his determination not to allow this aspect of the committee's work to break down before the voluntary compliance procedures were made largely obsolete by the dismantling of containment and oversight requirements. The legacy of "successful" voluntary compliance was valuable in itself, a precedent for future policymaking. Thus one notable finding of this study is its demonstration of the responsiveness of the NIH not simply to the biomedical researchers who were its primary clients but also to the pharmaceutical corporations and new genetic engineering firms, which wanted to assume the role of a voluntarily affected industry. The willingness of the NIH not to publicize Genentech's flaunting of the NIH guidelines and to keep its skirmishes with the rebellious firm under wraps (contrasting with its readiness to chastise the occasional unruly scientist) further confirms its responsiveness to the needs of the emerging genetic engineering industry.

The official image of this phase of NIH policymaking emphasized two features mandated by DHEW Secretary Califano in 1979: the inclusion of public interest members in the RAC and the pursuit of a program designed to test the possible hazards of genetic engineering. The policy process was portrayed as one of prudent development of a new field in which controls were lifted only when justified by evidence and approved by representatives of the public. Gilbert Omenn applauded "the cautious approach pursued by the NIH [which], despite the impatience of some affected scientists, served to rebuild public confidence and to broaden knowledge about the techniques among the scientific community."[72] Donald Fredrickson praised the "prudence [that] gained us entrance into new realms of beneficial understanding . . . and strengthened both the capacity and reputation of science for tending its broadest community responsibilities." And he commended the expanded RAC as "one of the most useful 'cultural innovations' to come out of the rDNA controversy."[73] The actual structure of RAC decisions and their relation to empirical evidence from 1976 onward reveal a very different image. Risk assessment not only lagged far behind advances in technique but was designed and used to soothe public concern rather than to test worst-case scenarios. Dissent from scientists and public interest representatives alike was skillfully contained through the design and management of the policy arena.

ELEVEN

Dismantling the Genetic Manipulation Advisory Group, 1979–1984

11.1 The Social and Political Setting

Until the end of 1978, the United Kingdom charted its own course in regulating genetic engineering, although not without some sensitivity to American decisions. At that point, however, the British and American systems diverged significantly, both in form and content. The British controls remained at the 1976 level, supported by regulations requiring notification of all genetic engineering work and a participatory policy-making system that had produced some major confrontations between the trade unions and private industry, and between the trade unions and the scientists. The American controls, on the other hand, were voluntary and likely to stay that way and were being substantially weakened in terms of oversight and containment requirements. In October 1978, the executive secretary of GMAG, anticipating (accurately) that the revisions would be approved, wrote: "The NIH categorizations have been relaxed beyond anything that would be acceptable in this country (I would guess). We . . . seem conservative by comparison: we now have legislative control and the discussions of categorization are unlikely to lead to dramatic changes."[1] When the EMBO committee repudiated GMAG's new risk assessment scheme in December 1978, it indicated that other European countries were likely to follow the American lead.

Furthermore the COGENE meeting at Wye College in April 1979 left no doubt that the Americans intended to dismantle their recombinant DNA controls, with the full support of the NIH. It was recognized that, in the words of a British official, there would be "tremendous pressure" on the European countries that had previously accepted GMAG's approach to "think in terms of the NIH guidelines." Indeed, by 1980 most countries had accepted the NIH approach as well as the successive revisions that it produced.[2]

Growing global competition in biotechnology intensified pressures to eliminate differences between the British and American systems. In the United Kingdom as in the United States, a declining economy characterized by loss of basic industries, spiraling inflation, and deep depression meant that development of new technology was widely seen as an important national priority. Keen attention was paid to the large equity investments being made by multinational corporations in small American genetic engineering firms. The *Economist* commented in December 1978 that "this time round Britain may identify the challenge at an early stage. It is not first in biotechnology but it is not far behind. . . . Britain's university scientists are right up with the leaders."[3]

The *Economist's* hope, that the United Kingdom would remain in the vanguard of biotechnology, was widely shared in government, industrial, and scientific circles. At the end of 1977, the House of Commons Select Committee on Science and Technology (a bipartisan parliamentary committee established to investigate aspects of British science and technology policy) convened a subcommittee to investigate the state of British genetic engineering. The committee's investigation was wideranging, including a field trip in March 1979 to meet with politicians, scientists, and corporate executives in the United States, and hearings from December 1978 to March 1979. The strongest and most persistent theme of the hearings was concern that British science and industry would be severely damaged if controls at home were more stringent than those abroad. A strongly worded memorandum from the CBI emphasized the "vital role of technology in improving Britain's competitive position in world markets" and insisted that the major factor inhibiting development of genetic engineering in the United Kingdom was the system of control.

> [This system] is much more restrictive than that in leading competitor nations, principally the United States, West Germany and Switzerland. There is likelihood that those leading chemical and pharmaceutical companies involved in genetic engineering research in the United Kingdom will carry out an increasing proportion of their activities and investment in other countries if this situation persists. Unnecessarily strict controls could dissuade other companies in Britain from diversifying into genetic engineering in spite of logical opportunities to do so which emerge as the technology advances. Furthermore, academic research work in the UK is becoming much less attractive than in the United States as a result of our stringent regulations; the likely outcome is that the most talented and entrepreneurial academic researchers will emigrate from the UK and continue their work abroad. Such a loss of leading scientific talent would have a highly detrimental effect on the quality of the basic research work carried out in Britain from which industrial applications could develop in the future.

To continue the British system, argued the CBI, would be to lose a "vital new 'heartland' technology" that could affect many other branches of British technology and industry.[4]

Government officials responsible for the future health of British science and technology were sensitive to these arguments. An Undersecretary for Science in the Department of Education and Science testified:

> I think it is necessary for GMAG to retain the confidence of industry, or there will be a danger that the work will go overseas, and we will lose opportunities in this country which we are very well equipped to take. . . . Other countries are catching up with us in their arrangements for regulating genetic manipulation. . . . They are not so likely to follow precisely our containment levels, but rather, to follow the more relaxed levels that are being promulgated now in the United States. So there are signs now—and I can only describe them as "signs"—that we may become somewhat isolated from scientific opinion and practice internationally, if we maintain things exactly as they are at the moment.[5]

The message was clear. Certainly it was reinforced by a visit paid by the British MPs to Genentech during their American field trip in March 1979, when preparations were already being made for production processes using genetic engineering.[6]

Nor were either the British government or influential sectors of the society eager to see international controls imposed on genetic engineering at this stage. A draft European Economic Community directive that proposed to develop uniform regulation of genetic engineering for European Economic Community member states received a cool appraisal from within the government on the grounds that it would require the United Kingdom to develop further regulations to cover the acquisition and use of the products of genetic engineering and introduce a further layer of bureacratic control at a point when the transition from research to application was under way.[7] The CBI strongly opposed the draft directive,[8] and the TUC, while in principle supporting international controls, was cool to the proposal on the grounds that it was being developed through consultations between government officials concerned primarily with research rather than with the health and safety of employees.[9]

With the election of the Thatcher government in May 1979, the work of the select committee came to an abrupt end. But the brief summary report issued by the dissolved committee sent a clear message to the new Parliament. Noting that GMAG's new risk assessment system had been strongly criticized by EMBO's Standing Advisory Committee on Recombinant DNA, that controls in the United States had been markedly relaxed, and that the subcommittee had "also heard from the NIH that further progressive relaxations of the guidelines were likely to take place over the next couple of years such that increasingly fewer areas of recombinant DNA work would be subject to regulation," the committee, while issuing a caveat that genetic engineering should be pursued safely, urged that "UK workers and industry not be placed at a competitive disadvantage."[10]

From the late 1970s onward, British science policy became increasingly selective and interventionist in character and directed toward fostering new fields with industrial potential (section 1.4). Biotechnology quickly became targeted for special attention. Early in 1979, a government committee was established by the Advisory Council for Applied Research and Development (ACARD) in conjunction with the Royal Society and the Advisory Board for the Research Councils (ABRC) to make recommendations for the industrial development of biotechnology. Chaired by Alfred Spinks, a former director of research for ICI, the committee included Sir William Henderson, a former secretary to the Agricultural Research Council (ARC) and the second chair of GMAG; Sir Austin Bide, chair of the Research and Technology Committee of the CBI; and Brian Hartley, professor of biochemistry at the Imperial College of Science and Technology and a founding member of Biogen—all with close ties to the industrial application of genetic engineering. Representatives of the general public and the trade unions were conspicuously excluded.[11]

After examining government support for biotechnology in the United States, West Germany, Japan, and France, the committee produced early in 1980 what was described in *Nature* as a "hard-hitting and interventionist" report that called for a policy of "technology push" (as opposed to "market pull") reflected in firm government commitment to strategic applied research.[12] This meant, among other things, substantial increases in support for research in biotechnology (a minimum of £3 million annually); coordination of industrial research and development among private industry, government research establishments, universities, and research associations; development of a limited number of "centers of excellence" in biotechnology; the training of a skilled work force; and the establishment by the National Enterprise Board and the National Research and Development Corporation (NRDC) of a national biotechnology company similar to Cetus or Biogen.[13]

Britain's relatively cautious regulatory policy for genetic engineering again received criticism. Warning that controls of recombinant DNA research in the United Kingdom should not be more restrictive than in other countries, the committee recommended that "GMAG and the HSE should continue as rapidly as possible to reduce constraints upon genetic manipulation experiments while maintaining an adequate degree of safety" and that "GMAG attends even more urgently than hitherto to the possible prejudicial consequences to British industry if constraints in the

United Kingdom on genetic manipulation are excessive compared with those of other countries."[14]

Further reports reinforced these proposals, insisting on the need to enable British biotechnology to compete on equal terms with other countries. A report of the House of Lords Select Committee on the European Communities issued in 1980 held that controls that were "more restrictive than those prescribed for the United States of America, Japan or [any] other advanced country would effectively distort competition and lead to re-siting of research, development, and production facilities" and that "nothing should be done to inhibit further progress" in the United Kingdom.[15] A Royal Society report on biotechnology and education, published in 1981, called for the training of about one thousand graduate biotechnologists over ten years.[16] A report of the Education, Science, and Arts Committee of the House of Commons called in 1982 for special measures to protect the research base for biotechnology and to encourage cooperation between government and industry.[17] An exception was a 1981 government White Paper on biotechnology, reflecting the "philosophy" of the Thatcher government, which argued that industry, not government, should be responsible for transforming genetic engineering into a commercially viable technology.[18] The report brought forth strong criticism from the universities and private industry, however, and was quietly shelved.[19]

As a result, the Thatcher government swallowed its scruples and proceeded to implement virtually all the recommendations of the Spinks report. It was recognized that, as one survey of British government policy for biotechnology during this period concluded, the government had "a role to play in stimulating coordination and awareness in addition to funding research and development and supporting educational policies."[20] Although support for the universities was being drastically cut back at this time, special funds were set aside for biotechnology by the University Grants Committee (UGC) and by the Research Councils. Several government committees were set up to coordinate the research and development effort. Measures were taken to facilitate the transfer of research advances into the market place and to induce industry investment in universities. In December 1980, a government-sponsored genetic engineering firm, Celltech, was created with capital from the British Technology Group, the Technology Development Corporation and private sources and involving a collaborative arrangement with the MRC. In addition to industry representatives, members of the board of directors included Sir Michael Stoker, vice president of the Royal Society; Sydney Brenner; and James Gowans, secretary of the MRC. By 1981, it was estimated that the British government was spending on the order of £25–30 million a year on biotechnology.[21]

At the same time, the Thatcher government pursued its preelection commitments to cut government spending and roll back regulation (section 1.4). The drive, launched with vociferous Tory support, to eliminate quangos directly hurt the Health and Safety Commission, whose budget was cut. Since the responsibilities of the commission were increasing, the resulting 11 percent reduction in staff meant that the commission increasingly restricted its concern to areas of high and known hazard. It was unlikely that much attention would be given to new technology such as genetic engineering, where hazards were conjectural.

The future of GMAG was also called into question. The new Secretary of Education, Mark Carlisle, was known to have a strong preference for policymaking on technical issues by expert committees and had little sympathy for union or lay representation. Neither as an experiment in lay participation in science policy nor as the creation of the previous Labour government was GMAG likely to appeal. Although GMAG was not slated for immediate elimination, calls for "quango culls" from Tory MPs and the government review of quangos boded ill.[22]

11.2 The New GMAG

Toward the end of 1978, as GMAG's first term drew to a close, there was a strong sense that the group's operations would eventually wind down. Leaders of industry portrayed the British controls as a "major constraint" on the development of the new genetic engineering industry, warning frequently after 1979 that the controls would drive industry abroad. One GMAG member who was a trade union representative and a molecular biologist active in the field commented on the change of attitude within the committee after 1979: "There was now considerable industrial involvement. . . . If we had continued to insist on a high level of safety measures, we were told time and time again that the industrial investment would move elsewhere, and this actually did happen in some cases." Scientists, moreover, were eager to get on with research. "The atmosphere [after 1978] shifted from one of willingness to be concerned to one of enormous hype—to some extent valid—about the power of the techniques—and I share in this. To be honest, I did not want to spend one day a week thinking about [the control of genetic engineering]. I wanted to be in the lab doing experiments."[23] And scientists who had originally dissented from the American consensus that the hazards were exaggerated were coming to accept this view.

At the same time, the Tory government's assault on regulation meant increasing difficulty in maintaining health and safety standards in the workplace. Moreover, the trade unions, having lost their basis of support in the government and steadily losing members to the unemployment lines, were in no position to resist this trend. At the local level, laboratory safety committees began to operate in a more casual manner. By 1982, only a minority of safety committees were meeting regularly, issuing minutes, and considering experimental protocols. Thus in the 1980s, the unions represented on GMAG, who had been most likely to support continued oversight and control of genetic engineering, were much less able to influence committee decisions. Furthermore, any effort to do so was resisted by the government. According to the same GMAG member, "The civil servants made it quite clear on every occasion that the situation had changed in Whitehall and that GMAG would not receive support. . . . There was no possibility of anyone [who supported the continuation of controls] raising this matter in an effective way because neither the scientific community nor the general public was disposed to regard this as a major issue." This member summarized the political context in which the second GMAG operated.

> I think it is hard to convey . . . the extent to which there was a shift, a gearing down, in the level of activity invested in this field, from 1978, when everything went to GMAG, when there was an enormous amount of interest and concern in this area, when the local safety committees met regularly, and when the entire system functioned at a very high profile, to the situation in 1982 when there was *no* interest in pursuing health and safety aspects of this area. . . . Anyone trying to maintain a high profile role for GMAG . . . was unable to do so. Basically, GMAG was at every stage of the game progressively being ignored and since it did not have political clout, there was no way to counter that. The balance of forces [affecting the operation of the committee] shifted totally.

All memberships on the committee automatically terminated at the end of 1978, and a majority of the membership changed. The chair, Sir Gordon Wolstenholme, stepped down and was replaced by Sir William Henderson, a former secretary of the ARC with a long involvement in the formation of British genetic engineering policy. Three new scientific members, three new members representing the public interest, and one new member representing management were appointed. The trade union representatives were all reappointed (table 11.1).

Although the interest-group composition of the committee remained the same, the changed membership shifted its style and direction. An important reason for the selection of Wolstenholme as the first chair of the committee was his ability as mediator and consensus producer. His appointment, seen at the time as an "inspired" choice, pleased those members of GMAG who believed in the value of participatory decision making. In 1980, Wolstenholme was described by representatives of the

Table 11.1 Membership of the Second U.K. Genetic Manipulation
Advisory Group, 1 January 1979

Sir William Henderson, Former Secretary to the Agricultural Research Council (chair)

Appointed as scientific and medical experts
J. B. Brooksby, Director, Animal Virus Research Institute, Pirbright, Surrey
J. Ingle, Scientific Adviser to the Secretary of the Agricultural Research Council, London
B. W. Langley, ICI Corporate Laboratory, Runcorn
M. H. Richmond, Professor, Department of Bacteriology, University of Bristol
J. H. Subak-Sharpe, Professor, Institute of Virology, University of Glasgow
P. M. B. Walker, Director, Medical Research Council, Mammalian Genome Unit,
    University of Edinburgh
P. Wildy, Professor, Department of Pathology, University of Cambridge

Appointed to represent the public interest
J. O. P. Chamberlain, Department of Epidemiology, Institute of Cancer Research, Royal
    Marsden Hospital, London
M. Kogan, Professor of Government and Social Administration, Brunel University,
    Uxbridge, Middlesex
J. Maddox, Director, Nuffield Foundation, London
C. M. Puxon, Barrister

Appointed to represent the interests of employees
D. C. Ellwood, Professor, Microbiological Research Establishment, Porton Down,
    Salisbury (Institute of Professional Civil Servants)
D. Haber, Association of Scientific, Technical, and Managerial Staffs, London
R. Owen, Trades Union Congress, London
R. Williamson, Professor, Department of Biochemistry, St. Mary's Hospital Medical
    School, London (ASTMS)

Appointed to represent the interests of management
Sir Frederick Dainton, Chairman, British Library Board and Chairman, National
    Radiological Protection Board, London
J. A. Gilby, Technical Director, Beecham Pharmaceuticals Division (U.K.), London

Source: Genetic Manipulation Advisory Group, *Second Report of the Genetic Manipulation
Advisory Group,* Cmnd. 7785 (December 1979), vii–x.

public interest and trade unions as fair, evenhanded, and able to bring
together those with widely differing interests in genetic manipulation.
Wolstenholme's GMAG spent long hours negotiating the content of the
minutes, the confidentiality of commercial information, and the right of
local committees to evaluate the merit of scientific research. But this
commitment to consensus building, seen by some as an asset, was
viewed by others, particularly those representing science and industry, as
a liability. These members saw GMAG's proceedings as unnecessarily
time-consuming, "getting out of hand," and threatening a growth area

of science and industry. Negative comparisons were made with the United States. As one industry representative put it: "GMAG flutters like a large butterfly. The United States goes at [the question of controls] like a bull." As pressures to relax genetic engineering controls increased, there can be little doubt that those responsible for GMAG now felt it necessary to select a chair for whom efficient movement toward relaxing controls would be the paramount concern.[24]

Henderson's strong commitment to facilitating the rapid development of genetic engineering in the United Kingdom was well known. As secretary of the ARC in 1973–74, he had been directly involved in the establishment of the Ashby committee and had been a member of the committee (section 3.5). In 1979, in addition to chairing GMAG, he was also serving on the Spinks committee. Styling himself a "benefits man not a hazards man," making it clear to the members of GMAG from the beginning of his chairmanship that the committee's responsibility for considering "potential hazards" should not inhibit the realization of the potential benefits,[25] he emphasized that there would be no delays in decisions. One of his first acts was to demonstrate that GMAG's decisions would not be delayed by union needs to consult with their constituency. While Wolstenholme tended to wait for the unions to consult, Henderson faced the unions down and proceeded. His ability to do so no doubt reflected confidence that his decisions would be backed by the minister.

Not surprisingly, Henderson's handling of GMAG evoked sharply differing reactions from members of GMAG, depending on their values and commitments. Scientists, representatives of industry and management, and the new public interest representatives all approved of Henderson's management of committee business. The isolated trade unionists saw him as "ruthless," "not evenhanded," and generally insensitive to union concerns. It is revealing that these qualities were not noted by any other group.[26]

A second influential appointment to GMAG was that of Peter Walker, professor of zoology and director of the MRC's Mammalian Genome Unit at the University of Edinburgh. Walker's views about the genetic engineering controversy, that the whole problem of genetic engineering hazards was virtually over and that the British controls should be brought into line with those in the United States, were well known. As a report in *Nature* noted, Walker had been "not uncritical of excessive control of recombinant DNA research."[27] Early in 1979, Walker was made chair of the new GMAG Technical Panel, composed largely of GMAG scientists and coopted experts. This group took on two key responsibilities: assessing applications for genetic engineering experiments and implementing the new risk assessment scheme. As chair of the group, Walker exerted considerable influence over the progressive relaxation of British controls (sections 11.3 and 11.4).[28]

Only one member's wish to remain on the committee was denied: Jerome Ravetz, reader in the history and philosophy of science at the University of Leeds, who during the first two years of GMAG's existence had focused on the question of hazard assessment and had pressed for combining implementation of the new risk assessment scheme with a substantial experimental program designed to define and measure recombinant DNA hazards. Within the committee, Ravetz found little active support for his position, although certain members were sympathetic to it. Outside the committee, Ravetz had also made his views known to Shirley Williams, the Minister for Education and Science. Significantly, Ravetz's perspective was seen as no longer useful. The message implicit in his nonreappointment was that risk assessment would be given a low priority on the new committee's agenda.[29]

Of the former public interest members, only John Maddox remained; three new public interest representatives—a professor of government, a barrister, and an epidemiologist—were appointed. At this stage in the committee's development and in the evolution of the issues associated with genetic engineering, the difficulties of absorbing knowledge of the issue and of prior policy decisions already assumed by the committee were daunting, especially to those drawn from other fields. All of the new public interest members acknowledged that they relied on the judgments of the scientists in the group to some extent—and two members relied on these judgments almost entirely. "I more or less accept what I am told," commented one public interest member. Another observed that in GMAG "one is, in a sense, an observer." The sense of some that GMAG's charge was being construed narrowly did not cause them to try to broaden the committee's agenda. In fact, in the political environment that developed in the United Kingdom from 1979 onward, any efforts to do so would almost certainly have been resisted.[30]

At the same time, communications between the MRC and NIH officials responsible for GMAG and the RAC, respectively, were improved. From 1980 onward, Keith Gibson and William Gartland, executive secretary of GMAG and of the RAC, respectively, communicated by telex. Thus the agendas of RAC meetings were immediately known in London, sometimes producing almost instant adjustments of the British controls.[31]

The new GMAG seemed fashioned to bring British policy into line with policies elsewhere, particularly the United States, and by the time Henderson resigned at the end of 1980 to become a member of the Science Council of Celltech, the British controls had been radically reduced (section 11.4). Henderson was replaced by Sir Robert Williams, the for-

mer chair of the Williams working party, who chaired GMAG until its
remaining functions were transferred to the Health and Safety Executive
in 1984.[32]

11.3 Implementing the New Risk Assessment Scheme

GMAG met almost every month in 1979 and 1980 and much less frequently thereafter, until its transfer to the Health and Safety Executive
in 1984. Previously contentious issues, such as the confidentiality of
proposals and the right of safety committees to address scientific merit,
were not allowed to disrupt the proceedings. GMAG operated, if uneas-
ily from the point of view of industry, with the confidentiality agreement worked out in 1978. Four major issues now occupied most of
the committee's agenda: implementing the new risk assessment scheme,
oversight of genetic manipulation, assessing large-scale fermentation
processes involving genetically engineered organisms, and deciding the
future of GMAG itself. The question that overshadowed all of these specific issues, however, was whether the United Kingdom should match
the relaxation of the American controls—a question that came up repeatedly, as the NIH progressively dismantled its controls.

Originally it was hoped that the new risk assessment scheme, initiated
by Sydney Brenner and developed by GMAG and its subcommittees in
1978, would be adopted by other countries in preference to the NIH
guidelines. But with the rejection of the scheme by the EMBO advisory
committee in December 1978 and cool responses from other European
countries, concern grew that the United Kingdom would be, in the
words of GMAG's executive secretary, left "out on a limb." [33]

Undoubtedly the new scheme's reception abroad colored the response
at home. Those skeptical about the need for controls grew more vocal,
arguing that the controls spelled disaster for the future of British biotechnology. At an open meeting organized by GMAG in December 1978,
reaction to the scheme was mixed. Although some scientists, particularly
those associated with GMAG, supported the scheme, others challenged
the need for continuing controls and questioned the practicality of implementing the scheme in the absence of firm data on hazards.[34]

The debate lit up the columns of *Nature*. An ad hoc committee of the
Royal Society, chaired by Oxford geneticist Walter Bodmer, warned that
"a lack of comparability between procedures in the United Kingdom and
elsewhere, especially in the United States, could seriously jeopardize this
country's research and development efforts in this important area of
work and in particular, threaten our contribution to extensive international cooperation in research in this field." [35] GMAG was urged to delay
action on the new scheme and to "assess the impact of the new NIH

Guidelines before any firm decisions about major changes in the approach to categorization are made." In letters to *Nature,* scientists dismissed the British system as "illogically restrictive." "Missing out on a revolution in biological technology" and "losing a generation of the best young scientists who wished to be trained in this area" became rallying cries for those who opposed continuing controls.[36] As one scientist summarized the situation, "There was an element of panic here because the British regulations were wildly out of line with anywhere else. People will put up with being marginally out of line but not [to that extent]. . . . We were in real trouble then. [It was] one moment in time when people were threatening to leave and meant it."[37]

Private industry also pressed for parity with controls elsewhere. The memo from the CBI to the House of Commons Select Committee on Science and Technology pointed to "several major discrepancies between the NIH and GMAG guidelines" and argued that "any substantial differences from the revised NIH guidelines would need to be justified on scientific and technical grounds."[38] A sympathetic editorial in *Nature* made the same point: "British firms will be penalized in the early stages of the potentially enormous business of biotechnology if GMAG's controls are any stiffer than those adopted voluntarily or by fiat elsewhere." The scheme's distinction between the safety of systems engineered to produce expression of a product and those that were not elicited the frank warning that "as industry clearly wants expression [of foreign genes], it is to industry's advantage that that factor is ignored."[39] Veiled threats that industry and science would leave the United Kingdom for less restrictive climates were made. That "scientists were threatening to leave and meant it" seems to have played a great part in GMAG's subsequent interpretations of its guidelines.[40]

The unions, on the other hand, remained committed to GMAG's controls, perhaps especially to the forms of oversight developed in the committee's first two years. Adding weight to the unions' contention that containment and oversight were required for safety was the escape of smallpox virus at the University of Birmingham in 1978, causing the death of a medical photographer. An official report on the incident, leaked to the press by the ASTMS in January 1979, served as a powerful reminder of the hazards not only of lax laboratory standards but also of lax oversight.[41] "All safety nets failed," announced a headline in *Nature.* A scientist favoring the weakening of British controls later observed: "The whole thing [the relaxation of British controls] was fouled up on the smallpox problem—which showed an 'expert' doing things an expert had no business doing."[42] A move to dismantle controls risked a major battle with the ASTMS and other unions.

Within GMAG, several came to see the new risk assessment scheme as the only way simultaneously to avoid public confrontation and to reduce containment levels for experiments. John Maddox wrote in a letter to *Nature* in March 1979:

> Making a bonfire of the regulations now will cause only trouble. *Surely it is in the interest of the scientific community that they should be dismantled by rational means.* GMAG's proposed way of working offers precisely such a possibility, provided that genetic manipulators play their part in gathering the necessary data. My own belief is that if the proposed system were adopted and willingly accepted, we should find that before the end of 1980 all but a handful of experiments were no more restricted than by the requirements of "good microbiological practice." Is not that a chance worth taking?[43]

A scientist closely associated with GMAG expressed the same view: "Critics [of the GMAG proposal] didn't realize that the scheme was a subtle British way of making things much easier. They didn't understand what was between the lines."[44]

What *was* between the lines emerged in incremental stages as decisions on the application of the new risk assessment scheme to the categorization of experiments and on oversight of these experiments, made by GMAG and by its new Technical Panel between 1979 and 1982, allowed the British controls to match the progressive relaxation of the NIH guidelines.

Soon after the Henderson-led GMAG began its term, it took a small step in this direction by exempting certain types of "self-cloning" experiments in which the genes from a given organism are cloned in the same organism—a step the NIH had taken in its revisions of December 1978. GMAG's exemption, announced in March 1979, was less permissive but served to reassure British scientists about the committee's intentions. As GMAG also stated at this time, it was "studying the NIH guidelines with a view to deciding whether there are procedures which could be adopted in the UK," and it "continued to advocate collaboration with other authorities and . . . to work towards some uniformity of procedure."[45]

That larger agenda surfaced as the Technical Panel turned its attention to implementing the new risk assessment scheme and to assessing genetic engineering applications.[46] These (closely related) responsibilities placed the Technical Panel in a pivotal role with respect to the evolution of British regulatory policy. Not only did the panel make the initial judgments concerning the validity of risk data submitted by scientists with their applications, but also it established the norms for the operation of the risk assessment scheme and decided when and how these norms were to be adjusted.[47]

One of the panel's first actions was to compare GMAG's requirements

Table 11.2 U.K. Genetic Manipulation Advisory Group Categorization Scheme,
ca. April 1979–August 1980

	Probability
Access	
Host systems such as wild-type *E. coli*	1
E. coli K12	10^{-3}
Approved disabled host/vector systems, e.g., χ1776	10^{-6}
Approved disabled host/vector systems, e.g., MRC1	10^{-9}
Expression	
Any DNA from which expression as a polypeptide is sought	1
Copy DNA inserted in a plasmid site from which expression is not sought	10^{-3}
Genomic DNA in a known plasmid expressing site	10^{-3}
Any DNA in a known plasmid where nonexpression has been demonstrated, depending on the sensitivity of the detection procedure	10^{-6}
Damage	
Expression of toxins, hormones, and similar biologically active molecules	1
Uncharacterized polypeptides of unknown biological activity	10^{-3}
Well-characterized polypeptides with no evidence of adverse biological activity, e.g., globin	10^{-6}
No polypeptide expressed	10^{-9}

Source: GMAG Note 11 (January 1980), Genetic Manipulation Advisory Group, *Third Report of the Genetic Manipulation Advisory Group,* Cmnd. 8665 (September 1982), 40–45.

with those of the revised NIH guidelines, issued in December 1978, confirming what its members already knew and many saw as a threat to British genetic engineering, that GMAG's containment requirements were significantly more stringent than those of the NIH. Recognition of this gap seems to have played a great part in GMAG's subsequent application of the risk assessment scheme. The details of that process are interesting because they show the malleability of the key variables of the scheme and the way in which these were adjusted to bring the British requirements into line with those across the Atlantic.

Under the guidance of the Technical Panel, GMAG substantially reduced the British controls and containment levels in two main stages. From March 1979 until July 1980, hazards were assessed according to standards that were formalized as GMAG Note 11, issued in January 1980 (table 11.2). A further drop in containment requirements followed in a second phase of assessment beginning in August 1980. These further reductions were formalized in GMAG Note 14, issued in September 1980 (table 11.3).[48] The Technical Panel also created a new category of containment in this period. Known as "good microbiological practice" (GMP), this category required fewer precautions than category I—

Table 11.3   U.K. Genetic Manipulation Advisory Group
Categorization Scheme, September 1980

	Probability
Access	
Wild-type *E. coli*	1
E. coli K12	10^{-3}
Approved disabled host-vector system, e.g., χ1776	10^{-6}
MRC8 with plasmid pAT 153	10^{-9}
Genetically manipulated DNA in tissue culture cells introduced as DNA that does not have the ability to infect or otherwise transfer to other cells	10^{-12}
Expression	
Expression in a system designed for maximum production	1
Insertion of copy DNA at an expressing site, with no special measures to increase expression	10^{-6}
Other systems that have minimal expression of eukaryotic genes in prokaryotic organisms	10^{-9}
Damage	
Expression of a high level of a toxic substance under conditions where the amount produced is likely to have a significant biological effect	1
Expression of a biologically active substance at a level where it might have a deleterious effect if it were completely absorbed and delivered to a target tissue	10^{-3}
Expression of a biologically active molecule at levels where it could not approach the normal body level or therapeutics dose (roughly 10 percent of the normal body level or therapeutic dose)	10^{-6}

Source: GMAG Note 14 (September 1980), Genetic Manipulation Advisory Group, *Third Report of the Genetic Manipulation Advisory Group,* Cmnd. 8665 (September 1982), 54–59.

essentially the basic requirements for any microbiological laboratory. Although there was some confusion over whether GMP counted as a formal containment category, in practice it would be used as a minimum standard when risks were assessed at 10^{-15} or lower.[49]

In notes 11 and 14, the Technical Panel established typical values for the three main parameters of the scheme, the probabilities for "access," "expression," and "damage."[50] These numbers were multiplied together to determine the overall probability of a harmful outcome. However, the values assigned to these quantities changed significantly between note 11 and note 14 (table 11.4). Such a change was not difficult to arrange, since the numbers employed to represent the hazards of experiments were merely reasonable guesses. ("Disabled" *E. coli,* for example, was given a million-to-one shot of establishing itself in the human intestinal tract; if expression was not sought, this was given a thousand-to-one chance of happening.) And the multiplication of these estimated probabilities to produce an overall hazard assessment that could be matched to contain-

Table 11.4   Comparison of British and American Containment Requirements
for the Human Insulin Gene, 1978–82

| Date | United Kingdom | | United States | |
	Source	Classification	Source	Classification
1976	Williams report (September)	EK1/category III/IV	NIH guidelines (June)	EK2P4
January 1979			NIH guidelines (December 1978)	EK2P2
April 1979	GMAG note 11 (formally issued January 1980)	EK1/category II/III MRC8/category I		
1980	GMAG note 14 (September)	EK1/category I/II EK1/GMP	NIH guidelines (January)	EK1P1
1982			NIH guidelines (April)	EK1: exempt; P1 recommended

Sources: Department of Health, Education, and Welfare, National Institutes of Health, "Guidelines for Research Involving Recombinant DNA Molecules," *Federal Register* 41 (7 July 1976): 27911–43; Department of Health, Education, and Welfare, National Institutes of Health, "Guidelines for Research Involving Recombinant DNA Molecules," *Federal Register* 43 (22 December 1978): 60108–31; Department of Health, Education, and Welfare, National Institutes of Health, "Guidelines for Research Involving Recombinant DNA Molecules," *Federal Register* 45 (29 January 1980): 6724–49; United Kingdom, *Report of the Working Party on the Practice of Genetic Manipulation,* Cmnd. 6600 (August 1976); GMAG Note 11 (issued January 1980) and GMAG Note 14 (issued September 1980), in Genetic Manipulation Advisory Group, *Third Report of the Genetic Manipulation Advisory Group,* Cmnd. 8665 (September 1982), 40–45, 54–59.

ment procedures assumed (with no basis in experiment) that hazard parameters were independent and that they included all sources of danger. As noted earlier, there was no significant interest among scientists in conducting the experimental work that could make hazard evaluation less hypothetical and containment assignment less a matter of arbitrary judgment. As Peter Walker later commented, "Initially, when people attempted to use [the scheme], it was impossible to see the justification for the figures they put down. Sometimes they worked it out a bit. So I'm afraid it was my object in life to simplify the scheme, to make it workable, to turn it into a set of rules to get a generally acceptable outcome."[51]

That such outcomes were deemed acceptable when they matched American standards can be seen from the committee's treatment of experiments designed to express particular proteins. The NIH guidelines did not require increased containment for such experiments; the Brenner system, in contrast, assumed that expression of a novel protein might—add significantly (by a factor of 10^3 to 10^6) to the potential hazard—a feature emphasized in GMAG's second report, issued in December

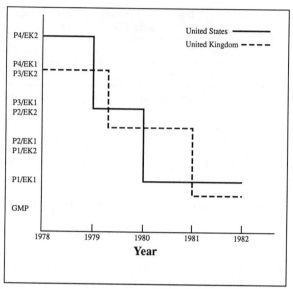

Fig. 11.1 Changes in American and British requirements for cloning the human insulin gene in *E. coli* K12, 1978–82.

1979.[52] To compensate for these higher assessments of hazard, the committee emphasized the use of new disabled systems of *E. coli* K12. One widely used system, MRC8, constructed by Sydney Brenner and a colleague at the MRC Laboratory of Molecular Biology in Cambridge, was assigned an access factor of 10^{-9} when used with a nonmobilizable plasmid.[53] The highest containment required for *any* work, including protein expression, with this system, was category II.[54]

By March 1979, GMAG produced categorizations of experimental hazard very close to those of the revised NIH guidelines. "It was not a very good system," Walker said. "We would have preferred the NIH system, which comes to much the same thing."[55] Indeed, comparison of the NIH and GMAG categorizations reveals only minor differences in containment requirements for the cloning of the human insulin gene, particularly where British use of disabled strains is allowed for (figure 11.1, table 11.4). Commenting on GMAG's work in December 1979, Henderson noted that GMAG had been "slightly embarrassed" when the NIH relaxed its controls in 1978 but that it was now "extremely difficult to detect the practical difference between the two systems." And he noted that, between July 1979 and December 1979, the percentage of experiments assigned to containment categories I and II had increased from 66 percent to 90 percent.[56] One scientist who had been involved in the development of the new risk assessment scheme commented in 1979, the scheme was "a peculiarly British way of saying 'OK, we're not changing anything but in fact, we're changing everything.'"[57]

Henderson's satisfaction with the achievement of parity between the British and American systems was short-lived, however. In January 1980, in response to the Rowe-Campbell proposal of April 1979, the NIH dropped containment levels for almost all work with E. coli K12, including the expression of the proteins of higher organisms in E. coli, to P1 (figure 11.1). The United Kingdom was left for a second time at a major disadvantage, particularly with respect to experiments closely linked to commercial applications.

The NIH-sponsored meeting at Pasadena in April 1980, organized under the NIH risk assessment program to evaluate the hazards of hormone-producing strains of E. coli K12 (section 10.1), provided an opportunity for GMAG to rethink the British requirements. GMAG was invited, and the question of who should represent GMAG became the object of a stormy struggle between the trade union representatives and Henderson. At a committee meeting in March 1980, the trade union representatives proposed a four-person delegation, one for each of the committee's constituencies.[58] Henderson, however, resisted the proposal, arguing that the Pasadena meeting would involve only experts and that those who represented GMAG should be scientists. When no resolution of the issue appeared to be at hand, he took the unusual step of referring the matter to the Department of Education and Science. The decision handed down to GMAG by the Thatcher government was that the committee would be represented only by the executive secretary of the committee, Keith Gibson. The report eventually transmitted to GMAG gave almost no indication of the actual range of dissent or uncertainty expressed at Pasadena. GMAG's third report could only conclude that "the NIAID workshop analysed the 'worst' possible case situation which may occur as a result of genetic manipulation work using E. coli with regard to the possible adverse effects of hormone producing strains of E. coli. It also discussed the possible occurrence of auto-antibodies or autoreactive cells due to the production of eukaryotic polypeptides and concluded that the public health risk would appear to be extraordinarily low."[59] The summary both obscured the actual range of dissent and uncertainty and omitted mention of the meeting's proposals for further experimentation and surveillance of the health of workers handling bacteria producing polypeptides (section 10.1). Such issues would undoubtedly have been noted by trade union representatives; their absence probably eased the way toward perceptions of reduced hazard.

Changes in GMAG's hazard assessments followed in July 1980. Comparison of note 11 (issued in January 1980) and note 14 (issued in September 1980) reveals important changes in assessment. The risk from expression of copy DNA was now set at 10^{-6}, a factor 10^3 lower than before, and damage factors for expression of biologically active substances were also reduced (in accordance with the official assessment of

the Pasadena conference). Taken together, these changes produced what
GMAG described as "a further radical revision of our procedures."[60] For
a second time, differences between British and American containment
requirements were minimized (table 11.4) and parity was achieved again
(figure 11.1). By this point, as John Maddox had anticipated, few experiments were categorized at more than GMP, and containment levels
for those experiments deemed scientifically or industrially important
were sufficiently low that any further changes were of much less
consequence.

11.4 Relaxing Oversight

Oversight requirements slid downward as well. Early in 1979, the Technical Panel proposed that all category I and II experiments could be initiated immediately after review and approval by local safety committees
and submission of details of the experiments to GMAG and to the Health
and Safety Executive without waiting for GMAG's approval. Apparently
uncontroversial, this was accepted by the committee at its meeting in July
1979 with the single modification that GMAG would continue to review
category II experiments involving expression.[61] With this exception,
British oversight requirements matched those operative in the United
States, where, under the December 1978 guidelines, experiments assigned containment could proceed after approval by an IBC, but were to
be registered with the NIH. When in January 1980 the NIH eliminated
national oversight for almost all experiments using *E. coli* K12 (with the
exception of experiments on the list of "prohibitions"), Henderson proposed a review of GMAG's oversight responsibilities. A subcommittee
of GMAG, consisting of one representative of each of the committee's
interest groups, was formed in response.

The key issue was whether central oversight of genetic engineering
work in categories I and II should continue. Peter Walker pressed for its
total exemption. The industry representatives wanted oversight of all genetic engineering work to be transferred to the Health and Safety Executive. The trade union and some public interest representatives wanted a
continuation of the status quo.[62]

At the June 1980 meeting of GMAG, the trade union representatives
proposed a compromise: category I and GMP experiments would be reviewed only at the local level, with GMAG being given a list of such
experiments annually; but procedures would continue unchanged for
categories II–IV and for large-scale work. Accepted virtually intact, this
proposal was formalized as GMAG Guidance Note 14.[63] At a press conference held to announce the new procedures, Henderson noted that
the changes would relieve GMAG of 90 percent of its responsibilities,

with only 2 percent of genetic engineering proposals requiring formal review.[64]

As in the United States, in the United Kingdom, these changes were represented as supported by new information and technical assessment. GMAG's third report, issued in September 1982, stated:

> In the USA the National Institutes of Health (NIH) made substantial changes to their guidelines [December 1978 and January 1980] at about the same time as we revised our categorization procedures [March 1979; July, 1980]. Most other countries follow the NIH guidelines but it was significant that, based on the published scientific data and information accumulating at that time, quite independent national advisory groups were able to reach similar conclusions and recommend an almost universal lowering of constraints on workers using this technique.[65]

In fact, this analysis shows that evaluating hazards without empirical data involved so much scope for arbitrary judgment that there was no reason for purely technical assessments of hazard to coincide and no reason to assume that systems expressing any protein whatsoever—hormones, oncogenic proteins, and so forth—were all equally safe. An editorial in *Nature* commented in October 1980:

> The most obvious deficiency of the past few years is that, faced with a problem of assessing the safety of a new and unexplored laboratory technique, the responsible committees (and the scientific community itself) did very little to provide data on which objective assessments of safety might be based. It is true that people have fed each other quantities of *E. coli* from time to time, and there have been a few half-hearted attempts to see whether genes incorporated into bacteria can find their way into the somatic cells of laboratory animals. Few opportunities have been seized for looking into the ways in which bacteria carrying foreign genes might actually damage their adventitious hosts. The customary explanation, during most of the past five years, is that the people most competent to carry out such experiments have been too preoccupied with exploiting the new techniques whose safety was called in question. With a few conspicuous (and honorable) exceptions, those with the most vivid interest in demonstrating that their techniques are as safe as they now appear have been content with arm-waving.[66]

Contrary to GMAG's claim that the two advisory systems acted independently of one another, arriving at the same conclusions on the basis of "published scientific data and [cumulative] information," this analysis shows that they interacted at several levels, especially through informal connections among scientists and through the links established between the NIH and the MRC. The downward trend of the NIH controls was closely watched in the United Kingdom, with British science and industry pressing GMAG to reach parity with the American standards when

differences existed. The unscientific haste with which British requirements were brought down to American levels should properly be ascribed to the intensity of British interests in not being "left behind" in
the "biotechnology race."

11.5 Closely Watched Trends: Regulating Industrial Processes

In August 1976, the Williams committee had considered large-scale, industrial processes to be "so remote" that it was unnecessary to address
their possible impact. Within two years, however, commercial interest in
industrial processes using genetically engineered organisms was growing
rapidly. From 1978 onward, the news media were noting the trends, announcing that industry was "start[ing] to do biology with its eyes open"
and that it was "time for bug valley." *The Economist* suggested that the
emerging genetic engineering industry had "striking parallels" with microelectronics, and that, like microelectronics, it would take off very rapidly. "The list of companies climbing on the biotechnology bandwagon
is growing," *The Economist* warned, noting that Hoffmann–La Roche,
Genentech, and Cetus were setting the pace of industrial development.
In March 1979, *Nature* reported Genentech's move to do large-scale
pilot projects (section 9.2) as well as the growing sense that many other
American companies were not far behind. The urgency conveyed by
such stories received reinforcement from the United Kingdom's leading
scientific society, the Royal Society, which in June 1979 hosted a meeting
of over 250 scientists, engineers, and industrialists to examine the prospects for producing substitute fuels, organic chemicals, drugs, food-
stuffs, and other products with genetic engineering and other forms of
biotechnology.[67] Industry organizations like the CBI were quick to seize
on these developments, reminding the government that more restrictive
controls would hold back development and cause a migration of activities
and investments to more congenial climates.[68]

The question of the precise coverage of large-scale processes by the
British regulations is complex. The definition of *genetic manipulation* used
by GMAG and the Health and Safety Executive covered the *construction*
of new combinations of DNA but not explicitly their subsequent *use*
(section 8.3). Guidance Note 4, first issued in May 1978, asserted that the
official definition did cover the subsequent *replication* of the products of
genetic manipulation, but this was not understood to include large-scale
uses of genetically manipulated organisms.[69] In its second report, issued
in December 1979, GMAG urged the Health and Safety Executive to
cover this gap.

> The term strictly does not extend to cover the processes by which the
> recombined product—be it a hybrid plasmid, a modified bacterial cell,
> a recombined virus, etc.—is cultured, either to produce a multiplication

of the new species or to secrete a product arising from genetic manipulation. The latter possibility is what many have in mind when considering the production of e.g. peptide hormones, or other protein products, by this means. The regulations do not include such "use" of the products of genetic manipulation but GMAG has advised HSE that the Regulations should be amended to include it; and in anticipation of such a change the Group has asked for the voluntary submission of brief information . . . of the nature of the work.[70]

Probably because they anticipated budget cuts, Health and Safety Executive officials were not keen either to engage in a further debate about the definition of genetic manipulation or to take on additional responsibilities, and the quango took no action. Consequently, GMAG's third report, issued in September 1982, simply reminded "users" of their obligations under the terms of the Health and Safety at Work Act, and at the same time noted that "although work involving 'use' of a genetically manipulated organism is not covered by the genetic manipulation regulations we consider our present terms of reference sufficiently wide to include this activity in our overall surveillance of the field and that the public still requires reassurance about the safety standards required for large scale work involving the use of genetically manipulated organisms."[71]

The debate over genetic engineering hazards had focused on the implications of new processes (chapter 9). Possible hazards involved in increasing their *scale* from laboratory cultures to the large volumes characteristic of fermentation processes received scant attention although the only official study of the potential impacts of large-scale processes, a report of the Commission of the European Communities issued in 1979, raised many questions.[72] GMAG's revised guidance note on large-scale use, issued in February 1982, also took cognizance of the possible pitfalls of industrial fermentation processes.

> Scale-up introduces an additional factor to those of access, expression and damage previously considered, namely the fact that an event considered so rare as to be negligible in work at a laboratory scale has to be taken into account when much larger volumes are involved. For example, even with a small volume such as 10 liters of culture of *E. coli* with the plasmid PAT 153 will contain approximately 10^{14} cells; therefore when integration of a plasmid into a host chromosome occurs at a finite frequency of 1 in 10^9 there will be approximately 10^5 such events in such a culture. Consideration should also be given to the relative viability and the degree of expression of these recombinants in the culture and to the survival of recombinants after spillage or disposal into the environment.[73]

The CBI and the British pharmaceutical industry, on the other hand, responded to such concerns by focusing on the growing scientific consensus, produced by the Bethesda, Falmouth, and Ascot meetings, that all genetic engineering hazards were minimal. In other words, that consensus, developed in response to the controversy over *laboratory* hazards, was now applied to *industrial* processes. These arguments, precisely paralleling those being used by industry representatives in the United States, were articulated in a memorandum from the CBI to GMAG.[74] Evoking the Falmouth conclusions, the CBI argued that "research has shown that it is extremely difficult to convert harmless micro-organisms into pathogens; present knowledge shows that the chances of this happening spontaneously are extremely remote." Thus the memorandum could also assert that "the problems with the use of genetically engineered micro-organisms are no different from those already well understood within the industry." It followed that the large-scale use of genetically engineered organisms could be safeguarded by existing industrial practices and that techniques of containment were well understood.[75]

The CBI also reminded GMAG that multinational corporations were unlikely to develop the new industry under hostile regulatory conditions.

> The application of recombinant DNA technology is a highly competitive international business. Major international companies are preparing plans for production using the technology, but the resources involved—the highly skilled scientists, engineers and technicians, the funds for development work and for longer-term investment and the specialist equipment—are highly mobile. World competition for these resources is intensive; they will migrate to those countries with the most suitable environment. One important determining factor here is the position adopted by regulatory authorities in individual nations with regard to the scaling-up of recombinant DNA technology.[76]

A major consideration was that high-containment conditions for industrial processes would greatly reduce profits from large-scale production. The CBI memorandum continued, "It must be stressed that in the majority of cases industry will wish to use genetically-manipulated micro-organisms which are already characterized at the lowest categorization because the costs involved might well render the whole process uncompetitive with alternative conventional ones and therefore uneconomic."

With similar arguments being made by industry officials on both sides of the Atlantic and with NIH and MRC officials in close contact with one another, the NIH with its RAC and the MRC with its GMAG made a series of parallel moves that produced virtual parity with respect to controls for industrial production.

The background and commitments of GMAG chair Henderson meant that he would be particularly adept in mediating this process. Experienced in the production of vaccines and as an industrial consultant, Henderson was sympathetic to the needs and perspective of private industry and "determined that we would not put ourselves ever in the position of justifiably being criticized for holding up industrial development of genetic engineering in this country." Henderson, citing his own experience with large-scale cultures of foot-and-mouth disease virus as an example of the safety of industrial fermentation, held that the hazards of large-scale processes were very much less than some anticipated.[77]

To address the scale-up question, Henderson reactivated the GMAG subcommittee on scale-up soon after his term as chair began, choosing to chair it himself. The new subcommittee consisted of scientists with expertise relevant to large-scale processes as well as representatives of industry, the trade unions, and the public interest.[78]

In November 1979, the subcommittee produced a guidance note for large-scale use of the products of genetic manipulation that differed significantly from the NIH approach, in which physical-containment requirements for large-scale work paralleled those used for laboratory manipulations. In contrast, GMAG did not develop distinct large-scale containment categories. Arguing ambiguously that "any risks involved [in using genetically manipulated organisms on a large scale] will not necessarily remain the same" as the risk of the original manipulation, the note concluded that "the different conditions of the industrial scale make it less appropriate to attempt to categorize proposals according to defined levels of containment."[79] That this approach reflected Henderson's own view is shown by his even stronger statement in a letter to the NIH in January 1980 that use of containment levels was "quite inappropriate," and his argument that "the very high cost of mounting the commercial operation of large-scale biotechnology . . . tends to ensure that the safety desirable for the growth of genetically manipulated organisms is likely to be achieved."[80] In other words, GMAG exempted large-scale processes from the new risk assessment scheme, deciding instead to assess applications individually, taking into account "the biological characteristics of the system and the physical characteristics of the plant combined with an appraisal of the competence of the management and the training and the skill of the operators" as well as the outcome of a site visit. These procedures were at once more detailed, in terms of scrutiny of each industrial plant and its safety committee, and more flexible than the approach required by the NIH.

GMAG's large-scale policy was soon being discussed in the United States, where pharmaceutical corporations were pushing the RAC to develop a large-scale containment category at the P1 level, lower than the two categories it was contemplating at P2-LS and P3-LS.[81] To NIH officials, Irving Johnson, vice president for research at Eli Lilly, argued that levels of containment lower than P2 were being approved in the United

Kingdom.[82] Letters from Henderson to William Gartland, executive secretary of the RAC, and from Keith Gibson, executive secretary of GMAG, to Gartland delineating GMAG's flexible scheme for large-scale applications were circulated to the RAC before its meeting in March 1980.[83] At that meeting, the RAC approved a set of large-scale containment categories that included a P1 category whose only expensive architectural feature would be the fermenter itself.[84]

As the RAC proceeded to relax genetic engineering controls during 1980 and 1981, requirements for large-scale containment higher than P1-LS became increasingly irrelevant. As the record of the RAC's reviews of large-scale applications shows, in the hawkish United States only P1-LS containment was required after the middle of 1980 for large-scale applications.[85] In addition, in September 1980, the RAC abandoned any effort to ensure compliance with its physical-containment requirements (section 10.2).

As the American controls eroded, the CBI repeated its warnings that British companies would be disadvantaged. The CBI's March 1981 memorandum urged an overhaul of the British large-scale procedures and particularly criticized GMAG's case-by-case review of large-scale work combined with site visits. Other areas of concern, as in the past, were the confidentiality of GMAG's review process and the lack of relevant expertise for the evaluation of proposals that occurred when those with conflicts of interest withdrew from decision making. The CBI insisted that the existing controls might "seriously affect the competitive position of industry in Britain." To avert these problems, the CBI proposed that the new risk assessment scheme (which by this point had been substantially relaxed, along with oversight requirements) should be applied to scaled-up processes—in other words, that large-scale work should be treated in the same way as laboratory work. (This position was, of course, contrary to that held by Henderson only a year previously, when containment requirements were more stringent.) Since work assigned to GMP or category I (and this included virtually all large-scale processes) could proceed without prior GMAG review, the CBI meant effectively to eliminate review of industry proposals.[86] When GMAG, chaired by Sir Robert Williams, considered the CBI's memorandum at its meeting in April 1981, a working group under Peter Walker was immediately constituted to meet with the CBI and draft a set of new proposals. Considered by GMAG at its meetings in June and September 1981, these were approved with minor modifications.[87]

GMAG's trade union members recognized the threat not only to controls for large-scale processes but, more broadly, to the continuation of the committee. "We are coming under heavy pressure on GMAG,"

Donna Haber warned Clive Jenkins in July 1981. "On the subject of large-scale work, industry is now wanting a situation where some of the work they do will not be notified to us at all. . . . It seems to me that we're losing the battle, and that if industry and the scientific establishment have their way, they'll win this one, and then eventually abolish GMAG altogether or confine it to a paper existence." [88]

The CBI had introduced two main issues: large-scale processes should be assessed in the same way and by the same criteria as laboratory processes (in other words, increases in volume made no difference to hazard); site visits and prior review by GMAG introduced unnecessary delays. The revised guidance note for large-scale processes proved favorable to the CBI in both respects. Although stating that large-scale work introduced risk factors different from those involved in laboratory work, industrial hazards were categorized in the same way as laboratory hazards. Furthermore, in harmony with the RAC's decision nine months earlier, GMAG now accepted responsibility only for the biological aspects of safety, not for the details of physical containment, and would make no more site visits. Finally, work categorized as either "exempt" or requiring GMP could proceed without prior review by GMAG. As the authors of the GMAG note stated, since most large-scale work was governed by "GMP" anyway, this would "allow scale-up to take place with notification alone, and without delay." [89]

There were immediate reverberations on the other side of the Atlantic. In the United States in 1981, all large-scale work was still being reviewed by the RAC on a case-by-case basis. In July 1981, Eli Lilly vice president Irving Johnson proposed to the NIH that large-scale work involving "exempt" organisms such as *E. coli* K12 and yeast be evaluated by the same criteria used for small-scale work (which would remove such processes from case-by-case review by the RAC and meant that oversight would exist only at the local level). A major reason cited by Johnson for the exemption of these industrial processes was "the Confederation of British Industry has already approached the GMAG with a similar proposal. . . . It is our understanding that it will be considered at the September 30 meeting of that group. We anticipate that GMAG will take positive action in order to assure British industry a competitive position in the evolution of this technology." [90] The RAC dutifully approved this request in September 1981.

Thus industry needs were secured on both sides of the Atlantic: at virtually the same moment, the RAC and GMAG dropped their requirements for case-by-case reviews of large-scale proposals. Containment requirements also converged, requiring at most category I. Communications between British and American government and industry officials, lubricated by reminders that industry should not be put at a competitive disadvantage, facilitated the downward course of regulation and a careful parity between the two systems.

11.6 Terminating GMAG

As controls were progressively relaxed, the questions in the background of GMAG's proceedings were not whether the committee would eventually be disbanded, but when and how. Long uncomfortable with GMAG's confidentiality arrangement as well as with delays in handling applications, the CBI was adamant in wanting to transfer the whole GMAG operation to the Health and Safety Executive.[91] "The sooner we get rid of GMAG the better," was the way the CBI position was summarized in 1980 by Jack Gilby, a Beecham Pharmaceuticals executive and representative of management on GMAG.[92] The trade unions were also open to the idea of transferring GMAG's functions to the Health and Safety Executive. In 1979, the trade union representatives proposed that both GMAG and a committee overseeing the use of dangerous pathogens, the Dangerous Pathogens Advisory Group, should come under the aegis of the Health and Safety Executive. However, the trade unions envisaged a far more substantial role for an advisory body to the Health and Safety Executive than did the CBI.[93] On the other hand, most scientists opposed a transfer of nonroutine decisions on genetic engineering to the Health and Safety Executive. GMAG under the Department of Education and Science, while unwelcome, was nevertheless seen as a committee responsive to scientists' needs. The Health and Safety Executive, on the other hand, was seen as detached from the scientific community. As an official in the MRC commented, "From an academic's point of view HSE is not always welcome. [It is seen] as interfering. [Academics] do not like anything which is going to restrict them at all."[94]

From the beginning of his chairmanship, Sir William Henderson forecast the eventual dissolution of the committee and was keen to facilitate that process. Independently of GMAG, Henderson briefed the Undersecretary of State for Science Neil MacFarlane in 1979, explaining that he could anticipate that by mid-1980 changes in GMAG's procedures would make it feasible to transfer most responsibilities to the Health and Safety Executive. Henderson suggested that the remaining GMAG activities could "nestle in what is a statutory function—an obligation, if you like, which exists anyway, regardless of GMAG. . . . We've provided the guidelines, we've been through all the stuff, we've given advice, and people know what to do. It's all case study now. You don't have to sit here and do all this stuff."[95]

According to Henderson, however, the Undersecretary "wasn't having any of this," seeing the move as premature: "His argument was he could see the necessity of having an advisory body, and further, he saw merit [in the fact] that the advisory body was not associated with the executive body [the Health and Safety Executive]. In other words, he preferred to see DES continuing to run GMAG, even though GMAG's workload—he could foresee—would change dramatically."[96] For the time being, GMAG was seen by the Department of Education and Science as one particular quango that was giving good value, actually protecting genetic engineering both from public criticism and from the possibly tougher standards that might have been enforced by the Health and Safety Executive.[97]

In March 1980, Henderson again raised the question of the transfer of the committee's functions to the Health and Safety Executive, this time at a GMAG meeting and in the context of the question of relaxing oversight. He proposed that responsibilities for category II work and below—which by this point meant merely receiving notification of it—be transferred to the Health and Safety Executive. Private industry, through its representative Jack Gilby, supported this proposal, but the trade union representatives vigorously opposed the idea of a piecemeal transfer. Furthermore, the Health and Safety Executive itself indicated that it was not anxious to add to its responsibilities at a time when reductions in its budget were contemplated.[98]

The future of GMAG continued to be debated within GMAG in 1980 without resolution. A call for the abolition of the committee by GMAG member John Subak-Sharpe did not receive general support.[99] Meanwhile, GMAG proceeded to reduce oversight and containment requirements. The recategorization of experiments and the weakening of controls for industrial processes radically reduced GMAG's workload: after June 1981, only three experiments were classified at levels higher than category II, and after December 1981, there were no further assignments to categories III and IV, and only two to category II. Since prior review and approval by GMAG was only required for categories II and above after September 1980, GMAG's review functions all but disappeared.

Nor did other issues associated with genetic engineering fill GMAG's increasingly thin agenda. An experimental program of hazard assessment was not given strong support by either of the departments involved in oversight of genetic engineering (section 8.3). Ministerial disinterest in this area was also reflected in the nonreappointment of the most active proponent of hazard assessment on the first GMAG. Moreover, scientists who were competent to do risk assessment were caught up in the pressures generated by the evolution of a new and competitive field. As one MRC official commented in 1980, "There's virtually *no* interest in [risk assessment] among scientists. It's all very well for people to recommend these things and for one or two people on GMAG to be fairly outspoken about it. But what it really comes down to is to get people to take a year or two out of their careers, and especially in this rather hot international race [this is unrealistic]."[100]

The emerging issues of release of genetically engineered organisms into the environment and of genetic engineering in humans were also addressed cursorily by GMAG but not actively engaged. The first was deemed too remote a possibility to require consideration of guidelines. In its third report, GMAG noted only that it would assess proposals on a case-by-case basis and maintain contact with two other government agencies with interests in this area, the Ministry of Agriculture, Fisheries, and Food and the Natural Environment Research Council.[101] The second, although provoking trade union interest, was merely passed to the MRC Committee of Inquiry into Human Fertilization and Embryology, where no action was taken. As GMAG chair Sir Robert Williams noted some years later, GMAG's consideration of gene therapy did not get "beyond the stage of long impassioned speeches, and then agreement not to do anything any more."[102]

GMAG's shrinking agenda and increasingly routine deliberations meant that the frequency of meetings dropped off sharply after 1980.[103] The question of the transfer of responsibility for GMAG's remaining functions from the Department of Education and Science to the Health and Safety Executive now began to receive serious consideration: by the end of 1981, almost no one involved in genetic engineering policy objected to the move. The CBI had pressed for this change all along; the trade unions saw it as advantageous, under the circumstances; the reservations of the scientific community dissipated with the recognition that public controversy or government oversight no longer threatened their independence.

Within the government, the Health and Safety Executive had always held that responsibilities for health and safety (given adequate funding to meet them) should fall under its authority. The MRC, on the other hand, had always been uncomfortable with its dual role of running GMAG and supporting research in genetic engineering. Once the scientific community no longer felt it necessary for the oversight of genetic engineering to be in the hands of the Department of Education and Science, the MRC eagerly shed its responsibility for GMAG.

Following the decision in the late summer of 1981 to allow most large-scale uses of genetically engineered organisms to proceed on notification, extended discussion of transferring oversight from the Department of Education and Science to the Health and Safety Executive took place among officials from both agencies. Since those involved were middle-level officials, the transfer had probably been already settled in principle.[104]

A draft document circulated in 1983 set forth four possible options for consideration: a continuation of the status quo; the formation of two separate committees, one under the Health and Safety Commission and one under the Department of Education and Science; the formation of a single committee jointly advisory to both the Health and Safety Executive and the Department of Education and Science; and the formation of a single advisory committee to the Health and Safety Commission.[105] Discussed by GMAG in November 1982, only the last was considered to be a serious option. Future restrictions on the proposed committee's range and powers were accepted with little debate: first, formal representation of the "public interest" was dispensed with; second, the new committee was restricted to health and safety dimensions of genetic engineering. Social and ethical issues were considered to be beyond its remit.

The committee that emerged a year later, the Advisory Committee on Genetic Manipulation (ACGM), consisted of eighteen members: five representing management and nominated by the CBI and the Committee of Vice-Chancellors and Principals of the British universities, five representing employees and nominated by the TUC, and eight scientists selected by the Health and Safety Executive in consultation with other government agencies and relevant interest groups. The committee was charged with advising the Health and Safety Executive on controls for genetic engineering to ensure that conditions in the workplace complied with the terms of the Health and Safety at Work Act.

The new committee was unlikely to undertake anything other than the routine oversight of genetic engineering work. The hazards issue, while unresolved, could not be reopened in the climate of the early 1980s: the battle for parity with the American controls had been won, and the threat of loss of science and industry abroad would be raised whenever the issued was broached again. Furthermore, it seemed unlikely that the Health and Safety Executive, lacking government support, uncertain of its future funding, under pressure from private industry to demonstrate that it would not impede the development of new industry, and understaffed to oversee facilities, would initiate policies that would result in more active or anticipatory forms of oversight of genetic engineering. Under the circumstances, it was unlikely that the agency would do anything other than implement existing controls. It was anticipated that the ACGM would operate exclusively in a "consultative" mode, considering only those issues brought to it by the Health and Safety Executive. Consequently, there was a strong sense, in 1984, that the ACGM's most important roles would be cosmetic and reactive: cosmetic in the sense that the committee would act, in the words of the chair, as a "comforter" to those with anxieties about genetic engineering; reactive in the sense that the committee would act only when anything problematic emerged.[106]

11.7 Achieving Parity

The ease with which concern about and control of recombinant DNA activities faded is no more to be ascribed in the United Kingdom to the resolution of the biohazard problem than it is in the United States. GMAG did not abrogate its responsibilities because genetic engineering had been proven utterly innocuous. Hazard investigation in the United Kingdom, as in the United States, remained at a rudimentary stage. The inability of GMAG to assign values to the key variables of Brenner's risk assessment program (transformed in 1978 into a mechanism for assessing hazards) underscored the extent of the problem.

GMAG, having diverged significantly both in policy and in discourse from the NIH controls, was brought back into line not by empirical or theoretical advances in understanding of the biohazard problem but by the impact of international competition. From 1979 onward, high-level government committees, sensitive to the threatened loss of industry and science to less restrictive locations, urged relaxation of the British controls. The Thatcher government's weakening of regulatory agencies and of trade unions removed most interests in conserving the original policy. A reshaped GMAG responded efficiently to the call for weakening containment and oversight requirements. The latitude in assigning values to the Brenner scheme's variables and in matching variables to containment requirements proved in 1979 to be a convenience, since it generally allowed scope for producing parity with the American controls. Application of the scheme to research proposals achieved that goal first in 1979, following the first revisions of the NIH controls, and again in 1980, following NIH approval of the essence of the Rowe-Campbell proposal.

In the case of large-scale proposals, the Brenner scheme was shelved as inapplicable in 1979, when it produced containment requirements that were costly for industry, and brought back into play in 1981, after its application had been revised to produce lower-containment requirements (generally GMP or category I). The British controls were somewhat weaker and certainly more flexible than those contemplated in the United States in the early months of 1980s. In this case, British flexibility was used as leverage in the United States by Eli Lilly to persuade the RAC to revise its more stringent industrial containment requirements.

Communications between British and American scientists, industrialists, and government officials paved the downward course of regulation and ensured parity between the two systems. The fear of being outcompeted elsewhere, skillfully used by scientists and industry officials to press for government action, was the primary leveler for government controls in each country.

TWELVE

Molecular Politics in a Global Economy

In the 1980s, the potential of molecular biology for reconstructing the natural world began to be realized. The pharmaceutical industry in the United States led the way with genetically engineered human insulin in 1982—the outcome of the Lilly-Genentech collaboration. Fourteen more genetically engineered drugs and vaccines were marketed in 1982–90.[1] Earlier forecasts of plentiful supplies of cheap drugs produced with genetic engineering were generally not confirmed by experience: Lilly's reasonably priced human insulin contrasted with the high prices of most other genetically engineered products. A year's supply of human growth hormone, for example, cost $10,000. As a government report issued in 1991 observed, such products were "no magic bullet" for pharmaceutical corporations—or, for that matter, for the consumers for whom they were designed. Revenues from genetically engineered drugs and vaccines for American companies in 1990 were $1.5 billion—a fraction of the projections of the 1970s.[2]

In other industries, advances were more modest. Nevertheless, ambitious visions for the future of genetic engineering persisted, fueled both by technical advances and by the need of firms to entice investors. The American pharmaceutical industry recorded over one hundred new drugs and vaccines under development in 1990. Many of these were said to be "market-driven"—products such as vaccines against AIDS, malaria, hepatitis, and other major diseases for which drugs were urgently needed. Others were described as "technology-driven"—products in search of a disease.[3] In agriculture, images of a brave new genetically engineered world were most vividly conjured by schemes to design "transgenic" cows and pigs to produce drugs in their milk, plants with resistance not only to pests but also to pesticides, bacteria and viruses designed to kill pests, and plant cells designed to produce substances previously obtained only through traditional extraction methods.[4] In medicine, approval of clinical tests of novel genetic therapies in which genetically modified cells were transferred to patients in efforts to replace diseased cells were depicted as potentially revolutionary approaches to curing major diseases such as cancer and AIDS.[5] Chemical and environmental applications—bacteria for cleaning up oil spills and toxic wastes or for producing chemicals from renewable resources, for example—developed more slowly, limited by low demand and the low price of oil.[6]

In contrast to the images of benign and obedient genetically modified plants, animals, and microbes for industry, agriculture, and medicine, those emerging from the military were ominous. In the 1980s, military reports began to portray genetic engineering as the source of a new biological warfare threat. The early fears voiced in the genetic engineering controversy but consistently avoided in policy decisions now received official recognition. As superpower tensions increased in the early 1980s, the development of biological weapons through genetic engineering and other biogenetic technologies was portrayed as an immediate menace to national security. "Biological warfare is not new, but it has a new face," warned a report to Congress in 1986.[7] By the beginning of the 1990s, with military scenarios refocusing from the Cold War to Third World conflicts, biotechnology was placed on a list of technologies deemed "critical" for maintaining national security.[8]

Claims by science and industry about the need to remain competitive with civilian developments elsewhere and warnings from the military about the need to be prepared to meet new security threats kept public and private investments in genetic engineering in particular, and biotechnology in general, flowing bountifully. By fiscal year 1987, annual investment in biotechnology research and development by the U.S. government amounted to $3 billion.[9] Governments elsewhere similarly emphasized support for biotechnology research and development. In 1987, biotechnology support by the British, French, and Japanese governments amounted to 0.015, 0.0215, and 0.05 percent of gross national product—figures comparable to the U.S. percentage of 0.058.[10]

The private sector also made huge investments in biotechnology. In 1987, American biotechnology companies and major corporations invested $2 billion in research and development.[11] By the end of the 1980s, the international character of this investment was striking. A government examination of the biotechnology industry chaired by Robert Reich (later President Clinton's Secretary of Labor) concluded in 1990: "The increasingly global economy . . . makes it more difficult to view industrial competitiveness [from the point of view of the nominal location of a corporation]. Many companies actively investing in biotechnology are multinational, conducting research, manufacturing, and marketing throughout the world. These companies contribute to the economies of nations other than the one in which they are headquartered."[12] The ability of large corporations to move investments and operations around the globe, enhanced by the rapidity of information transfer through electronic systems, meant that it became increasingly difficult to define "national competitiveness" in biotechnologies such as genetic engineering. Many of the research, equity, and marketing deals struck in the 1980s involved a foreign corporation, as did some notable mergers and acquisitions. A large part of Genentech, for example, was acquired by the huge Swiss multinational Hoffmann–La Roche in 1990.[13] Science and technology developed by national governments were deployed in an "increasingly seamless global market."[14]

Legacies of the Genetic Engineering Policies of the Early 1980s

As investments poured into genetic engineering and applications emerged, the legacies of the policy deliberations in the United States and the United Kingdom over the previous decade were weak controls justified by the claim that the field posed "no extraordinary hazard." Two developments—release of genetically engineered organisms into the environment and attempts to anticipate, in the name of biological defense, the use of genetic engineering for biological weapons—are of special interest because they produced problems that might have challenged the consensus of the early 1980s.

The release of genetically engineered organisms into the environment negated one of the original principles for their safe handling—that they should be contained. The NIH guidelines prohibited deliberate dissemination until 1982, when all prohibitions were lifted. In the early 1980s, research requiring such release triggered an important scientific debate, on the possible impacts of genetically modified organisms, that divided roughly along disciplinary lines.[15] Molecular and microbiologists tended to reduce the problem to analysis of the genetic constitution of an organism and of the introduced genes and to emphasize the capacity of molecular genetics to predict and control behavior. Plant molecular biologist Winston Brill, for example, argued that release of organisms into the environment would not pose serious ecological problems unless either the organism or the introduced genes indicated problems.[16] In contrast, ecologists tended to focus on the complexity and interdependence of living things and possible nonlinearity of their interactions. A group of ecologists responding to Brill's defense of the safety of new agricultural practices based on genetic engineering emphasized the unpredictability of the properties of a genetically altered organism and of its interactions with other organisms in an ecosystem. They concluded that detailed, case-by-case testing would be required for each introduced novel organism.[17]

Similar positions were elaborated in reports issued by the National

Academy of Sciences and the Ecological Society of America.[18] A National Academy of Sciences report of 1987 produced by a committee composed mainly of molecular and microbiologists and notable for its brevity and simplicity concluded that genetically engineered organisms posed "no unique hazards," and that risks were comparable to those associated with releasing unmodified organisms or to those modified by traditional techniques.[19] On the other hand, a report issued by the Ecological Society of America warned that "the diversity of organisms that will be modified, functions that will be engineered, and environments that will receive altered organisms makes ecological risk evaluation complex," that "precise genetic characterization does not ensure that all ecologically important aspects of the phenotype can be predicted for the environments into which an organism will be introduced," that while many engineered organisms might be less fit than the parent organism, important exceptions might arise, and that transfer of genes from modified organisms to fitter organisms, including pests, in an ecosystem, needed to be addressed.[20]

In the fiercely antiregulatory climate of the 1980s, neither the Reagan administration nor Congress was inclined to restrict such applications through new regulatory initiatives, although existing law was widely seen as inadequate and ill-adapted to covering the full range of applications contemplated.[21] Instead the Reagan administration established the Biotechnology Science Coordinating Committee (BSCC), a high-level interagency body under the White House Office of Science and Technology Policy, to coordinate existing controls for plants, animals, and microbes.[22] Formally charged with sharing scientific information and coordinating and guiding regulatory policy for genetic engineering, the BSCC sought informally to realize the role ascribed to it by one government official, to "unshackle industry in pursuit of biotechnology" by encouraging agencies to limit the scope of rule making under their statutes.[23] An early move, for example, was to attempt to encourage regulatory agencies to limit screening to certain subcategories of genetically altered organisms.[24] When the BSCC ended in disarray in 1990 after confrontations between the EPA and White House officials over the scope of EPA regulations, the committee's functions were assumed by President Bush's Council on Competitiveness, chaired by Vice President Quayle.[25] By this point, official discussion of possible hazards of genetic engineering had vanished, replaced by calls for eliminating "unneeded regulatory burdens for all phases of developing new biotechnology products— laboratory and field experiments, product development, and eventual sale and use."[26] Through continued deregulatory pressure, the Reagan and Bush administrations assured that the discursive practice established during the first decade of genetic engineering policy, according to which the technology would be considered safe until proved dangerous, would apply to future uses of the technology. By February 1991, more than two hundred field trials of genetically engineered plants and microorganisms had taken place.[27]

The British response to the problem of release of genetically modified organisms was shaped by the Council of the European Communities, which in May 1990 issued a directive requiring that plans for such release be screened and approved by the relevant national agency.[28] Thus all member states of the European Economic Community were required to establish case-by-case screening of genetically engineered organisms—an approach that was in principle more cautious than those encouraged by the Reagan and Bush administrations. In addition, sales of products were contingent upon acceptance by member states, thus raising the possibility of delays or denials of approval by states with stringent controls. The provisions of part 6 of the British Environmental Protection Act of 1990, which implemented the European Community directive, introduced an elastic criterion for evaluating applications. "Best available technique not entailing excessive cost," written into the Environmental Protection Act despite strong opposition, introduced a balancing of possible costs against possible benefits into reviews. Although the British regulations were more comprehensive than their American counterparts, they were also sufficiently elastic to be adapted to regulatory trends.[29]

The second development that might have affected the established discursive practice was the possible use of genetic engineering for biological warfare. Biological weapons were formally banned by the 1972 Biological Weapons Convention, to which the United States and many other nations were parties. But the terms of this treaty were weak; the specifications of its first article, formulating the principal obligation not to possess biological weapons, were vague, providing a loophole through which weapons exploration might be pursued in the name of defense.[30] Moreover, the treaty had no mechanism for verification of compliance with its terms. As concerns about hazards produced inadvertently through civilian application of genetic engineering evaporated in the early 1980s, concerns about hazards produced deliberately through construction of biological weapons escalated. When superpower tensions intensified in the early 1980s, military reports elaborated weapons applications—the possibility of transferring virulence factors or genes encoding toxins into common bacteria such as E.coli;[31] rapid production of large quantities of toxins, which were hundreds of times more lethal than chemical weapons;[32] production of large quantities of bioregulators that affect physiological processes in minute amounts and of novel toxins.[33]

Perceptions of military threats from the Soviet Union and the developing world fostered a radically different discourse concerning possible impacts of genetic engineering—one that claimed that genetically engineered organisms could be highly dangerous and destructive for humans and for the environment. A year after the U.S. Office of Science and Technology Policy asserted that "experience . . . has mitigated many of the concerns about the risk [of genetic engineering]," President Reagan's Chemical Warfare Review Commission concluded that, using advances in genetic technology, "biological weapons . . . could be designed to spread disease and kill indiscriminately in wider fashion, and thus are adaptable for targets far from one's own population. They could infect enormous areas."[34] While military agencies insisted on the possibility of emerging hazards from genetic engineering, civilian institutions insisted that civilian uses could pose "no unique hazards."[35] While military agencies urged preparation for every conceivable scenario, civilian agencies weakened regulatory controls on the grounds that there was "no cause for alarm."[36]

These military and civilian discourses, with their differing constructions of the potential of biogenetic technologies for novel forms of ecological and human disruption, evolved along separate paths, supported by separate bodies of law and practice. The boundary separating the appraisals of genetic engineering hazards incorporated in these discourses, resting on the slender logic of the assertion that what might happen by design could never happen by accident, was maintained intact. That the science of ecology, with its emphasis on the unpredictability of complex ecosystems, implicitly threatened to unravel that logic was never confronted. Indeed, the RAC's consensus that genetic engineering was incapable of generating unusual hazards supported approval of military applications for the cloning of genes encoding potent toxins (substances far more powerful than those used for chemical weapons).[37] When a move to institute some degree of control over military research was made in 1982 in the form of an RAC proposal to include in the NIH guidelines a prohibition on the use of genetic engineering for weapons construction, the committee sidestepped the issue, asserting with tranquil confidence that international law could adequately cope with the weapons problem.[38] Ominous projections in the military sector had no effect on the process of deregulation in the civilian sector.

In summary, with the exception of occasional eruptions of public concern and despite the emergence of military interest in the field, genetic engineering was relieved of the stigma of hazard in the 1980s. The predominant political concern became the promotion of biotechnology, not its control. The controversy surrounding the biohazard issue all but faded away.

The Development of American and British Policy: An Appraisal

In this context, the history of policymaking in the previous decade was reconstructed. The policy effort aimed at prevention of harm that followed the Berg committee's call for caution in 1974 was now seen as an episode that was over, leaving only a trace. In the words of one observer, the Asilomar conference and the debate it inspired might "rate a footnote in the history of science."[39] In the 1980s, the received interpretation of this transformation of both scientific and public perception was that the issue had been definitively resolved. That "the fears of the early days" had been found "groundless" was widely aired.[40] In the formulations of the press, "the hazards once thought to be associated with this research are now considered to have been exaggerated."[41] The "episode" was cast as a war between truth and error, rational analysis and irrational alarmism, in which rationality and truth (that is, "new data" revealing "over-estimation of risk") ultimately won.[42]

In contrast, this study has demonstrated that large uncertainties about the nature of genetic engineering hazards persisted throughout this period of policymaking—uncertainties that expanded over time as the technology advanced. The processes being assessed were immensely complex, a complexity multiplied by deregulation (which eliminated restrictions to specific host-vector systems that simplified risk assessment) and by advances in technique. Any claim that a judgment concerning the safety of the entire field of operations could be made on purely technical grounds is called into question by the multiplicity of positions to be found in the transcripts of the various conferences called to assess hazards or in the proceedings of the RAC. In response to the reassessment of the hazards that began to take place in 1976–78, Robert Sinsheimer wrote that "the whole structure of graded risk implicit in the guidelines floats at some level dependent upon investigatorial confidence rather than documented fact."[43] The complexity of the field being assessed and the novelty of its techniques, while constrained within broad limits (as far as anyone knew, no monsters had emerged from the test tubes of the genetic engineers), provided scope for major differences in perception and opinion within the United States and the United Kingdom (as reflected, for example, in the later debate between molecular biologists and ecologists on the possible impact of release of genetically modified organisms) and between the United States and the United Kingdom (as reflected in differences in the meaning of *hazard* in the policy processes in each country).

Attempts to explain the closure of the controversy surrounding genetic engineering on technical grounds exclude the possibility that policies, as well as the assessments of hazard that provided their justifications, were affected by socioeconomic contexts and by interests in securing certain kinds of policy outcomes. In contrast, this study has been designed to leave these possibilities open, and the attempt to explain the closure of the technical debate is seen as part of the larger project of examining socioeconomic contexts, the formation and expression of governmental and sectoral interests in genetic engineering, the operation of interests in creating and shaping policy arenas, and the effects of interests, expressed as policies and discursive practices.

Within this framework, both the similarities and the differences between the American and British policies require explanation. In particular, any alternative to the received view must address, first, why the American and British policies so closely resembled one another until 1976; second, why they diverged from 1976 until the end of 1978; and third, why they began to converge again from 1979 onward, becoming indistinguishable for most practical purposes after 1982.

The advent of genetic engineering, as a novel and potentially powerful technoscientific resource, activated diverse and often conflicting interests in its development: the scientific community, universities, executive branches of governments, private industry, labor unions, and pressure groups all responded in ways shaped by their own specific goals and institutional needs. As chapter 1 shows, the interests of scientists, the universities where most scientists pursued research, and the government agencies responsible for funding research were formed under the influence of the policies for government-sponsored science that evolved in the years following the Second World War. In the United States, the growing squeeze on scientific resources in the 1960s and calls for "accountability" from politicians accentuated the utilitarian emphasis that had always characterized American science. In the United Kingdom, a long-standing tradition of protecting scientific autonomy broke down as difficult economic conditions encouraged moves toward selective intervention and an increasing emphasis on practical results. The potential of genetic engineering to generate both theoretical and practical advances meant that the field was immediately perceived at many levels in both countries—by the scientists who were responsible for the early techniques, by universities pressed by reduced funding, and by funding agencies—as a highly desirable prospect. In the United States, in particular, since no organized political ties of scientists supported other commitments, the interests of researchers, universities, and the NIH adminstration formed a strongly bound, mutually reinforcing system of interests that had especially powerful effects. In the United Kingdom, the allegiance of some scientists to trade unions introduced a commitment to laboratory safety and participatory policymaking (in addition to interest in development) that would register during the second phase of policymaking.

The private sector was also immediately alert to the potential of genetic engineering to generate new products and modes of production. While the field was in its infancy and its long-term implications remained uncertain, however, major corporations were for the most part content to keep watching briefs on developments in both scientific and political arenas and were not actively engaged either in investing in the field (section 2.3) or in attempting to influence government policy.

In responding to questions of hazard and social use raised by genetic engineering, scientists and the government agencies that sponsored their research acted in a political climate shaped by the diverse movements of the 1960s and early 1970s that aimed at protecting workers, the public, and the environment from unanticipated effects of science and technology. These movements produced major regulatory legislation in both countries in the 1970s. Thus the initial phase of policymaking occurred in a political context in which unrestrained deployment of science and technology might be challenged. Leaders of the scientific communities in each country were sensitive to the possibility that new advances could touch off moves to regulate research. Alvin Weinberg asked at the CIBA conference held in London in 1971 (attended by two of the future chairs of British policy committees, Eric Ashby and Gordon Wolstenholme), "[I]f the unaccredited public becomes involved in debate on matters as close to the boundary between science and trans-science as the direction of biological research, is there some danger that the integrity of the Republic of Science will be eroded?" For such leaders, the Pandora's box that might be opened by genetic engineering was not a box of monsters but a box of monster regulations forced on science by a public that feared science (section 3.3). To be "seen to be acting" in order to forestall moves to control genetic engineering from outside science was certainly as important as the action itself.

During the first few years of genetic engineering (from 1972 to 1975), few outside the biomedical research sector registered the misgivings being aired by some close to the field. At this point, the principal actors were the communities of molecular biologists within which the techniques were being developed, government agencies with responsibilities for research in molecular biology, and scientific leaders and organizations concerned primarily with the advancement of science. Thus the immediate influences on policy—the American one in the hands of the National Academy of Sciences and the NIH and the British one in the hands of the Advisory Board for the Research Councils within the Department of Education and Science—were the scientific communities and government agencies with close links to the development being appraised.

While the early policies in each country were produced by different institutions—the one private and the other governmental—their resemblance is not surprising. Scientific leaders in both countries maintained close contact. Paul Berg, chair of the committee established under the auspices of the National Academy of Sciences and cochair of the Asilomar conference, and Eric Ashby, chair of the working group convened by the Advisory Board for the Research Councils, communicated regularly as policy decisions were being made. Those government agencies closely associated with biomedical research—the NIH and the Department of Education and Science and its Research Councils—played supportive roles, establishing the expert committees their scientist-clients desired, reminded, when necessary, of the leverage top scientists could exert by refusing to be bound by controls. (See, e.g., section 4.4.) Before the Asilomar conference in February 1975, these communities established policy goals and a supporting discursive practice: the genetic engineering "problem" would be defined as a technical problem, that of developing suitable safety precautions to contain possible hazards—a definition that served to reinforce the central role of the biomedical research community in policymaking and to exclude others with different experiences and perspectives. The inscription of this reduction in discourse into formal charges to policymaking bodies marginalized social and ethical questions regarding genetic engineering, making it impossible to achieve their sustained consideration.

The Asilomar conference was organized to cement the technical discourse, and it largely succeeded in doing so: the few dissenting voices that struggled to express concerns about possible social uses of genetic engineering or the implications of dangerous research were silenced or found those concerns transformed into elements of the technical agenda (chapter 3). As the NIH moved to develop voluntary controls under the guidance of a committee of peers, the expectation among British scientists at the Oxford meeting in the summer of 1975 was that a similar arrangement would be established for their own research (section 4.6).

Until this point, the similarities between the American and British processes can be explained in terms of the influences of communities of scientists who maintained close trans-Atlantic contact. (These ties make it tempting to depict these communities as a single "community," although I argue below that the description is misleading.) No other sector appears to have played a substantial role in the formation of these early policies and discursive practices. This explanation suggests a simple reductionist formula for the entire course of policymaking: the influence of an international scientific elite.[44] But this explanation fails to account for the character of later policies, and for differences between the American and British policies. Furthermore, it takes the interests of scientists as given, rather than as social formations that require explanation.

In the second phase of policymaking, during which the American RAC

and the British Williams working party and GMAG proceeded to formulate controls for genetic engineering, what promised to be a straight-
forward implementation of the Asilomar paradigm was disrupted, on the
American side, by the growing public controversy that surrounded the
NIH process and, on the British side, by the entry of the trade unions
into the issue. The American controversy is important for revealing a
great deal about the scientific and industrial interests at stake. Most visibly, the response of the American biomedical research community to
regulatory legislation (especially to legislation that would have expanded
participation in policymaking) revealed American scientists' intense commitment to maintaining control of policymaking and intense opposition
to expanding participation in policymaking (a primary goal of pressure
groups such as Friends of the Earth, the Environmental Defense Fund,
the Natural Resources Defense Council, Science for the People, and the
newly formed Coalition for Responsible Genetic Research). For leaders
of the biomedical research community, the entry of the public into the
hazards debate represented a most threatening turn in "molecular politics." Equally important, the controversy reveals the network that linked
industrial, scientific, and university leaders, and NIH administrators in
their commitments to maintaining voluntary controls administered by
the NIH. Finally, the controversy shows how preferred practice (voluntary controls) was linked politically to the promotion of new discourses
about hazards ("new evidence" revealing "exaggeration" of hazard) and
hazard assessment (experiments aimed at testing "containment" rather
than "worst-case" scenarios) and disseminated to the rank and file at the
laboratory bench (chapter 5). As those present at the Enteric Bacteria
Meeting had agreed, it was essential to convince the public that the hazards had been exaggerated. "Then a lot of the public thing will go away."
The dissemination of the report of the Falmouth conference in the summer of 1977 initiated this exercise in public persuasion. As a *New York
Times* headline announced, there would be "No Sci-Fi Nightmare after
All" (sections 5.3 and 5.6).

   That the American practice that evolved at the end of the controversy
was not at all logical or natural is shown by its differences from the policy
that evolved in the United Kingdom after 1976: in a political setting in
which a new and influential group of political actors (the trade unions)
entered the policy arena, a different response to the problem was deemed
appropriate. In contrast to the American legislative fracas, the British
developed in relative calm a policy and a policy instrument (GMAG) that
incorporated some of the elements of the contested American legislation.
At this point, the two policies and discursive practices diverged. While
American policy conserved and formalized the original vision of the Berg
committee—voluntary controls developed by a committee of peers—

the British moved toward regulation of genetic engineering, with requirements developed by a participatory committee. While American scientists focused on hazards outside the laboratory and argued that no "epidemic pathogens" could emerge from genetic engineering (chapter 5), British scientists focused on hazards inside the workplace, reacting with some skepticism to the American arguments (chapter 7). The differences are underscored by the vehemence with which the American biomedical research community attacked Edward Kennedy's proposal for a free-standing, GMAG-like commission to regulate genetic engineering (chapter 6). This is not to say that all sides in the United Kingdom welcomed GMAG, or that GMAG was as balanced as superficial appearances might suggest: GMAG's formation and operation revealed a great deal about the resistances inside and outside the government to this experiment in participatory policymaking and the built-in constraints that determined just how far this committee might go in taking British science away from the path being charted across the Atlantic.

The response of the American biomedical research community to the genetic engineering legislation certainly demonstrated both its political power as an "affected industry" and the corresponding political weakness of environmental and other public interest organizations. Given the uncertainties in the evidence available in 1977, it was remarkable that scientists could orchestrate a series of conferences and achieve a change in consensus that could be used to justify dropping the legislation. But the derailing of legislation was not the work of the biomedical research community alone. While scientists like Wallace Rowe, Harlyn Halvorson, James Watson, David Baltimore, Norton Zinder, and Stanley Cohen played influential roles, they had the active support of the universities and the quiet but enormously persuasive support of industry (chapter 6). At the moment when genetic engineering research had begun to indicate its industrial potential (the fall of 1977), the low-key opposition of industry to legislation undoubtedly swayed members of Congress. The unison with which all three sectors opposed regulatory legislation, drowning the concerns of environmental groups with the argument that nothing less than a good part of the future of American science and industry was at stake, was probably decisive. Thus the structure of international industrial competition entered as a major factor in shifting congressional interest from regulation of genetic engineering to its promotion (chapter 6). If genetic engineering had provided only a set of powerful research techniques without also offering the basis for novel forms of chemical, pharmaceutical, and agricultural production, it seems unlikely that Congress would have been so easily persuaded by such slight scientific arguments to drop its interest in regulation. Congress has proven tough with science in other circumstances.

The British turn toward regulation and participatory decision making may be explained in terms of the active engagement of the British trade unions with the genetic engineering issue from 1975 onward, the presence of a Labour government that was responsive to union interests, the fact that substantial numbers of laboratory workers were union members, and the preexistence of a regulatory framework (the Health and Safety at Work Act) and a regulatory agency (the Health and Safety Commission) with a mandate to control laboratory research (chapter 4). That many British scientists belonged to unions was also important. What seemed radical heresy to American scientists—that laboratory workers would be consulted about safety issues and that they would be actively represented in local and national decisions—was calmly accepted by many of their British colleagues. The determined focus of American scientists on hazards outside the laboratory was not acceptable in the United Kingdom. As one scientist recalled, "In the first instance, our concern was the health and safety of the people at work." Despite constant exchanges among British and American scientists, political and cultural differences, rooted in differing histories and organizational structures, meant that the two communities did not always act as a single, homogeneous interest group, operating across national boundaries.

Limits on the extent of the British divergence from the American policy paradigm are also instructive. With the important exception that participation of representatives of laboratory and industrial workers was seen as essential, GMAG, like the RAC, focused almost exclusively on formulating and overseeing safety measures for the pursuit of genetic engineering. Other questions were not allowed to disturb GMAG's primary agenda: the review of applications for experiments. This became the imperative, driving the committee's agenda in the same way that it drove the RAC's agenda in the United States. Furthermore, the Secretary for Education and Science and the Department of Education and Science civil servants responsible for selecting members and setting agendas had no intention of allowing the group to unduly dull the United Kingdom's competitive edge in the new field. When tensions erupted on GMAG, especially over the question of the confidentiality of industry secrets, GMAG's managers soon moved to reshape the committee and to weaken trade union influence (section 8.3). International competition in science and industry therefore set limits on the United Kingdom's experiment in participatory policymaking.

At the beginning of the third phase of policymaking examined in this study, the American scientific community emerged from the struggle over genetic engineering legislation victorious and in control of the twin discourses of hazards and hazard assessment. But it would be a mistake to conclude that the final phase of policy formation examined here, in which controls were largely dismantled, is simply the story of the further influence of a powerful scientific elite. Science itself was undergoing a major social transformation; a far-reaching swing toward deregulation was also occurring.

Toward the end of the 1970s, American science policy, shaped by a determined campaign by industrial and financial interests, aimed at forging strong links between science and industry and at deregulation. High-technology fields like genetic engineering were targeted for additional government support, and tax, budget allocation, and patent policies stimulated huge corporate and venture capital investments in university and private research and development, thus artificially heating the already hot competition (section 2.5). Furthermore, these policies achieved rather precisely the goal for which they had been designed—that is, a new alignment between science and industry (section 1.4). As developers of patents on basic techniques and the organisms produced by these techniques, as recipients of corporate grants for research, and as cosponsors of new biotechnology companies, leading research universities established major interests in the industrial development of the field, and so did research scientists as equity holders, advisers, consultants, and executives for new genetic engineering companies and multinational corporations. As competition accelerated, pressures to pursue rapid development intensified and multiplied. In the late 1970s and early 1980s, corporate representatives and scientists alike framed the genetic engineering "problem" in terms of a "race" in which the NIH controls presented a major "handicap." Scientific interests in genetic engineering had been profoundly restructured by this point.

The intensity of this transformation was probably unique to the United States. In the United Kingdom, where venture capital did not flow as freely and relatively few small firms were established, scientists' roles were not transformed to the same extent. Perception of competition was largely directed outward, toward a need not to be penalized by requirements significantly stricter than those of other nations. British policy was mainly reactive, not proactive (chapter 11).

In the United States, in the charged climate of the late 1970s and early 1980s, the interests of several sectors converged on the goal of weakening the NIH controls. Large corporations wanted relief from the NIH requirements for RAC review of proposals for new host-vector systems and large-scale processes, given the possibilities such reviews opened up for leaks of sensitive industry information to competitors or employees. What industry wanted (and eventually got) were reviews at the local level, by committees appointed by the industry in question (sections 7.7 and 9.2). The small genetic engineering firms like Genentech, under great competitive pressures to secure products and patents, supported these goals but also needed to keep capital costs low. Genentech's strategy was to play hardball with the NIH, threatening to ignore the NIH guidelines unless these were reduced, thus embarrassing the agency, which had no legal control over the company's actions but hardly wanted its inability to force compliance to become the subject of further congressional scrutiny. Academic scientists, who were experiencing considerable discrimination as a result of the double standard the NIH controls enforced with respect to research in industry and academe, pressured the NIH by threatening to move their research abroad, to countries with weaker controls (chapters 7 and 9).

The fine structure of the RAC advice and NIH decisions in this period in which the NIH guidelines were dramatically weakened illuminates the relative influences of science and industry. In analyzing this process, it should be remembered that the RAC did not have direct industrial representation; only a few members of the committee in the period from 1979 to 1982 were known to have ties to pharmaceutical corporations and genetic engineering firms. Industry representatives put forth their most important goals and needs to government officials offstage, in private meetings, rather than in the formal policy arena.

Where academic and industry interests converged (for example, on the goal of reducing safety requirements), their effects within the RAC were powerful, overwhelming the few voices within the committee, mainly those of public interest representatives, that still advocated caution (section 9.7). The drops in containment requirements after 1979 were so precipitous that they might almost be described as a free-fall. In the absence of any effective countervailing force operating inside or outside the RAC, the discursive practice established at Asilomar ("containment of unknown hazards") was swiftly abandoned as the burden of proof was transferred from science and industry to the public, and the discourse of "no extraordinary hazard" prevailed.

Where these interests diverged (notably, on voluntary compliance procedures for industry and on the replacement of the NIH controls by a voluntary code of practice), industry had less influence within the RAC. These splits provided the few public interest representatives on the committee with their only significant leverage in pressing for caution. I have argued that these cases are especially interesting tests of the politics of the NIH. In response to these issues, the NIH director chose to play an activist role, ensuring that the RAC appointed in 1979 was shaped to contain dissent (the supposedly "participatory" RAC was more accurately a committee in which limits to participation were calculated and defined in advance), narrowing committee choices to alternatives acceptable to the NIH, and skillfully working to ensure that the RAC would grant what industry desired: weak controls, mainly overseen by the industry itself at the local level, with strong protection for trade secrets. A telling characteristic of the NIH policy shaped in the early 1980s was that the only *criminal* penalty associated with the NIH controls was that for release of proprietary information, not the release of the genetically engineered organisms whose dissemination the controls were supposed to restrain. In response to the second split between academic science and industry—on the question of whether the controls should be entirely dismantled or kept formally on the books—the committee's agenda was again shaped to respond to industry's interest in retaining the formal appearance of controls; scientists who had pressed for complete deregulation had to be content with somewhat less than that. In the end, the RAC's managers proved to be more responsive to the needs of the pharmaceutical industry and the new genetic engineering firms than to the desires of the academic scientists who were the agency's primary clients (sections 10.2 and 10.3).

The British response to the precipitous dismantling of the NIH controls further reinforces the argument that the reversal of the Asilomar paradigm was accomplished for political, not technical, reasons and that a fundamental requirement was to satisfy the needs of industry. By 1979, the British controls, influenced by union commitments to laboratory safety, were significantly more cautious than those in the United States. But under tough economic conditions in which British preeminence in science was perceived as a major economic resource, governmental responsiveness to the calls from industry and science to match the American deregulation transcended party affiliations: it was James Callaghan's Labour government, not the Conservative government of Margaret Thatcher, that took the first actions to weaken the United Kingdom's regulatory policy by establishing the Spinks committee to recommend policies for promoting biotechnology and by appointing Sir William Henderson, whose commitment to facilitating rapid development of genetic engineering was well known, to chair GMAG in 1979. Thatcher's government went further, weakening the major countervailing force favoring caution, the trade unions, and cutting support for the agency responsible for implementing controls, the Health and Safety Executive. Under these conditions, and with the malleable Brenner scheme in place, the British controls followed the American ones on their downward course (sections 11.3 and 11.4, figure 11.1). As industrial needs became more prominent, industry representatives in both countries pressed their respective governments to maintain parity, lubricating the downward slide with predictions of imminent relaxation on the other side of the Atlantic. The fear of being placed at a competitive disadvantage by weaker regulations elsewhere, skillfully used by industry representatives as well as leading scientists, proved a powerful leveler. The final dismantling of GMAG and its replacement by the tripartite ACGM, working primarily in a consultative rather than policymaking mode, responded precisely to the CBI's concerns about leaks of commercially significant gene-splicing secrets.

In summary, this analysis shows that there is no simple, one-dimensional explanation of the rise and fall of genetic engineering controls in the United States and the United Kingdom. Neither technical arguments nor the interests of a scientific elite communicating across national boundaries can explain the course of policymaking in each country. Although a group of prominent scientists played an important role in establishing the initial policy paradigm in 1975, and although American scientists were influential in generating technical justifications for derailing legislation and weakening controls, later events reveal the effects of other sectors, above all the pervasive influence of private industry. Industry influence was expressed at three main levels: in shaping the twin science policies of science-industry "synergy" (to use Edward David's word) and deregulation of the late 1970s and early 1980s; in forging, through major investments, the strong links between academic science and industry that transformed large parts of molecular biology; and in exerting pressure to achieve the specific elements of genetic engineering policy—protection of commercial secrets and deregulation in all but name—that it desired. No other sector operated with nearly the same range or clout.

The analysis also points to the importance of the socioeconomic structures in which industry and other sectors operated. The structure of American biomedical research, in which the primary source of support is serial-competitive funding, made it difficult for scientists or their universities to tolerate anything that might retard research. Such effects were initially less evident in the United Kingdom, where a dual system of support acted as a brake on competitive pressures, but selectivity and a squeeze on funding had similar effects from the late 1970s onward. In addition, the links formed between science and industry meant that the conditions of industrial competition were transmitted to and expressed at the laboratory bench. Finally, and in the end the most powerful factor, the structure of global industrial competition, transmitted by corporations that are free to move personnel and capital across national boundaries, set nations in competition to attract and keep new sources of innovation, industry, and employment. Scientists, industrialists, and government bureaucrats pursued their interests with respect to government controls on genetic engineering, but they did so under political and economic conditions that were not theirs to choose.

In summary, the rise and fall of genetic engineering controls reveals a great deal about the political economy of science, that is, the mix of political and economic interests affecting the disposition of science in a world economy characterized by the increasing globalization of production, trade, and finance. In particular, it reveals both the strength of the pressures that could be exerted on government policy when industrial and scientific development was seen to be jeopardized as well as the force of the competitive relations under which industry and science operated. An irony of the history of genetic engineering policy is that the same competitive structures that allowed scientists, industrialists, and heads of government agencies to exert powerful leverage in policy arenas also imposed on their activities conditions that were not theirs to choose.

This does not mean that these competitive structures are permanent fixtures in the global landscape or that the interests and policies are completely determined by them. Undoubtedly, competitive conditions were experienced by the primary actors as important constraints on their options, but it is important to recognize that these options were not entirely determined. American scientists, the front-runners in genetic engineering, might have chosen to accept the need for strict regulation and effective risk assessment; a minority of them did so. Governments might have pursued the possibility of uniform international controls. This would certainly have been a novel and radical development; international competition is assumed and fostered when states interact over scientific and technological development. As environmental and other global problems loom large however, new political structures are clearly required to cope with the leverage that transnational industrial interests can exert and that pushes regulatory controls down to the lowest common denominator, ultimately discriminating against the cautious. The steps being taken by the European Economic Community to develop uniform regulatory standards are an example (albeit limited to Europe) of an alternative, transgovernmental approach to regulation that may be less susceptible to the pressures of international competition.

The skeptical reader may ask whether this analysis really matters. After all, it may be argued that no problematic hazards have emerged from genetic engineering and that the established consensus reached in the late 1970s has in fact been borne out by the experience of the 1980s and early 1990s. A great deal of virtually unregulated cloning has been conducted in laboratories and factories, and genetically engineered organisms have been released both intentionally and, one may assume, unintentionally in countries all over the world. As far as anyone knows (and of course, this qualification is especially poignant in view of the deliberate decision in 1976 *not* to do worst-case analysis of genetic engineering hazards), none of this activity has produced harmful effects, at least in the short term.

Two responses to such arguments may be made. In the first place, these arguments are restricted in the same way as the genetic engineering "problem" itself in the early stages of policymaking. "Impacts" are defined narrowly in terms of health or environmental effects. Questions of social use of genetic engineering, which were deliberately set aside during these early stages and never reengaged, are not perceived as problems. Yet plans are proceeding apace, with few real restraints, to explore chemical and biological warfare applications of genetic engineering and to increase the resistance of agricultural plants to herbicides and produce a large variety of genetically modified plants and microbes for release into the environment.[45]

Even if the narrow terms of chosen policies are accepted, the precedents established by the evolution of genetic engineering policy are problematic. The original policies of the United States and the United Kingdom, although framed narrowly, were unusual in attempting to forestall the emergence of unknown hazards from a novel form of technology and in requiring a degree of international cooperation for their success. The abandonment of those policies signified a return to laissez-faire development of technology driven primarily by the interests of its funders and creators and by the conditions of international industrial and scientific competition.

Those responsible for these decisions no doubt felt that they had no choice but to weaken genetic engineering controls, that otherwise their research, their firm, or their country would be "left behind" in the genetic engineering race. This study shows that this compulsion is the product of social and economic structures that intensify competition for products, patents, modes of production, and markets. The history of genetic engineering policy may prove valuable in revealing the interests and structures that produce the appearance of inevitable, determined, ever-quickening technological development; and it may suggest why new forms of international cooperation are required to redirect the development of science and technology toward the fulfillment of human needs.

Excerpts from Transcript of the Enteric Bacteria Meeting, National Institutes of Health, Bethesda, Maryland, 31 August 1976

[Morning Session on Nature of Hazards Posed by Recombinant DNA Research] [1]

—I would like to try to introduce what I think this meeting is all about. It is basically an *ad hoc* group of consultants to try to give us a crash course in all of the things you need to know about infectious disease and were afraid to ask, and a refresher course in everything from molecular structure to gastroenterology. You are all aware of recombinant DNA—the molecular politics of it—those from Boston are acutely aware of the multi-levels of concern—of how the reaction has flared up to the heights of glory and fanned by a number of Nobel Prize winners. . . .

. . . Other general comments . . . We are just floundering for ideas, for coherence . . . for a mode of analysis of what might be going on if people have unlimited access to doing recombinant DNA experiments. Are the super-critics, the "sky-is-falling" people, justified in the fear of saying [that] we are creating new forms of death, as the front page of the *New York Times* supplement last week said?

Is there any way that one can imagine that an epidemic of virulent disease can emerge without having a string of remote probabilities [that] this, and this, and this, happen? . . .

Out of the infinite universe of combinations that DNA recombinant research can involve, the guidelines have narrowed it down, it seems to me, to a very advanced level. Of all the bacteria in the universe, we're only really talking about one particular bacterium with options to find parallel ones that are as laboratory restricted. Okay, so *E. coli* K-12 is really the focus of . . . these experiments. No other organism is presently considered as "licensed" under the guidelines. . . .

I thought . . . if we could talk first or chiefly about what are the possible ways whereby the researcher might generate dangers to the public health, to the eco-system in any way by working with K-12 and its plasmids and lambdaphage systems. Our agenda is the general problem of what could happen under worst [possible] cases of *E. coli* research. Secondly, we are doing an experiment as the very first stage of some risk evaluations—looking at certain aspects of . . . insertion [of polyoma virus DNA] into *E. coli*. We would like to have some advice on this experiment, about the interpretation of it and particularly the design of any experiments involved in colonization of animals. The third area that might be worth talking about is the eukaryotic vectors of the polyoma-SV40 type. I don't think that the public concern is at all focused on these yet. But once you start saying, "Well, I am going to create a new tumor virus, one that produces dozens of different kinds of tumors," all you have to do is let this get into the front page of the *New York Times,* or the *Boston Globe,* or something, and you're in trouble.

I think it very important to recognize that what we are talking about . . . is . . . in part a hardcore analysis of molecular genetics and pathogenetic mechanisms of enteric bacteria, transfer of plasmids, the biology of phages, the evolutionary pressures against concerted DNAs, population dynamics and so on. But equally, we always have the problem of PR and of dealing with the public and to have the public understand what is being done, what is not being done, what is real, and what is not real. . . .

What I would like to eventually come up with is . . . a set of ideas . . . saying . . . what you would like to see as informed clinicians, molecular geneticists . . . before you would say, "Well, gee, that looks safe; I can't really see any reason not to go ahead with that type of experimentation." . . .

—One general idea that I wanted to mention is that this form of argumentation that has occupied the public, as far as I know, none of it has come from infectious disease people. It has always been people who have expertise in some other area of science. . . . Part of the agenda today is to get you guys involved and get your voices heard, and maybe if the "Infectious Disease Society of America" comes out and says, "By God, if it's just insertion you are talking about, nobody is worried about this mechanism." That carries a tremendous amount of weight, at least to me. If I could say . . . to the prophets of doom: "Look, these guys have come out and said there is nothing to worry about here, so let's really start and get on with more serious business." That's what I hope we can accomplish. . . .

—[I]t's an important accomplishment if we can all at least put into a lower probability category, the production of epidemic disease by *E. coli* K-12 that has been genetically manipulated over the years in many, many, many different ways. I gather that it is the consensus of the people in Infectious Diseases that they know of no reports or, in their personal experience, have no evidence for epidemic disease, which seems to be much more important than infections of laboratory workers.

—I make that as a major condition. I am really not as concerned about laboratory workers as long as the infection is restrained in our midst. Introducing new things in the eco-system, into populations [outside] the lab, that is what I worry about. A case in the investigator—the technicians, is bad, but that's not the major question. . . .

—It seems to me that what everybody is worried about is naturally happening by some method. . . . it is not a practical thing to make *E. coli* into the kind of organism that we want. . . . The question is that this kind of epidemic [spread of *E. coli*] just doesn't happen and isn't happening in our society largely because of sanitation. We don't have house rats and we don't house fleas and we don't have lice and we don't eat shit and that's what it comes down to. . . . And, if that is the answer, then I think we should stop worrying about all this secondary [transfer of DNA inserted into *E. coli* to organisms outside the laboratory] and maybe make the additional point that physical containment does a lot of good because, basically, your average split-level house is a physical containment facility. . . .

—[Could *E. coli* bacteria containing a specific gene spread in human hosts?]

—In otherwise normal people, it would be very difficult. In abnormal situations, in a hospital environment, antibiotic feeding, strange diseases of the GI tract, it becomes theoretically possible. . . .

—Can you make an epidemic? Suppose you had to make an epidemic. . . . could you do it? . . .

—The question that I think you are talking about is making an epidemic with *E. coli* K-12 or with any organism [at all]. Are you talking about putting something, a new product, into *E. coli* K-12; it gets into one person in the lab as an accident and that person goes outside and craps. That goes into the water. Water happens to have typhoid in it and that new product that you made in *E. coli* K-12 is then transferred to typhoid.

—Right.

—Will that make an epidemic?

—It makes people sick.

—No. That's not fair. What I want to know is: . . . Can you make an organism so virulent that it will make an epidemic in Washington? That's what I want to hear, an organism that will colonize and spread.

—I think the point is that your K-12 could be carrying a new product that is quiescent as a genetic entity. It's got virulence, but is not expressing itself. As soon as by some accident of nature it then leaves that environment and gets superimposed. . . .

—Can you arrange this accident? Can you think of a circumstance in which you could make it spread? This is really the heart of the issue. Can it be done?

—I can't answer that.

—There are large-scale epidemics that have occurred over months at a time. In Riverside, California, for example, there was an epidemic involving 20,000 people in which there was continual contamination of the water supply by the sewage and it was a *Salmonella typhimurium* epidemic. I think in the summer of 1931 or so there was an epidemic in the Pick Congress Hotel that caused amebiasis that went on for two or three nights. . . .

—If [critics of the NIH controls] say, "Look at the risk you are putting me under of getting your epidemic Andromeda disease," somehow we have to communicate to them, "Listen, lady. You are under a risk of developing cervical carcinoma . . . and that risk is finite and real and horrible and if we do these experiments we can guarantee that, within generations, at least, we will reduce that risk. How about putting up with this risk?" . . .

—You can't promise with recombinant DNA where the results are likely to come. . . . I guess we all agree [that recombinant DNA research] should all go on—that's why we are here.

—These claims have already been made in the press and it doesn't do any good whatsoever.

—It doesn't sell! Science is under a very serious attack.

—But where is it coming from?

—From ourselves! One has to be very careful about the tack one uses and should not say, "Well, gee, we have been doing much more dangerous experiments for years." That's murder! You have to use a very positive approach.

—But the positive approach has unfortunately so far been unavailable. . . . Part of the problem is that [critics of the NIH controls] have also scared some of our scientific colleagues who don't know anything about bacteriology. I think what is really important right now . . . is to get out the word you can't make epidemics. . . .

—We have a serious political disease. You have to be careful in these arguments that you don't spread it to other people. The big danger about the argument: "But look! Something else is much more dangerous than what we do already," is that the "something else," all of a sudden, gets in with a big bag of red tape. . . .

—Once you write something down and give it to lawyers, it has a life of its own. That's really serious and we are seeing that happening here. I personally believe that, politically, the recombinant DNA business is lost in this country. I think it is hopeless; I absolutely believe that. I am going to fight to the last man, but I think it is hopeless. The real problem is [that] not all science should go down with recombinant DNA. At a certain point, I think we have to abandon ship. I am serious. Otherwise, what's going to happen is that everything is going to go down. All of genetics which I care more about than recombinant DNA. . . .

—Why I got you here is that I think that if somebody acquires data that convinces important people, they'll say, "It's a bunch of nonsense; you cannot change E. coli; you've tried, so and so has done this until he is blue in the face, and I can't see and a thousand other Infectious Disease people can't see any danger in working with a Salmonella donor into E. coli and a Drosophila into E. coli."

—Well, who do you have to impress? How does it come to pass that I have to write an application to do a standard genetic cross?

—That's really where it's at. Grass roots outrage at some point carries a lot of weight. When I see how NIH will react to one letter from somebody that says, "If you don't file an environmental impact statement before you issue these guidelines or we are going to court." I have seen Building 1 just go, "Uh." A few people can have tremendous impact and I think [grass roots lobbying] can be used [by scientists] in the other direction. But part of the problem . . . is this business of really basic principles in how infectious diseases are made. The point . . . is that the ingredients for infectious disease with E. coli K-12 are simply not there and the number of unknowns that you have to specify is very large and each probability is very small. You multiply them together and you come out with nothing. . . .

—You are going on the argument that people say you are going to create drastic epidemics, everybody bleeding to death.

—Right, but that's what people are being scared with; that is what the other side is winning with. They are not winning with the idea that a few lab technicians or a few scientists are going to get sick. They don't care about that. Nobody cares about that.

—I think the intermediate ground is where it's at. If you introduce something that persists at a low level and from time to time, for an indefinite period of time, does cause disease.

—The way [to address this is] to really separate these out and somehow to try to get the word going around that informed people are really not worried about epidemics; that there may be problems of low grade endemicity and, depending on what's created, in special cases, serious endemicity—the *Botulinus,* the growth hormone producing *E. coli.* To me, those are frightening.

—The Mayor of Cambridge doesn't know the difference. What the Mayor of Cambridge is worried about, besides not being re-elected the next time, is the possibility of an epidemic. . . . Serious arguments are about this kind of low level thing, but in terms of the PR you have to hit epidemics, because that is what people are afraid of and if we can make a *strong* argument about epidemics and make it stick, then a lot of the public thing will go away.

—But are we here really to talk about the other [possibilities for generating hazards]? . . . How are you going to recognize when low level passage [of DNA from one organism to another] has occurred is one of the central issues: To be able to decide what the risk is at *any* level. . . .

—I would rather get on to easier things. . . . If somebody takes every clone of *E. coli* carrying a recombinant DNA . . . and if somebody put these into mice by all kinds of routes and can't kill a mouse no matter what he does, I would think George Wald [Nobel laureate and a leading critic of the NIH controls] would have to be impressed.

—I think that is what you have to deal with. It may not mean a thing, but that is very easy to do. It's molecular politics, not molecular biology and I think we have to consider both, because a lot of science is at stake. . . .

[Afternoon Session on Evaluation of Hazards]

—If I were to draw up for the Recombinant Committee or for NIAID a formal proposal for a program to evaluate risks, whether it's a political or scientific exercise . . . I would include the following elements: 1) Whether it's worthwhile doing; 2) if it is, what is the best way to go about it. . . .

—It seems to me, there are two kinds of experiments. . . . One kind of experiment involves learning something about what's going on and those experiments require the standard sort of thing that we are interested in. The second . . . is a class of experiment whose aim is to convince people that some of their worst fears won't happen very easily. Now, one experiment in the latter case that I would argue is definitely worthwhile doing is to try every way you can, germ-free or otherwise, to make an epidemic of serious disease with cloned DNA. Just to try to do it. I agree that it makes no sense. Just titer the survival of mice that have been injected with this material and do that for a while and say, "Look, there's no epidemic!" And the experiment gets published in the *New York Times,* not in the *Journal of Molecular Biology.* That's what ought to be done, that kind of thing. A kind of thing that will never live to be a famous experiment in the history of science; it's like all other experiments that end up in the front of the *New York Times.* It's something that captures the imagination of reporters. I would argue as strong as possible that, in the short term, that's what is needed.

—That's what I'm talking about. Let's get rid of the most outlandish hypotheses that are the cause of the hysteria. What can we do to convince ourselves and other reasonable people. . . .

—Do you think a negative experiment is going to make the front page of the *Times*?

—I think if it was handled properly, yes.

—I would like to carry that one step further, because this is something that crossed my mind when I was reading your proposal with the polyoma system. One of the thoughts I had was why not do this experiment under P4 containment, but use a known mouse pathogen; use a *Typhimurium*. This experiment is going to be done in an isolated facility under severe restrictions and so forth. Here you have an organism (*S. typhimurium*) which cuts across the gut barrier. It gets inside the epithelial cells, delivers your polyoma to an epithelial cell. Here is an organism that goes beyond the mucosa and goes into a Peyer's patch. It goes into the liver. It goes into the spleen. It's delivering your "biohazard," and you can see what happens. If you get a negative result with *this,* I think this carries much more weight emotionally against the extremists. A negative result with *E. coli* where you don't know anything about the colonization and "pathogenicity" of K-12 may be meaningless. . . .

—Your shotgun piece of DNA is carried to the liver and may express itself, giving you a positive result that says, "Yes, we have an absolute hazard." . . . I am not saying that everyone should do this experiment. But you have been charged with doing these kinds of experiments. Take the opportunity to do a good experiment.

—The experiment is *ipso facto* illegal.

—Well, all right, I agree, at the moment.

—K-12 is the only vector-host that we can use.

—Fine. But you have had some sort of dispensation. I think that you could ask for it. Present this rationale. I think that if this is positive, then you have something to work with. . . .

—I think that all of the factors that we are arguing about: pathogenicity, colonization, toxin production, will be present in a *Typhimurium*-mouse system. There are base line studies that have already been done. And one can work with it and with a natural disease. . . .

—You could attack many parameters. But at the same time, *Typhimurium* is a pathogen of mice. The argument is *E. coli* K-12 is no kind of pathogen of anything. The point is this: Do you expect that some of the survivors of the *Salmonella* get tumors? That is the question. . . .

—But is *Typhimurium* more real than K-12?

—In the mouse, yes, it is a natural disease . . .

—No, but the point is, the analogy to the human isn't with *Salmonella typhi.* No one is proposing that we clone with *Pasteurella pestis* [plague].

—You are arguing two different things. I am not arguing in favor of using *Typhimurium* as the particular cloning host. All I am saying is that for the first few experiments, you are trying to assess not whether this host is good or bad, but you are trying to assess the reality of creating a biological hazard. We could use something better than K-12.

—No. What I would like to assess is the reality of creating a risk in *E. coli* K-12. . . .

—If you got a negative result with K-12, and a positive result with *Salmonella,* what it would tell you is that there is something wrong with the K-12 system that you do not understand.

—That is why it was chosen.

—Okay. I agree with you. But we are trying to find out, first of all, whether you could, indeed, make a hazard.

—Yeah, but what you are doing is looking at a super germ warfare model. You want to get a bad pathogen and make it worse. . . .

—We still have to talk about *E. coli* as your vehicle-host.

—For the final use in various experimentation—Yes. I am not challenging the use of *E. coli.* All I am trying to drive at is that if you do all of these experiments in *E. coli* K-12 in man or mouse or whatever you choose and you consistently come out with negative results, all you can say is the results are negative. We will not have any evidence of any kind of biohazard. . . .

—That may be the problem with K-12. You would have to do too many things to make K-12 pathogenic so that it is a slick *New York Times* kind of experiment. . . .

—Well, we've been here for three hours, and what I hear people saying is that these things are not going to hurt anybody, so let's prove it. I don't think that is the proper starting place. I think we should ask ourselves, are there any possibilities in recombinant DNA experimentation that worry us as being conceivable dangers. I think that there are. The [insertion of the gene for] *Botulinus* [toxin into *E. coli* K-12] is the extreme case, and I don't want to see us throw that away. We keep assuming that we are going to get a negative here, and a negative there. But if you put in *Botulinus,* we are not going to get a negative. . . .

—There may not be a toxic product. Take rheumatic fever, for example. Presumably that's an immunologic reaction against the streptococcal cell wall which somehow has been modified to cause a cross reaction between streptococcal cell wall and the heart. By putting DNA in, you change the cell wall of an organism so the immunologic reaction to it causes a cross reaction with human tissue.

—It is certainly one mechanism that has been proposed [for] producing a product which is autoallergenic. . . .

—I am concerned about the idea of changing the host [from *E. coli* to *S. typhi* for risk assessment experiments]. I understand that scientifically it makes some sense, but it seems to me that there are two arguments against it. One is that the problem about carrying out the experiment and getting [permission to do it] may not be completely non-trivial. The second thing is that if you tip the balance very close to pathogenicity, then you are also tipping it in a particular direction. That, in fact, the way that recombinant DNA can help [the] bacterium, may not be the way in which you, *a priori,* have chosen [that is,] to make it merely pathogenic. . . . One of the things I would like to see documented is that this millionfold business with *E. coli* K-12 [that the bacteria are able to contain the inserted DNA] is really there. . . . I think that's very important and it seems to me that if you find [that exposing test animals to naked DNA] works and that it doesn't work at all with *E. coli* [containing the same DNA], that difference is a documentable difference and is the difference we want. That, politically, is going to help if it's true. I don't regard the thing as prejudged with polyoma. I could see it happening either way with polyoma, easily. But at the same time, the fact that *E. coli* is not going to become an epidemic pathogen because you put in any DNA, I believe that, too. I think that you can't get around that and by using known pathogens, it seems to me we go politically in the wrong direction even though scientifically it does make more sense. . . .

—My concern is that if you . . . put a particular factor into [*E. coli*], you may clone something virulent. But you don't detect it because that strain has got how many strikes against it? . . . You could have a biohazard sitting in that K–12 and not know it and be led to believe that it is a safe system.

APPENDIX B

New Data on Recombinant DNA Hazards Addressed in Relation to the Rowe-Campbell Proposal

I. Effects of bacteria carrying tumor virus genes
  A. Main sources of new information
    1. Mark A. Israel et al., "Molecular Cloning of Polyoma Virus DNA in *Escherichia coli:* Plasmid Vector System," *Science* 203 (2 March 1979): 883–87.
    2. Hardy W. Chan et al., "Molecular Cloning of Polyoma Virus DNA in *Escherichia coli:* Lambda Phage Vector System," *Science* 203 (2 March 1979): 887–92.
    3. Mark A. Israel et al., "Interrupting the Early Region of Polyoma Virus DNA Enhances Tumorigenicity," *Proceedings of the National Academy of Science* 76 (August 1979): 3713–16.
    4. Mark A. Israel et al., "Molecular Cloning of Polyoma Virus DNA in *Escherichia coli:* Oncogenicity Testing in Hamsters," *Science* 205 (14 September 1979): 1140–42.
    5. These experiments were designed to test one of the original hazard scenarios: the possibility that splicing tumor virus DNA into *E. coli* bacteria might transform the bacteria into vehicles for spreading cancer in human or animal populations. These experiments used a rodent model and the virus used as the source of DNA was polyoma virus, which infects and causes tumors in rodents. The strain of *E. coli* used in the experiments was the weakened strain, $\chi 1776$. The phage and plasmid vectors were also those designed for maximum biological containment.
    6. The phage, plasmids, and bacteria bearing the polyoma virus DNA were tested (1) for their capacity to produce polyoma infections in mice, and (2) for their capacity to produce tumors in newborn hamsters.
    7. Two forms of polyoma DNA were used in these experiments: single copies, or monomers; and double copies, or dimers, in which two copies of the polyoma genome were joined end to end.
  B. Results
    1. Positive results from the infection and tumor induction tests performed in this series of experiments had quite different impli-

cations. Infection required that the entire polyoma genome was taken up and was able to function in the cells of the test animal. Thus, positive results from assays for infection meant that the viral genome had been excised from the phage, plasmid, or host-vector system into which it had been introduced and had entered the cells of the test animal. Tumor induction, on the other hand, required only the functioning of a single gene and did not depend on the transfer of the intact viral genome into the cells of the host animal.

2. The results are summarized in tables 9.6–9.8. In experiments 1 and 2, no cases of infection were observed in mice when a single copy of polyoma DNA was used. However, when double copies (dimers) were used, infections were observed in two of the five mice injected with phage DNA and in two of the five mice injected with DNA prepared from *E. coli* infected with phage into which two copies of the polyoma DNA had been inserted. In these cases, the level of infection was comparable to that observed with free polyoma DNA and roughly 10^4-fold less infectious than polyoma virus itself.

3. The investigators also showed that the polyoma DNA retained its function when it was joined to a plasmid or a phage. When plage and plasmids bearing polyoma DNA were cleaved with the same restriction enzyme used in their construction, the polyoma DNA was freed, and it was shown that it would still cause infection.

4. In experiment 3, tumor induction was observed in three out of sixteen hamsters injected with lambda phage containing the double copy of polyoma DNA (a level comparable to the level observed with free polyoma DNA) and two out of twenty-seven hamsters injected with plasmids containing a single copy of polyoma DNA. Tumor induction was not observed with *E. coli* bacteria carrying plasmids bearing one copy of polyoma DNA. *E. coli* bacteria carrying phage bearing one or two copies of DNA were apparently not tested.

5. In experiment 4, it was shown that cleavage of polyoma DNA with a variety of restriction enzymes markedly enhanced the tumor-forming potential of the DNA. In one case, 100 percent of the animals tested with the fragments of the polyoma DNA developed tumors.

6. The experiments were incomplete in several ways: only one route of exposure was tested, and tests were not run with plasmids containing two copies of polyoma DNA.

C. Further investigation advocated? Even though not all combinations were tested, the authors do not specifically comment on a need to complete the experiments. They note only (at the end of experiment 3) that the experiments "are in need of extension to other host-vector combinations."

D. Use in policy documents[1]
   1. Rowe: Studies "negated" the possibility that *E. coli* could act as a "vehicle for the delivery of viral genomes."

2. NIAID Ad Hoc Working Group: Not addressed.
3. RAC Background Document: Experiments demonstrated the following general results: (1) Polyoma DNA contained in plasmid pBR322 is not infectious. (2) Polyoma DNA contained in plasmid pBR322 and introduced into χ1776 is not infectious. (3) Polyoma DNA introduced into lambda phage in single copies is not infectious. (4) Polyoma DNA introduced into lambda phage in double copies is infectious, but the experiments "appear to indicate" that the frequency of occurrence of dimeric polyoma DNA is "rather low." (5) Polyoma DNA contained in plasmid pBR322 is tumorigenic but less tumorigenic than the polyoma DNA itself. (6) Polyoma DNA contained in lambda phage is only tumorigenic if double copies of polyoma DNA are introduced into the phage.
4. NIH Decision Document: States only that "the Rowe-Martin polyoma experiments . . . provide valuable data on the question of the likelihood of an animal virus recombinant DNA insert being transferred out of an *E. coli* K12 host cell used in its propagation into a eukaryotic cell" (69241).

II. Survival of weakened *E. coli* systems in human intestinal tract
   A. Main source of new information: Stuart Levy and Bonnie Marshall, "Survival of *E. coli* Host-Vector Systems in the Human Intestinal Tract," *Recombinant DNA Technical Bulletin* 2 (July 1979): 77–80. Tested (1) survival of EK2 strains of *E. coli*; (2) in vivo transfer of plasmid pBR322 from an EK2 strain to aerobic bacteria in the intestinal tract in four human subjects.
   B. Results
      1. The EK2 strain without a plasmid was not recovered from any of the human subjects.
      2. The EK2 strain carrying a common plasmid *was* recovered from the feces of the human subjects for an average time of eighty-two hours after ingestion—a result the researchers found "surprising" and "unexpected." They also noted that it was five hundred times higher than the survival rate required for a weakened strain of *E. coli* K12.
      3. No in vitro transfer of the plasmid pBR322 from the EK2 strain of *E. coli* to another strain of *E. coli* was found.
   C. Further investigation advocated?
      1. Survival of other EK2 strains should be tested.
      2. Studies of the transfer of plasmids to endogenous bacteria in the human host were deemed to be "critically important."
   D. Use in policy documents
      1. Rowe: No discussion of this specific result. In general, it is claimed that "K12 itself cannot become established in nature; only transfer of the recombinant molecule to another host need be considered" (pt. 6, p. 8). Rowe predicts that such transfer will be "highly unlikely."
      2. NIAID Ad Hoc Working Group: (1) Recognition that Levy shows that "there is some survival of the disarmed strain after it is fed to

humans" (3). (2) Informal consensus that, regardless of the precise results obtained from experiments on transfer of plasmids, the frequency of transfer from EK1 and EK2 strains to more robust strains will be low (53).

3. RAC Background Document: (1) Emphasizes the inability of EK2 strains to colonize the intestinal tracts of the four human subjects. (2) Recognizes the survival of the EK2 strain carrying plasmid pBR322 for eighty-two hours. (3) Argues that this EK2 strain was unable to transfer plasmids to other bacteria.

4. NIH Decision Document: Not addressed.

III. Survival of *E. coli* K12 strains in mice

A. Main source of new information: Rolf Freter et al., "Testing of Host-Vector Systems in Mice," *Recombinant DNA Technical Bulletin* 2 (July 1979): 68–76. Tested several strains of *E. coli* strains isolated from mice and humans and EK1 and EK2 strains of *E. coli* K12 for their ability to survive passage through the intestinal tract in mice.

B. Main results

1. All strains tested survived passage to some extent.

2. Strains of *E. coli* isolated from mice and humans survived effectively, one strain being recovered in larger numbers than those fed to mice.

3. EK1 strains survived passage more effectively than EK2 strains. Freter concluded that EK2 strains "are killed at a moderate rate within the gut and for this reason cannot persist *in vivo* for prolonged periods of time" (70). In contrast, Freter concluded that EK1 strains "pass through the gut without significant net multiplication or death" (72).

4. EK2 strains were unable to implant in the intestines of germ-free mice or conventional mice, but a bile-resistant mutant of χ1776 was able to do so.

5. EK1 strains were able to implant in germ-free mice and in less than 1 percent of mice with conventional flora.

6. Strains of *E. coli* isolated from mice and humans were able to implant effectively.

7. While the EK1 strains of *E. coli* had difficulty implanting, they were able to survive in the gut for many hours.

8. Preliminary results on the transfer of lambda phage from one K12 strain to another suggested that this was inefficient.

C. Further investigation advocated?

1. Implantation of weakened host-vectors systems in the intestinal microflora.

2. Factors determining the rate of in vivo transfer of plasmids and phages used as vectors. (Note: investigator assumes that weakened strains will be used for genetic engineering.)

D. Use of results in policy debate

1. Rowe: Not addressed.

2. NIAID Ad Hoc Working Group: Discussion of significance of results inconclusive.

3. RAC Background Document: Addressed. Noted: a) ability of all *E. coli* strains to survive in the gut; b) inability of EK2 strains to implant in germ-free mice; c) inefficient in vivo transfer "shown" by Freter's experiment on the transfer of lambda phage.

4. NIH Decision Document: Not addressed.

IV. Survival of weakened EK2 strains of *E. coli* K12 in sewage treatment plants

   A. Main sources of new information: B. P. Sagik and C. A. Sorber, "The Survival of Host-Vector Systems in Domestic Sewage Treatment Plants," *Recombinant DNA Technical Bulletin* 2 (July 1979): 55–61. Tested EK2 strains and lambda phage vector for survival in sewage treatment. Results were circulated in 1979 prior to publication.

   B. Results

      1. One strain (DP50supF) multiplied in primary sewage treatment.

      2. Both EK2 strains survived for several days in sewage treatment, as did a phage, charon 4a.

   C. Further investigation advocated?

      1. Disinfection of effluent from sewage treatment plants with goal of reducing EK2 vectors and hosts to "acceptable" levels.

      2. Survival of EK2 hosts and vectors in sewage sludge.

      3. Reasons for ability of DP50 supF to survive.

      4. Transmission of plasmid and phage vectors to indigenous organisms in waste waters.

   D. Use in policy documents

      1. Rowe: No discussion.

      2. NIAID Ad Hoc Working Group: Emphasis on low survival rate *after* sewage treatment rather than survival *during* sewage treatment and potential for transmission.

      3. RAC Background Document: Not addressed.

      4. NIH Decision Document: Not addressed.

V. Release and survival of *E. coli* K12 in laboratory procedures

   A. Source: Mark A. Chatigny et al., "Studies on Release and Survival of Biological Substances Used in Recombinant DNA Laboratory Procedures," *Recombinant DNA Technical Bulletin* 2 (July 1979): 62–67.

   B. Results

      1. Laboratory manipulation of liquid cultures produces aerosols and particles that contaminate the surrounding area.

      2. Both EK1 and EK2 strains survived on laboratory surfaces, although EK2 strains survived less effectively.

   C. Further investigation advocated? Testing of survival of organisms released by other laboratory manipulations.

   D. Use in policy documents

      1. Rowe: Not addressed.

      2. NIAID Ad Hoc Working Group: Not addressed.

      3. RAC Background Document: Not addressed.

      4. NIH Decision Document: Not addressed.

VI. Monitoring of laboratory workers

   A. Main source of new information: M. Richmond, "Summary of Conclusions from Feeding and Monitoring Experiments," report to CO-

GENE Risk Assessment Subcommittee, Royal Society, London, 30 March 1979 (Rowe, pt. 3). A single paragraph summarizes results of studies of workers in Bristol, London, and Seattle. Details of the study of the Bristol workers are provided in V. Petrocheilou and M. H. Richmond, "Absence of Plasmid or *Escherichia coli* K12 Infection among Laboratory Personnel Engaged in R-Plasmid Research," *Gene* 2 (1977): 323–27. Workers were monitored for acquisition of the K12 strain and plasmids resistant to antibiotics that some of the strains carried.

B. Results

    1. In general, the studies showed that workers did not acquire the strains or the plasmids used in these laboratories.

    2. Fecal samples taken in the Bristol study of five laboratory workers were never shown to test positive for the *E. coli* strains with which they worked.

C. Further investigation advocated? No.

D. Use in policy documents

    1. Rowe: Cites results of Bristol study.

    2. NIAID Ad Hoc Working Group: Richmond study noted.

    3. RAC Background Document: Richmond study cited.

    4. NIH Decision Document: Cited at length as evidence that *E. coli* K12 does not infect laboratory personnel (69239–40).

VII. Survival of *E. coli* carrying insulin sequences in the intestinal tracts of rats and mice

A. Main source of new information: Paul Burnett (Eli Lilly Research Laboratories) to Wallace Rowe, 8 May 1979, ORDAR, doc. 670. Letter provides one table giving preliminary results of feeding experiments in which the survival of *E. coli* strains carrying insulin sequences in the intestinal tracts of rats and mice was measured.

B. Results

    1. Table shows that *E. coli* K12 strains could persist up to eight days in the intestinal tracts of "conventional" rats and mice, up to twenty-nine days in germ-free mice, and up to fifty days in germ-free rats.

    2. Experimenter describes the loss of the strain in conventional rodents as "rapid" and in germ-free rodents as "somewhat longer."

C. Further investigation advocated? Experimenter notes that these data are "preliminary."

D. Use in policy documents

    1. Rowe: Not addressed.

    2. NIAID Ad Hoc Working Group: Not addressed.

    3. RAC Background Document: Not addressed.

    4. NIH Decision Document: Cites experimenter's conclusion that, when *E. coli* K12 carrying recombinant DNA plasmids containing insulin gene sequences were fed to conventional rats and mice, they were "lost very rapidly."

VIII. Growth competition between *E. coli* K12 and the same organism containing a variety of recombinant DNA molecules

A. Main source of new information: A single table of unpublished data provided by Donald Brown, Carnegie Institution of Washington, Baltimore.

B. Results: Eventually, the *E. coli* bacteria without the recombinant DNA always outgrew the same bacteria with recombinant DNA, although in some cases this took many generations.

C. Further investigation advocated? No.

D. Use in policy documents

　　1. Rowe: Cited as evidence that "recombinants do not have positive selection" (pt. 6, p. 9).

　　2. NIAID Ad Hoc Working Group: Not addressed.

　　3. RAC Background Document: Cited as evidence that addition of recombinant DNA does not increase potential for survival.

　　4. NIH Decision Document: Cited as evidence that addition of recombinant DNA does not increase potential for survival (69240).

IX. Survival of *E. coli* K12 bacteria containing fragments of yeast DNA from a shotgun procedure

A. Main source of new information: Unpublished results of experiment by David Botstein, Hardy Chan, Malcolm Martin, and Wallace Rowe, cited in David Botstein to Donald Fredrickson, 16 October 1979, in NIH, *Recombinant DNA Research* 5:354–55, and in ORDAR, doc. 677.

B. Results

　　1. Fragments of yeast introduced into *E. coli* K12 were injected into mice. Authors concluded that introduction of these fragments did not increase the virulence of *E. coli* K12 for the mice.

　　2. Authors' conclusion: "Shot-gun cloning of the *S. cerevisiae* genome and/or subgenomic fragments into *E. coli* did not yield any clone with increased virulence for mice or with increased ability to adapt to mouse virulence as compared with K12 carrying a non-recombinant plasmid" (2).

C. Further investigation advocated? No.

D. Use in policy documents

　　1. Rowe: Not addressed.

　　2. NIAID Ad Hoc Working Group: Not addressed.

　　3. RAC Background Document: "Shot-gun cloning of the *S. cerevisiae* genome and/or subgenomic fragments into *E. coli* did not increase virulence for mice" (suppl., 4).

　　4. NIH Decision Document: Cited authors' conclusion (69239).

X. Autoimmune effects

A. Main sources of new information (all from NIH, *Recombinant DNA Research* 5 [March 1980]).

　　1. Jonathan Beckwith (Department of Microbiology and Molecular Genetics, Harvard Medical School) to William Gartland (Office of Recombinant DNA Activities, NIH), 26 October 1978 and 16 May 1978 (241–42, 268–69).

　　2. Wallace Rowe to Frank Dixon, 22 March 1979 and to George Cahill, 20 April 1979 (239–40).

    3. Baruj Benacerraf (Department of Pathology, Harvard medical School) to Wallace Rowe, 29 March 1979 (243–44).

    4. Norman Talal (Department of Medicine, School of Medicine, University of California, Berkeley) to Wallace Rowe, 5 April 1979 (247).

    5. Frank Dixon (Scripps Clinic and Research Foundation) to Wallace Rowe, 2 April 1979 (245–46).

    6. George Cahill (Joslin Diabetes Foundation) to Wallace Rowe, 39 April 1979 (248–49).

    7. William Paul and Richard Asofsky (NIAID) to Wallace Rowe, 18 May 1979 (270–75).

    8. Philip Patterson to Wallace Rowe, 15 May 1979 (265–67).

B. Information

    1. The question in this case was whether bacteria designed to make proteins closely related to human proteins could trigger an immune response in humans in which the immune system mistakenly turned its defenses against the body's own tissues. Jonathan Beckwith, in a letter to the NIH in October 1978, argued that the possibility of such an autoimmune effect should be considered particularly in the case of bacteria designed to express novel proteins on their outer cell surfaces rather than inside the bacterial cell.

    2. Letters 2–6 above were solicited by Wallace Rowe in response to Beckwith's letter (1). The information generated was entirely theoretical in nature; no experiments were designed to test the risk of generating autoimmune responses in individuals exposed to organisms carrying recombinant DNA. Responses 2–6 do not represent a uniform position on this question. Responses 2–5 indicated that, in general, autoimmune effects were unlikely, although response 4 noted that such effects could be tested. Response 6 concluded that bacteria producing human proteins might constitute a potential risk, especially if the bacteria were able to propagate and spread from individual to individual. Little risk was expected for strains of *E. coli* that were unable to transfer genes to more robust strains. This response concluded that the "requirements for enfeeblement" of *E. coli* K12 constituted "a very adequate safeguard" against these hazards.

C. Further investigation advocated?

    1. Response 4: Yes.

    2. Response 6: implied for strains of *E. coli* K12 other than the enfeebled strains.

D. Use in policy documents

    1. Rowe: Assessment delivered in a single sentence: "The concept of production of immunologic disease by bacteria producing eukaryotic proteins is considered by knowledgeable immunologists to be without sufficient substance to generate concern about a public health problem" (pt. 6, p. 9).

    2. NIAID Ad Hoc Working Group: Not addressed.

3. RAC Background Document: Not addressed.
4. NIH Decision Document: Selective quotation giving the impression that responses 2–6 concluded without qualification that "the probability of recombinant DNA experimentation leading to auto-immune disease was remote" (69241).

NOTES

The notes use the following abbreviations for archival sources.

HHP   Harlyn O. Halvorson personal papers
MRCUW   Modern Records Centre, University of Warwick, Coventry, United Kingdom
ODNIH   Office of the Director, National Institutes of Health files, Bethesda, Md.
ORDAR   Office of Recombinant DNA Activities, National Institutes of Health records, Bethesda, Md.
OSDHEW Office of the Secretary, Department of Health, Education, and Welfare files, Washington, D.C.
RDHC   Recombinant DNA History Collection, MC100, Institute Archives and Special Collections, MIT Libraries, Cambridge, Mass.
RWP   Robert Williamson personal papers

Introduction

1. For an insightful and wide-ranging account of the debate among American historians, see Peter Novick, *That Noble Dream: The Objectivity Question and the American Historical Profession* (Cambridge: Cambridge University Press, 1989).

2. Thomas Kuhn, *The Structure of Scientific Revolutions* (Chicago: Chicago University Press, 1962).

3. Steven Shapin, "History of Science and Its Sociological Reconstructions," *History of Science* 20 (1982): 157–211. References to feminist history of science in this period are included in *Feminism and Science,* ed. Nancy Tuana (Bloomington: Indiana University Press, 1989), 229–39.

4. Elizabeth Fox-Genovese, "Placing Women's History in History," *New Left Review* 133 (1982): 6–19; Joan Wallach Scott, "History in Crisis? The Others' Side of the Story," *American Historical Review* 94 (June 1989): 680–92.

5. John E. Toews, "Intellectual History after the Linguistic Turn: The Autonomy of Meaning and the Irreducibility of Experience," *American Historical Review* 92 (1987): 885; Novick, *That Noble Dream,* 541–46; Frederic Jameson, *The Prison-House of Language* (Princeton, N.J.: Princeton University Press, 1972).

6. Paul Rabinow and William Sullivan, "The Interpretive Turn: A Second Look," in *Interpretive Social Science: A Second Look,* ed. Paul Rabinow and William Sullivan (Berkeley: University of California Press, 1987), 7.

7. For reviews of these approaches, see Shapin, "History of Science"; Jan Golinski, "The Theory of Practice and the Practice of Theory: Sociological Approaches to the History of Science," *Isis* 81 (1990), 492–505; Andrew Webster, *Science, Technology, and Society* (London: Macmillan, 1991), 15–32.

8. David Bloor, *Knowledge and Social Imagery* (London: Routledge and Kegan Paul, 1976), 1–19.

9. Shapin, "History of Science," 164.

10. Trevor J. Pinch and Wiebe E. Bijker, "The Social Construction of Facts and Artifacts: Or How the Sociology of Science and the Sociology of Technology Might Benefit Each Other," in *The Social Construction of Technological Change,* ed. Wiebe E. Bijker, Thomas P. Hughes, and Trevor J. Pinch (Cambridge, Mass.: MIT Press, 1987), 17–50.

11. Steven Shapin, "Following Scientists Around," *Social Studies of Science* 18 (1988): 533–50.

12. Robert Young, "Science *Is* Social Relations," *Radical Science Journal* 5 (1977): 65–129.

13. Hilary Rose and Steven Rose, eds., *The Political Economy of Science: Ideology of/in the Natural Sciences* (London: Macmillan, 1976); Hilary Rose and Steven Rose, eds., *The Radicalization of Science* (London: Macmillan, 1976).

14. David Noble, *America by Design: Engineers as Agents and Victims of Capital* (New York: Knopf, 1977).

15. David Dickson, *The New Politics of Science* (New York: Pantheon, 1984).

16. See, e.g., Carolyn Merchant, *The Death of Nature: Women, Ecology, and the Scientific Revolution* (San Francisco: Harper and Row, 1980); Margaret Rossiter, *Women Scientists in America: Struggles and Strategies to 1940* (Baltimore: Johns Hopkins University Press, 1982); Evelyn Fox Keller, *Reflections on Gender and Science* (New Haven: Yale University Press, 1985); Sandra Harding, *The Science Question in Feminism* (Ithaca: Cornell University Press, 1986); Ludi Jordanova, *Sexual Visions: Images of Gender in Science and Medicine between the Eighteenth and Twentieth Centuries* (Brighton: Harvester Wheatsheaf, 1989); Mary Jacobus, Evelyn Fox Keller, and Sally Shuttleworth, eds., *Body/Politics: Women and the Discourses of Science* (New York: Routledge, Chapman, and Hall, 1990); Donna Haraway, *Simians, Cyborgs, and Women: The Reinvention of Nature* (New York: Routledge, Chapman, and Hall, 1991).

17. For critical commentaries on the dominant movements within the social studies of science, see Dorothy Nelkin, "Science Studies in the 1990s," *Science, Technology, and Human Values* 14 (Summer 1989): 305–11; Susan Douglas, "Technology and Society," *Isis* 81 (1990): 80–83; Brian Martin, "The Critique of Science Becomes Academic," *Science, Technology, and Human Values* 18 (Spring 1993): 247–59; Langdon Winner, "Upon Opening the Black Box and Finding It Empty: Social Constructivism and the Philosophy of Technology," *Science, Technology, and Human Values* 18 (Summer 1993): 362–78.

18. See Sal Restivo, "Modern Science as a Social Problem," *Social Problems* 35 (June 1988): 206–25.

19. Donald Fredrickson, "Science and the Cultural Warp: RDNA as a Case Study," paper presented at the annual meeting of the American Association for the Advancement of Science, January 1982. This view is held widely within the scientific community, and elements of it have influenced some accounts of the history of genetic engineering policy.

20. Steven Lukes, *Essays in Social Theory* (New York: Columbia University Press, 1977), 6–7.

21. Ibid., 3.

22. Steven Lukes, *Moral Conflict and Politics* (Oxford: Clarendon Press, 1991), ch. 6.

23. For discussion of the debates among political scientists, see, e.g., Peter Saunders, *Urban Politics: A Sociological Interpretation* (London: Hutchinson, 1979), 33–48. For debates among historians and sociologists of science, see, e.g., Steve Woolgar, "Interests and Explanation in the Social Study of Science," *Social Studies of Science* 11 (1981): 365–94; Barry Barnes, "On the 'Hows' and 'Whys' of Cultural Change," *Social Studies of Science* 11 (1981): 481–97; Donald MacKenzie, "Interests, Positivism, and History," *Social Studies of Science* 11 (1981): 498–503.

24. Steven Lukes, *Power: A Radical View* (London: Macmillan, 1974), 24.

25. Jurgen Habermas, "Toward a Theory of Communicative Competence," *Inquiry* 13 (1970): 360–75; William Connally, "On 'Interests' in Politics," *Politics and Society* 2 (1972): 459–77; Lukes, *Power*, 34–35.

26. E.g., Robert Dahl, *A Preface to Democratic Theory* (Chicago: University of Chicago Press, 1957).

27. Peter Bachrach and Morton Baratz, "Two Faces of Power," *American Political Science Review* 56 (1962): 947–52; Peter Bachrach and Morton Baratz, "Decisions and Non-decisions," *American Political Science Review* 57 (1963): 632–42; Peter Bachrach and Morton Baratz, *Power and Poverty: Theory and Practice* (New York: Oxford University Press, 1970); E. E. Schattschneider, *The Semi-sovereign People* (New York: Holt, Rinehart, and Winston, 1960).

28. Schattschneider, *Semi-sovereign People,* 71.

29. Bachrach and Baratz, "Two Faces of Power," 948.

30. E.g., Raymond Wolfinger, "Nondecisions and the Study of Local Politics," *American Political Science Review* 65 (1971): 1063–80.

31. Bachrach and Baratz, *Power and Poverty,* 49–50. See also the sensitive appraisal in Frederick Frey, "Comment: On Issues and Non-issues in the Study of Power," *American Political Science Review* 65 (1971): 1081–1101.

32. Lukes, *Power,* 23–24.

33. Ibid., 21–22.

34. Michel Foucault, *Power/Knowledge: Selected Interviews and Other Writings, 1972–1977,* ed. Colin Gordon (New York: Pantheon Books, 1980), 211. Hereafter pages given in text.

35. Michel Callon and John Law, "On Interests and Their Transformation: Enrolment and Counter-enrolment," *Social Studies of Science* 12 (1982): 622.

36. Nancy Hartsock, "Foucault on Power," in Linda J. Nicholson, *Feminism/Postmodernism* (New York: Routledge, Chapman and Hall, 1990), 169–70.

37. Shapin, "Following Scientists Around."

38. Rabinow and Sullivan, "Interpretive Turn," 21.

39. Cf. Martin Rudwick, *The Great Devonian Controversy* (Chicago: University of Chicago Press, 1985), 454. Rudwick discusses the nature of science, but his arguments can be applied reflexively to the nature of history.

40. As Richard Falk argues in a different connection, the openness of such debate must ultimately be relied upon ("Cultural Foundations for the International Protection of Human Rights," in *Human Rights in Cross-cultural Perspective: A Quest for Consensus*, ed. Abdullahi Ahmed An-Na'im [Philadelphia: University of Pennsylvania Press, 1992], 44–64).

41. Scott, "History in Crisis?" 681.

1. Social Interests in Promoting and Controlling Science and Technology

1. Charles W. Eliot, "Academic Freedom," *Science* 26 (5 July 1907): 11.

2. James Shannon, "The Advancement of Medical Research: A Twenty-Year View of the Role of the National Institutes of Health," *Journal of Medical Education* 42 (1967): 102.

3. For examinations of the role of the Rockefeller Foundation in promoting molecular biology, see Edward Yoxen, "Life as a Productive Force: Capitalizing the Science and Technology of Molecular Biology," in *Science, Technology, and the Labor Process*, ed. Les Levidow and Bob Young (London: CSE Books, 1981), 66–122; Edward Yoxen, "Giving Life a New Meaning: The Rise of the Molecular Biology Establishment," in *Scientific Establishments and Hierarchies: Sociology of the Sciences*, ed. N. Elias, H. Martins, and R. Whitley (Dordrecht: D. Reidel, 1982), 123–43; Pnina Abir-am, "The Discourse of Physical Power and Biological Knowledge in the 1930s: A Reappraisal of the Rockefeller Foundation's 'Policy' in Molecular Biology," *Social Studies of Science* 12 (1982): 341–82; Robert E. Kohler, *Partners in Science: Foundations and Natural Scientists, 1900–1945* (Chicago: Chicago University Press, 1990); Lily E. Kay, *The Molecular Vision of Life: Caltech, the Rockefeller Foundation, and the Rise of the New Biology* (Oxford: Oxford University Press, 1993).

4. Alex Roland, "Science and War," *Osiris* 2 (1986): 263–67.

5. Shannon, "Advancement of Medical Research," 99.

6. As Lily Kay notes, however, the molecular approach to analyzing biological problems, which the Rockefeller Foundation played a key role in initiating, was disseminated through multiple funding channels in the postwar years (*Molecular Vision of Life*, 273–74).

7. For a survey of studies of postwar sponsorship and support for research, see Margaret Rossiter, "Science and Public Policy since World War II," *Osiris* 1 (1985): 273–94.

8. James Shannon, "The Advancement of Medical Research," 99.

9. Donald Swain, "The Rise of a Research Empire: NIH, 1930–1950," *Science* 138 (14 December 1962): 1233–37.

10. Daniel Kevles, "The National Science Foundation and the Debate over Postwar Research Policy, 1942–1945: A Political Interpretation of Science: The Endless Frontier," *Isis* 68 (1977): 5–26.

11. Organization for Economic Co-operation and Development, *Reviews of National Science Policy: United States* (Paris: OECD, 1968), 37.

12. Roland, "Science and War," 266.

13. For a detailed account of the development of American health policy in the postwar years, see Steven P. Strickland, *Politics, Science, and Dread Disease: A Short History of United States Medical Research Policy* (Cambridge, Mass.: Harvard University Press, 1972); Steven P. Strickland, "Integration of Medical Research and Health Policies," *Science* 173 (17 September 1971): 1093–1103.

14. Strickland, "Integration of Medical Research and Health Policies," 1094–95.

15. Charles Kidd, "The NIH Phenomenon," *Science* 179 (19 January 1973): 272.

16. *NIH Factbook: Guide to National Institutes of Health Programs and Activities* (Chicago: Marquis Who's Who, 1976), 99; U.S. National Science Foundation, Division of Science Resources Studies, "Federal Funds for Research and Development: Detailed Historical Tables: Fiscal Years 1955–1987" (n.d.), table A, 3–9; U.S. National Science Foundation, *Federal Funds for Science*, vols. 8–15 (Washington D.C.: U.S. Government Printing Office, 1959–67).

17. National Science Foundation, "Federal Funds for Research and Development," 5–6.

18. Daniel S. Greenberg, *The Politics of Pure Science* (New York: New American Library, 1967), 276ff.

19. Strickland, *Politics, Science, and Dread Disease*, 169–77.

20. Robert Semple, "President Orders a Medical Review," *New York Times*, 28 June 1966, 35.

21. Elizabeth Brenner Drew, "The Health Syndicate: Washington's Noble Conspirators," *Atlantic*, December 1967, 78.

22. Donald Fredrickson, "Health and the Search for New Knowledge," *Daedalus* 106 (Winter 1977): 165.

23. Strickland, *Politics, Science, and Dread Disease*, ch. 12; Richard Rettig, *Cancer Crusade: The Story of the National Cancer Act of 1971* (Princeton: Princeton University Press, 1977), chs. 4, 5.

24. National Science Foundation, "Federal Funds for Research and Development," 10–12, 179.

25. *NIH Factbook*, 101–2.

26. Strickland, *Politics, Science, and Dread Disease*, 236.

27. Ibid., 254–55.

28. Rettig, *Cancer Crusade*, 12.

29. Organization for Economic Cooperation and Development, *Reviews of National Science Policy*, 95.

30. See, e.g., Department of Health, Education, and Welfare, *Report of the President's Biomedical Research Panel* (Washington, D.C.: U.S. Government Printing Office, 20 April 1976), 23, 28.

31. Shannon, "Advancement of Medical Research," 102. Data on federal support for research in universities show a substantially enlarged proportion of the university budget coming from the government (Department of Health, Education, and Welfare, Education Office, *Higher Education Finances: Selected Trends and Summary Data* [June 1968], table 1, 3).

32. Spencer Klaw, *The New Brahmins: Scientific Life in America* (New York: William Morrow, 1968), 114–16.

33. Organization for Economic Cooperation and Development, *The Research System: Comparative Survey of the Organization and Financing of Fundamental Research* (Paris: OECD, 1974), 3:49.

34. Department of Health, Education, and Welfare, *Report of the President's Biomedical Research Panel,* appendix C, 6.

35. National Science Foundation, "Federal Funds for Research and Development," 201.

36. See, e.g., the view of Surgeon General William H. Stewart, quoted in Greenberg, *Politics of Pure Science,* 287.

37. Klaw, *New Brahmins,* 123; Greenberg, *Politics of Pure Science,* 151–52.

38. Greenberg, *Politics of Pure Science,* 151–52.

39. Shannon, "Advancement of Medical Research," 103.

40. Jerome Ravetz, *Scientific Knowledge and Its Social Problems* (Oxford: Oxford University Press, 1971), 44–47.

41. Harold Orlans, *The Effects of Federal Programs on Higher Education: A Study of 36 Universities and Colleges* (Washington, D.C.: Brookings Institution, 1962), 201–3.

42. Max Heirich, "Why We Avoid the Key Questions: How Shifts in the Funding of Scientific Inquiry Affect Decision Making about Science," in *The Recombinant DNA Debate,* ed. Steven P. Stich and David A. Jackson (Englewood Cliffs, N.J.: Prentice Hall, 1979), 240–41. On the merging of basic and applied research, see also Klaw, *New Brahmins,* 172.

43. Heirich, "Why We Avoid the Key Questions," 240.

44. For accounts of the development of the British economy after World War II, see Keith Smith, *The British Economic Crisis: Its Past and Future* (Harmondsworth, Middlesex: Penguin, 1984), ch. 1; Sidney Pollard, *The Wasting of the British Economy,* 2d ed. (Beckenham, Kent: Croon Helm, 1984), ch. 3.

45. A. H. Halsey and M. A. Trow, *The British Academics* (Cambridge, Mass.: Harvard University Press, 1971), 60–64.

46. Ibid., 63.

47. Hilary Rose and Stephen Rose, *Science and Society* (Harmondsworth, Middlesex: Penguin, 1970), ch. 5.

48. U.K. Council for Science Policy, *First Report,* Cmnd. 3007 (May 1966), 18.

49. Smith, *British Economic Crisis,* 87–94.

50. Ibid., 81; Norman Vig, *Science and Technology in British Politics* (Oxford: Pergamon Press, 1968), ch. 3.

51. Vig, *Science and Technology in British Politics,* 69.

52. Quoted in Rose and Rose, *Science and Society,* 93. I thank Colin Presswood for reminding me of Wilson's metpaphor.

53. Ibid., 106.

54. Halsey and Trow, *British Academics,* 140–44.

55. For a description of the United Kingdom's economic problems at this time, see, e.g., Smith, *British Economic Crisis,* ch. 4.

56. Council for Scientific Policy, *First Report,* 3, 11.

57. Lord Rothschild, *A Framework for Government Research and Development,* Cmnd. 4814 (1971); for discussion of the Rothschild proposals and their reception, see Roger Williams, "Some Political Aspects of the Rothschild Affair," *Science Studies* 3, no. 1 (1973): 31–46; Philip Gummett, *Scientists in Whitehall* (Manchester: Manchester University Press, 1980), 195–202.

58. Rothschild, *Framework for Government Research and Development.*

59. Gummett, *Scientists in Whitehall,* 205.

60. Robert Gilpin, *American Scientists and Nuclear Weapons Policy* (Princeton, N.J.: Princeton University Press, 1962), 24–25.

61. Ibid., 155–58.

62. Charles Schwartz, "The Movement vs. the Establishment," *Nation,* 22 June 1970, 749.

63. Hilary Rose and Stephen Rose, "The Radicalization of Science," in *The Radicalization of Science,* ed. Hilary Rose and Stephen Rose (London: Macmillan, 1976), 18–24.

64. Jonathan Beckwith, "The Scientist in Opposition in the United States," in *The Biological Revolution: Social Good or Social Evil?* ed. Watson Fuller (London: Routledge and Kegan Paul, 1972), 296–97.

65. Allan Schnaiberg, *The Environment from Surplus to Scarcity* (New York: Oxford University Press, 1980), 27–28; for insightful treatments of aspects of the consumer safety, environmental, and occupational health movements, see Allan Schnaiberg, "Politics, Participation, and Pollution: The 'Environmental Movements,'" in *Cities in Change: Studies on the Urban Condition,* ed. J. Walton and D. E. Carns (Boston: Allyn and Bacon, 1973), 605–27; Jack Walker, "Setting the Agenda in the U.S. Senate: A Theory of Problem Selection," *British Journal of Political Science* 7 (1977): 423–45; Charles Noble, *Liberalism at Work: The Rise and Fall of OSHA* (Philadelphia: Temple University Press, 1986).

66. Robert O. Herrmann, "The Consumer Movement in Historical Perspective," in *Consumerism: Search for Consumer Interest,* ed. David A. Aaker and George S. Day (New York: Free Press, 1971), 15; Walker, "Setting the Agenda in the U.S. Senate," 432–33.

67. Schnaiberg, *Environment from Surplus to Scarcity,* 27–28.

68. Annual statistics compiled by the Bureau of Labor Statistics, cited in *1970 Congressional Quarterly Almanac,* 675; Noble, *Liberalism at Work,* 62–63; Samuel Bowles, David M. Gordon, and Thomas E. Weisskopf, *Beyond the Wasteland* (Garden City, N.Y.: Anchor Press, 1984), 93.

69. U.K. Committee on Safety and Health at Work, *Safety and Health at Work: Report of the Committee, 1970–1972* (the Robens report), Cmnd. 5034 (July 1972), table 103, pp. 3, 161–63.

70. Nicholas Ashford, *Crisis in the Workplace: Occupational Disease and Injury* (Cambridge, Mass.: MIT Press, 1976), 47, 88.

71. This interpretation is developed persuasively by Jack Walker in "Setting the Agenda in the U.S. Senate," 433–35.

72. Noble, *Liberalism at Work,* 81.

73. Walker, 436ff.

74. Robert Cameron Mitchell and J. Clarence Davies, III, "The United States Environmental Movement and Its Political Context: An Overview," Resources for the Future Discussion Paper, May 1978, fig. 1.

75. For examples of this perception, see Richard Falk, *This Endangered Planet: Prospects and Proposals for Human Survival* (New York: Random House, 1971); David Sills, "The Environmental Movement and Its Critics," *Human Ecology* 3, no.1 (1975): 23.

76. For discussion of the significance of the environmental and occupational health and safety legislation enacted in the late 1960s and early 1970s, see, e.g., Richard Andrews, *Environmental Policy and Administrative Change: Implementation of the National Environmental Policy Act* (Lexington, Mass.: Lexington Books, 1976); Sally Fairfax, "A Disaster for the Environmental Movement," *Science* 199 (17 February 1978): 743–49; Ashford, *Crisis in the Workplace;* Noble, *Liberalism at Work.*

77. For discussion of the importance of these laws for providing public access to governmental processes, see Dorothy Nelkin, "Public Participation in Decisions about Science and Technology in the United States," report to the OECD Science Policy Directorate, December 1977.

78. Roger C. Crampton and Barry B. Boyer, "Citizen Suits in the Environmental Field: Peril or Promise?" *Ecology Law Quarterly* 2, no.3 (1972): 407–29. For a contrary view, see Fairfax, "Disaster for the Environmental Movement," 743–49.

79. Alexander Capron, book review, *Southern California Law Review* 48 (1974): 571–72; Technology Assessment Act, 2 *U.S.C.* sec. 1862 (1976). For a critique of the technology assessment movement and the implementation of the Technology Assessment Act, see Dickson, *New Politics of Science,* 232–43.

80. On the limitations of citizen representation on advisory boards, see Nelkin, "Public Participation."

81. Francis Sandbach, *Environmental Ideology and Policy* (Oxford: Basil Blackwell, 1980), 17–18.

82. Quoted in Jeremy Bugler, *Polluting Britain: A Report* (Harmondsworth, Middlesex: Penguin, 1972), 11.

83. Howard A. Scarrow, "The Impact of British Domestic Air Pollution Legislation," *British Journal of Political Science* 2 (1972): 261–82.

84. Roger Williams, *Government Regulation of the Occupational and General Environments in the United Kingdom, the United States, and Sweden* (Science Council of Canada Background Study no. 40, October 1977), 37–38.

85. Ibid.

86. Timothy O'Riordan, *Environmentalism* (London: Pion, 1976), 238.

87. United Kingdom, *The Protection of the Environment,* Cmnd. 4373 (May 1970).

88. Royal Commission on Environmental Pollution, *Third Report: Pollution in Some British Estuaries and Coastal Waters,* Cmnd. 5054 (September 1972), 2–3.

89. Royal Commission on Environmental Pollution, *Fifth Report: Air Pollution Control: An Integrated Approach,* Cmnd. 6371 (January 1976).

90. Committee on Safety and Health at Work, *Safety and Health at Work,* 1.

91. Ibid., 1, 12, 22, 42–46.

92. Quoted in Theo Nichols and Pete Armstrong, *Safety or Profit: Industrial Accidents and the Conventional Wisdom* (Bristol: Falling Wall Press, 1973), 4.

93. United Kingdom, *Parliamentary Debates* (Commons), 5th series, 21 May 1973, 857:67, col. 1.

94. A. Ian Glendon and Richard T. Booth, "Worker Participation in Occupational Health and Safety in Britain," *International Labour Review* 121, no. 4 (1982): 409–10.

95. Health and Safety at Work Act 1974, ch. 37.

96. This provision was later amended by the Employment Protection Act of 1975 to require that safety representatives be chosen only from recognized trade unions.

97. John W. Leopold and P. B. Beaumont, "Joint Health and Safety Committees in the United Kingdom: Participation and Effectiveness: A Conflict?" *Economic and Industrial Democracy* 3 (1982): 270–71.

98. Thomas Ferguson and Joel Rogers, "The Reagan Victory: Corporate Coalitions in the 1980 Campaign," in *The Hidden Election: Politics and Economics in the 1980 Presidential Campaign,* ed. Thomas Ferguson and Joel Rogers (New York: Pantheon, 1981), 13–17; David Dickson and David Noble, "By Force of Reason: The Politics of Science and Technology Policy," in Ferguson and Rogers, *Hidden Election,* 260–61; Bowles, Gordon, and Weisskopf, *Beyond the Wasteland,* ch. 5.

99. National Science Board, *Science Indicators 1980* (Washington, D.C.: U.S. Government Printing Office, 1981), 32–33, 235.

100. See, e.g., Smith, *British Economic Crisis,* ch. 1; Pollard, *Wasting of the British Economy,* chs. 1–3.

101. Philip Handler, address to the Biochemical Society in London on 17 December 1969, quoted in "Sombre Greeting from Abroad," *Nature* 224 (27 December 1969): 1250

102. Alvin Weinberg, "In Defense of Science," *Science* 167 (9 January 1970): 141.

103. Peter Medawar, "On 'the Effecting of All Things Possible,'" *New Scientist,* 4 September 1969, 467.

104. Philip Handler, "Science's Continuing Role," *BioScience* 20 (15 October 1970): 1103.

105. John Maddox, "On Which Side Are the Angels?" *Nature* 224 (27 December 1969): 1242.

106. Handler, "Science's Continuing Role," 1104.

107. National Academy of Sciences, Ad Hoc Committee, *Applied Science and Technological Progress* (Washington, D.C.: National Academy of Sciences, 1969), quoted in Dickson, *New Politics of Science,* 233.

108. National Academy of Sciences, *Technology: Processes of Assessment and Choice* (Washington, D.C.: National Academy of Sciences, 1969), 14, 79, 84; see also comments on the public's relation to science in *Civilization and Science: In Conflict or Collaboration?* ed. Gordon E. W. Wolstenholme and Maeve O'Connor (Amsterdam: Elsevier, 1972); see section 4.3.

109. On the corporate response to the environmental, consumer safety, and workplace safety movements, see Dickson and Noble, "By Force of Reason,"

260–312; Dickson, *New Politics of Science,* chs. 1, 2, 6; Noble, *Liberalism at Work,* ch. 4.

110. David Rockefeller, "Corporate Tasks for 70's: Social Action Not Words," *Commercial and Financial Chronicle* 215 (6 January 1972): 69, emphasis in the original.

111. Leonard Silk and David Vogel, *Ethics and Profits: The Crisis of Confidence in American Business* (New York: Simon and Schuster, 1976), 52–53, 57.

112. E.g., Rockefeller, "Corporate Tasks for 70's," 69.

113. Dickson, *New Politics of Science,* 64; National Science Board, *Science Indicators: The 1985 Report* (Washington, D.C.: U.S. Government Printing Office, 1985), 108, and table 5-20.

114. Philip Abelson, "Academic Science and Industry," *Science* 183 (29 March 1974): 1251.

115. Noble, *Liberalism at Work,* 100–103; Dickson and Noble, "By Force of Reason," 266–70.

116. Dickson and Noble, "By Force of Reason," 270–72.

117. Ibid., 276–79; see also speech by MIT economist Karl Kaysen at a conference given by the Committee on Economic Development in 1976, quoted in the same pages.

118. Dickson, *New Politics of Science,* 87–88; Dickson and Noble, "By Force of Reason," 284–87.

119. Edward David, "Science Futures: The Industrial Connection," *Science* 203 (2 March 1979): 837–40.

120. Dickson, *New Politics of Science,* 35–36.

121. For an account of science policy in the Carter administration by its chief architect, see Frank Press, "Science and Technology in the White House, 1977–1980: Part 1," *Science* 211 (9 January 1981): 139–45.

122. Dickson, *New Politics of Science,* 284.

123. For an account and appraisal of the way in which the Carter administration balanced the interests of industry, science, unions, and environmentalists in its science policy, see ibid., 282–86.

124. W. Bowman Cutter, executive associate director of the Office of Management and Budget, prepared presentation, in *R&D in an Inflationary Environment: Federal R&D, Industry and the Economy, Universities, Intergovernmental Science: Colloquium Proceedings, June 19–20, 1980,* ed. Albert H. Teich, Gail J. Breslow, and Ginger F. Payne (Washington, D.C.: American Association for the Advancement of Science, 1980), 42; Dickson, *New Politics of Science,* 124–25.

125. Dickson, *New Politics of Science,* 40–41.

126. For accounts of the impact of the Reagan administration's deregulatory policies on environmental policy, see, e.g., Jonathan Lass, Katherine Gillman, and David Sheridan, *A Season of Spoils: The Story of the Reagan Administration's Attack on the Environment* (New York: Pantheon, 1984).

127. On this development, see Dickson, *New Politics of Science,* 288–89.

128. Quoted in Claude E. Barfield, *Science Policy from Ford to Reagan: Change and Continuity* (Washington, D.C.: American Enterprise Institute for Public Policy Research, 1982), 43.

129. See, e.g., remarks of Glenn Schleede, executive associate director of the Office of Management and Budget, in ibid., 41.

130. Colin Norman, "Is Reaganomics Good for Technology?" *Science* 213 (21 August 1981): 843–45; U.S. Congress, Office of Technology Assessment, *Commercial Biotechnology: An International Analysis* (Washington, D.C.: U.S. Government Printing Office, 1984), 278.

131. National Science Board, *Science Indicators: The 1985 Report,* 82–84, 261–62; Office of Technology Assessment, *Commercial Biotechnology,* 278.

132. Barbara Culliton, "Harvard and Monsanto: The $23-Million Alliance," *Science* 195 (25 February 1977): 759–63.

133. "Business and the Universities: A New Partnership," *Business Week,* 20 December 1982, 58–62.

134. National Science Board, *Science Indicators: The 1985 Report,* 262.

135. Dickson, *New Politics of Science,* 73–74.

136. National Science Board, *Science Indicators: The 1985 Report,* 85.

137. Pollard, *Wasting of the British Economy,* 61–62.

138. Health and Safety Commission, *Report 1977–1978* (London: HMSO, 1978), *Report 1978–1979* (London: HMSO, 1979), and *Report 1979–1980* (London: HMSO, 1980).

139. U.K. Advisory Board for the Research Councils, *First Report for the Research Councils,* Cmnd. 7367 (February 1979), appendix 5, 20; U.K. Advisory Board for the Research Councils, *Third Report of the Advisory Board for the Research Councils, 1976–1978,* Cmnd. 7467 (February 1979), appendix 3, 26.

140. Advisory Board for the Research Councils, *Second Report for the Research Councils,* Cmnd. 6430 (March 1976), 13–15.

141. Advisory Board for the Research Councils/University Grants Committee, *Report of a Joint Working Party on the Support of University Scientific Research,* Cmnd. 8567 (June 1982), 8–11.

142. Advisory Board for the Research Councils, *Third Report,* 26.

143. Medical Research Council, *Annual Report, 1973–74,* 6.

144. Advisory Board for the Research Councils, *Third Report,* 21.

145. Clive Booth, "Higher Education and Research Training," in *U.K. Science Policy: A Critical Review of Policies for Publicly Funded Research,* ed. Maurice Goldsmith (London: Longman, 1984), 136.

146. At the time of ACARD's establishment, the membership included six representatives from private industry, two from public enterprises, one trade unionist, two academics, and the chair of the ABRC (*New Scientist* 72 [9 December 1976]: 572).

147. Advisory Council for Applied Research and Development, *Industrial Innovation* (London: HMSO, 1978), 7–10, 13–15.

148. Advisory Council for Applied Research and Development/Advisory Board for the Research Councils, *First Joint Report,* Cmnd. 8957 (July 1983), 6.

149. For an insightful account of science policy developments during the Thatcher years, see Martin Ince, *The Politics of British Science* (Brighton, Sussex: Wheatsheaf Books, 1986).

150. Ibid., 39.

151. Advisory Board for the Research Councils/University Grants Committee, *Report of a Joint Working Party*, 2.

152. Advisory Board for the Research Councils, *The Science Budget: A Forward Look* (London: Department of Education and Science, 1982). For a similar view, see Advisory Council for Applied Research and Development/Advisory Board for the Research Councils, *Improving Research Links between Higher Education and Industry* (London: HMSO, June 1983), 45.

153. Advisory Council for Applied Research and Development/Advisory Board for the Research Councils, *First Joint Report*.

154. David Vogel, *National Styles of Regulation: Environmental Policy in Great Britain and the United States* (Ithaca: Cornell University Press, 1986), 22.

155. United Kingdom, *Report on Non-departmental Public Bodies,* Cmnd. 7797 (January 1980), 18.

156. Peter Hennessey, "Savings of £11.6 m. Is Expected in Loss of 246 'Quangos' after Review of 2,117 Organizations," *Times,* 17 January 1980, 6; Chairman of the Health and Safety Commission, William J. Simpson, to Secretary of Employment, James Prior, 21 May 1980, published in *Compress* (journal of the Health and Safety Executive), June 1980, 1–2; Judy Redfearn, "Occupational Safety: Cuts Ahead," *Nature* 286 (21 August 1980): 751.

157. Philip Armstrong, Andrew Glyn, and John Harrison, *Capitalism since World War II* (London: Fontana, 1984), 418–23.

158. Edward David, "To Prune, Promote, and Preserve," *Science* 217 (13 August 1982): 589.

2. The Social Transformation of Recombinant DNA Technology, 1972–1982

1. The selection of about sixty scientific papers judged to be critical for the development of recombinant DNA technology was based on references in scientific reviews and news articles, interviews, and cross-references.

2. For accounts of the development of molecular biology in the 1950s and 1960s, see Garland Allen, *Life Science in the Twentieth Century* (Cambridge: Cambridge University Press, 1975); Robert H. Haynes and Philip C. Hanawalt, eds., *The Molecular Basis of Life* (San Francisco: Freeman, 1968); David Freifelder, ed., *Recombinant DNA* (San Francisco: Freeman, 1978).

3. On the transmission of drug resistance, see Tsutomu Watanabe, "Infectious Drug Resistance," *Scientific American,* December 1967, 19–27; Royston C. Clowes, "The Molecule of Infectious Drug Resistance," *Scientific American,* April 1973, 18–27. On the restriction enzymes, see Herbert Boyer, "DNA Restriction and Modification Mechanisms in Bacteria," *Annual Review of Microbiology* 25 (1971): 153–76.

4. See Philip Handler, ed., *Biology and the Future of Man* (New York: Oxford University Press, 1970); James D. Watson and John Tooze, *The DNA Story: A Documentary History of Gene Cloning* (San Francisco: Freeman 1981), 551–52; Gunther Stent, *The Coming of the Golden Age: A View of the End of Progress* (New York: Natural History Press, 1969). Stent went so far as to claim that "the gene problem was solved" (57). For others, however, what remained unknown about

the functioning of genes, particularly in higher organisms, led to more circumspect conclusions.

5. Cetus Corporation, "Special Report," ca. 1975. This unpublished report on the commercial potential of genetic engineering was circulated privately in the second half of 1975.

6. For examinations of the provenance of the key concepts of molecular biology and its reductionist thrust, and of the role of the Rockefeller Foundation in promoting the new field, see Yoxen, "Life as a Productive Force"; Yoxen, "Giving Life a New Meaning"; Abir-am, "Discourse of Physical Power"; Kohler, *Partners in Science;* Kay, *Molecular Vision of Life.* While the reasons for the role played by the Rockefeller Foundation in reordering biological problems and focusing attention on molecular approaches remain controversial, there is broad agreement that this role was influential.

7. Edward L. Tatum, "A Case History in Biological Research," *Science* 129 (26 June 1959): 1714.

8. Sydney Brenner, "New Directions in Molecular Biology," *Nature* 248 (26 April 1974): 786.

9. Robert L. Sinsheimer, "Genetic Engineering: The Modification of Man," *Impact of Science on Society* 20 (1970): 279–90. For a review of the debate on genetic engineering in this period, see U.S. Library of Congress, Congressional Research Service, "Genetic Engineering: Evolution of a Technological Issue," report prepared for the House Subcommittee on Science and Astronautics, 92d Cong., 2d sess., November 1972.

10. Joshua Lederberg, "Genetics," in *Encyclopaedia Britannica Yearbook of Science and the Future 1969* (Chicago: Encyclopaedia Britannica, 1968), 321.

11. James Danielli, "Artificial Synthesis of New Life Forms," *Bulletin of the Atomic Scientists* 28 (December 1972): 20–24.

12. Joshua Lederberg, statement to the Conference of the Committee on Disarmament, Geneva, 17 August 1976, UN Doc. CCD/312 (27 August 1970).

13. Robert J. Bazell, "Molecular Biology: Corporate Citizenship and Potential Profit," *Science* 174 (15 October 1971): 275–76.

14. For details of the founding of the company and the initial conception of its goals, see Roger Lewin, "Modern Biology at the Industrial Threshold," *New Scientist,* 5 October 1978, 18–19; see also Cetus Corporation, "Special Report."

15. Ronald Cape, transcript of interview with Charles Weiner, 19 April 1978, 9, RDHC. ʾ

16. Cetus Corporation, prospectus, 6 March 1981.

17. For reviews of this research, see Theodore Friedmann and Richard Roblin, "Gene Therapy for Human Genetic Disease? Proposals for Genetic Manipulation in Humans Raise Difficult Scientific and Ethical Problems," *Science* 175 (3 March 1972): 949–55; J. Leslie Glick, "Reflections and Speculations on the Regulation of Molecular Genetic Research," *Annals of the New York Academy of Sciences* 265 (1976): 179.

18. Friedmann and Roblin, "Gene Therapy for Human Genetic Disease?" 952–953.

19. Robert L. Sinsheimer, "Recombinant DNA," *Annual Review of Biochemistry* 46 (1977): 415–16; Friedmann and Roblin, "Gene Therapy for Human Genetic Disease?" 950–51.

20. K. L. Agarwal et al., "Total Synthesis of the Gene for an Alanine Ribonucleic Acid from Yeast," *Nature* 227 (4 July 1970): 27–34; Sinsheimer, "Recombinant DNA," 416–17.

21. Interview with Robert Helling, 24 September 1982.

22. Joshua Lederberg, "Genetics of Bacteria," grant application to the National Institutes of Health, no. AI 05160-11, 20 December 1967, Joshua Lederberg personal papers.

23. Ibid.

24. Peter Lobban and A. Dale Kaiser, "Enzymatic End-to-End Joining of DNA Molecules," *Journal of Molecular Biology* 78 (1973): 453–71; David Jackson, Robert Symons, and Paul Berg, "Biochemical Method for Inserting New Genetic Information in DNA of Simian Virus 40: Circular SV40 Molecules Containing Lambda Phage Genes and the Galactose Operon of *Escherichia coli*," *Proceedings of the National Academy of Sciences of the United States of America* 69 (1972): 2904–9.

25. Peter Lobban, "The Generation of Transducing Phage *in Vitro*," essay for third Ph.D. examination, Stanford University, 6 November 1969, Peter Lobban personal papers.

26. Interview with David Jackson, 15 July 1982.

27. Jackson, Symons, and Berg, "Biochemical Method."

28. Lobban and Kaiser, "Enzymatic End-to-End Joining of DNA Molecules."

29. Interview with Robert Helling, 24 September 1982.

30. Herbert Boyer, "Host-Controlled Modification and Restriction of DNA," grant application to the National Institute of General Medicine, no. GM 14378-04, December 1968, obtained through a request under the Freedom of Information Act.

31. Interview with Robert Helling, 24 September 1982.

32. Stanley N. Cohen, "The Manipulation of Genes," *Scientific American,* July 1975, 28.

33. Janet E. Mertz and Ronald W. Davis, "Cleavage of DNA by R Restriction Endonuclease Generates Cohesive Ends," *Proceedings of the National Academy of Sciences of the U.S.A.* 69 (1972): 3370–74.

34. Joe Hedgpeth, Howard M. Goodman, and Herbert W. Boyer, "DNA Nucleotide Sequence Restricted by the RI Endonuclease," *Proceedings of the National Academy of Sciences of the U.S.A.* 69 (1972): 3448–52.

35. Interview with Robert Helling, 24 September 1982.

36. Mertz and Davis, "Cleavage of DNA," 3374.

37. Stanley N. Cohen et al., "Construction of Biologically Functional Bacterial Plasmids *in Vitro*," *Proceedings of the National Academy of Sciences of the U.S.A.* 70 (November 1973): 3240–44.

38. Annie C. Y. Chang and Stanley N. Cohen, "Genome Construction between Bacterial Species *in Vitro*: Replication and Expression of *Staphylococcus* Plasmid Genes in *Escherichia coli*," *Proceedings of the National Academy of Sciences of the U.S.A.* 71 (1974): 1033; Cohen, "Manipulation of Genes," 29–31.

39. John F. Morrow et al., "Replication and Transcription of Eucaryotic DNA in *Escherichia coli,*" *Proceedings of the National Academy of Sciences of the U.S.A.* 71 (1974): 1743–47; Cohen, "Manipulation of Genes," 31.

40. Sydney Brenner, "Evidence for the Ashby Working Party," paper submitted to the Working Party on the Experimental Manipulation of the Genetic Composition of Micro-organisms, 26 September 1974, RDHC.

41. Cetus Corporation, "Special Report."

42. Brenner, "Evidence for the Ashby Working Party."

43. Gene Bylinsky, "Industry Is Finding More Jobs for Microbes," *Fortune,* February 1974, 96–102.

44. "Getting Bacteria to Manufacture Genes," *San Francisco Chronicle,* 21 May 1974.

45. Victor K. McElheny, "Animal Gene Shifted to Bacteria, Aid Seen to Medicine and Farm," *New York Times,* 20 May 1974, 61.

46. For references to bacteria factories, see, e.g., Bylinsky, "Industry Is Finding More Jobs for Microbes"; McElheny, "Animal Gene Shifted to Bacteria"; Roger Lewin, "The Future of Genetic Engineering," *New Scientist,* 17 October 1974, 166.

47. Interview with Ronald Cape, 19 May 1982.

48. David Fishlock, "ICI Launches £40,000 Project on 'Genetic Engineering,'" *Financial Times,* 3 October 1974.

49. Miranda Robertson, "ICI Puts Money on Genetic Engineering," *Nature* 251 (18 October 1974): 564–65.

50. U.S. patent no. 4,237,224, granted to Cohen and Boyer and assigned to Stanford University in December 1980, for the method of inserting foreign DNA into *E. coli.*

51. Eleanor Lawrence, "Nuts and Bolts of Genetic Engineering," *Nature* 263 (28 October 1976): 726–27; Robert Williamson, "First Mammalian Results with Genetic Engineering," *Nature* 260 (18 March 1976): 189–90.

52. "Insulin Gene Transferred to Bacterium," *Chemical and Engineering News,* 30 May 1977, 4; Charles Weissmann, "The Cloning of Interferon and Other Mistakes," in *Interferon 3,* ed. I. Gresser (New York: Academic Press, 1981), 101–34, esp. 101–2; Jean L. Marx, "Nitrogen Fixation: Prospects for Genetic Manipulation," *Science* (6 May 1977): 638–41; Peter R. Day, "Plant Genetics: Increasing Crop Yield," *Science* 197 (30 September 1977): 1334–39.

53. Vickers Hershfield et al., "Plasmid Col E1 as a Molecular Vehicle for Cloning and Application of DNA," *Proceedings of the National Academy of Sciences of the U.S.A.* 71 (1974): 3455–59; Francisco Bolivar et al., "Construction and Characterization of New Cloning Vehicles: 2. A Multipurpose Cloning System," *Gene* 2 (1977): 95–113; Sinsheimer, "Recombinant DNA," 423–24.

54. Marjorie Thomas, John R. Cameron, and Ronald W. Davis, "Viable Molecular Hybrids of Bacteriophage Lambda and Eucaryotic DNA," *Proceedings of the National Academy of Sciences of the U.S.A.* 71 (1974): 4579–83; Alain Rambach and Pierre Tiollais, "Bacteriophage Having Eco RI Endonuclease Sites Only in the Nonessential Region of the Genome," *Proceedings of the National Academy of Sciences of the U.S.A.* 71 (1974): 3927–80; Noreen E. Murray and Kenneth Murray, "Manipulation of Restriction Targets in Phage to Form Receptor Chro-

mosomes for DNA Fragments," *Nature* 251 (1974): 476–81; Sinsheimer, "Recombinant DNA," 423–24.

55. Maria Szekely, "Two Approaches to Gene Synthesis," *Nature* 263 (23 September 1976): 277–78.

56. Ibid.

57. Jean L. Marx, "Molecular Cloning: Powerful Tool for Studying Genes," *Science* 191 (1976): 1160–62; Sinsheimer, "Recombinant DNA," 435.

58. Sinsheimer, "Recombinant DNA," 419, 428.

59. See, e.g., Michael Grunstein and David S. Hogness, "Colony Hybridization: A Method for the Isolation of Cloned DNAs That Contain a Specific Gene," *Proceedings of the National Academy of Sciences of the U.S.A.* 72 (1975): 3961–65; Marx, "Molecular Cloning," 1161; Sinsheimer, "Recombinant DNA," 428. For a later refinement of early screening techniques, which allowed screening over very high density bacterial colonies, see Douglas Hanashan and Matthew Meselson, "Plasmid Screening at High Colony Density," *Gene* 10 (1980): 63–67.

60. F. Sanger and A. R. Coulson, "A Rapid Method for Determining Sequences in DNA by Primed Synthesis with DNA Polymerase," *Journal of Molecular Biology* 94 (1975): 441–48; Allan M. Maxam and Walter Gilbert, "A New Method for Sequencing DNA," *Proceedings of the National Academy of Sciences of the U.S.A.* 74 (1977): 560–64. For reviews of these methods, see Walter Gilbert, "DNA Sequencing and Gene Structure," *Science* 214 (18 December 1981): 1305–12; Gena Bari Kolata, "DNA Sequencing: A New Era in Molecular Biology," *Science* 192 (14 May 1976): 645–47.

61. Herbert L. Heyneker et al., "Synthetic *lac* Operator DNA Is Functional *in Vivo*," *Nature* 263 (28 October 1976): 748–52; Richard H. Scheller et al., "Chemical Synthesis of Restriction Enzyme Recognition Sites Useful for Cloning," *Science* 196 (8 April 1977): 177–80.

62. John Atkins, "Expression of a Eukaryotic Gene in *Escherichia coli*," *Nature* 262 (22 July 1976): 256–57.

63. Brenner, "Evidence for the Ashby Working Party."

64. Annie C. Y. Chang et al., "Studies of Mouse Mitochondrial DNA in *Escherichia coli*: Structure and Function of the Eucaryotic-Procaryotic Chimeric Plasmids," *Cell* 6 (1975): 231–44.

65. Kevin Struhl, John R. Cameron, and Ronald W. Davis, "Functional Genetic Expression of Eukaryotic DNA in *Escherichia coli*," *Proceedings of the National Academy of Sciences of the U.S.A.* 73 (1976): 1471–75.

66. Joshua Lederberg, personal communication, 30 October 1982.

67. Cetus Corporation, "Special Report."

68. Ronald Cape, transcript of interview with Charles Weiner, 19 April 1978, 24, RDHC; Genentech, Inc., prospectus, 14 October 1982, 4; Nicholas Wade, "Guidelines Extended but EPA Balks," *Science* 194 (15 October 1976): 304.

69. Nicholas Wade, "Gene Splicing Company Wows Wall Street," *Science* 210 (3 October 1980): 506–7; interview with Ronald Cape, 19 May 1982.

70. David Fishlock, "Controlling the Hazards of Genetic Engineering," *Financial Times*, 5 October 1974.

71. Imperial Chemical Industries, memo, 2 July 1976, Imperial Chemical Industries press files, Runcorn, United Kingdom.

72. Roger Lewin, "Genetic Engineers Ready for Stage Two," *New Scientist,* 14 October 1976, 86–87.

73. Roy J. Britten and Eric H. Davidson, "Gene Regulation for Higher Cells: A Theory," *Science* 165 (25 July 1969): 349–57.

74. J. Sambrook, "Adenovirus Amazes at Cold Spring Harbor," *Nature* 268 (14 July 1977): 101–4. Graham Chedd gives an accessible and entertaining account of this discovery in "Genetic Gibberish in the Code of Life," *Science 81,* November 1981, 50–55. It was later discovered that some introns coded for proteins.

75. On editing, see John Abelson, "RNA Processing and the Intervening Sequence Problem," *Annual Review of Biochemistry* 48 (1979): 1035–69; on complexity, see Gunther Stent, "DNA," *Daedalus* 99 (1970): 926.

76. Interview with Robert Helling, 24 September 1982.

77. Keiichi Itakura et al., "Expression in *Escherichia coli* of a Chemically Synthesized Gene for the Hormone Somatostatin," *Science* 198 (9 December 1977): 1056–63.

78. Genex Corporation, prospectus, 29 September 1982.

79. Weissmann, "Cloning of Interferon," 103.

80. "A Commercial Debut for DNA Technology," *Business Week,* 12 December 1977, 128–32; "Genetic Engineering: Taking the Plunge," *Economist,* 16 July 1977, 91.

81. U.S. Congress, Office of Technology Assessment, *Commercial Biotechnology: An International Analysis* (Washington, D.C.: U.S. Government Printing Office, 1984), 276; David Dickson, "Genentech Makes a Splash on Wall Street," *Nature* 287 (23 October 1980): 669–70.

82. "Where Genetic Engineering Will Change Industry," *Business Week,* 22 October 1979, 160.

83. Biogen, N.V., prospectus, 22 March 1983, 14.

84. Office of Technology Assessment, *Commercial Biotechnology,* 92; Lois S. Peters, "Industries and Universities in the Context of Biotechnology: A Working Paper," paper presented at a conference sponsored by the British Science Museum and the European Association for the Study of Science and Technology, 30 March–1 April 1984, 18 and table 2. The latter study concludes that "most key scientists of new biotechnology firms were associated with universities and that only a small number were drawn from corporations."

85. Genentech, prospectus, 22.

86. Quoted in Robert Cooke, "Clone Business: It's Growing Fast, It's Growing Fast," *Boston Globe,* 25 June 1978, 1, 10–11; "Commercial Debut for DNA Technology"; Lewin, "Modern Biology at the Industrial Threshold," 18.

87. Cooke, "Clone Business," 10; "Industry Starts to Do Biology with Its Eyes Open," *Economist,* 2 December 1978, 95–102.

88. "Industry Starts to Do Business with Its Eyes Open"; Cooke, "Clone Business"; Jeremy Campbell, "All Aboard the Gene Machine," *Evening Standard* (London), 14 December 1977, 17.

89. *Diamond v. Chakrabarty, United States Reports* 447 (1980): 303.

90. For reviews of these developments, see Roger Lewin, "Do Jumping Genes Make Evolutionary Leaps?" *Science* 213 (7 August 1981), 634–36; L. E. Orgel and F. H. C. Crick, "Selfish DNA: The Ultimate Parasite," *Nature* 284 (7 April 1980): 604–7; Miranda Robertson, "Gene Families, Hopeful Monsters, and the Selfish Genetics of DNA," *Nature* 293 (1 October 1981): 333–34; Roger Lewin, "Tale of the Orphaned Genes," *Science* 212 (1 May 1981): 530; Stanley N. Cohen and James Shapiro, "Transposable Genetic Elements," *Scientific American*, February 1980, 40–49; B. D. Hall, "Mitochondria Spring Surprises," *Nature* 282 (8 November 1979): 129–30; Evelyn Fox Keller, "McClintock's Maize," *Science* *81*, October 1981, 55–58; Donald Brown, "Gene Expression in Eukaryotes," *Science* 211 (13 February 1981): 667–74; James Darnell, "Variety in the Level of Gene Control in Eukaryotic Cells," *Nature* 297 (3 June 1982), 365–70.

91. However much the new findings may have challenged accepted thinking about DNA, some of these discoveries had a direct and constructive effect on related areas of biology and medicine. The new data on gene structure generated new approaches to the diversity of antibodies and to the molecular basis of cancer. For reviews of these developments, see Jean L. Marx, "Antibodies," *Science* 212 (29 May 1981): 1015–17; Philip Leder, "The Genetics of Antibody Diversity," *Scientific American*, May 1982, 102–15; Miranda Robertson, "Clues to the Genetic Basis of Cancer," *New Scientist*, 9 June 1983, 688–91; Robert A. Weinberg, "A Molecular Basis of Cancer," *Scientific American*, November 1983, 126–42. For an analysis, see Joan Fujimura, "The Molecular Bandwagon in Cancer Research," *Social Problems* 35 (1988): 261–83.

92. Interview with David Jackson, 15 July 1982; Walter Gilbert, "Useful Proteins from Recombinant Bacteria," *Scientific American*, April 1980, 74–94; Thomas M. Roberts and Gail D. Lauer, "Maximizing Gene Expression on a Plasmid Using Recombination *in Vitro*," *Methods in Enzymology* 68 (1979): 473–82.

93. Interview with David Jackson, 15 July 1982.

94. "Biotechnology Becomes a Gold Rush," *Economist*, 13 June 1981, 81–86; Craig R. Johnson, "Genetic Engineering: Why the Microbe Makers Got Stung," *New York Times*, 14 March 1982, sec. 3, p. 2.

95. J. Leslie Glick, "Impact of Recombinant DNA Technology on the Economy," in *Biomedical Research Institutions, Biomedical Funding, and Public Policy*, ed. H. H. Fudenberg (New York: Plenum, 1982), 640–73.

96. Ann Crittenden, "The Gene Machine Hits the Farm," *New York Times*, 6 June 1981, sec. 3, pp. 1, 20.

97. E. F. Hutton, *Biotechnology*, no. 2 (March 1980): 8.

98. Paul Jacobs, "'Glamour Stock' Could Help Cancer Patients," *Los Angeles Times*, 21 January 1980, sec. 1, pp. 3, 16.

99. E. F. Hutton, *Biotechnology*, 8.

100. Marjorie Sun, "Insulin Wars: New Advances May Throw Market into Turbulence," *Science* 210 (12 December 1980): 1225–28; Charles J. Elia, "Battle for the Synthetic Human Insulin Market between Lilly and Danish Firm Is Intensifying," *Wall Street Journal*, 7 January 1981, 47.

101. Katherine Bouton, "Academic Researchers and Big Business: A Delicate Balance," *New York Times Magazine*, 11 September 1983, 63.

102. For critiques of the information conveyed in this new conference, see

Spyros Andreopoulos, "Gene Cloning by Press Conference," *New England Journal of Medicine* 302 (27 March 1980): 743–46; Ray Goodell, "The Gene Craze," *Columbia Journalism Review* 19 (November–December 1980): 41–45; Nicholas Wade, "Cloning Gold Rush Turns Basic Biology into Big Business," *Science* 208 (16 May 1980): 668–93. Charles Weissmann, in "The Cloning of Interferon and Other Mistakes," asserted that Biogen's claims about the significance of the experiment were properly qualified by Walter Gilbert, but these qualifications do not seem to have registered with members of the press who covered this announcement. See, e.g., Jacobs, "'Glamour Stock' Could Help Cancer Patients," 3, 16.

103. Office of Technology Assessment, *Commercial Biotechnology*, 100–101.

104. "Where Genetic Engineering Will Change Industry," 160–72.

105. The numbers are approximate. They were obtained from a list of biotechnology companies given in Office of Technology Assessment, *Commercial Biotechnology*, 67–70. Those companies not engaged in recombinant DNA technology were eliminated from the list, using the annual "Guide to Biotechnology Companies" published in *Genetic Engineering News*, November–December 1982, 6–19.

106. J. Leslie Glick, "The Biotechnology Industry," paper for the American Law Institute and the American Bar Association, Committee on Continuing Professional Education, October 1981, J. Leslie Glick personal papers.

107. Office of Technology Assessment, *Commercial Biotechnology*, 92.

108. Ibid., 104–6.

109. Ibid., 103; Cetus Corporation, prospectus, 4, 10, 11; Genentech, Inc., prospectus, 14 October 1980, 10, 21; Genex Corporation, prospectus, 14. I thank my student Margaret Coe for pointing out the high degree of dependence of biotechnology firms on research and development contracts from multinational corporations.

110. Cf. Office of Technology Assessment, *Commercial Biotechnology*, 107–8. Because of the long-term disadvantages of research and development contracts, biotechnology firms, as they matured, attempted to find other types of support that promised higher profits and more effective control over the technology. For example, successful firms increasingly preferred to enter into joint research and development ventures in which equity capital was provided by both partners. See, e.g., Genentech, *1983 Annual Report*, 25.

111. Derek Bok, "Business and the Academy," *Harvard Magazine*, May–June 1981, 25.

112. Zsolt Harsanyi, "Biotechnology: Its Impact on Genetic Sciences: The View from the Investment Community," paper presented at the annual meeting of the American Association for the Advancement of Science, New York, 17 May 1984, 2.

113. See, e.g., Barbara Culliton, "The Hoechst Department at Mass General," *Science* 216 (11 June 1982): 1200–1203; David E. Sanger, "Business Rents a Lab Coat and Academia Hopes for the Best," *New York Times*, 21 February 1982, sec. 5, p. 7.

114. Nicholas Wade, "Recombinant DNA: Warming Up for Big Pay-Off," *Science* 206 (9 November 1979): 663–65; Brian Hartley, "The Biology Business: The Bandwagon Begins to Roll," *Nature* 283 (10 January 1980): 122;

Wade, "Cloning Gold Rush"; Robert Reinhold, "There's Gold in Them Thar Recombinant Genetic Bits," *New York Times*, 22 June 1980, sec. 4, p. 8.

115. Quoted in Dickson, "Genentech Makes Splash on Wall Street," 669–70.

116. Wade, "Gene Splicing Company Wows Wall Street," 505–7.

117. Thomas J. Lueck, "Cetus Charting a Broad Course: Plans to Use Biotechnology in Many Fields," *New York Times*, 5 June 1981, 25.

118. Office of Technology Assessment, *Commercial Biotechnology*, 281; "GENdex: A Biotechnology Stock Monitoring Tool," *Genetic Engineering News*, January–February 1984, 34.

119. Glaucon, "Biology Loses Her Virginity," *New Scientist*, 18/25 December 1980, 826.

120. "Biotechnology Becomes a Gold Rush."

121. Ibid.; Office of Technology Assessment, *Commercial Biotechnology*, 270, 283.

122. See, e.g., "Biotechnology Becomes a Gold Rush"; Glaucon, "Biology Loses Her Virginity," 826.

123. Lawrence K. Altman, "U.S. Unit Backs Human Insulin for the Market: Gene Splicing Drug Gets First F.D.A. Approval," *New York Times*, 30 October 1982, 1, 16.

124. Office of Technology Assessment, *Commercial Biotechnology*, 98, 102.

125. John Walsh, "Biotechnology Boom Reaches Agriculture," *Science* 213 (18 September 1981): 1339–41; William Marbach et al., "The Bust in Biotechnology," *Newsweek*, 26 July 1982, 73.

126. "A New Weapon against Japan: R&D Partnerships," *Business Week*, 8 August 1983, 42; Judith Johnson, "Biotechnology: Commercialization of Academic Research," Issue Brief no. IB81160, Congressional Research Service, Library of Congress, July 1982, 3; Office of Technology Assessment, *Commercial Biotechnology*, 278–79, 292–94; Hal Lancaster, "R&D Tax Shelters Catching Fire, but Potential Abuses Cause Concern," *Wall Street Journal*, 14 September 1981, sec. 2, p. 33.

127. Details of these arrangements are given in Office of Technology Assessment, *Commercial Biotechnology*, 416–21.

128. David E. Sanger, "Harvard Shuns the Apple but Doesn't Step on the Serpent," *New York Times*, 23 November 1980, sec. 5, p. 7; Nicholas Wade, "Gene Gold Rush Splits Harvard, Worries Brokers," *Science* 210 (21 November 1980): 878–79.

129. "A $10 Million Genetic Project," *San Francisco Chronicle*, 15 September 1981, 54; David Dickson, "Backlash against DNA Ventures?" *Nature* 288 (20 November 1980): 203.

130. Robert Walgate, "Brussels Asked to Spend £16M on Biotechnology," *Nature* 283 (10 January 1980): 125–26; Robert Walgate, "United Kingdom: Biotechnology Report Urges £10 Million Programme to Match Competitors'," *Nature* 283 (24 January 1980): 324–325; Stephanie Yanchinski, "France Entices Its Biotechnologists into Industry," *New Scientist*, 25 March 1982, 767; Peter Newmark, "France: Biotechnology Company Planned," *Nature* 285 (15 May 1980): 124; "UK Plant Biotechnology: ARC Joins In," *Nature* 298 (29 July 1982): 412.

131. Winston Williams, "Midwest Steps Up Effort to Attract Industry," *New*

York Times, 16 March 1982, sec. 1, p. 1, sec 4, p. 6; Kim McDonald, "North Carolina Governor Urges States to Support Scientific Research at Academic Institutions," *Chronicle of Higher Education,* 13 January 1982, 9.

132. Margaret A. Coe, "Multinational Corporations and Genetic Engineering: The Midwives of an Industry," paper for course on the history of genetic engineering, Residential College, University of Michigan, 1983.

133. Precise data for total corporation investments in genetic engineering have not been collected, but it is clear simply from the sum of documented investments in the original biotechnology firms, university research, and in-house research and development that the amount is comparable to the total NIH support for genetic engineering in 1977–82, $686 million.

134. Perceptively described by Barbara Culliton, "Biomedical Research Enters the Marketplace," *New England Journal of Medicine* 304 (14 May 1981): 1195–1201; and by Charles Weiner, "Relations of Science, Government, and Industry," in *Science, Technology, and the Issues of the Eighties: Policy Outlook,* ed. Albert H. Teich and Ray Thornton (Boulder, Colo.: Westview, 1982), 71–97.

135. Anne Keatley, "Knowledge as Real Estate," *Science* 222 (18 November 1983): 717; Johnson, "Biotechnology," 3.

136. Paul Berg, quoted in Sharon Begley, "The DNA Industry," *Newsweek,* 20 August 1979, 53; DeWitt Stetten, "The DNA Disease" (letter), *Nature* 297 (27 May 1982): 260.

137. Nicholas Wade, "University and Drug Firm Battle over Billion-Dollar Gene," *Science* 209 (26 September 1980): 1492–94; Hal Lancaster, "Rights to Life: Profits in Gene Splicing Bring the Tangled Issue of Ownership to Fore," *Wall Street Journal,* 3 December 1980, 1, 15; Culliton, "Biomedical Research Enters the Marketplace," 1200.

138. Examples of university guidelines are given in Office of Technology Assessment, *Commercial Biotechnology,* 574–85. David Dickson, in *The New Politics of Science,* argues that one of the main conferences held to discuss the social and ethical issues associated with the growing corporate presence in the universities was not intended to foster open public discussion of the issues (since the meeting was closed) but rather "to act as a forum for moving toward a consensus on both goals and strategy . . . for the university and corporate leadership" (57–58).

139. Andreopoulos, "Gene Cloning by Press Conference," 744.

140. Bronson, "Bacteria Induced to Produce Insulin," 17.

141. Jeff Wheelwright, "Boom in the Biobusiness," *Life,* May 1980, 57–58.

142. Zsolt Harsanyi, personal communication, 26 September 1984.

143. Donald Kennedy, "Health Research: Can Utility and Quality Co-exist?" speech given at the University of Pennsylvania, December 1980.

144. Charles Weiner, "Universities, Professors, and Patents," *Technology Review,* February–March 1986, 33–43; Sheldon Krimsky, "The Corporate Capture of Academic Science and Its Social Costs," paper presented at Genetics and the Law: The Third National Symposium, conference sponsored by the American Society of Law and Medicine and the Boston Schools of Medicine, Law, and Public Health, 1 April 1984; Martin Kenny, *Biotechnology* (New Haven: Yale University Press, 1986).

145. Harsanyi, "Biotechnology," 2.

3. The Emergence and Definition of the Genetic Engineering Issue, 1972–1975

1. Robert L. Sinsheimer, "Recombinant DNA: On Our Own," *BioScience* 26, no. 10 (1976), 599.

2. Martin Pollock, transcript of BBC program "Scientifically Speaking," 26 September 1974, British Broadcasting Corporation Script Library, London.

3. Frederick Neidhardt, quoted in University of Michigan news release, 25 February 1976, University of Michigan News and Information Services Collection, Bentley Historical Library, University of Michigan.

4. Department of Health, Education, and Welfare, *Report of the President's Biomedical Research Panel,* 23.

5. Sydney Brenner, statement on BBC-2 television program, Horizon series, "Genetic Roulette," 1 April 1977, British Broadcasting Corporation Script Library, London.

6. Colin Leys, *Politics in Britain: From Labourism to Thatcherism* (London: Verso, 1989), 140–48.

7. John Hall, "The Union Boss," *New Scientist,* 26 October 1978, 266–67.

8. Association of Scientific, Technical, and Managerial Staffs, "Guide to Health and Safety at Work" (ca. 1977), 2.

9. ASTMS documents from this period emphasize this position: in a briefing note for an address by Clive Jenkins in 1978, Robert Williamson wrote that "ASTMS supports the development of a strong (and safe) science-based industry" (briefing note for Clive Jenkins's talk to ASTMS conference on recombinant DNA, 27 October 1978, MRCUW, MSS.79/AS/3/8/55-58).

10. For details, see Susan Wright, "Evolution of Biological Warfare Policy, 1945–1990," in *Preventing a Biological Arms Race,* ed. Susan Wright (Cambridge, Mass.: MIT Press, 1990), 26–68.

11. House Committee on Appropriations, *Hearings: Department of Defense Appropriations for 1963,* 87th Cong., 2d sess., 10 January–13 October 1963, 177; House Committee on Appropriations, *Hearings: Department of Defense Appropriations for 1970,* 91st Cong., 1st sess., 3 January–23 December 1970, 129.

12. Proceedings of the Conference on Chemical and Biological Warfare, sponsored by the American Academy of Arts and Sciences and the Salk Institute, 25 July 1969, in House Commitee on Foreign Affairs, *Hearings on Chemical-Biological Warfare: U.S. Policies and International Effects,* 91st Cong., 1st sess., 18 November–19 December 1969, 492–96.

13. Ibid., 87.

14. Joshua Lederberg, remarks at an informal meeting of the Conference of the Committee on Disarmament, 5 August 1970.

15. Interview with Matthew Meselson, 13 November 1987.

16. Spencer Weart, "Scientists with a Secret," *Physics Today,* February 1976, 23–30. Weart shows that, in 1939, some physicists, like Leo Szilard, believed that the discovery of fission would lead to the construction of weapons many times more powerful than existing bombs, whereas others, like Enrico Fermi and Niels Bohr, expressed doubts about this possibility. See also Daniel Kevles, *The Phy-*

sicists: The History of a Scientific Community in Modern America (New York: Vintage, 1979), 324.

17. Proceedings of the Conference on Chemical and Biological Warfare, 494.

18. Wright, "Evolution of Biological Warfare Policy," 37–43.

19. For discussion of the weaknesses of the Biological Weapons Convention, see Susan Wright and Robert L. Sinsheimer, "Recombinant DNA and Biological Warfare," *Bulletin of the Atomic Scientists* 39 (November 1983): 20-26; Richard Falk, "Inhibiting Reliance on Biological Weaponry: The Role and Relevance of International Law," in *Preventing a Biological Arms Race,* ed. Susan Wright (Cambridge, Mass.: MIT Press, 1990), 241–66.

20. Maddox, "On Which Side Are the Angels?"

21. Eric Ashby, "Science and Antiscience," *Sociological Review Monograph 18: The Sociology of Science,* ed. Paul Halmos (Keele, United Kingdom: University of Keele, 1972).

22. Steven Cotgrove, "Anti-science," *New Scientist,* 12 July 1973, 82–84.

23. Wolstenholme and O'Connor, *Civilization and Science,* 73, 113, 121.

24. Lily Kay argues that this was no accident, that the Rockefeller Foundation in particular fostered in the 1930s and 1940s a vision of rational social control that attracted leading researchers in genetics and biochemistry (*Molecular Vision of Life,* ch. 1).

25. Hermann Muller, "Means and Aims in Human Genetic Betterment," in *The Control of Human Heredity and Evolution,* ed. T. M. Sonneborn (New York: Macmillan, 1965), 104–10.

26. Rollin D. Hotchkiss, "Portents for a Genetic Engineering," *Journal of Heredity* 56 (1965): 199.

27. Robert L. Sinsheimer, "The End of the Beginning," *Engineering and Science* 33 (December 1966): 10.

28. From the discussion in "Evaluation of Applications to Man," in *The Control of Human Heredity and Evolution,* ed. T. M. Sonneborn (New York: Macmillan, 1965), 123. See also Salvador Luria, "Directed Genetic Change: Perspectives from Molecular Genetics," in the same volume, 3–4.

29. Hotchkiss, "Portents for Genetic Engineering," 201.

30. James Shapiro, Larry Eron, and Jonathan Beckwith, "More Alarms and Excursions," *Nature* 224 (27 December 1969): 1337.

31. Salvador Luria, "Modern Biology: A Terrifying Power," *Nation,* 20 October 1969, 406–9.

32. James D. Watson, "Moving toward the Clonal Man," *Atlantic,* May 1971, 53.

33. Leon Kass, "The New Biology: What Price Relieving Man's Estate?" *Science* 174 (19 November 1971): 785.

34. Philip Handler, "Science and the Federal Government: Which Way to Go?" *Federation Proceedings* 29 (May–June 1970): 10902.

35. Philip Abelson, "Anxiety about Genetic Engineering," *Science* 173 (23 July 1971): 225.

36. Bernard Davis, "Prospects for Genetic Intervention in Man: Control of Polygenic Traits Is Much Less Likely Than Cure of Monogenic Diseases," *Science* 170 (18 December 1970): 1279–83.

37. Joshua Lederberg, "Egg Transplants: Not the End of the World," *Washington Post*, 20 June 1970, B2.

38. Nicholas Wade, *The Ultimate Experiment: Man-Made Evolution* (New York: Walker and Company, 1977), 29–33; John Lear, *Recombinant DNA: The Untold Story* (New York: Crown, 1978), 32–36, 54–59; Sheldon Krimsky, *Genetic Alchemy* (Cambridge, Mass.: MIT Press, 1982), 39–69.

39. See Krimsky, *Genetic Alchemy*, 59, for data on support for the viral oncology program in current dollars.

40. Alfred Hellman, Michael N. Oxman, and Robert Pollack, eds., *Biohazards in Biological Research* (Cold Spring Harbor, N.Y.: Cold Spring Harbor Laboratory, 1973), 353. See also Wade, *Ultimate Experiment*, 29–30, who quotes Emmett Barkley, head of biological safety at the NIH: "Most people working with tumor viruses have been exposed to some extent."

41. James A. Rose to scientific director, NIAID (John R. Seal), 13 December 1972, RDHC.

42. Wade, *Ultimate Experiment*, 30. For further details of concerns about the SV40-adeno hybrids, see Krimsky, *Genetic Alchemy*, 39–57.

43. James A. Rose to scientific director, NIAID (John R. Seal), 13 December 1972, RDHC.

44. United Kingdom, *Report of the Committee of Inquiry into the Smallpox Outbreak in London in March and April 1973* (the Cox report), Cmnd. 5626 (June 1974).

45. Department of Health and Social Security, *Report of the Working Party on the Laboratory Use of Dangerous Pathogens* (the Godber report), Cmnd. 6054 (May 1975).

46. In other words, these "nondefective" viruses are able to grow in cells in tissue culture on their own, as opposed to "defective" viruses, which require that the cells be simultaneously infected with a nondefective virus.

47. Quoted in Krimsky, *Genetic Alchemy*, 42.

48. For details, see ibid., 47–57.

49. Andrew Lewis to the Committee on Recombinant DNA, National Academy of Sciences, 29 November 1974, quoted in Wade, *Ultimate Experiment*, 32.

50. "A Two-Edged Sword," *Nature* 240 (10 November 1972): 73–74.

51. For details of the first reactions to Berg's proposal to insert DNA from the virus SV40 into *E. coli K12*, see Lear, *Recombinant DNA*, 21–38; Krimsky, *Genetic Alchemy*, 24–38.

52. Maxine Singer and Dieter Soll, "Guidelines for DNA Hybrid Molecules," *Science* 181 (21 September 1973): 1114.

53. Brenner, "Evidence for the Ashby Working Party."

54. Paul Berg to Irving P. Crawford, 9 December 1974, RDHC.

55. Bernard Davis to Paul Berg, 5 September 1974, RDHC.

56. For discussion of scientific speculation concerning the latter scenario, see Krimsky, *Genetic Alchemy*, 119–20.

57. H. Williams Smith, "Survival of Orally Administered *E. coli* K12 in Alimentary Tract of Man," *Nature* 255 (5 June 1975): 500–502; E. S. Anderson, "Viability of, and Transfer of a Plasmid from, *E. coli* K12 in the Human Intestine," *Nature* 255 (5 June 1975): 502–4.

58. Roy Curtiss to Paul Berg et al., memo on potential biohazards of recombinant DNA molecules, 6 August 1974, RDHC.

59. Maxine Singer to Hans Kornberg, 6 June 1974, RDHC.

60. Brenner, "Evidence for the Ashby Working Party."

61. Bernard Dixon, "Biological Research (1)," *New Scientist,* 25 October 1973, 236.

62. Brenner, "Evidence for the Ashby Working Party."

63. Robert L. Sinsheimer, "Recombinant DNA Research (Genetic Engineering): Social Implications and Potential Hazards," UCLA Distinguished Lecture Series, 1976–77, 22 November 1976, Robert L. Sinsheimer personal papers.

64. Joshua Lederberg to Robert Stone, 23 August 1974, RDHC.

65. Joshua Lederberg to Martin Kaplan, 23 September 1974, RDHC.

66. Stanley Cohen to Paul Berg, 27 August 1974, RDHC.

67. Bernard Davis to Paul Berg, 5 September 1974, RDHC; see also Waclaw Szybalski to William Reznikoff, 25 October 1974, RDHC.

68. Several articles that appeared at this time treated the possible hazards of genetic engineering as a subcategory of a larger issue of laboratory hazards posed by biological techniques: see, e.g., Nicholas Wade, "Microbiology: Hazardous Profession Faces New Uncertainties," *Science* 182 (9 November 1973): 566–67; Edward Ziff, "Benefits and Hazards of Manipulating DNA," *New Scientist* (25 October 1973): 274–75; Robin Weiss, "Virological Hazards in Routine Procedures," *Nature* 255 (5 June 1975): 445–47.

69. Lear, *Recombinant DNA,* 71.

70. Ibid., 76.

71. Paul Berg, interview with Rae Goodell, 17 May 1975, 33, 48, RDHC.

72. For details, see Lear, *Recombinant DNA,* 76–82; Krimsky, *Genetic Alchemy,* 81–96.

73. Quoted in Krimsky, *Genetic Alchemy,* 82–83.

74. Drafts of Berg committee's letter, RDHC.

75. Paul Berg et al., "Potential Biohazards of Recombinant DNA Molecules," *Science* 185 (26 July 1974): 303; Paul Berg et al., "Potential Biohazards of Recombinant DNA Molecules," *Proceedings of the National Academy of Sciences of the U.S.A.* 71 (July 1974): 2593–94; "NAS Ban on Plasmid Engineering," *Nature* 250 (19 July 1974): 175.

76. Lear, *Recombinant DNA,* 87–88; interview with Andrew Lewis, 24 September 1980.

77. David Baltimore, transcript, National Academy of Sciences press conference, 18 July 1974, RDHC.

78. See, e.g., Victor McElheny, "Genetic Tests Renounced over Possible Hazards," *New York Times,* 18 July 1974, 1, 46; Stuart Auerbach, "Halt in Genetic Work Urged," *Washington Post,* 18 July 1974, A1, A8.

79. These recommendations were noted by the *New York Times* but received little attention in other accounts.

80. Department of Health, Education, and Welfare, National Institutes of Health, "Advisory Committee: Establishment of Committee," *Federal Register* 39 (6 November 1974), 39306.

81. David Bennett, Peter Glasner, and David Travis, *The Politics of Uncertainty: Regulating Recombinant DNA Research in Britain* (London: Routledge and Kegan Paul, 1986), 17; see also Hans Kornberg to Maxine Singer, 6 May 1974, RDHC; Maxine Singer to Hans Kornberg, 6 June 1974, RDHC; Hans Kornberg to Maxine Singer, 6 July 1974, RDHC; Hans Kornberg to Paul Berg, 6 July 1974, RDHC.

82. Interview with Hans Kornberg, 5 June 1979; Hans Kornberg to Maxine Singer, 6 July 1974, RDHC; Hans Kornberg to Paul Berg, 6 July 1974, RDHC.

83. Interview with Hans Kornberg, 5 June 1979.

84. Hans Kornberg to Maxine Singer, 6 July 1974, RDHC; Hans Kornberg to Paul Berg, 6 July 1974, RDHC.

85. Interview with Sydney Brenner, 11 April 1980.

86. A meeting at the MRC held in June 1974 between Tony Vickers, an MRC official responsible for research in virology, and approximately twelve scientists funded by the MRC, including Sydney Brenner and John Subak-Sharpe, director of the MRC Virology Unit at the University of Glasgow, may have been one point at which interest in separating genetic engineering from dangerous pathogens was conveyed to MRC officials.

87. Interview with Eric Ashby, 10 January 1977.

88. Interview with Sydney Brenner, 11 April 1980.

89. For example, three pieces of testimony submitted to the Ashby committee and letters in *Nature* on the proposed moratorium.

90. See, e.g., Michael Stoker, "Molecular Dirty Tricks Ban," E. S. Anderson, "The Indiscriminate Use of Antibiotics Has Exerted More Pressure on the Bacterial Population Than Could Be Wielded by All Research Workers in the Field Put Together," and Ken Murray, "Alternative Experiments?" all in *Nature* 250 (26 July 1974): 278–80.

91. Brenner, "Evidence for the Ashby Working Party."

92. BBC television program, "Controversy," autumn 1974, draft transcript by Audio Overload, British Broadcasting Corporation Script Library, London.

93. Sydney Brenner to Paul Berg, 24 September 1974, RDHC.

94. Eric Ashby to Paul Berg, 18 September 1974, RDHC.

95. Paul Berg to Eric Ashby, 12 September 1974, RDHC.

96. Interview with Eric Ashby, 10 January 1977.

97. United Kingdom, *Report of the Working Party on the Experimental Manipulation of the Genetic Composition of Micro-organisms,* Cmnd. 5880 (January 1975).

98. Eric Ashby to Paul Berg, 5 November 1974, RDHC.

99. The Asilomar conference has been the subject of a number of carefully crafted eyewitness accounts as well as academic analyses. News reports include Stuart Auerbach, "And Man Created Risks," *Washington Post,* 9 March 1975, B1, B4; Horace F. Judson, "Fearful of Science: Who Shall Watch the Scientists?" *Harper's,* June 1975, 70–76; Colin Norman, "Berg Conference Favors Use of Weak Strains," *Nature* 254 (6 March 1975): 6–7; Michael Rogers, "The Pandora's Box Congress," *Rolling Stone,* 19 June 1975; Nicholas Wade, "Genetics: Conference Sets Strict Controls to Replace Moratorium," *Science* 187 (4 March 1975): 931–35; Janet Weinberg, "Asilomar Decision: Unprecedented Guidelines for Gene-Transplant Research," *Science News,* 8 March 1975, 148–49, 156; "Decision at

Asilomar," *Science News,* 22 March 1975, 194–96. The Asilomar proceedings have also been treated in several books, including Krimsky, *Genetic Alchemy,* Lear, *Recombinant DNA,* Michael Rogers, *Biohazard* (New York: Knopf, 1977), and Wade, *The Ultimate Experiment.* The account of the conference given in this chapter draws on these materials as well as on the documents and other records, particularly tapes, generated by the meeting.

100. The conference was supported by the NIH and the National Science Foundation, but these government agencies had no responsibility for its organization.

101. Paul Berg to E. S. Anderson, 13 October 1974, RDHC.

102. Paul Berg, interview with Rae Goodell, 17 May 1975, 77–78, RDHC.

103. Baron was suggested by Hermann Lewis, head of the Cellular Biology Section of the National Science Foundation, and was an associate of Maxine Singer (Maxine Singer to Paul Berg, 26 November 1974, RDHC).

104. Lear, *Recombinant DNA,* 115–22; Krimsky, *Genetic Alchemy,* 105.

105. Asilomar Conference Program, 24–27 February 1975, reprinted in Watson and Tooze, *The DNA Story,* 41–42.

106. References to drafts and the final version of the Ashby report occur in the following correspondence: Sydney Brenner to Paul Berg, 20 November 1974, RDHC; Paul Berg to Eric Ashby, 9 December 1974, RDHC; E. S. Anderson to Stanley Cohen, 24 January 1975, RDHC.

107. Robert L. Sinsheimer, "Troubled Dawn for Genetic Engineering," *New Scientist,* 16 October 1975, 150.

108. Judson, "Fearful of Science," 70, 75; Wade, "Genetics: Conference Sets Strict Controls to Replace Moratorium," 931–34.

109. David Baltimore, opening remarks at Asilomar conference, on tape of the conference proceedings, RDHC.

110. Sinsheimer, "Troubled Dawn for Genetic Engineering," 150.

111. J. Michael Bishop et al., "Preparation and Use of Recombinant Molecules Involving Animal Virus Genomes," n.d., RDHC; Royston C. Clowes et al., "Proposed Guidelines on Potential Biohazards Associated with Experiments Involving Genetically Altered Microorganisms," 24 February 1975, RDHC; Sydney Brenner et al., "Cloned Eukaryotic DNA" (ca. 1975), RDHC. The account given here addresses characteristics of the reports related to the policy process. For further details, see Krimsky, *Genetic Alchemy,* ch. 9.

112. Clowes et al., "Proposed Guidelines," 3.

113. Ibid., 10.

114. Brenner et al., "Cloned Eukaryotic DNA," 5.

115. Bishop et al., "Recombinant Molecules."

116. Ibid., 2.

117. Rogers, "The Pandora's Box Congress."

118. Clowes et al., "Proposed Guidelines."

119. Science for the People, Genetic Engineering Group, "Open Letter to the Asilomar Conference on Hazards of Recombinant DNA" [1975], RDHC.

120. Wade, "Genetics: Conference Sets Strict Controls," 934.

121. Wade, *Ultimate Experiment,* 48.

122. Wade, "Genetics: Conference Sets Strict Controls," 932, 934.

123. Clowes et al., "Proposed Guidelines," appendix C, RDHC.

124. Norman, "Berg Conference Favours Use of Weak Strains," 6–7.

125. Roy Curtiss III to Committee on Recombinant DNA, n.d., cited in Lear, *Recombinant DNA,* 132–33.

126. Rogers, *Biohazard,* 76.

127. Alexander Capron, "Legal Analysis and Response to Recombinant DNA Hazards," 26 February 1975, RDHC.

128. Roger Dworkin, "Legal Liability of Investigators and Institutions in the Event of Proximate or Remote Injury Arising out of Work with Recombinant DNAs," quoted in Rogers, *Biohazard,* 81.

129. Wade, "Genetics: Conference Sets Strict Controls," 935.

130. "Provisional Statement of the Conference Proceedings," 27 February 1975, RDHC.

131. Rogers, *Biohazard,* 84.

132. Ibid., 89–90.

133. Tape of Asilomar conference made by David Perlman, RDHC.

134. Paul Berg et al., "Asilomar Conference on Recombinant DNA Molecules," *Science* 188 (6 June 1975): 994.

135. Ibid., 991.

136. Details of the implementation of such a code by local biohazard committees and by the agencies responsible for funding research were contained in the provisional statement but were dropped from the final statement, presumably because the particular model described applied to the United States but not necessarily to other countries.

137. Judson, "Fearful of Science," 72; Roger Lewin, "Ethics and Genetic Engineering," *New Scientist,* 17 October 1974, 16.

4. Initiating Government Controls in the United States and the United Kingdom, 1975–1976

1. Stanley Cohen, testimony, in Senate Committee on Labor and Public Welfare, Subcommittee on Health, *Hearing: Genetic Engineering, 1975,* 94th Cong., 1st sess., 22 April 1975, 4–5.

2. E.g., Stanley Falkow to William Gartland, 19 September 1975, ORDAR; Nicholas Wade, "Recombinant DNA: NIH Group Stirs Storm by Drafting Laxer Rules," *Science* 190 (21 November 1975): 767–69.

3. DeWitt Stetten to Paul Primakoff, 6 October 1975, ORDAR.

4. Philip Leder to the director, NIH, 10 February 1976, ORDAR.

5. Senate Committee on Labor and Public Welfare, Subcommittee on Health, *Hearing: Genetic Engineering,* 94th Cong., 1st sess., 22 April 1975, 1–35.

6. Edward Kennedy, "Remarks at the Harvard School of Public Health," 9 May 1975, RDHC.

7. Maxine Singer to Willard Gaylin, 30 April 1975, RDHC.

8. See, e.g., DeWitt Stetten to Charles Thomas, 12 November 1975, ORDAR.

9. Maxine Singer to Paul Berg and David Baltimore, 16 September 1974, RDHC.

10. Department of Health, Education, and Welfare, National Institutes of Health, "Advisory Committee: Establishment of Committee," *Federal Register* 39 (6 November 1974): 39306. This announcement reprinted purposes formulated in the original charter established a month earlier (Secretary of Health, Education, and Welfare, Charter, Recombinant DNA Molecule Program Advisory Committee, 7 October 1974, ORDAR).

11. Paul Berg to Robert S. Stone, 10 December 1974, RDHC.

12. Interview with Leon Jacobs, 24 September 1980; Krimsky, *Genetic Alchemy,* 156–59.

13. DeWitt Stetten, transcript of interview with Rae Goodell, 3 August 1976 and 2 September 1976, RDHC, quoted in Krimsky, *Genetic Alchemy,* 156–57.

14. Department of Health, Education, and Welfare, National Institutes of Health, Recombinant DNA Molecule Program Advisory Committee, minutes of meeting (hereafter cited as RAC minutes), 28 February 1975, 2, ORDAR.

15. Interview with Leon Jacobs, 24 September 1980. Jacobs emphasized that this was his own view and did not necessarily represent the NIH position.

16. Jane Setlow to DeWitt Stetten, 4 September 1975, ORDAR.

17. For accounts of these meetings, see Wade, "Recombinant DNA: NIH Group Stirs Storm by Drafting Laxer Rules," *Science* 190 (21 November 1975): 767–69; "Recombinant DNA: NIH Sets Strict Rules," *Science* 190 (19 December 1975): 1175–79.

18. RAC minutes, 28 February 1975.

19. For discussion of the assumptions behind the phylogenetic ordering of recombinant DNA hazards, see Krimsky, *Genetic Alchemy,* 184–86.

20. Wade, *Ultimate Experiment,* 70.

21. Paul Berg to DeWitt Stetten, 2 September 1975, ORDAR.

22. Stanley Falkow to D. Vesley, 17 November 1975, RDHC.

23. Roy Curtiss to DeWitt Stetten, 13 August 1975, ORDAR. Curtiss announced in this letter that he was "very much opposed to the concept of the EK2 biological containment."

24. Rolf Freter to John Seal, 7 August 1975, ORDAR.

25. Sydney Brenner, remarks at the RAC meeting in La Jolla, Calif., in December 1975, quoted in Wade, *Ultimate Experiment,* 95.

26. Stanley Falkow, draft revision to the NIH guidelines, 6 November 1975, ORDAR. Falkow was apparently adapting a passage in a letter from DeWitt Stetten to Paul Berg, dated 9 September 1975: "The hazards of the experiment that has never been done may be guessed at, intuited, speculated about, or voted upon, but they can not be known in the total absence of any actuarial data."

27. DeWitt Stetten to L. J. Heere, 15 September 1975, ORDAR.

28. For details, see Wade, "Recombinant DNA: NIH Group Stirs Storm by Drafting Laxer Rules," 767–69.

29. Stanley Falkow to DeWitt Stetten, 7 August 1975, ORDAR.

30. Roy Curtiss to DeWitt Stetten, 13 August 1975, ORDAR.

31. Richard Goldstein et al. to DeWitt Stetten, 27 August 1975, ORDAR.

32. Paul Berg to DeWitt Stetten, 2 September 1975, ORDAR.

33. Margaret Duncan, Richard Goldstein, Cristian Orrego, and Paul Primakoff to DeWitt Stetten, 24 November 1975, in Department of Health, Education,

and Welfare, National Institutes of Health, *Recombinant DNA Research* (Bethesda, Md.: NIH Office of Recombinant DNA Activities, August 1976), 1:351, hereafter cited as NIH, *Recombinant DNA Research*.

34. Leon Jacobs to Members, NIH Program Advisory Committee on DNA Recombinant Molecules (*sic*), 16 October 1975, ORDAR.

35. In a letter to Donald Brown, David Hogness described the new subcommittee as "consist[ing] entirely of those who thought the guidelines approved last July at Woods Hole were too lax" (17 November 1975, RDHC).

36. DeWitt Stetten to Charles Thomas, 12 November 1975, ORDAR.

37. Stanley Falkow to Elizabeth Kutter, 6 November 1975, RDHC.

38. Maxine Singer to Elizabeth Kutter, 15 October 1975, RDHC.

39. Sydney Brenner, statement at meeting of the Recombinant DNA Molecule Program Advisory Committee, 4–5 December 1975, unofficial transcript of proceedings (hereafter cited as RAC meeting, 4–5 December 1975, unofficial transcript), 120, ORDAR.

40. RAC minutes, 4–5 December 1975, 3.

41. Rogers, *Biohazard,* 159–76; Wade, "Recombinant DNA: NIH Sets Strict Rules," 1175–79; Wade, *Ultimate Experiment,* 87–98; RAC meeting, 4–5 December 1975, unofficial transcript.

42. Rogers, *Biohazard,* 174; RAC meeting, 4–5 December 1975, unofficial transcript, 584–85.

43. Department of Health, Education, and Welfare, National Institutes of Health, "Proposed Guidelines for Research Involving Recombinant DNA Molecules" (January 1976), in NIH, *Recombinant DNA Research* 1:71–133.

44. Wade, *Ultimate Experiment,* 81.

45. Margaret Duncan et al. to DeWitt Stetten, 24 November 1975, 350–71.

46. Paul Primakoff to DeWitt Stetten, 25 September 1975, ORDAR.

47. Margaret Duncan et al. to DeWitt Stetten, 24 November 1975, 351.

48. Stanley Falkow to Paul Primakoff, 8 October 1975, RDHC; Wallace Rowe to Paul Primakoff, 8 October 1975, RDHC; DeWitt Stetten to Paul Primakoff, 6 October 1975, ORDAR.

49. Paul Berg to Donald Fredrickson, 17 February 1976, in NIH, *Recombinant DNA Research* 1:466–70.

50. Paul Berg et al., "Potential Biohazards of Recombinant DNA Molecules," *Science* 185 (26 July 1974): 303.

51. Sherwood L. Gorbach, ed., "Proceedings from a Workshop on Risk Assessment of Recombinant DNA Experimentation with *Escherichia coli* K12," *Journal of Infectious Diseases* 137 (1978): 611–714. For analysis of this meeting, see chapter 6.

52. RAC minutes, 4–5 December 1975, 25; RAC meeting, 4–5 December 1975, unofficial transcript, 686–97, ORDAR. Brenner's proposal for the experiment is recorded on page 686.

53. See, e.g., Department of Health, Education, and Welfare, National Institutes of Health, "Proposed Guidelines for Research Involving Recombinant DNA Molecules," January 1976, in NIH, *Recombinant DNA Research* 1:113–16.

54. Harrison Echols to DeWitt Stetten, 17 September 1975, ORDAR; Harri-

son Echols to Elizabeth Kutter, 31 October 1975, RDHC; Margaret Duncan, Richard Goldstein, et al., to DeWitt Stetten, 24 November 1975, ORDAR.

55. Department of Health, Education, and Welfare, National Institutes of Health, "Proposed Guidelines for Research Involving Recombinant DNA Molecules," January 1976, in NIH, *Recombinant DNA Research* 1:113.

56. RAC meeting, 4–5 December 1975, unofficial transcript, 101–4.

57. For detailed description of the views of Chargaff and Sinsheimer, see Wade, *Ultimate Experiment,* 99–103.

58. Chargaff was also known for his acerbic criticism of the reductionist and applied character of molecular biology, and of its competitive ethos: see Pnina Abir-am, "From Biochemistry to Molecular Biology: DNA and the Acculturated Journey of the Critic of Science Erwin Chargaff," *History and Philosophy of the Life Sciences* 2 (1980): 3–60.

59. Robert Sinsheimer to the Board of Regents, University of Michigan, 12 May 1976, Robert L. Sinsheimer personal papers.

60. Erwin Chargaff, "On the Dangers of Genetic Meddling," *Science* 192 (4 June 1976): 938–40.

61. Sinsheimer, "Troubled Dawn for Genetic Engineering," 150.

62. Chargaff, "On the Dangers of Genetic Meddling"; Robert Sinsheimer to Donald Fredrickson, 5 February 1976, Robert L. Sinsheimer personal papers.

63. Erwin Chargaff, "Profitable Wonders," *Sciences,* August–September 1975, 21.

64. Science for the People, Genetics and Society Group, "Proposals on Research Involving Gene Manipulation" (circulated in 1975), RDHC.

65. "Politics and Genes," *Newsweek,* 12 January 1976, 50–52.

66. DeWitt Stetten, transcript of interview with Rae Goodell, 3 August and 2 September 1976, RHDC, 69, cited in Krimsky, *Genetic Alchemy,* 170.

67. Roy Curtiss to DeWitt Stetten, 22 September 1975, ORDAR.

68. Donald Fredrickson, comments at RAC meeting, 2 April 1976, transcribed from tape of meeting, ORDAR.

69. Department of Health, Education, and Welfare, National Institutes of Health, "Proceedings of a Conference on NIH Guidelines for Research on Recombinant DNA Molecules," public hearings held at a meeting of the director's advisory committee, 9–10 February 1976, in NIH, *Recombinant DNA Research,* comments of Donald Fredrickson, 1:150, 318–19.

70. Donald Fredrickson, comments at RAC meeting, 2 April 1976, transcribed from tape of meeting, ORDAR.

71. For a transcript of the hearing and correspondence associated with it, see NIH, *Recombinant DNA Research* 1:140–551.

72. NIH, *Recombinant DNA Research* 1:322–23.

73. Robert Sinsheimer to Donald Fredrickson, 12 February 1976, reprinted in NIH, *Recombinant DNA Research* 1:443. The probability arguments used by biologists were also criticized in detail by Arthur Schwartz in "Analysis of the Probability Estimates Used in Establishing the Effectiveness of Disarming *E. coli* by Gene Deletion," Department of Mathematics, University of Michigan, 1976.

74. Peter Hutt to Donald Fredrickson, 20 February 1976, in NIH, *Recombinant DNA Research* 1:482–83. A similar argument was made by the RAC's nonscien-

tist member, Emmette Redford (Redford to Donald Fredrickson, 26 March 1976, in NIH, *Recombinant DNA Research* 1:502–6).

75. Richard Andrews to Donald Fredrickson, 16 April 1976 in NIH, *Recombinant DNA Research* 1:525–27.

76. Ibid., 317–18.

77. Department of Health, Education, and Welfare, National Institutes of Health, "Decision of the Director, National Institutes of Health, to Release Guidelines for Research on Recombinant DNA Molecules," 23 June 1976, in *Federal Register* 41 (7 July 1976): 27902–29811.

78. EMBO was formed in 1963 and supported early in its history by the government of Israel and a major grant from the Volkswagen Foundation. After 1974, it was also supported by the governments of many European nations (Peter Newmark, "European Molecular Biology," *Nature* 273 [18 May 1978]: 182–83).

79. Charles Weissmann to DeWitt Stetten, 18 February 1976, ORDAR.

80. Nicholas Wade, "Recombinant DNA: The Last Look before the Leap," *Science* 192 (16 April 1976): 237.

81. RAC minutes, 1–2 April 1976, attachment 2, "Selected Issues for Committee Review."

82. Compare Department of Health, Education, and Welfare, National Institutes of Health, "Proposed Guidelines for Research Involving Recombinant DNA Molecules" (January 1976), in NIH, *Recombinant DNA Research* 1:71–133, and Department of Health, Education, and Welfare, National Institutes of Health, "Guidelines for Research Involving Recombinant DNA Molecules," *Federal Register* 41 (7 July 1976): 27911–43. For an account of the committee's deliberations, see Wade, "Recombinant DNA: The Last Look before the Leap," 236–38. For detailed analysis focusing on the technical content of the guidelines, see Krimsky, *Genetic Alchemy,* ch. 13.

83. Donald Fredrickson, comments at RAC meeting, 2 April 1976, transcribed from tape of meeting, ORDAR.

84. RAC minutes, 1–2 April 1976, 3.

85. Associate Director for Program Planning and Evaluation to Director, NIH, 9 April 1976, in NIH, *Recombinant DNA Research* 1:423.

86. Associate Director for Program Planning and Evaluation to Director, NIH, "Summary of Your Meeting with Private Industry, June 2," 4 June 1976, in NIH, *Recombinant DNA Research* 1:129–30.

87. Cornelius W. Pettinga to Donald S. Fredrickson, 4 June 1976, RDHC.

88. Department of Health, Education, and Welfare, National Institutes of Health, "Recombinant DNA Research: Guidelines," *Federal Register* 41 (7 July 1976): 27911–43.

89. "Forever Amber on Manipulating DNA Molecules?" *Nature* 256 (17 July 1975): 155.

90. Interview with Sydney Brenner, 11 April 1980.

91. Ibid. On this point see also Bennett, Glasner, and Travis, *Politics of Uncertainty,* 28–30.

92. Interview with Sydney Brenner, 11 April 1980.

93. Bennett, Glasner, and Travis, *Politics of Uncertainty,* 36–39.

94. Roger West, "Meeting of A.S.T.M.S. Molecular Biologists in Edinburgh," ca. January 1975, RWP.

95. Williamson, "Briefing Paper for ASTMS," 11 February 1975, MRCUW, MSS.79/AS/3/8/57.

96. Robert Williamson, "Briefing Notes to A.S.T.M.S. re 'Genetic Engineering'," 25 January 1976, RWP.

97. E.g., Association of Scientific, Technical, and Managerial Staffs, "Evidence to the TUC on the [Health and Safety Executive] Consultative Document and the Williams Committee Report on the Practice of Genetic Manipulation," ca. August 1976, RWP.

98. Gavin Drewry, "Legislation," in *The Commons Today,* ed. S. A. Walkland and Michael Ryle (Glasgow: Fontana, 1979), 87–117.

99. For detailed discussions of these characteristics, reflecting a general consensus on the salient differences between the British and American political systems, see, e.g., Stephen T. Early, Jr., and Barbara B. Knight, *Responsible Government: American and British* (Chicago: Nelson-Hall, 1981), chs. 5, 6; Drewry, "Legislation"; Richard Rose, "Still the Era of Party Government," *Parliamentary Affairs* 36 (Summer 1983): 282–99; Leys, *Politics in Britain,* ch. 13.

100. Sydney Brenner, "Evidence for the Ashby Working Party."

101. R. A. Bird (ASTMS National Officer) to All Branches, memo on genetic engineering, ca. January–February 1975, RWP.

102. "Forever Amber on Manipulating DNA Molecules?" 155.

103. Interview with Sir Robert Williams, 10 February 1977.

104. Bennett, Glasner, and Travis, *Politics of Uncertainty,* 43.

105. United Kingdom, *Report of the Working Party on the Practice of Genetic Manipulation,* Cmnd. 6600 (London: HMSO, August 1976), 3. Hereafter, cited as the *Report on Genetic Manipulation.*

106. R. A. Bird to Robert Williamson, 4 December 1975, RWP.

107. United Kingdom, *Parliamentary Debates* (Commons), 902 (1975–76), "Written Answers to Question," 8 December 1975, 43–44.

108. *Report on Genetic Manipulation,* 19. For details of testimony, see Bennett, Glasner, and Travis, *Politics of Uncertainty,* 54–56.

109. Interview with Robert Williamson, 21 April and 7 May 1980.

110. Ibid.

111. "Why Are We Waiting?" *Nature* 262 (22 July 1976): 244.

112. Department of Education and Science, "Controls Recommended for Experiments in Genetic Engineering," press release, 25 August 1976; Health and Safety Commission, "Compulsory Notification of Proposed Experiments in the Genetic Manipulation of Micro-organisms," press release, 25 August 1976, Health and Safety Executive Press Office, London.

113. European Molecular Biology Organization, Standing Advisory Committee on Recombinant DNA, *Report and Recommendations,* second meeting, London, 18–19 September 1976, EMBO files, Heidelberg, Germany.

114. *Report on Genetic Manipulation,* 3.

115. Bennett, Glasner, and Travis, *Politics of Uncertainty,* 66.

116. Second meeting of the EMBO Standing Advisory Committee on Recombinant DNA, *Report and Recommendations,* 2.

117. *Report on Genetic Manipulation,* 5.

118. For press coverage noting similarities between the British and American containment proposals, see Eleanor Lawrence, "Genetic Manipulation: Guidelines Out," *Nature* 263 (2 September 1976): 4–5; "Genetic Engineering Gets British Go-Ahead," *New Scientist,* 2 September 1976, 475.

119. *Report on Genetic Manipulation,* 13–16.

120. Bennett, Glasner, and Travis, *Politics of Uncertainty,* 68.

121. John Subak-Sharpe, A. R. Williamson, and J. Paul to Shirley Williams, 13 October 1976, in Watson and Tooze, *DNA Story,* 320–21.

122. On two occasions in 1976, first in its evidence to the Williams committee and second in its comments on the Health and Safety Executive consultative document submitted in the fall of 1976, the TUC proposed that the advisory body for genetic manipulation should be established within the Health and Safety Executive.

123. Interview with William Henderson, 30 April 1980.

124. Department of Education and Science, "Controls Recommended."

125. Tony Vickers, "Flexible DNA Regulation: The British Model," *Bulletin of the Atomic Scientists* 34 (January 1978): 5.

126. Health and Safety Commission, *Consultative Document: Compulsory Notification of Proposed Experiments in the Genetic Manipulation of Micro-organisms* (London: HMSO, 1976).

127. Recipients of the draft regulations included other relevant government departments, local authority associations, scientific societies, trade unions, organizations representing higher education, and the CBI.

128. Health and Safety Commission, Consultative Document: Compulsory Notification of Proposed Experiments, appendix B.

129. Association of Scientific, Technical, and Managerial Staffs, "Evidence to the TUC on the Consultative Document and the Williams Report on the Practice of Genetic Manipulation," ca. August 1976, MRCUW, MSS.79/AS/3/8/57.

130. S. J. Pirt, letter to the editor, *Times,* 19 October 1976, p. 13, col. 5.

131. Michael Ashburner, "An Open Letter to the Health and Safety Executive," *Nature* 264 (4 November 1976): 2.

132. *New Scientist,* 28 October 1976, 221, and 18 November 1976, 372.

133. Ashburner, "Open Letter to the Health and Safety Executive," 2. See also John Subak-Sharpe, A. R. Williamson, and J. Paul to Shirley Williams, 13 October 1976, in Watson and Tooze, *DNA Story,* 320–21.

134. John Locke, "An Open Reply from the Director of the Executive," *Nature* 264 (4 November 1976): 3.

135. Bennett, Glasner, and Travis, *Politics of Uncertainty,* 23.

136. Department of Education and Science, "Genetic Manipulation Advisory Group Membership Announced," press release, 8 December 1976.

137. Norman Hardyman, testimony before the House of Commons Select Committee on Science and Technology, 7 February 1979, in House of Commons, Select Committee on Science and Technology, *Recombinant DNA Research: Interim Report* (London: HMSO, 1979), 89.

138. Interview with Shirley Williams, 13 June 1978.

139. Interview with Norman Hardyman, 25 April 1980.

140. Gordon Wolstenholme to Susan Wright, 1 June 1977.

141. Gordon Wolstenholme to Donald Fredrickson, 18 October 1976, Office of the NIH Director files.

142. Norman Hardyman, testimony before the House of Commons Select Committee on Science and Technology, 88.

143. Interview with Gordon Wolstenholme, 15 April 1980.

144. The meeting had been scheduled for and was held on December 16, but because of the absence of the trade union members, it was designated as a preliminary meeting, whose decisions would need to be confirmed at the following one.

145. Shirley Williams, testimony before the House of Commons Select Committee on Science and Technology, 7 March 1979, in House of Commons, Select Committee on Science and Technology, *Recombinant DNA Research: Interim Report,* 155.

146. Interview with Gordon Wolstenholme, 15 April 1980.

147. Interview with Shirley Williams, 13 June 1978; interview with Norman Hardyman, 25 April 1980.

148. Interview with Norman Hardyman, 25 April 1980; Norman Hardyman to Susan Wright, 26 August 1992.

149. House of Commons, Select Committee on Science and Technology, *Recombinant DNA Research: Interim Report,* 89, 161.

150. Edward Yoxen, "Regulating the Exploitation of Recombinant Genetics," in *Directing Technology: Policies for Promotion and Control,* ed. Ron Johnston and Philip Gummett (London: Croom Helm, 1979), 231.

151. The question of military use was not addressed either by the RAC or by the Williams committee. Industrial use was raised on the RAC by one of its lay members and was addressed by the Williams committee but was set aside in both cases. The Williams committee acknowledged, however, that industrial use was an issue that policymakers might address at a later stage.

152. See, e.g., Department of Health, Education, and Welfare, National Institutes of Health, "Decision of the Director," 27902–11.

153. Donald Fredrickson, comments at RAC meeting, 27–28 April 1978, transcribed from tape of meeting, ORDAR.

154. Berg et al., "Potential Biohazards of Recombinant DNA Molecules," 303.

155. Department of Health, Education, and Welfare, National Institutes of Health, "Advisory Committee: Establishment of Committee," 39306.

156. Second meeting of the EMBO Standing Advisory Committee on Recombinant DNA, *Report and Recommendations.*

157. Issues associated with the proliferation of recombinant DNA techniques were raised by Sydney Brenner at the December 1975 meeting of the RAC but were not taken up by the committee.

158. Schattschneider, *Semi-sovereign People,* 71.

5. Defusing the Controversy

1. Robert Cooke, "Experimental Biologists Surveyed: Controlled DNA Research Favored," *Boston Globe,* 28 March 1977, 1, 8; Robert Cooke, "DNA

Question: Kind of Control," *Boston Globe,* 29 March 1977, 5; Robert Cooke, "MDs Worried Least by DNA Research," *Boston Globe,* 31 March 1977, 23.

2. Department of Health, Education, and Welfare, National Institutes of Health, "Guidelines for Research Involving Recombinant DNA Molecules," *Federal Register* 41 (7 July 1976): 27911.

3. Cooke, "Experimental Biologists Surveyed."

4. Krimsky, *Genetic Alchemy,* ch. 22.

5. Barbara Culliton, "Recombinant DNA," *Science* 193 (23 July 1976): 300–301; Krimsky, *Genetic Alchemy,* ch. 22.

6. City of Cambridge, Commonwealth of Massachusetts, Cambridge Experimentation Review Board, "Guidelines for the Use of Recombinant DNA Molecule Technology in the City of Cambridge," recommendations and findings submitted to the commissioner of health and hospitals (21 December 1976) and to the city manager (5 January 1977).

7. Ibid.

8. Sheldon Krimsky, Anne Baeck, and John Bolduc, *Municipal and States Recombinant DNA Laws: History and Assessment,* report prepared for the Boston Neighborhood Network, June 1982; Sheldon Krimsky, "A Comparative View of State and Municipal Laws Regulating the Use of Recombinant DNA Molecules Technolgy," *Recombinant DNA Technical Bulletin* 2 (November 1979): 121–33. Controls were subsequently enacted in Emeryville (April 1977), Maryland (February 1978), Berkeley (September 1977), New York State (February 1978), Princeton (March 1978), and Amherst (October 1978).

9. Edward M. Kennedy and Jacob K. Javits to Gerald Ford, 19 July 1976, in NIH, *Recombinant DNA Research* (March 1978), 2:158–60.

10. Interagency Committee on Recombinant DNA Research, minutes, 4 November 1976, and charter, in NIH, *Recombinant DNA Research* 2:168–75, 181–83.

11. Edward M. Kennedy, opening statement before the Senate Subcommittee on Health hearings on recombinant DNA research and the NIH guidelines, 22 September 1976, Office of Senator Edward Kennedy.

12. See, e.g., Francine Robinson Simring, "The Double Helix of Self-Interest," *Sciences* (May–June 1977): 10–13, 27.

13. "Petition of the Environmental Defense Fund, Inc., and Natural Resources Defense Council, Inc., to the Secretary of Health, Education, and Welfare to Hold Hearings and Promulgate Regulations under the Public Health Service Act Governing Recombinant DNA Activities," 11 November 1976, in NIH *Recombinant DNA Research* 2:325–45.

14. William Bennett and Joel Gurin, "The Science That Frightens Scientists: The Great Debate over DNA," *Atlantic,* February 1977, 43–62; "Decoding the Genes," *New York Times,* 13 October 1976, 42.

15. Liebe Cavalieri, "New Strains of Life—or Death," *New York Times Magazine,* 22 August 1976; "Fruits of Gene-Juggling: Blessing or Curse?" *Medical World News,* 4 October 1976, 45; "Progress or Peril? Gene Transplants Stir Communities' Fears, Scientists Are Split," *Wall Street Journal,* 28 September 1976, 1.

16. "Decoding the Genes," *New York Times,* 13 October 1976, 42. For discussion, see Diana Dutton, *Worse Than the Disease* (Cambridge: Cambridge University Press, 1988), 174–225.

17. National Academy of Sciences, *Research with Recombinant DNA* (Washington, D.C.: National Academy of Sciences, 1977).

18. Photograph in Watson and Tooze, *DNA Story,* 134; Lear, *Recombinant DNA,* 166.

19. Roger Lewin, "Total Ban Sought on Genetic Engineering," *New Scientist,* 10 March 1977, 571; "U.S. Genetic Engineering in a Tangled Web," *New Scientist,* 17 March 1977, 640–41.

20. Roger G. Noll and Paul A. Thomas, "The Economic Implications of Regulation by Expertise: The Guidelines for Recombinant DNA Research," in National Academy of Sciences, *Research with Recombinant DNA,* 262–77.

21. See, e.g., House Committee on Science and Technology, Subcommittee on Science, Research, and Technology, *Science Policy Implications of DNA Recombinant Molecule Research,* 95th Cong., 2d sess., March 1978, ix.

22. For reviews of these studies, see Shapin, "History of Science," and Harry Collins, "An Empirical Relativist Programme in the Sociology of Scientific Knowledge," in *Science Observed,* ed. Karin Knorr-Cetina and Michael Mulkay (London: Sage Publications, 1983), 85–113.

23. For discussion of some examples, see Shapin, "History of Science," 186–94.

24. For example, in his detailed analysis of the development of inertial guidance technology for missiles, Donald MacKenzie notes debates over the need for missile accuracy in Congress and within the U.S. navy and air force but does not examine in detail the politics of these debates, the political reasons for their resolution, and the impact of the particular forms that resolution took ("Missile Accuracy: A Case Study in the Social Processes of Technological Change," in *The Social Construction of Technological Systems,* ed. Wiebe E. Bijker, Thomas P. Hughes, and Trevor J. Pinch [Cambridge, Mass.: MIT Press, 1987], 195–222; *Inventing Accuracy: A Historical Sociology of Nuclear Missile Guidance* [Cambridge, Mass.: MIT Press, 1990]).

25. Schattschneider, *Semi-sovereign People,* 35.

26. Ibid., 71.

27. The support of university administrators and biomedical researchers for NIH controls of the policy process is reflected in the strength of their opposition to the transfer of responsibility to another body; see chapter 7.

28. Interview with Malcolm Martin, 22 November 1983.

29. Participants at the Ascot meeting were asked to identify themselves and their interests in recombinant DNA technology: see Department of Health, Education, and Welfare, National Institutes of Health, U.S.-EMBO Workshop to Assess the Containment Requirements for Recombinant DNA Experiments Involving the Genomes of Animal, Plant, and Insect Viruses, 27–29 January 1978, transcript (hereafter cited as U.S.-EMBO Workshop), 1–5. A list of participants is given in Department of Health, Education, and Welfare, National Institutes of Health, "U.S.-EMBO Workshop to Assess the Risks for Recombinant DNA Experiments Involving the Genomes of Animal, Plant, and Insect Viruses," *Federal Register* 43 (31 March 1978): 13748–55.

30. Quoted in Krimsky, *Genetic Alchemy,* 215–16.

31. "The Stretching of Consensus in Assessment of Hazards of DNA Research," ca. July 1978, MRCUW, MSS.79/AS/3/8/56.

32. Enteric Bacteria Meeting, 31 August 1976, transcript, 2, 39–40, 42, ORDAR.

33. U.S.-EMBO Workshop, 16.

34. E.g., Bachrach and Baratz, *Power and Poverty,* 3–51.

35. Enteric Bacteria meeting, transcript, 6. Hereafter pages given in text. Excerpts from the transcript of this meeting are in appendix A.

36. RAC minutes, 13–14 September 1976, 9–10.

37. Sherwood L. Gorbach, "Recombinant DNA: An Infectious Disease Perspective," *Journal of Infectious Diseases* 137 (May 1978): 615–16.

38. E.g., H. Williams Smith, "Is It Safe to Use *Escherichia coli* K12 in Recombinant DNA Experiments?" *Journal of Infectious Diseases* 137 (May 1978): 655–60.

39. See Rolf Freter et al., "Possible Effects of Foreign DNA on Pathogenic Potential and Intestinal Proliferation of *Escherichia coli,*" *Journal of Infectious Diseases* 137 (May 1978): 624–29. Freter noted that "[feeding] experiments tell us little about the ability of bacteria to grow in the human intestine. It has been known for some time that the feeding of bacteria, especially those grown under the usual laboratory conditions, rarely results in implantation because the normal intestinal flora is antagonistic to the growth of invaders" (626).

40. Samuel B. Formal and Richard B. Hornick, "Invasive *Escherichia coli,*" *Journal of Infectious Diseases* 137 (May 1978): 641–44.

41. Later research carried out with the benefit of more precise knowledge of these properties and with genetic manipulation techniques showed that the "invasiveness" of *E. coli* K12 could be substantially enhanced: see Philippe J. Sansonetti, Dennis J. Kopecko, and Samuel B. Formal, "*Shigella sonnei* Plasmids: Evidence That a Large Plasmid Is Necessary for Virulence," *Infection and Immunity* 34 (1981): 75 and 35 (1982): 852.

42. See, e.g., E. S. Anderson, "Plasmid Transfer in *Escherichia coli,*" *Journal of Infectious Diseases* 137 (May 1978): 686–87.

43. Bruce Levin to Donald Fredrickson, 29 July 1977, ORDAR.

44. Jonathan King, "Recombinant DNA and Autoimmune Disease," *Journal of Infectious Diseases* 137 (May 1978): 663–66.

45. Sherwood Gorbach to Donald Fredrickson, 14 July 1977, ORDAR.

46. Bruce Levin to Donald Fredrickson, 29 July 1977, ORDAR; Jonathan King and Richard Goldstein to Participants, Cold Spring Harbor Phage Meetings, 22 August 1977, Jonathan King personal papers.

47. Interview with Sydney Brenner, 6 May 1980.

48. Interview with member of GMAG, April 1980.

49. Paul Berg to Donald Fredrickson, 13 October 1977, in NIH, *Recombinant DNA Research* (September 1978), appendix A, 3:12–14; John Tooze, testimony, meeting of the Advisory Committee to the Director, NIH, on the Proposed Revision of the NIH Guidelines on Recombinant DNA Research, 15–16 December 1977, transcript, *Recombinant DNA Research* 3:306, 351.

50. U.S.-EMBO Workshop, 39. Hereafter pages given in text.

51. See also ibid., 1010, 1009, 1016.

52. Interview with Robin Weiss, 18 June 1979.

53. *Federal Register* 43 (31 March 1978): 13749.

54. Ibid., 13751.

55. Comments at RAC meeting, 27–28 April 1978, personal notes.

56. Interview, 1979; anonymity requested.

57. Roy Curtiss to Paul Berg et al., 6 August 1974, RDHC.

58. Roy Curtiss to Donald Fredrickson, 12 April 1977, ORDAR.

59. Roy Curtiss, "Biological Containment and Cloning Vector Transmissibility," *Journal of Infectious Diseases* 137 (May 1978): 668–75.

60. Jaap Jelsma and W. A. Smit, "Risks of Recombinant DNA Research: From Uncertainty to Certainty," in *Impact Assessment Today,* ed. Henk A. Becker and Alan L. Porter (Utrecht: Uitgeverij Jan van Arkel), 715–40.

61. Stanley N. Cohen, "Recombinant DNA: Fact and Fiction," *Science* 195 (18 February 1977): 654–57.

62. Shing Chang and Stanley N. Cohen, "*In Vivo* Site-Specific Genetic Recombination Promoted by Eco RI Restriction Endonuclease," *Proceedings of the National Academy of Sciences* 74 (November 1977): 4811–15.

63. Stanley Cohen to Donald Fredrickson, 6 September 1977, ORDAR.

64. Roy Curtiss to Donald Fredrickson, 12 April 1977, 8, ORDAR.

65. Gorbach, "Recombinant DNA," 615.

66. Gorbach, "Proceedings from a Workshop," 704.

67. It was proposed that the first two questions be investigated both in human volunteers and in germ-free mice, used as a model system.

68. Enteric Bacteria Meeting, transcript, 45–46, ORDAR. Hereafter pages given in text.

69. "Risk Assessment Protocols," 707.

70. E.g., Victor Cohn, "Scientists Now Downplay Risks of Genetic Research," *Washington Post,* 18 July 1977, A1. The Cohn article did not mention criticism of the Gorbach letter. News coverage of such criticism did not appear until October 1977: Joanne Omang, "Genetic Research Enthusiasts Said Misleading Public," *Washington Post,* 18 October 1977, C3; Robert Cooke, "Gene Researchers Accused of Exaggerating Safety Factor," *Boston Globe,* 12 October 1977.

71. Cohn, "Scientists Now Downplay Risks of Genetic Research."

72. Philip Abelson, "Recombinant DNA," *Science* 197 (19 August 1977): 721.

73. See, e.g., Nicholas Wade, "Gene Splicing: Senate Bill Draws Charges of Lysenkoism," *Science* 197 (22 July 1977): 348–50.

74. "DNA Research: Not So Dangerous after All?" *Time,* 17 August 1977, 56; cf. "The DNA Furor: Tinkering with Life," *Time,* 18 April 1977, 32–45; see also "Recombinant DNA Debate Three Years On," *Nature* 268 (21 July 1977): 185.

75. Victor Cohn, "DNA Research Control Dims," *Washington Post,* 28 September 1977, A1, A4.

76. Herbert Boyer and S. Nicosia, eds. *Genetic Engineering,* proceedings of the International Symposium on Genetic Engineering: Scientific Developments and Practical Applications, Milan, 29–31 March 1978 (Amsterdam: Elsevier, 1978), 224.

77. Ibid., 254.

78. Ibid., 282. Cf. concerns about the hazards of routine biological techniques, aired privately before the advent of genetic engineering, section 4.3.

79. Peter Newmark, "WHO Looks for Benefits from Genetic Engineering," *Nature* 272 (20 April 1978): 663–64.

80. Bernard D. Davis, "The Recombinant DNA Scenarios: Andromeda Strain, Chimera, and Golem," *American Scientist* 65 (September–October 1977): 547–55.

81. See, e.g., Bernard Davis, "Evolution, Epidemiology, and Recombinant DNA," *Science* 193 (6 August 1976): 442; Bernard Davis/Robert Sinsheimer, "Discussion Forum: The Hazards of Recombinant DNA," *Trends in the Biological Sciences* 1 (August 1976): N178–80; Davis, "Recombinant DNA Scenarios," 547–55; "The DNA Research Scare," *Wall Street Journal,* 5 April 1978.

82. James Watson, "In Defense of DNA," *New Republic,* 25 June 1977, 11–14. See also James Watson, "An Imaginary Monster," *Bulletin of the Atomic Scientists,* May 1977, 19–20; James Watson, "Remarks on Recombinant DNA," *Co-Evolution Quarterly* 14 (Summer 1977): 40–41; James Watson, "The Nobelist vs. the Film Star," *Washington Post,* 16 May 1978, D1–2; James Watson, "Trying to Bury Asilomar," *Clinical Research* 26 (April 1978): 113–15.

83. Rene Dubos, "Genetic Engineering," *New York Times,* 21 April 1977, A25.

84. Lewis Thomas, "Notes of a Biology Watcher: The Hazards of Science," *New England Journal of Medicine* 296 (10 February 1977): 324–27.

85. Peter B. Medawar, "The DNA Scare: Fear and DNA," *New York Review of Books,* 27 October 1977, 15–20.

6. Derailing Legislation, 1977–1978

1. Treating the formation and deployment of the new discourse in separate chapters does not mean that the two processes were in any way distinct: using the arguments about hazard in the congressional and NIH arenas was an integral part of the processes' formation and legitimation.

2. Gilbert S. Omenn, "Government as a Broker between Private and Public Institutions in the Development of Recombinant DNA Applications," in *Proceedings of the Battelle Conference on Genetic Engineering,* ed. Melissa Kenberg (Seattle: Battelle Seminars and Studies Program, 1981), 1:38.

3. Ibid., 38–39.

4. Interviews with committee members in the period February–July 1978; permission to cite by name not sought.

5. Interview with John Finklea, director of the National Institute of Occupational Safety and Health, 26 July 1978.

6. U.S. Environmental Protection Agency, EPA Public Interest Group Meeting on Recombinant DNA Research, minutes, 15 February 1977, presentation by Dr. Thomas S. Bath, EPA Science Advisory Board, 12–13, EPA Historical Document Collection, USEPA Headquarters, Washington, D.C.

7. Interagency Committee on Recombinant DNA Research, "Interim Report," 15 March 1977, appendix 3: "Regulation of Recombinant DNA Research in Laboratories," in NIH, *Recombinant DNA Research* 2:307–18.

8. Interagency Committee on Recombinant DNA Research, minutes, 23 November 1976, in NIH, *Recombinant DNA Research* 2:240–49. Hereafter cited as IAC minutes.

9. IAC minutes, 25 February, 10, 14, and 29 March 1977, 253–61, 266–75, 352–60.

10. IAC minutes, 25 February 1977, 256; Victor Cohn, "Drug Industry Seeks to Alter U.S. Rules on Genetic Studies," *Washington Post,* 20 November 1976, A3. Representatives of Abbott Laboratories, the American Type Culture Collection, Cetus Corporation, CIBA-Geigy Corporation, Du Pont, General Electric, Eli Lilly, Merck, Sharp and Dohme, Monsanto, the Upjohn Company, Wyeth Laboratories, Searle Laboratories, Pfizer, and the Pharmaceutical Manufacturers Association attended the meeting with Commerce Department officials (list of participants at meeting, 19 November 1976, John Finklea files, National Institute of Occupational Safety and Health, Rockville, Md.). No further information is available from the Department of Commerce about this meeting (Jordan Baruch to Susan Wright, 4 December 1979, in response to a request under the Freedom of Information Act).

11. IAC minutes, 25 February 1977, 256.

12. Memo of John F. Finklea, transmitted to Joseph G. Perpich, NIH Associate Director for Program Planning and Evaluation, 4 March 1977, John Finklea files.

13. IAC minutes, 25 February 1977, 256; U.S. Environmental Protection Agency, EPA Public Interest Group Meeting on Recombinant DNA Research, 15 February 1977.

14. IAC minutes, 25 February 1977, 256–57.

15. Helen Whitley to Donald Fredrickson, 7 March 1977, HHP.

16. Harlyn Halvorson to Donald Fredrickson, 8 March 1977, HHP.

17. IAC minutes, 25 February 1977, 253–61.

18. Ibid., 258.

19. Ibid., 10 and 14 March 1977, 268–69.

20. Ibid., 266–75.

21. Ibid., 25 March 1977, 357.

22. Ibid.; IAC minutes, 10 and 14 March 1977, 268.

23. Ibid., 25 February 1977, 257. It should be noted that, given the composition of the meeting with biomedical researchers, which included at least one critic of the NIH controls, there was probably some dissent on the question of preemption.

24. Ibid., 10 and 14 March 1977, 267.

25. "Interim Report of the Federal Interagency Committee on Recombinant DNA Research: Suggested Elements for Legislation," 15 March 1977, in *Recombinant DNA Research* 2:279–301; IAC minutes, 29 March 1977, 279–318, 352–65.

26. "A Bill to Regulate Activities Involving Recombinant Deoxyribonucleic Acid," draft, 21 March 1977, circulated to the IAC by NIH associate director for program planning and evaluation, 23 March 1977, John Finklea files.

27. Marion Suter to the Chairman of the Council on Environmental Quality, Charles Warren, memo, 1 April 1977.

28. IAC minutes, 29 March 1977, 352–60.

29. Burke Zimmerman, "Recombinant DNA Regulatory Legislation: A Discussion of Issues and Analysis of Legislative Alternatives," ca. April 1977, Burke K. Zimmerman personal papers. Zimmerman was a member of the staff of Paul Rogers in 1977–78.

30. Quoted in Barbara Culliton, "Recombinant DNA Bills Derailed: Congress Still Trying to Pass a Law," *Science* 199 (20 January 1978): 274–77.

31. S. 1217, Subcommittee Print 2, 27 April 1977, 25.

32. Donald Cox to Harlyn Halvorson, 3 March 1977, RDHC.

33. Ralph Arlinghaus et al. to Members of Congress, 29 June 1977, RDHC.

34. "Philip Handler on Recombinant DNA Research," *Chemical and Engineering News,* 9 May 1977, 3.

35. Harlyn Halvorson, interview with Aaron Seidman, 10 and 22 May 1978, RDHC. ASM's request to Berg for representation at the Asilomar conference was not accepted (Donald Cox, Chairman, Public Affairs Committee, American Society of Microbiology, to Paul Berg, 3 December 1974, HHP).

36. Harlyn Halvorson, interview with Aaron Seidman, 10 and 22 May 1978, RDHC.

37. Interview with Allan Fox, 1 March 1978.

38. Frank Young to Harlyn Halvorson, 3 August 1977, HHP.

39. Interview with Harlyn Halvorson, 1 December 1979.

40. Culliton, "Recombinant DNA Bills Derailed"; interview with Harlyn Halvorson, 1 December 1979.

41. Culliton, "Recombinant DNA Bills Derailed."

42. Harlyn O. Halvorson to Paul Rogers and Edward Kennedy, mailgram, 9 May 1990, RDHC.

43. Harlyn O. Halvorson, "ASM on Recombinant DNA," *Science* 196 (10 June 1977): 1154.

44. In addition to the ASM, the other societies represented on the ISCBM were the American Institute of Biological Sciences, the American Society for Medical Technology, the American Society of Allied Health Professions, the Association of American Medical Colleges, the Federation of American Societies for Experimental Biology, and the National Society for Medical Research. The Federation of American Societies for Experimental Biology encompassed the American Physiological Society, the American Society of Biological Chemists, the American Society for Pharmacology and Experimental Therapeutics, the American Association of Pathologists, the American Institute of Nutrition, and the American Association of Immunologists.

45. Culliton, "Recombinant DNA Bills Derailed"; Harlyn O. Halvorson to Edwin H. Lennette, 15 September 1977, HHP.

46. The resolution was adopted on 26 April 1977. It was reprinted in the *Bulletin of the Atomic Scientists,* December 1977, 9.

47. Walter Gilbert, "Recombinant DNA Research: Government Regulation," *Science* 197 (15 July 1977): 206.

48. David Dickson, "Friends of DNA Fight Back," *Nature* 272 (20 April 1978): 664–65.

49. Interview with Nan Nixon, 14 November 1979.

50. Harlyn Halvorson, phone log, HHP.

51. Herbert W. Boyer, Coralee Stevens Kuhn, Robert Swanson, reports filed with the secretary of the U.S. Senate and the clerk of the House of Representatives pursuant to the federal Regulation of Lobbying Act for the second and third quarters, 1977, House of Representatives, Office of Records and Registration.

52. "DNA Regulation Bill Hits Roadblock Again," *Congressional Quarterly,* 27 May 1978, 1331–35; interview with Nan Nixon, 14 November 1979.

53. Friends of the Earth, "Legislative Alert," ca. May 1977, Friends of the Earth files, Washington, D.C.

54. E.g., interview with Olga Grkawac (office of Edward Markey), 2 December 1979.

55. For an insider's account of the turns in the legislative road, see Burke K. Zimmerman, "Science and Politics: DNA Comes to Washington," in *The Gene-Splicing Wars: Reflections on the Recombinant DNA Controversy,* ed. Raymond A. Zilinskas and Burke K. Zimmerman (New York: Macmillan, 1986), 33–54. Zimmerman, a member of representative Paul Rogers's staff, generally reinforces the details given below. However, his account underestimates the understanding that united sectors of the biomedical research community in efforts to derail legislation, despite appearances of differences between "moderates" and "extremists."

56. Harlyn Halvorson, interview with Aaron Seidman, 10 and 22 May 1978, RDHC. Because Zimmerman's previous position had been as a staff biologist for the Environmental Defense Fund, he was initially viewed with suspicion by some biologists, but apparently his manner of interacting with the members of the biomedical research community dispelled these reservations.

57. Harlyn O. Halvorson to Charles Yanofsky, 26 May 1977, HHP.

58. Harlyn Halvorson, phone log, HHP; interview with Ray Thornton, 4 December 1979.

59. House Committee on Science and Technology, Subcommittee on Science, Research, and Technology, *Hearings: Science Policy Implications of DNA Recombinant Molecule Research,* 95th Cong., 1st sess., 29–31 March, 27–28 April, 3–5 and 25–26 May, 7–8 September 1977.

60. Interview with Harlyn Halvorson, 1 December 1979.

61. Harlyn O. Halvorson to Members of Congress, 28 July 1977, HHP.

62. Interview with Harlyn Halvorson, 1 December 1979.

63. "Kennedy Opens Door to Discussion of His Bill to Regulate DNA Research," *FASEB Newsletter* 10 (July 1977): 1–2; see also Harlyn Halvorson, interview with Aaron Seidman, 10 and 22 May 1977, RDHC.

64. Gaylord Nelson, "Remarks on Recombinant DNA Act, S. 1217 (Amendment No. 754)," *Congressional Record,* 2 August 1977, S. 13312–19. Nelson also attached a *New York Times* article that described attacks on the legislation by various groups of biologists at the Gordon conferences and by Americans for Democratic Action, the Curtiss letter, the Gorbach letter, and an article by James Watson (Walter Sullivan, "Legislating the Laboratories," *New York Times,* 31 July 1977, E9).

65. Interview with Halsted Holman, 6 November 1979.

66. Wade, "Gene Splicing," 348–50.

67. National Board, Americans for Democratic Action, "Recombinant DNA and Government Control of Research," 26 June 1977, RDHC. The ADA National Board later passed a second resolution moderating its position and "going on record as supporting the concept of safeguarding the public through DNA legislation" ("DNA Research," 17–18 September 1977, Americans for Democratic Action papers, Washington, D.C.). By the time the second motion was approved, however, the lobbying effort against the Kennedy legislation had virtually succeeded.

68. Harlyn Halvorson, interview with Aaron Seidman, 10 and 22 May 1978, RDHC; Harlyn Halvorson, phone log, HHP.

69. Adlai Stevenson, "Recombinant DNA Legislation," *Congressional Record,* 22 September 1977, 30457–60.

70. Edward M. Kennedy, speech to the Association of Medical Writers, 27 September 1977, New York, reprinted in Watson and Tooze, *DNA Story,* 173–74; Victor Cohn, "Scientists' Lobby Successful: DNA Research Control Dims," *Washington Post,* 28 September 1977, A1, A4. For discussion of Cohen's claim, see section 6.4.

71. Frank Young to Robert Watkins, 10 August 1977, HHP.

72. Harlyn Halvorson, telephone log, ca. September 1977, HHP.

73. Interview with Nan Nixon, 14 November 1979.

74. Harlyn O. Halvorson, "Recombinant DNA Legislation: What Next?" *Science* 198 (28 October 1977): 357.

75. Stanley Cohen to Harley Staggers, 7 October 1977, reprinted in Watson and Tooze, *DNA Story,* 187–88.

76. Harlyn Halvorson, interview with Aaron Seidman, 10 and 22 May 1978, RDHC; interview with Burke Zimmerman, 13 November 1979.

77. Harlyn Halvorson, interview with Aaron Seidman, 10 and 22 May 1977; interview with Nan Nixon, 14 November 1979; interview with Burke Zimmerman, 13 November 1979.

78. E.g., letter from Robert Acker (on behalf of the ISCBM) to Harley Staggers, 1 November 1977, RDHC. This letter does not allude to the Rogers legislation directly but states that the council "has hesitated to recommend legislation to regulate recombinant DNA research because recent scientific evidence supports our contentions that risks have been moderated" and that it recommends "minimal interim legislation" that would include protection of proprietary information, protection of "free scientific inquiry," and preemption of local controls.

79. For an overview of the course of the DNA legislation in 1978, see Elizabeth Wehr, "DNA Regulation Bill Hits Roadblock Again," *Congressional Quarterly,* 27 May 1978, 1331–35. Bills H.R. 11192 and S. 1217, amendment 1713, are reprinted in NIH, *Recombinant DNA Research* 3, appendix B.

80. Senate Committee on Commerce, Science, and Transportation, Subcommittee on Science, Technology, and Space, *Hearings: Regulation of Recombinant DNA Research,* 95th Cong., 1st sess., 2, 8, and 10 November 1977, 13–14, 36–37.

81. On the irregularity of the somatostatin announcement, see David Perlman. "Scientific Announcements" (letter), *Science* 198 (25 November 1977): 782.

82. Senate Committee on Commerce, Science, and Transportation, Subcommittee on Science, Technology, and Space, *Hearings: Regulation of Recombinant DNA Research,* 95th Cong., 1st sess., 2, 8, and 10 November 1977, 37.

83. Ibid., 13.

84. Ibid., 16.

85. House Committee on Science and Technology, Subcommittee on Science, Research, and Technology, *Report: Science Policy Implications of DNA Recombinant Molecule Research,* 95th Cong., 2d sess., March 1978.

86. *Regulation of Recombinant DNA Research,* 125.

87. House Committee on Interstate and Foreign Commerce, *Report: Recombinant DNA Act,* 95th Cong., 2d sess., 24 March 1978, 35–43.

88. Edward Kennedy et al. to Joseph Califano, 1 June 1978, and Joseph Califano to Edward Kennedy, 12 September 1978, reprinted in NIH, *Recombinant DNA Research* (December 1978), 4:26–28.

89. Stevenson's position is discussed further in chapter 10.

90. See, e.g., House Committee on Science and Technology, Subcommittee on Science, Research, and Technology, *Report: Science Policy Implications of DNA Recombinant Molecule Research,* 95th Cong., 2d sess., March 1978.

7. Revising the National Institutes of Health Controls, 1977–1978

1. William Whelan, testimony, Department of Health, Education, and Welfare, Public Hearing on Proposed Revised Guidelines on Recombinant DNA Research, 15 September 1978, in NIH, *Recombinant DNA Research* 4:127.

2. John Tooze, "Harmonizing Guidelines: Theory and Practice," in *Genetic Engineering,* ed. Herbert W. Boyer and S. Nicosia (Amsterdam: Elsevier, 1978), 281.

3. U.S. Interagency Committee on Recombinant DNA Research, "Report on International Activities," November 1977, in NIH, *Recombinant DNA Research* 2:415–16.

4. Stanley Falkow to Donald Fredrickson, 19 April 1978, in NIH, *Recombinant DNA Research,* appendix A, 3:284–86.

5. Stanley Cohen to Donald Fredrickson, 15 May 1978, in NIH, *Recombinant DNA Research,* appendix A, 3:306–8.

6. John Douglas, "U.S. Geneticists Look to Europe for Research Facilities," *Nature* 275 (21 September 1978): 170.

7. Richard Saltus, "Genetic Research: Quiet Move to Europe," *San Francisco Examiner,* 29 October 1978, 1, 9.

8. House Committee on Science and Technology, Subcommittee on Science, Research, and Technology, *Science Policy Implications,* 9.

9. See also Julian Davies to Donald Fredrickson, 19 April 1978, in NIH, *Recombinant DNA Research,* appendix A, 3:21–22.

10. Nicholas Wade, "Recombinant DNA: NIH Rules Broken in Insulin Gene Project," *Science* 197 (30 September 1977): 1342–45; Nicholas Wade, "A Tangled Tale from the Biology Classroom," *Science* 200 (5 May 1978): 516–17; David Dickson, "NIH Confirms Violation of Recombinant DNA Research Guidelines," *Nature* 273 (4 May 1978): 5.

11. Roy Curtiss to Donald Fredrickson, 12 April 1977, ODNIH.

12. Roy Curtiss III to John Littlefield, 14 March 1977, ORDAR.

13. Working copy of 1976 guidelines and proposed revisions, circulated in May 1977, ORDAR. See, e.g., proposed revisions for containment of insect DNA, III-7, compared with the 1976 conditions, III-18.

14. David Hogness to Donald Fredrickson, 30 June 1977, ORDAR, doc. 268.

15. Department of Health, Education, and Welfare, National Institutes of Health, "Recombinant DNA Research: Proposed Revised Guidelines," *Federal Register* 42 (27 September 1977): 49596–49609.

16. Ibid., 49597.

17. Office of the Director, NIH, "Background on the Proposed Revisions," November 1977, 28–34, ORDAR.

18. Details of the meeting, including a transcript, are given in NIH, *Recombinant DNA Research* 3:178–496. Hereafter pages given in text.

19. Ibid., e.g., 233, 245, 268–69, 425–30.

20. Chang and Cohen, "*In Vivo* Site-Specific Genetic Recombination," 4811–15.

21. Senate Committee on Commerce, Science, and Transportation, Subcommittee on Science, Technology, and Space, *Recombinant DNA Research and Its Applications*, 95th Cong., 2d sess., August 1978.

22. Victor Cohn, "Discoverer Would Lift Curbs on DNA," *Washington Post*, 17 December 1977, A1.

23. Harold Schmeck, "Rules on DNA Studies Viewed as Too Strict," *New York Times*, 18 December 1977, L19.

24. See Nicholas Wade, "Gene-Splicing Rules: Another Round of Debate," *Science* 199 (6 January 1978): 30–33; Schmeck, "Rules on DNA Studies Viewed as Too Strict"; Victor Cohn, "Amid Latest Gene Research Flap, Easing of Rules Is Eyed," *Washington Post*, 16 December 1977, A2.

25. Transcript of proceedings, 309. See also the remarks of David Gelfand (Genentech), in NIH, *Recombinant DNA Research* 3:309.

26. On the question of potential problems posed for industry by the ten-liter limit, see John Adams, "Recombinant DNA Research: The Drug Industry Viewpoint," *Society for Industrial Microbiology News*, January 1978, 3–5.

27. NIH, *Recombinant DNA Research* 3:222.

28. Ibid., 416, 418. For further statements of the industry position at this time, see C. W. Pettinga (Lilly) to Donald S. Fredrickson, 7 November 1977, and C. Joseph Stetler to Donald S. Fredrickson, 14 November 1977, in NIH, *Recombinant DNA Research*, appendix A, 3:36–47, 62–69.

29. George S. Gordon to Sidney R. Galler and Robert B. Ellert, "Meeting to Discuss Possible Approaches to Development to Private-Sector Voluntary Compliance with NIH Guidelines for Recombinant DNA Research, under Surveillance by the Department of Commerce," memo, ca. December 1977, obtained through a Freedom of Information Act request, Department of Commerce.

30. This point was also made by John Adams in an address to the annual meeting of the Society for Industrial Microbiology, East Lansing, Michigan, 25 August 1977. See Adams, "Recombinant DNA Research," 3–5.

31. Gordon, "Meeting to Discuss Possible Approaches."

32. Paul Berg to Donald Fredrickson, 13 October 1977, in NIH, *Recombinant DNA Research,* appendix A, 3:12–14.

33. David Baltimore to Donald Fredrickson, 18 October 1977, in ibid., 21–22.

34. Ann Skalka to Donald Fredrickson, 13 October 1977, in ibid., 9–11.

35. Paul Berg to Donald Fredrickson, 13 October 1977, in ibid., 13.

36. Ibid. 3:306, 351.

37. U.S.-EMBO Workshop, 9–10.

38. Interview with Robin Weiss, 8 June 1984.

39. Ibid.

40. Department of Health, Education, and Welfare, National Institutes of Health, "U.S.-EMBO Workshop," 13751.

41. The transcript records that thirteen of the original twenty-three participants and two of the chairs were making contributions toward the end of the meeting.

42. See, e.g., U.S.-EMBO Workshop, 283, 1017.

43. Harold S. Ginsberg and Malcolm A. Martin, report of a meeting of the Virus Working Group, 6–7 April 1978, ORDAR, doc. 391.

44. Wallace P. Rowe to Recombinant DNA Advisory Committee, "Memorandum on Report of the Working Group on Virus Guidelines," 14 April 1978, ORDAR, doc. 382.

45. Boyer and Nicosia, *Genetic Engineering,* 6, 253.

46. Newmark, "WHO Looks for Benefits from Genetic Engineering," 663–64.

47. House Committee on Science and Technology, Subcommittee on Science, Research, and Technology, *Report: Science Policy Implications,* VII–IX.

48. Donald Fredrickson, introductory remarks at RAC meeting, 27–28 April 1978, transcribed from tape of meeting, ORDAR.

49. Remarks of DeWitt Stetten, ibid.

50. The committee did, however, exempt research using many combinations of DNA, on the grounds that these combinations were derived either from a single organism, or from organisms deemed to exchange DNA through natural processes. For details, see RAC minutes, 27–28 April 1978, 5–8, and attachments III and IV.

51. Ibid., 19–20.

52. Ibid., 28.

53. Ibid., 12.

54. This point was made at the meeting by Daniel Weiss of the National Academy of Sciences, who observed that "industry wants to be protected from private law suits" (RAC meeting, 27 April 1978, Wright notes).

55. Harlyn Halvorson, telephone log, 2 May 1978.

56. Edward M. Kennedy to Donald S. Fredrickson, 2 May 1978, in NIH, *Recombinant DNA Research* 3:556–58.

57. Department of Health, Education, and Welfare, National Institutes of Health, "Recombinant DNA Research: Proposed Revised Guidelines," *Federal Register* 43 (28 July 1978): 33042–33178.

58. Ibid., 33087.

59. Joseph A. Califano to Edward M. Kennedy, 12 September 1978, in NIH, *Recombinant DNA Research* 4:27–28.

60. Harlyn Halvorson, testimony, Department of Health, Education, and Welfare, Public Hearing on the Proposed Revised Guidelines, 16 September 1978, 114–20.

61. Paul Berg to Donald Fredrickson, 20 September 1978, in NIH, *Recombinant DNA Research,* appendix A, 4:208–14.

62. David Baltimore to Donald Fredrickson, 9 August 1978, in ibid., 7.

63. Ronald Davis to Donald Fredrickson, 15 September 1978, in ibid., 159–61.

64. The main elements of the critics' position were presented at the DHEW hearing and in the subsequent meeting on 18 October 1978 with the DHEW committee.

65. Jonathan King to Joseph Califano, 22 September 1978, in NIH, *Recombinant DNA Research,* appendix A, 4:295–302.

66. Marcia Cleveland and Louis Slesin, "Comments of the Natural Resources Defense Council, Inc., on the Proposed Revision of the National Institutes of Health's Guidelines on Recombinant DNA Research," 19 September 1978, in ibid., 175–207.

67. Ibid., 295.

68. Ibid., 176.

69. Ibid., 301.

70. Philip Bereano, "Testimony on Proposed Revised Guidelines for Recombinant DNA Research," 15 September 1978, in NIH, *Recombinant DNA Research* 4:314–38.

71. Ibid., appendix A, 4:200.

72. See, e.g. ibid., appendix A, 4:180–81, 204, 207.

73. See, e.g. testimony of Leslie Dach, Science Associate, Environmental Defense Fund, and Pamela Lippe, Friends of the Earth, at the DHEW Public Hearing on the Proposed Revised Guidelines, 15 September 1978, in ibid. 4:102–10, 169–78.

74. Robert Cooke, "New DNA Guidelines Coming," *Boston Globe,* 18 October 1978.

75. Adlai E. Stevenson to Joseph A. Califano, 13 October 1978, in NIH, *Recombinant DNA Research,* appendix A, 4:414–27.

76. Robert Swanson to Donald Fredrickson, 9 October 1978, circulated to NIH officials on 16 October 1978, ODNIH.

77. Vagueness in interpretation of the limits to disclosure under the terms of the Freedom of Information Act was discussed in the report of the Senate Subcommittee on Science, Technology and Space, "Recombinant DNA Research and Its Applications," 36–40.

78. Excerpts from minutes, meeting of DHEW committee with representatives of the PMA, 13 October 1978, OSDHEW.

79. Donald S. Fredrickson to C. Joseph Stetler, 19 October 1978, ODNIH.

80. Joseph Califano, remarks at commencement exercises, University of Michigan, 18 December 1977, in Michigan Historical Collections, Bentley Historical Library, Chief Marshall's papers, box 8.

81. Department of Health, Education, and Welfare, Committee on Proposed Revised Guidelines for Recombinant DNA Research, minutes, 18 October 1978, OSDHEW.

82. Maxine Singer, "Spectacular Science and Ponderous Progress," *Science* 203 (5 January 1979): 9.

83. Department of Health, Education, and Welfare, National Institutes of Health, "Guidelines for Research Involving Recombinant DNA Molecules," *Federal Register* 43 (22 December 1978): 60108–31, reprinted in NIH, *Recombinant DNA Research* 4:30–53.

84. David Dickson, "U.S. Extends Recombinant DNA Controls to Private Industry," *Nature* 276 (21/28 December 1978): 744; David Dickson, "DNA Critics Appointed to Advisory Committee," *Nature* 277 (11 January 1979): 83.

85. Joseph Califano to Douglas Costle, 15 December 1978, in NIH, *Recombinant DNA Research* 4:57–58. Simultaneously the FDA commissioner, Donald Kennedy, issued a notice of the FDA's intent to implement this directive (Department of Health, Education, and Welfare, Food and Drug Administration, "Recombinant DNA: Intent to Propose Regulations," *Federal Register* 43 [22 December 1978]: 60134–35, reprinted in NIH, *Recombinant DNA Research* 4:55–56).

86. Department of Health, Education, and Welfare, National Institutes of Health, "Recombinant DNA Research: Revised Guidelines," *Federal Register* 43 (22 December 1978): 60080–81, reprinted in NIH, *Recombinant DNA Research* 4: 3–4.

87. NIH, *Recombinant DNA Research* 4:4, 5.

88. Ibid., 4.

89. Ibid., 3.

90. Donald Fredrickson, "Science and the Cultural Warp: DNA as a Case Study," paper presented at the annual meeting of the American Association for the Advancement of Science, Washington, D.C., 7 January 1982.

8. Operating the Genetic Manipulation Advisory Group, 1977–1978

1. The small Genetic Engineering Group formed in 1977 by the BSSRS appears to have been the only public interest organization that followed British policymaking in some detail. Its activities were largely limited to written critiques of British genetic engineering policy. By its own acknowledgment, the BSSRS group, while articulate, lacked resources and influence. See British Society for Social Responsibility in Science, Genetic Engineering Group, "Genetic Engineering: Life in the Margins of the Politics of Science," January 1979, mimeo, British Society for Social Responsibility in Science, London.

2. Sir Gordon Wolstenholme to Donald Fredrickson, 9 August 1977, ODNIH.

3. Second meeting of the EMBO Standing Advisory Committee on Recombinant DNA, *Report and Recommendations*, 2–6.

4. European Science Foundation, "Recommendations of the European Science Foundation's Ad Hoc Committee on Recombinant DNA Research (Genetic Ma-

nipulation)," in European Science Foundation, *Report 1976* (Strasbourg: European Science Foundation, 1976), appendix B and 8–12.

5. Robert Walgate, "GMAG Wants to Stay On," *Nature* 273 (25 May 1978): 25.

6. David Dickson, "GMAG: Stormy Weather Ahead?" *Nature* 271 (5 January 1978): 5–6.

7. "Plenty for GMAG to Do," *Nature* 276 (2 November 1976): 1.

8. Robert Freedman, "Gene Manipulation: A New Climate," *New Scientist,* 27 July 1978, 268–69.

9. John Maddox, "A Way to Safety without Brakes," *Financial Times,* 23 May 1978.

10. R. H. Pritchard, "Recombinant DNA Is Safe," *Nature* 273 (29 June 1978): 696.

11. Genetic Manipulation Advisory Group, *First Report of the Genetic Manipulation Advisory Group,* Cmnd. 1215 (May 1978), 19.

12. Sydney Brenner, "Six Months in Category IV," *Nature* 276 (2 November 1978): 2–3.

13. "Genetic Engineering," *Medical World,* November–December 1978, report on ASTMS conference on genetic engineering, 27 October 1978, 10.

14. Nicholas Wade, "New Smallpox Case Seems Lab Caused," *Science* 201 (8 September 1978): 893; Lawrence McGinty, "Smallpox Laboratories: What Are the Risks?" *New Scientist,* 4 January 1979, 8–14; "Ignorance Is Never Bliss," *Nature* 277 (11 January 1979): 75–76.

15. "GMAG Debates New Controls on Genetic Engineering," *New Scientist,* 4 January 1979, 4.

16. Interview with Sydney Brenner, 6 May 1980.

17. Interviews conducted in April and May 1980 with ten members appointed to GMAG in 1976.

18. Interview with Derek Ellwood, 21 April 1980.

19. Interview with Gordon Wolstenholme, 15 April 1980.

20. GMAG, *First Report,* May 1978, 21–22, 32.

21. Dickson, "GMAG: Stormy Weather Ahead?" 6.

22. See, e.g., Hall, "Union Boss," 266–68.

23. Interview with Robert Williamson, 21 April 1980.

24. Interview with Jerome Ravetz, 14 June 1979.

25. Interview with Sir Gordon Wolstenholme, 25 April 1980.

26. Interviews with Jerome Ravetz, 14 June 1979, and with Gordon Wolstenholme, 15 April 1980.

27. GMAG, *First Report,* 37–39.

28. Bennett, Glasner, and Travis, *Politics of Uncertainty,* 85.

29. Interview with Gordon Wolstenholme, 15 April 1980.

30. Interview with Jerome Ravetz, 18 April 1980.

31. One member of GMAG subsequently called its operation a "cosmetic exercise," but that description should not be taken to indicate that the exercise was empty: GMAG's procedures were widely acknowledged as having some impact on safety: see Jerome Ravetz, "Recombinant DNA Research: Whose Risks?" in

The Merger of Knowledge and Power: Essays in Critical Science (London: Mansell Publishing, 1990), 75–78.

32. Ibid.

33. Interview with Gordon Wolstenholme, 15 April 1980.

34. Interview with Jerome Ravetz, 18 April 1980.

35. Bennett, Glasner, and Travis, *Politics of Uncertainty,* 84–85.

36. Health and Safety Commission, *Genetic Manipulation: Regulations and Guidance Notes* (London: HMSO, 1978), 2.

37. GMAG, *First Report,* 17.

38. GMAG Note 4, "Transfer and Use of Recombinant Material," in GMAG, *First Report,* 49–51.

39. Department of Health, Education, and Welfare, National Institutes of Health, "Guidelines for Research," *Federal Register* 41 (7 July 1976): 27920.

40. ASTMS memos from this period show that the union was aware of the problem and wanted to amend the Health and Safety Executive regulations to cover use: see, e.g., Donna Haber to Clive Jenkins, 24 October 1977, and Clive Jenkins to Lionel Murray, 25 May 1978, MRCUW, MSS 79/AS/3/8/58.

41. GMAG Note 6, "Health Monitoring," in GMAG, *First Report,* 53–59.

42. Department of Health, Education, and Welfare, National Institutes of Health, "Guidelines for Research" (7 July 1976), 27920.

43. Ibid.

44. GMAG, *First Report,* 22. The Health and Safety Commission's Regulations on Safety Representatives and Safety Committees came into force on 1 October 1978.

45. Ibid., 42–44.

46. Three such issues were cited by many members of the committee in interviews in 1980: using "scientific merit" as a criterion for evaluating genetic engineering experiments, procedures for protecting the confidentiality of industrial proposals, and the siting of certain category III laboratories.

47. On this point, see Bennett, Glasner, and Travis, *Politics of Uncertainty,* 51–52, 116–17.

48. GMAG Note 7, "Information and Advice on the Completion of Proposal Forms for Centres (Part A) and Projects (Part B)," March 1978, GMAG Records, MRC, London.

49. Interview with Robert Williamson, 21 April 1980.

50. Interview with Jerome Ravetz, 16 November 1978.

51. John Subak-Sharpe to J. H. Morris, 21 August 1978, John Subak-Sharpe personal papers.

52. GMAG Note 7, March 1978, revised May 1979.

53. G. S. A. Szabo, "Patents and Recombinant DNA," *Trends in the Biochemical Sciences* 2 (November 1977): N246; GMAG, *First Report,* 24–26.

54. The following account is based on documents deposited by the ASTMS in the MRCUW, MSS.79/AS/3/8/57, and on interviews with members of GMAG in 1980.

55. Interview with Bernard Langley, 17 April 1980.

56. See, e.g., memoranda submitted by the British pharmaceutical industry and the CBI, appendixes 3 and 5, in House of Commons, Select Committee on

Science and Technology, *Second Report: Recombinant DNA Research: Interim Research* (3 April 1979).

57. On this point, see also Bennett, Glasner, and Travis, *Politics of Uncertainty,* 85.

58. British Society for Social Responsibility in Science, Genetic Engineering Group, "Genetic Engineering Report: Too Many Questions Unasked," press release, 18 May 1978, London.

59. Interview with John Maddox, 2 April 1980.

60. Interview with Jerome Ravetz, 18 April 1980.

61. Ibid.

62. Interview with Gordon Wolstenholme, 25 April 1980.

63. Interview with Derek Ellwood, 9 May 1980.

64. Ibid.

65. J. R. Hunter, D. C. Ellwood, A. Atkinson, and P. J. Greenaway, "Bile Salt Sensitivity of *Escherichia coli* χ1776," *Nature* 275 (7 September 1978): 70–71.

66. H. R. Smith and B. Rowe, "Testing of Disabled Strains of *Escherichia coli* K12," in GMAG, *Third Report of the Genetic Manipulation Advisory Group,* Cmnd. 8665 (September 1982), 132–35.

67. Interview with Derek Ellwood, 21 April 1980.

68. The account that follows is based on interviews in 1978–80 with Sydney Brenner and others who made important contributions to the development of the new risk assessment scheme.

69. Interview with John Maddox, 2 April 1980.

70. Sydney Brenner, paper submitted to GMAG in July 1978, in GMAG, *Second Report of the Genetic Manipulation Advisory Group,* Cmnd. 7785 (December 1979), 77–90.

71. Ibid.

72. This paragraph is taken from an appendix to Brenner's paper that was not published in GMAG, *Second Report.*

73. Interview with Derek Ellwood, 21 April 1980.

74. Interview with Peter Walker, 2 May 1980.

75. Jerome Ravetz to Gordon Wolstenholme, 13 January 1978, Jerome Ravetz personal papers.

76. Interview with John Maddox, 2 April 1980.

77. See, e.g., David Dickson, "NIH May Loosen Recombinant DNA Research Guidelines," *Nature* 273 (18 May 1978): 179.

78. European Science Foundation, Genetic Engineering Liaison Committee, minutes, 22–23 May 1978, GMAG Records, MRC, London.

79. Interview with Peter Rigby, 2 April 1980.

80. Transcript of London Weekend Television program, part 1, "Genetic Engineering: How Safe Is It?" 5 November 1978, London Weekend Television files.

81. Interview with Jerome Ravetz, 14 June 1979.

82. Peter Rigby, "Examples of the Application of the New Approach," manuscript, ca. June 1978, Sydney Brenner personal papers.

83. Bennett, Glasner, and Travis, *Politics of Uncertainty,* 139.

84. John Subak-Sharpe, comments, conference organized by the Committee

on Genetic Experimentation and the Royal Society, in *Recombinant DNA and Genetic Experimentation,* ed. Joan Morgan and William J. Whelan (Oxford: Pergamon Press, 1979), 237–38.

85. Interview with Mark Richmond, 23 April 1980.

86. Much supporting evidence for the way in which Brenner's original scheme was simplified is provided by interviews with those involved in the development of the scheme, from the paper eventually published by GMAG in *Nature* ("Genetic Manipulation: New Guidelines for UK," *Nature* 276 [9 November 1978]: 104–8), and from the examples of application of the scheme worked out by Peter Rigby in June 1978 and circulated to GMAG ("Examples of the Application of the New Approach").

87. Document written circa July 1978, quoted in an interview in 1980 (anonymity requested). The same point was made by Frances Rolleston, director of special programs for the Medical Research Council of Canada, in a commentary in *Nature* 281 (25 October 1979): 626–27.

88. Interview with Mark Richmond, 23 April 1980.

89. Genetic Manipulation Advisory Group, "Genetic Manipulation," 104–8.

90. Jerome Ravetz to Keith Gibson, 30 October 1978, Jerome Ravetz personal papers.

91. Interview with Donna Haber, 15 June 1979.

92. Transcript of London Weekend Television program, part 2, "Interview with Shirley Williams on the Risks," 5 November 1978, London Weekend Television files.

93. Interview with Derek Ellwood, 21 April 1980.

94. GMAG, "Genetic Manipulation," 102–8.

95. "Call for International Control on Genetic Experiments," *Guardian,* 9 November 1978; "Genetic Manipulation: New Guidelines," *Times,* 10 November 1978; "Rational Risk Assessment for Genetic Engineering," *New Scientist,* 9 November 1978, 421.

96. Eleanor Lawrence, "Bacteriologists Lobby GMAG's First Public Meeting," *Nature* 277 (4 January 1979): 3.

97. "Genetic Engineering: How Safe Is It?"

98. "Now Reason Can Prevail," *Nature* 276 (9 November 1978): 103; European Molecular Biology Organization, Standing Advisory Committee on Recombinant DNA, *Report,* fifth meeting, London, 2–3 December 1978, EMBO files.

99. Fifth meeting of the EMBO Standing Advisory Committee on Recombinant DNA, *Report.*

9. Dismantling the National Institutes of Health Controls

1. Michel Foucault, "Truth and Power," in Foucault, *Power/Knowledge,* 133.

2. U.S. Congress, Office of Technology Assessment, *Impacts of Applied Genetics: Micro-organisms, Plants, and Animals* (Washington, D.C.: U.S. Government Printing Office, 1981), 43.

3. "The Promise of Gene-Splicing," *New York Times,* 19 January 1980, 22.

4. Department of Defense, "Annual Report on Chemical Warfare and Biological Research Programs (October 1, 1978–September 30, 1979), 30 November 1979," *Congressional Record,* 5 August 1980, S10852–68.

5. Department of Defense, *Annual Report on Chemical Warfare and Biological Research Programs, FY 1980* (15 December 1980), sec. 2, p. 4, Office of the Assistant to the Secretary of Defense for Atomic Energy, files.

6. Advertisement for contracts, *Science* 209 (12 September 1980): 1282; "DoD Offers Dollars for Far-out Research Ideas in Molecular Detection," *McGraw-Hill's Biotechnology Newswatch* 1 (21 September 1981): 1–2.

7. Wright, "Evolution of Biological Warfare Policy," 43–60.

8. Senate Committee on Commerce, Science, and Transportation, Subcommittee on Science, Technology, and Space, *Hearing: Industrial Application of Recombinant DNA Techniques,* 96th Cong., 2d sess., 20 May 1980, 26; see also testimony of Steven Turner, president of Bethesda Research Laboratories, 48.

9. Ibid., 23–24.

10. Ibid., 32, 37.

11. Department of Health, Education, and Welfare, minutes of meeting between DHEW General Counsel Peter Libassi and representatives of the PMA, 25 May 1979, OSDHEW.

12. Hal Lancaster, "Gene Splicers Ponder Mass Production: Will Fermenting Pose Major Problems?" *Wall Street Journal,* 28 August 1981, 28.

13. Robert Swanson to Donald Fredrickson, 9 October 1978, ODNIH.

14. William Gartland to Robert A. Swanson, 9 January 1979, ODNIH.

15. William Gartland, "Telephone Conversation with Dr. Robert Swanson," memo, 14 March 1979 (copies to Donald Fredrickson, Joseph Perpich, Bernard Talbot, Burke Zimmerman, Ruth Kirschstein, Richard Krause), ODNIH.

16. Department of Health, Education, and Welfare, minutes of meeting between DHEW General Counsel Peter Libassi and representatives of the PMA, 25 May 1979, OSDHEW.

17. "Insulin Research Raises Debate on DNA Guidelines," *New York Times,* 29 June 1979, A18.

18. Robert Locke, "His Bugs May Alter Life" (AP story), *Ann Arbor News,* 28 January 1979, F1, F5.

19. E.g., Sheldon Samuels (director of health and safety for the AFL-CIO) to Donald Fredrickson, 23 May 1980, in NIH, *Recombinant DNA Research* (April 1981), 6:290.

20. Interview with Eula Bingham, 19 June 1986.

21. Interview with Anthony Robbins, 28 April 1986.

22. Interview with Robert Williamson, 21 April 1980.

23. K. Sargeant and C. G. T. Evans, *Hazards Involved in the Industrial Use of Micro-organisms* (Brussels-Luxembourg: Commission of the European Communities, 1979).

24. Joseph Califano to the Assistant Secretary for Health and the Director, National Institutes of Health, 15 December 1978, *Federal Register* 43 (22 December 1978): 60082.

25. Michael Stoker, "Introduction and Welcome," in *Recombinant DNA and Genetic Experimentation,* ed. Joan Morgan and William J. Whelan (Oxford: Pergamon Press, 1979), ix–xx. Hereafter pages given in text.

26. "Molecular Biology: Suffering from Shock," *Nature* 278 (12 April 1979): 587.

27. Edward Yoxen, *The Gene Business: Who Should Control Biotechnology?* (New York: Harper and Row, 1983), 45.

28. Ibid. Setlow's remark was edited out of the official transcript of the Wye meeting.

29. Ibid.

30. "Molecular Biology: Suffering from Shock."

31. Interview with Robin Weiss, 18 June 1979.

32. Ibid.

33. Ibid.

34. Secretary of Health, Education, and Welfare, Charter, Recombinant DNA Advisory Committee, 30 June 1978, ORDAR; Secretary of Health and Human Services, Charter, Recombinant DNA Advisory Committee, 26 June 1980, ORDAR. Cf. Secretary of Health, Education, and Welfare, Charter, Recombinant DNA Molecule Program Advisory Committee, 7 October 1974, ORDAR.

35. Fredrickson, "Science and the Cultural Warp."

36. Secretary of Health and Human Services, Recombinant DNA Advisory Committee charter, 26 June 1989, ORDAR, doc. 923.

37. Department of Health, Education, and Welfare, minutes of meeting with representatives of the PMA, 13 October 1978, OSDHEW.

38. Interview with John Adams, 13 December 1979.

39. Interview with Rick Cotton, 26 November 1985.

40. Interview with Sheila Pires, 26 November 1985.

41. Ibid.

42. One of the scientists referred to by Singer was Jonathan King, who, like Baltimore, was a professor of biology at MIT and had played a leading role as a critic of NIH policy.

43. Singer, "Spectacular Science and Ponderous Progress," 9.

44. At least three of the members of the RAC appointed in 1979 had ties as consultants or equity owners with genetic engineering and pharmaceutical firms: David Baltimore (Collaborative Research), Milton Zaitlin (Plant Genetics), and Frank Young (Miles Laboratories). At least three RAC members appointed in 1980 and 1981 also had industry ties: Winston Brill (Cetus), King Holmes (Genex), and David Martin (Genentech). I thank Sheldon Krimsky for providing this information.

45. Interviews with nonscientist members of the RAC, June–July 1986.

46. Senate Subcommittee on Science, Technology, and Space, *Hearing: Industrial Applications of Recombinant DNA Techniques,* 62.

47. For membership of the "kitchen RAC," see Donald S. Fredrickson, "The Recombinant DNA Controversy: The NIH Viewpoint," in *The Gene-Splicing Wars: Reflections on the Recombinant DNA Controversy,* ed. Raymond A. Zilinskas and Burke K. Zimmerman (New York: Macmillan, 1986), 25.

48. Donald Fredrickson, introductory remarks recorded at the fourteenth meeting of the RAC, 15–16 February 1979, ORDAR.

49. Interview with Elena Nightingale, 7 June 1986.

50. William Gartland, quoted in Susan Wright, "The Recombinant DNA Advisory Committee," *Environment* 21 (April 1979): 3.

51. Ibid., 2–5.

52. RAC minutes, 21–23 May 1979.

53. The isolation of the public members of the RAC has also been noted by historian of science Charles Weiner, a frequent observer of the committee in the 1970s and 1980s: see his "Relations of Science, Government, and Industry."

54. Department of Health, Education, and Welfare, National Institutes of Health, "Recombinant DNA Research: Proposed Actions under Guidelines," part 6, "Proposed Exemption under I-E-5 for Experiments Involving EK1 and EK2 Host-Vector Systems," *Federal Register* 44 (13 April 1979): 22314–16.

55. Appendix B provides details of the sources of this information, the main experimental results achieved, proposals for further investigation, and the use of this information in policy documents. Table 9.5 summarizes this material.

56. "Risk Assessment Protocols for Recombinant DNA Experimentation," *Journal of Infectious Diseases* 137 (May 1978): 704–8; John E. Nutter, Chief OSFR, NIAID, "Major Recommendations of Ad Hoc NIAID Working Group on Risk Assessment," 4 September 1979, in NIH, *Recombinant DNA Research* (March 1980), 5:319–20; Ad Hoc Working Group on Risk Assessment, NIAID, 30 August 1979, transcript, ORDAR.

57. V. Petrocheilou and M. H. Richmond, "Absence of Plasmid or *Escherichia coli* K12 Infection among Laboratory Personnel Engaged in R-Plasmid Research," *Gene* 2 (1977): 323–27. The Petrocheilou-Richmond results were summarized by Richmond in a brief report to the Risk Assessment Subcommittee of COGENE in March 1979 and circulated to the RAC by Wallace Rowe as part of ORDAR, doc. 670.

58. Paul Burnett (Eli Lilly Laboratories) to Wallace Rowe, 8 May 1979, ORDAR; Donald Brown, Carnegie Institution of Washington, Baltimore, Md., table of unpublished data on growth competition between *E. coli* K12 and the same organism containing recombinant DNA molecules, circulated to the RAC by Wallace Rowe as part of ORDAR, doc. 670 on 4 May 1979; Hardy Chan, David Botstein, et al., "Testing of *E. coli* K12 Carrying a Eukaryotic Shotgun Mixture for Altered Pathogenicity," ORDAR, doc. 677, circulated to the RAC by Wallace Rowe on 10 May 1979.

59. Wallace Rose to Frank Dixon, 29 March 1979, reprinted in NIH, *Recombinant DNA Research* 5:239–40.

60. Jonathan Beckwith to William Gartland, 26 October 1978, ORDAR, doc. 616, in NIH, *Recombinant DNA Research* 5:241–42.

61. Stuart Levy and Bonnie Marshall, "Survival of *E. coli* Host-Vector Systems in the Human Intestinal Tract," *Recombinant DNA Technical Bulletin* 2 (July 1979): 77–80.

62. Rolf Freter et al., "Testing of Host-Vector Systems in Mice," *Recombinant DNA Technical Bulletin* 2 (July 1979): 68–76.

63. Levy and Marshall, "Survival of *E. coli* Host-Vector Systems," 80.

64. Roy Curtiss to Donald Fredrickson, 11 May and 4 October 1979, in NIH, *Recombinant DNA Research* 5:260–64, 339–40.

65. See, e.g., comment of Susan Gottesman in RAC minutes, 6–7 September 1979, 13.

66. Jonathan Beckwith to William Gartland, 26 October 1978, ORDAR, doc. 616, in NIH, *Recombinant DNA Research* 5:241–42.

67. See, e.g., Frank J. Dixon to Wallace Rowe, 2 April 1979, reprinted in NIH, *Recombinant DNA Research* 5:245–46.

68. William E. Paul and Richard Asofsky to Wallace Rowe, 18 May 1979, reprinted in NIH, *Recombinant DNA Research* 5:270–75.

69. RAC minutes, 6–7 September 1979, 11.

70. Paul Berg to Donald Fredrickson, 17 December 1979, in NIH, *Recombinant DNA Research* 5:505.

71. Roy Curtiss III to Donald Fredrickson, October 4, 1979.

72. Minutes of RAC meeting, 15–16 January 1977, 9.

73. Mark A. Israel et al., "Molecular Cloning of Polyoma Virus DNA in *Escherichia coli:* Plasmid Vector System," *Science* 203 (2 March 1979): 883–87; Hardy W. Chan et al., "Molecular Cloning of Polyoma Virus DNA in *Escherichia coli:* Lambda Phage Vector System," *Science* 203 (2 March 1979): 887–92.

74. In the case of the *E. coli*–phage-dimer combination, the researchers stated that "the extremely low efficiency of lambda phage production in the absence of aerobic conditions, as would be the case in the tissues of the animal, makes it quite likely that the same reduction in activity would have been seen" (Mark A. Israel et al., "Molecular Cloning of Polyoma Virus DNA: Oncogenicity Testing in Hamsters," *Science* 205 [14 September 1979]: 1142).

75. Robert Rohwer, personal communication, 1979.

76. Mark A. Israel et al., "Interrupting the Early Region of Polyoma Virus DNA Enhances Tumorigenicity," *Proceedings of the National Academy of Sciences* 76 (August 1979): 3714.

77. Chan et al., "Molecular Cloning," 892.

78. Israel et al., "Molecular Cloning . . . : Plasmid Vector System," 887; Chan et al., "Molecular Cloning," 892.

79. Chan et al., "Molecular Cloning," 892.

80. Israel et al. "Molecular Cloning . . . : Oncogenicity Testing in Hamsters," 1142.

81. Ibid., 1141, 1142.

82. Chan et al., "Molecular Cloning," 892; Israel et al., "Molecular Cloning . . . : Oncogenicity Testing in Hamsters," 1142.

83. Eliot Marshall, "Gene Splicers Simulate a 'Disaster,' Find No Risk," *Science* 203 (23 March 1979): 1223.

84. Harold Schmeck, "Experiment Indicates Little Risk In Some Gene-Splicing Research," *New York Times,* 2 March 1979, A13. See also "Infection Risk Small in Some DNA Experiments," *Chemical and Engineering News,* 12 March 1979, 7.

85. Jonathan King, Ethan Signer, David Ozonoff, and Stuart Newman to the Editors, *Science,* 11 April 1979, Jonathan King personal papers. This letter was

rejected for publication by *Science* on 21 January 1980. A shorter version, signed by King and Signer, was published in the *New York Times* on 3 May 1979.

86. Barbara Rosenberg and Lee Simon, "Recombinant DNA: Have Recent Experiments Assessed All the Risk?" *Nature* 282 (20/27 December 1979): 773.

87. King and Signer, letter to the *New York Times*, 3 May 1979.

88. Indeed, the results of the experiments on tumor induction in hamsters published by Rowe in September 1979 confirmed this prediction: compare King et al., unpublished letter to *Science*, 11 April 1979, and Israel et al., "Molecular Cloning . . . : Oncogenicity Testing in Hamsters," 1140–42.

89. In more technical terms, the lambda vector used in the Rowe-Martin experiments, known as λgtWES.λB, contained three amber mutations so that it was able to grow only in strains possessing the amber suppressor gene. Furthermore, it was not able to lysogenize in χ1776.

90. Cecil R. Smith and Malcolm A. Martin, "In Vivo Biological Activity of Polyoma-pBR322 Recombinants Cloned in WT *Escherichia coli* or *Escherichia coli* K12," unpublished paper, 1982, 4.

91. Sheldon Krimsky to Donald Fredrickson, 24 September 1979, in NIH, *Recombinant DNA Research* 5:328–39.

92. Wallace P. Rowe, "Statement to the NIH Recombinant DNA Advisory Committee," May 1979, ORDAR, doc. 670.

93. Meeting of RAC, 21–23 May 1979, transcript of personal tape recording.

94. David Baltimore and Allan Campbell, "Proposal to Convert the NIH Guidelines into a Non-regulatory Code of Standard Practice and the Reduce the Recommended Levels for Some Experiments," 11 February 1981, ORDAR, doc. 994.

95. RAC minutes, 21–23 May 1979, 15.

96. See, e.g., Kan Agarwal et al. to Donald Fredrickson, 14 June 1979, in NIH, *Recombinant DNA Research* 5:279–83.

97. Meeting of the Ad Hoc Working Group on Risk Assessment, NIAID, 30 August 1979, transcript, 2. Hereafter pages given in text.

98. The arguments associated with this method of measuring transfer events are analyzed further in Stuart Newman, "Down the Drain: A Short History of Recombinant DNA Risk Assessment" (1982), photocopy, New York Medical College, Valhalla. Newman draws attention to the repeated attempts by some participants to link the agreement on using a robust strain of *E. coli* for risk assessment with a claim for the safety of research with *E.coli* K12, and notes that the transcript shows no agreement on such a linkage.

99. Nutter, "Major Recommendations," in NIH, *Recombinant DNA Research* 5:319–20.

100. Department of Health, Education, and Welfare, National Institutes of Health, "Recombinant DNA Research: Proposed Actions under Guidelines," part 1, "Proposed Exemption for *E. coli* K12 Host-Vector Systems," *Federal Register* 44 (31 July 1979): 145; *E. coli* K-12 Host-Vector Systems Working Group, "Background Document for Proposed Exemption of *E. coli* K12 Host-Vector Systems from the Recombinant DNA Guidelines," ORDAR, doc. 708, emphasis in the original.

101. RAC minutes, 6–7 September 1979, 10.

102. Cf. statements of Rowe and Baltimore, ibid., 9, 12, and the discussion at the Bethesda meeting, registered in transcript, Enteric Bacteria meeting, 31 August 1976, transcript, 52–66.

103. RAC minutes, 6–7 September 1979, 7.

104. Ibid., 12.

105. Details of this vote are, for: David Baltimore, Winston Brill, Francis Broadbent, Allan Campbell, Werner Maas, Richard Novick, Ray Thornton, LeRoy Walters, Luther Williams, Milton Zaitlin; against: Karim Ahmed, Richard Goldstein, Susan Gottesman, Sheldon Krimsky; abstained: James Mason (Susan Wright, personal notes on meeting).

106. Adlai E. Stevenson to Patricia Harris, 21 November 1979, in NIH, *Recombinant DNA Research* 5:374–75.

107. Department of Health, Education, and Welfare, National Institutes of Health, "Proposed Guidelines for Research Involving Recombinant DNA Molecules," *Federal Register* 44 (30 November 1979): 69210–34.

108. Department of Health, Education, and Welfare, National Institutes of Health, "Recombinant DNA Research: Proposed Actions under Guidelines," *Federal Register* 44 (30 November 1979): 69241.

109. Richard Novick, quoted in Susan Wright, "Recombinant DNA Policy: From Prevention to Crisis Intervention," *Environment* 21 (November 1979): 37.

110. David Dickson, "U.S. Expected to Exempt Most Recombinant DNA Experiments from Federal Regulation," *Nature* 281 (13 September 1979): 90; Nicholas Wade, "Major Relaxations in DNA Rules," *Science* 205 (21 September 1979), 1238; Ann Roark, "Easing of DNA Rules Draws Fire from Scientists," *Chronicle of Higher Education,* 29 October 1979), 15; Harold Schmeck, "Exemptions from U.S. Restrictions Are Urged for Most DNA Research," *New York Times,* 8 September 1979, 1, 26; "Less Peril in Gene Splicing," *New York Times,* 21 September 1979, A26; Victor Cohn, "US Lifts Most Curbs on Gene Experiments," *Washington Post,* 30 January 1980, A1, A4; "DNA: Risks and Guidelines," *Washington Post,* 4 February 1980, A24; "NIH Relaxes Recombinant DNA Guidelines," *Chemical and Engineering News,* 4 February 1980, 7; Roger Lewin, "US to Relax Controls on Genetic Engineering," *New Scientist,* 13 September 1979, 787.

10. Dismantling the National Institutes of Health Controls but Preserving Quasi-regulation, 1980–1982

1. E.g., "The Promise of Gene-Splicing," *New York Times,* 19 January 1980, 22; "The Genetic Payoff," *Wall Street Journal,* 22 January 1980, 20; Wade, "Cloning Gold Rush," 668–93.

2. RAC minutes, 5–6 June 1980, 30–31.

3. Department of Health, Education, and Welfare, National Institutes of Health, "Guidelines for Research Involving Recombinant DNA Molecules," *Federal Register* 45 (29 January 1980): 6724–49.

4. Rowe, "Statement to the NIH Recombinant DNA Advisory Committee."

5. David Baltimore, testimony, Senate Committee on Commerce, Science, and Transportation, Subcommittee on Science, Technology, and Space, *Hearing: Industrial Applications of Recombinant DNA Techniques,* 64.

6. Harold S. Ginsberg to Donald Fredrickson, 21 December 1979, in NIH, *Recombinant DNA Research* 5:575.

7. Bernard Davis to Donald Fredrickson, 28 December 1979, in ibid., 618–19.

8. Data drawn from RAC minutes for February, May, September, and December 1979 and March, June, and September 1980.

9. Richard Krause, "Report on Proceedings of the Recombinant DNA Risk Assessment Workshop," Pasadena, Calif., 11–12 April 1980, ORDAR, doc. 898.

10. Transcript of Final Plenary Session, Report on Proceedings of the Recombinant DNA Risk Assessment Workshop, Pasadena, California, 11–12 April 1980, ORDAR, doc. 898.

11. "Significant Recommendations of the Pasadena Workshop," c. June 1980, ORDAR, doc. 898A; NIH, minutes of RAC meeting, 5–6 June 1980, 12–14, reprinted in NIH, *Recombinant DNA Research* 6:112–14.

12. Transcript of Final Plenary Session, 42–43.

13. See, e.g., "Evaluation of Risks Associated with Recombinant DNA Research," *Federal Register* 46 (4 December 1981): 59385–92, reprinted in NIH, *Recombinant DNA Research* 7 (December 1982), 274–81; section 11.3.

14. Diana Dutton, "Results of a Survey of Institutional Biosafety Committees in California," 22 September 1980, ORDAR, doc. 937. For the final report of this study, see Diana B. Dutton and John L. Hochheimer, "Institutional Biosafety Committees and Public Participation: Assessing an Experiment," *Nature* 297 (6 May 1982): 11–15. The results of a later study of safety procedures in six industrial genetic engineering facilities conducted by a NIOSH team were reported in Larry J. Elliott et al., "Industrial Hygiene Characterization of Commercial Applications of Genetic Engineering and Technology," CDC/NIOSH Report 131.17, and in Philip J. Landrigan, John M. Harrington, and Larry J. Elliott, "The Biotechnology Industry," in *Recent Advances in Occupational Health,* ed. John M. Harrington (Edinburgh: Churchill Livingstone, 1984), 2:3–13.

15. Quoted by Paul Berg in a letter to William Gartland, 22 April 1981, in NIH, *Recombinant DNA Research* 7:591–92.

16. Paul Berg to Maxine Singer, 29 July 1980, in NIH, *Recombinant DNA Research* 6:291.

17. National Institute of Allergy and Infectious Diseases, Institutional Biosafety Chairpersons' Meeting, 24–25 November 1980, Washington, D.C., edited transcript of Plenary Session 3, 9, 16, 18–19, 34–35, ORDAR, doc. 977.

18. RAC meeting, 8–9 January 1981, personal notes.

19. L. J. Lazear to William Gartland, 14 May 1981, in NIH, *Recombinant DNA Research* 7:596.

20. Department of Health, Education, and Welfare, National Institutes of Health, "Recombinant DNA Research: Proposed Actions under Guidelines," *Federal Register* 46 (20 March 1981): part 2, 17995.

21. Baltimore and Campbell, "Proposal to Convert the NIH Guidelines."

22. Letters of Norton Zinder, Maxine Singer, and Paul Berg to William Gartland in April 1981, in NIH, *Recombinant DNA Research* 7:581, 586–87, 591–92.

23. Recombinant DNA Advisory Committee, Working Group on Revision of the Guidelines, National Institutes of Health, RAC minutes, attachment 3, 10–11 September 1981.

24. See, e.g., David Baltimore's position (recorded in RAC minutes, 10–11 September 1981, 12), that review by IBCs was a "serious obstruction of science."

25. Ibid., 21–25.

26. Ray Thornton, statement at the meeting of the RAC, ibid., attachment V.

27. Department of Health, Education, and Welfare, National Institutes of Health, "Recombinant DNA Research; Proposed Revised Guidelines," *Federal Register* 46 (4 December 1981): 59368–59425.

28. Department of Health, Education, and Welfare, National Institutes of Health, "Recombinant DNA Research: Proposed Actions under Guidelines," *Federal Register* 46 (7 December 1981): 59734–37.

29. Stuart Newman, "The 'Scientific' Selling of rDNA," *Environment* 24 (July–August 1982): 56.

30. Ditta Bartels, "Oncogenes: Implications for the Safety of Recombinant DNA Work," *Search* 14 (April–May 1983): 88–92; Stephanie Yanchinski, "Oncogene Researchers May Run Cancer Risk," *New Scientist* (25 August 1983): 530; Ditta Bartels, "Genetische Manipulation an Krebsgenen," in *Die ungeklärten Gefahrenpotentiale der Gentechnologie,* ed. Regine Kollek, Beatrix Tappeser, and Gunter Altner (Munich: J. Schweitzer Verlag, 1986), 70–83; Ditta Bartels, "Escape of Cancer Genes?" *New Scientist* (30 July 1987): 52–54.

31. The last report of the risk assessment program appeared in 1982 (Department of Health and Human Services, National Institutes of Health, "Program to Assess the Risks of Recombinant DNA Research: Proposed Second Annual Update," *Federal Register* 47 [7 December 1982]: 55104–9, reprinted in NIH, *Recombinant DNA Research* 8 [May 1986], 58–63).

32. Numbers are based on attendance records included in the minutes of RAC meetings for 1979–82.

33. Califano later made similar requests to Bob Bergland, Secretary of Agriculture, and Ray Marshall, Secretary of Labor (Joseph Califano to Bob Bergland, 26 February 1979, OSDHEW; Joseph Califano to Ray Marshall, 9 July 1979, OSDHEW). See Federal Interagency Committee on Recombinant DNA Research, minutes, 17 July 1979, in *Recombinant DNA Research* 5:134. Bergland's reply indicates that the Department of Agriculture might use its authority "to the extent any products of recombinant DNA research or organisms containing recombinant DNA molecules could be classified as plant pests or plant pathogens." Neither agency took immediate action.

34. David Dickson, "US Drug Companies Push for Changes in Recombinant DNA Guidelines," *Nature* 278 (29 March 1979): 385–86.

35. Excerpts from minutes, meeting of DHEW with representatives of the PMA, 13 October 1978, OSDHEW. This meeting is described in chapter 8.

36. Department of Health, Education, and Welfare, National Institutes of Health, "Guidelines for Research Involving Recombinant DNA Molecules" (proposal for voluntary compliance by private institutions), n.d., discussed at meeting of DHEW general counsel Peter Libassi with representatives of the NIH and of public interest and labor organizations, 25 May 1979, ORDAR.

37. RAC meeting, 21–23 May 1979, transcript of personal tape recording.

38. Department of Health, Education, and Welfare, minutes of meeting between DHEW general counsel Peter Libassi and the PMA, 25 May 1979, OSDHEW. At this meeting, the PMA representatives also proposed corporate representation on the RAC, authorization of IBCs to approve new host-vector systems, and even more stringent control of sensitive industry information.

39. RAC minutes, 6–7 September 1979, 16.

40. IAC minutes, 17 July 1979, in NIH, *Recombinant DNA Research* 5:132–43.

41. Department of Health, Education, and Welfare, National Institutes of Health, "Proposed Supplement to the NIH Guidelines for Recombinanat DNA Research," *Federal Register* 44 (3 August 1979), 45868.

42. Comment of Melvin Myers, NIOSH, RAC minutes, 6–7 September 1979.

43. "Proposed Settlement," 45868–69.

44. Donald Fredrickson, remarks to RAC, 6–7 September 1979, transcript of personal tape recording.

45. Ibid. The statements of David Baltimore and Sheldon Krimsky were not recorded in the RAC minutes.

46. National Institutes of Health, Office of Recombinant DNA Activities, "Recombinant DNA Large-Scale Proposals Approved by NIH as of November 3, 1981," ORDAR.

47. RAC meeting, 6–7 September 1979, transcript of personal tape recording.

48. RAC meeting, 6–7 March 1980, transcript of personal tape recording.

49. Ibid.

50. Ibid.

51. RAC meeting, 5–6 June 1980, transcript of personal tape recording.

52. Comments of Ray Thornton and LeRoy Walters, RAC minutes, 6–7 December 1979, 24.

53. RAC meeting, 6–7 March 1980, transcript of personal tape recording.

54. RAC minutes, 6–7 March 1980, 55.

55. Adlai Stevenson to Patricia Roberts Harris, 21 November 1979, in NIH, *Recombinant DNA Research* 5:374–75.

56. Eula Bingham to Donald Fredrickson, 17 December 1979, in NIH, *Recombinant DNA Research* 6:342–44.

57. Interview with Gilbert Omenn, 12 May 1986.

58. E.g., Fredrickson's response to a question concerning health surveillance in industry (RAC minutes, 5–6 June 1980, 7).

59. Donald Fredrickson, remarks to the RAC on voluntary compliance procedures, in ibid., 4–6.

60. Industrial Practices Subcommittee of the Federal Interagency Advisory Committee on Recombinant DNA Research, minutes, 18 June 1980, attachment H, in NIH, *Recombinant DNA Research* 6:437–53. The same arguments were made by Irving Johnson and David Baltimore in testimony to the Senate Subcommittee on Science, Technology, and Space a month earlier.

61. RAC meeting, 5–6 June 1980, personal notes.

62. Irving Johnson to William Gartland, 20 July 1981, ORDAR, doc. 1027.

63. RAC minutes, 10–11 September 1981, 31–35; Department of Health and Human Services, National Institutes of Health, "Recombinant DNA Research: Actions under the Guidelines," *Federal Register* 46 (30 October 1981): 52981. See also section 12.4.

64. Allan Campbell to Donald Fredrickson, 4 January 1982, in NIH, *Recombinant DNA Research* 7:632–34.

65. Harvey S. Price to William Gartland, 1 February 1982, in ibid. 7:753–57.

66. Michael Ross to William Gartland, 1 February 1982, in ibid. 7:750–52.

67. William N. Lipscomb to William Gartland, 28 January 1982, in ibid. 7:710.

68. Paul Berg to William Gartland, 22 January 1982, in ibid. 7:689–90.

69. Comment of Mark Saginor, RAC minutes, 8–9 February 1982, 15.

70. Ibid., 14.

71. Revised guidelines based on the RAC's advice were issued in August 1982 (Department of Health and Human Services, National Institutes of Health, "Guidelines for Research Involving Recombinant DNA Molecules," *Federal Register* 47 (27 August 1982): 38048–68.

72. Omenn, "Government as a Broker," 38.

73. Fredrickson, "Science and the Cultural Warp," 31, 18.

11. Dismantling the Genetic Manipulation Advisory Group, 1979–1984

1. Tony Vickers to Susan Wright, 10 October 1978.

2. Interview with Keith Gibson, 28 April 1979 and 11 June 1979; Genetic Manupulation Advisory Group, *Third Report of the Genetic Manipulation Advisory Group*, Cmnd. 8665 (September 1982), 8.

3. "Industry Starts to Do Biology with Its Eyes Open," *Economist*, 2 December 1978, 95.

4. House of Commons, Select Committee on Science and Technology, *Second Report*, 218, 221.

5. Ibid., 92.

6. Telephone conversation with David Dickson, ca. April 1979.

7. For the response of Department of Education and Science officials, see Hillary Benn to Clive Jenkins, 25 January 1979, MRCUW, MSS.79/AS/3/8/55-58, and U.K. Department of Education and Science, "Explanatory Memorandum on European Community Legislation 5899/70," 25 July 1979, MRCUW, MSS.79/AS/3/8/55-58.

8. See, e.g., Confederation of British Industry, "Policy Issues Arising from Genetic Engineering," in House of Commons, Select Committee on Science and Technology, *Second Report*, 224.

9. Trades Union Congress, "Memorandum," in ibid., 250–51. A European Economic Community directive was not adopted until 1990.

10. Select Committee on Science and Technology, *Second Report*, viii.

11. For further analysis of the British government's promotion of biotechnology, see Edward Yoxen, "Assessing Progress with Biotechnology," in *Science and Technology Policy in the 1980s and Beyond*, ed. Michael Gibbons, Philip Gummett, and Bhalchandar Udgaonkar (London: Longman, 1984), 207–24.

12. Walgate, "Biotechnology Report Urges £10 Million Programme," 324–25.

13. Advisory Council for Applied Research and Development/Advisory Board for the Research Councils and the Royal Society, *Biotechnology: Report of a Joint Working Party* (London: HMSO, March 1980).

14. Ibid., 43.

15. House of Lords, Select Committee on the European Communities, *Genetic Manipulation (DNA)* (report on proposal for a council directive establishing safety measures against the conjectural risks associated with recombinant DNA work with minutes of evidence), sess. 1979–80, 4 March 1980.

16. Royal Society, "Biotechnology and Education: Report of a Working Group," 1981, Royal Society, London.

17. House of Commons, Education, Science, and Arts Committee, *Biotechnology: Interim Report on the Protection of the Research Base in Biotechnology*, sess. 1981–82, 27 July 1982.

18. United Kingdom, *Biotechnology*, Cmnd. 8177 (March 1981).

19. "Scientists Slam White Paper on Biotechnology," *New Scientist*, 12 March 1981, 660; Madeleine Vaquin, "Biotechnology in Britain," draft contract report prepared for the U.S. Office of Technology Assessment, October 1981, 3–4.

20. Vaquin, "Biotechnology in Britain," i.

21. "Science after the Cuts," *New Scientist*, 13 August 1981, 396–97; "Mighty Microbe?" *New Scientist*, 13 November 1980, 411; "Britain's Biotechnology Company Takes Off," *New Scientist*, 13 November 1980, 413.

22. Robin McKie, "Britain's Shadow Minister Believes in Experts," *Nature* 278 (29 March 1979): 63; United Kingdom, *Report on Non-departmental Public Bodies;* Michael Hatfield, "Measures of 'Quango Cull' May Be Severe," *Times*, 12 January 1980, 3; George Clark, "Champion Hunter of Quangos Offers His Death List of 707," *Times*, 25 September 1980, 4.

23. Interview with Robert Williamson, 7 June 1984.

24. Interviews with officials in the Department of Education and Science, the Department of Health and Human Services, and the MRC, and with members of GMAG, April and May 1980.

25. Interview with William Henderson, 30 April 1980; see also Bennett, Glasner, and Travis, *Politics of Uncertainty*, 102.

26. Interviews with trade union representatives on GMAG, April and May 1980; see also Donna Haber to Clive Jenkins, 4 June 1979, MRCUW, MSS.79/AS/3/8/58.

27. Robert Walgate, "Genetic Manipulation: Britain May Exempt 'Self-Cloning,'" *Nature* 277 (22 February 1979): 589.

28. In September 1979, the Technical Panel was reconstituted, with a broader membership, as a Technical Subcommittee, assuming the responsibilities of both the panel and the former Safe Vectors Subcommittee.

29. Jerome Ravetz to Keith Gibson, executive secretary of GMAG, 30 October 1978; Ravetz to Shirley Williams, 8 November 1978, Jerome Ravetz personal papers.

30. Interviews with public interest representatives, GMAG, April and May 1980.

31. Interview with William Henderson, 30 April 1980.

32. Further minor changes in the composition of GMAG occurred at the end of 1981, when all the memberships terminated at the end of the committee's second two-year term and few members opted not to be reappointed.

33. Interview with Keith Gibson, 11 June 1979.

34. Lawrence, "Bacteriologists Lobby GMAG's First Public Meeting," 3; "GMAG Debates New Controls on Genetic Engineering," 4.

35. "Flaws in GMAG's Guidelines," *Nature* 277 (15 February 1979): 509–10, statement issued by the Council of the Royal Society and prepared by an ad hoc group chaired by Walter F. Bodmer.

36. Alan R. Williamson, "GMAG Should Look to NIH" (letter), *Nature* 277 (1 February 1979): 346.

37. Interview with Robert Williamson, 13 May 1980. Acquiring valid risk assessment data, as we have seen, was apparently already a dead issue as far as GMAG was concerned.

38. Confederation of British Industry, "Policy Issues Arising from Genetic Engineering," in Select Committee on Science and Technology, *Second Report*, 222.

39. "Now Reason Can Prevail," *Nature* 276 (9 November 1978): 103.

40. Interview with Robert Williamson, 13 May 1980.

41. "It Takes a Death: We Publish Professor Shooter," *Medical World*, January 1979, 2–13; "All Safety Nets Failed, Says Shooter," *Nature* 277 (11 January 1979): 78–79; Nigel Hawkes, "Science in Europe: Smallpox Death in Britain Challenges Presumption of Laboratory Safety," *Science* 203 (2 March 1979): 855–56.

42. Interview with Mark Richmond, 23 April 1980.

43. John Maddox, "GMAG: NIH Guidelines Are Not the Way," *Nature* 278 (1 March 1979): 10, emphasis added. Note the presumption, early in 1979, that controls should be eliminated.

44. Interview with Robin Weiss, 18 June 1979.

45. GMAG, *Second Report*, 40–41.

46. Initially, researchers were given a choice of submitting their proposals under the Williams scheme or under the new scheme, and they were encouraged to provide evidence that would enable their experiments to be reassessed. After a trial period, which ended in the summer of 1979, the new scheme replaced the Williams categorizations entirely.

47. Bennett, Glasner, and Travis, *Politics of Uncertainty*, 105.

48. GMAG Note 11, "The GMAG Categorisation Scheme," in GMAG, *Third Report*, 40–45; GMAG Note 14, "Revised Guidelines for the Categorisation of Recombinant DNA Experiments," in GMAG, *Third Report*, 54–59.

49. GMAG, *Third Report*, 54.

50. For discussion of the development of these categories during the evolution of the risk assessment scheme, see section 9.4.

51. Interview with Peter Walker, 2 May 1980.

52. GMAG, *Second Report*, 4.

53. Judy Redfearn, "UK Genetic Engineering Regulations Relax Gradually," *Nature* 282 (20/27 December 1979): 769; GMAG, *Third Report*, 126.

54. GMAG, *Third Report*, 55.

55. Interview with Peter Walker, 2 May 1980.

56. Redfearn, "UK Genetic Engineering Regulations Relax Gradually," 769.

57. Interview with Robin Weiss, 18 June 1979.

58. Interview with Donna Haber, 13 May 1980.

59. GMAG, *Third Report*, 6.

60. Ibid.

61. GMAG Note 11.

62. Interview with William Henderson, 30 April 1980.

63. Bennett, Glasner, and Travis, *Politics of Uncertainty*, 110.

64. "Still Looser UK Guidelines," *Nature* 287 (25 September 1980): 265–66.

65. GMAG, *Third Report*, 6.

66. "Will the DNA Guidelines Wither Away?" *Nature* 287 (9 October 1980): 474.

67. "Industry Starts to Do Biology with Its Eyes Open," 95–96; Pearce Wright, "Time for Bug Valley," *New Scientist,* 5 July 1979, 27–29; Dickson, "US Drug Companies Push," 385–86.

68. Confederation of British Industry, "Issues Arising from Scale-up of Recombinant DNA," 26 March 1981, ORDAR, doc. 1027, circulated to GMAG before its meeting on 3 April 1981. See also Confederation of British Industry, "Policy Issues Arising from Genetic Engineering," 221–22.

69. GMAG Note 4, in GMAG, *First Report,* 49–51.

70. GMAG, *Second Report,* 13. A similar position was stated in GMAG Note 8, reprinted in the same report (38–41).

71. GMAG, *Third Report,* 7.

72. Sargeant and Evans, *Hazards Involved in the Industrial Use of Micro-organisms.* For discussion of the analysis given in this report, see section 9.3.

73. GMAG Note 12 (revised February 1982), "Large-Scale Use of Genetically Manipulated Organisms," in GMAG, *Third Report,* 46–49.

74. Cf. arguments of Irving Johnson before the Senate Subcommittee on Science, Technology, and Space, the IAC, and RAC in 1980 (chapter 9).

75. Confederation of British Industry, "Issues Arising from Scale-up of Recombinant DNA."

76. Ibid.

77. Select Committee on Science and Technology, *Second Report,* 39.

78. Interview with William Henderson, 30 April 1980; William Henderson to William Gartland, 31 January 1980, ORDAR, doc. 850; Select Committee on Science and Technology, *Second Report,* 39.

79. GMAG Note 12.

80. William Henderson to William Gartland, 31 January 1980, ORDAR, doc. 850.

81. See, e.g., James D. Punch and John H. Coats (cochairs, Upjohn Recombinant DNA Advisory Committee) to Donald Fredrickson, 25 January 1980, ORDAR, doc. 835; Irving S. Johnson (vice president, Eli Lilly and Company) to Donald S. Fredrickson, 30 November 1979, ORDAR, doc. 840.

82. William Henderson to William Gartland, 31 January 1980, ORDAR, doc. 850. Henderson refers to a letter from Gartland to Keith Gibson, executive secretary of GMAG, in which Gartland mentions "comments by Dr. Irving Johnson

about levels of containment lower than 'P2-LS' being approved in the UK for large scale work."

83. William Henderson to William Gartland, 31 January 1980, ORDAR, doc. 850; Keith Gibson to William Gartland, 29 January 1980, ORDAR, doc. 851.

84. Department of Health, Education, and Welfare, National Institutes of Health, "Recombinant DNA Research: Physical Containment Recommendations for Large-Scale Uses of Organisms Containing Recombinant DNA Molecules," *Federal Register* 45 (11 April 1980): 24968–71.

85. Elizabeth A. Milewski, "The NIH Guidelines for Research Involving Recombinant DNA Molecules," in *Recombinant DNA Products: Insulin, Interferon and Growth Hormone,* ed. Arthur P. Bolton (Boca Raton, Fla.: CRC Press, 1984), table 6.

86. Confederation of British Industry, "Issues Arising from Scale-up of Recombinant DNA."

87. GMAG Note 12 (revised February 1982).

88. Donna Haber to Clive Jenkins, 30 July 1981, MRCUW, MSS.79/AS/ 3/8/58.

89. Ibid.

90. Irving Johnson to William Gartland, 20 July 1981, ORDAR, doc. 1027.

91. Confederation of British Industry, "Policy Issues Arising from Genetic Engineering," 223.

92. Interview with Jack Gilbey, May 1980.

93. Interview with Donna Haber, 15 June 1979.

94. Interview with Keith Gibson, 11 June 1979.

95. Interview with William Henderson, 30 April 1980.

96. Ibid.

97. Bennett, Glasner, and Travis, *Politics of Uncertainty,* 111.

98. Ibid., 113.

99. Ibid., 114–15.

100. Interview with Keith Gibson, 28 April 1980.

101. GMAG, *Third Report,* 10.

102. Interview with Robert Williams, 15 June 1984.

103. Only eight meetings were held from 1981 until the last meeting of the committee in 1984 (interview with Keith Gibson, 28 April 1980).

104. Bennett, Glasner, and Travis, *Politics of Uncertainty,* 121.

105. "Reconstitution of the Genetic Manipulation Advisory Group (GMAG)," report to Ministers and the Health and Safety Commission by officials of the Department of Education and Science and the Health and Safety Executive, April 1983, GMAG records, MRC, London, 5.

106. Interview with Robert Williams, 14 June 1984.

12. Molecular Politics in a Global Economy

1. U.S. Congress, Office of Technology Assessment, *Biotechnology in a Global Economy* (Washington, D.C.: U.S. Government Printing Office, 1991), 77.

2. Ibid., 47, 77, 80, 84. Cf. the projection of J. Leslie Glick in 1982 for a $40 billion market for genetically engineered products by the year 2000 (section 3.5).

3. Ibid., 78–81.

4. Ibid., ch. 6; Barnaby Feder, "The 'Pharmers Who Breed Cows That Can Make Drugs," *New York Times*, 9 February 1992, F9.

5. See, e.g., Andrew Pollack, "Gene Therapy Gets the Go-Ahead," *New York Times*, 14 February 1992, C1, C4.

6. Office of Technology Assessment, *Biotechnology in a Global Economy*, chs. 7, 8.

7. U.S. Department of Defense, Department of the Army, "Biological Defense Program," report to the House Committee on Appropriations, May 1986, Office of the Assistant to the Secretary of Defense for Atomic Energy files, chs. 1, 13.

8. U.S. Department of Defense, *Critical Technologies Plan* (Washington, D.C.: U.S. Government Printing Office, 1990); U.S. Department of Defense, *Report of the National Critical Technologies Panel* (Washington, D.C.: U.S. Government Printing Office, 1991). For further discussion and analysis of these proposals, see Susan Wright, "Prospects for Biological Disarmament," *Transnational Law and Contemporary Problems* 2 (Fall 1992): 453–92.

9. U.S. Congress, Office of Technology Assessment, *New Developments in Biotechnology: 4. U.S. Investment in Biotechnology* (Washington, D.C.: U.S. Government Printing Office, 1988), 36, 80.

10. These percentages are calculated using data provided in Office of Technology Assessment, *Biotechnology in a Global Environment*, appendix C.

11. Office of Technology Assessment, *New Developments in Biotechnology: 4*, 80.

12. Ibid., 19.

13. Ibid., 45–69.

14. Joan Spero, "Guiding Global Finance," *Foreign Policy* 73 (1988–89): 114–34.

15. For details of these projects and the controversies surrounding them, see Sheldon Krimsky, "Release of Genetically Engineered Organisms into the Environment," in Sheldon Krimsky and Alonzo Plough, *Environmental Hazards: Communicating Risks as a Social Process* (Dover, Mass.: Auburn House Publishing Company, 1988), 75–129.

16. Winston Brill, "Safety Concerns and Genetic Engineering in Agriculture," *Science* 227 (25 January 1985): 381–84.

17. Robert K. Colwell et al., "Genetic Engineering in Agriculture," *Science* 229 (12 July 1985): 111–12.

18. See, e.g., National Academy of Sciences, Committee on the Introduction of Genetically Engineered Organisms into the Environment, *Introduction of Recombinant DNA–Engineered Organisms into the Environment: Key Issues* (Washington, D.C.: National Academy Press, 1987); National Academy of Sciences, Committee on Scientific Evaluation of the Introduction of Genetically Modified Microorganisms and Plants into the Environment, *Field Testing Genetically Modified Organisms: Framework for Decisions* (Washington, D.C.: National Academy Press, 1989); James M. Tiedje et al., "The Planned Introduction of Genetically Engineered Organisms: Ecological Considerations and Recommendations," *Ecology* 70 (1989): 298–315.

19. National Academy of Sciences, *Introduction of Recombinant DNA–Engineered Organisms into the Environment*, 22. A further report by two NAS committees that included several ecologists was more nuanced: *Field Testing Genetically Modified Organisms*.

20. Tiedje et al., "Planned Introduction of Genetically Engineered Organisms," 298–99.

21. For discussion of the details of relevant statutes, see Valerie Fogleman, "Regulating Science: An Evaluation of the Regulation of Biotechnology Research," *Environmental Law* 17, no. 2 (1987): 183–273; Gary Marchant, "Modified Rules for Modified Bugs: Balancing Safety and Efficiency in the Regulation of Deliberate Release of Genetically Engineered Microorganisms," *Harvard Journal of Law and Technology* 1 (Spring 1988): 163–208; Sidney Shapiro, "Biotechnology and the Design of Regulation," *Ecology Law Quarterly* 17, no. 1 (1990): 1–70. Margaret Mellon, *Biotechnology and the Environment* (Washington, D.C.: National Wildlife Federation, 1988), 37–49. For criticism of the use of existing law, see, e.g., Jack Doyle, Director, Agriculture and Biotechnology Project, Environmental Policy Institute, to Ron Evans, Office of Toxic Substances, EPA, 16 May 1989, cited in Shapiro, "Biotechnology and the Design of Regulation," 22–23.

22. U.S. Office of Science and Technology Policy, "Coordinated Framework for Regulation of Biotechnology: Establishment of the Biotechnology Science Coordinating Committee," *Federal Register* 50 (14 November 1985), 47, 174–76; U.S. Office of Science and Technology Policy, "Coordinated Framework for Regulation of Biotechnology," *Federal Register* 51 (26 June 1986): 23302–93. The administration's approach was also implicitly supported by a report of the congressional Office of Technology Assessment, which chose to ignore the possibility of new legislation, focusing instead on the application of existing statutes (Office of Technology Assessment, *New Developments in Biotechnology: 3. Field-Testing Engineered Organisms: Genetic and Ecological Issues* [Washington, D.C.: U.S. Government Printing Office, 1988], 25–29).

23. Letter from W. Walsh, State Department, to Presidential Science Adviser George Keyworth, 27 May 1983, quoted in Fogleman, "Regulating Science," 232 n. 250.

24. Shapiro, "Biotechnology and the Design of Regulation," 27.

25. For details of the operation of the Council on Competitiveness, see Christine Triano and Nancy Watzman, *All the Vice President's Men* (Washington, D.C.: OMB Watch and Public Citizen's Congress Watch, 1991).

26. U.S. President's Council on Competitiveness, *Report on National Biotechnology Policy* (February 1991), 11.

27. Ibid., 14.

28. Council of the European Communities, "Council Directive of 23 April 1990 on the Deliberate Release into the Environment of Genetically Modified Organisms," *Official Journal of the European Communities* 117 (8 May 1990): 15–27. For an account of the development of the EC directive, see Gordon Lake, "Scientific Uncertainty and Political Regulations: European Legislation on the Contained Use and Deliberate Release of Genetically Modified (Micro) Organisms," *Project Appraisal* 6 (March 1991): 7–15.

29. For analysis, see Les Levidov and Joyce Tait, "Risk Regulation: Release of Genetically Modified Organisms: Precautionary Legislation," *Project Appraisal* 7 (June 1992): 93–105.

30. Wright and Sinsheimer, "Recombinant DNA and Biological Warfare," 20–26; Richard Falk, "Inhibiting Reliance on Biological Warfare," in *Preventing a Biological Arms Race,* ed. Susan Wright (Cambridge, Mass.: MIT Press, 1990), 241–66.

31. U.S. Department of Defense, *Annual Report on Chemical Warfare and Biological Defense Programs, FY 1980,* sec. 2, p. 4; U.S. Department of Defense, *Annual Report on Chemical Warfare and Biological Defense Programs, FY 1981,* sec. 2, p. 16.

32. Frank B. Armstrong, A. Paul Adams, and William H. Rose, "Recombinant DNA and the Biological Warfare Threat," report commissioned by U.S. Army Dugway Proving Ground TECOM 8-CO-513-FBT-021 (May 1981).

33. U.S. Army Medical Research Institute of Infectious Diseases, "Biological Defense: Functional Area Assessment: Overview," 4 January 1985. For details, see Susan Wright, "Evolution of Biological Warfare Policy"; Susan Wright and Stuart Ketcham, "The Problem of Interpreting the U.S. Biological Defense Research Program," in *Preventing a Biological Arms Race,* ed. Susan Wright (Cambridge, Mass.: MIT Press, 1990), 26–68, 169–96.

34. U.S. Office of Science and Technology Policy, "Proposal for a Coordinated Framework for Regulation of Biotechnology," *Federal Register* 49 (31 December 1984): 50856; U.S. Chemical Warfare Review Commission, *Report of the Chemical Warfare Review Commission* (Washington, D.C.: U.S. Government Printing Office, 1985), 70.

35. National Academy of Sciences, *Introduction of Recombinant DNA–Engineered Organisms,* 6.

36. E.g., U.S. Congress, Office of Technology Assessment, *New Developments in Biotechnology: 3,* 4.

37. See, e.g. RAC minutes, 6 February 1984.

38. Ibid., 28 June 1982.

39. Nicholas Wade, "The Roles of God and Mammon in Molecular Biology," in *From Genetic Experimentation to Biotechnology: The Critical Transition,* ed. William J. Whelan and Sandra Black (New York: John Wiley, 1982), 207.

40. Peter Day, "Engineered Organisms in the Environment: A Perspective on the Problem," in *Engineered Organisms in the Environment,* ed. Harlyn O. Halvorson, David Pramer, and Marvin Rogul (Washington, D.C.: American Society for Microbiology, 1985), 5.

41. Norton Zinder, "A Personal View of the Media's Role in the Recombinant DNA War," in *The Gene-Splicing Wars: Reflections on the Recombinant DNA Controversy,* ed. Raymond A. Zilinskas and Burke K. Zimmerman (New York: Macmillan, 1986), 117.

42. Donald Fredrickson's retrospective account, "Science and the Cultural Warp," nicely conveys this sense.

43. Robert Sinsheimer to Donald Fredrickson, 4 January 1978, in NIH, *Recombinant DNA Research,* appendix A, 3:179.

44. I thank one of the anonymous readers of the manuscript of this book for proposing this as a possible explanation.

45. With respect to biological warfare, see Susan Wright, ed., *Preventing a Biological Arms Race;* with respect to agricultural applications, see Rebecca Goldburg, Jane Rissler, Hope Shand, and Chuck Hassebrook, *Biotechnology's Bitter Harvest: Herbicide Tolerant Crops and the Threat to Sustainable Agriculture* (Washington, D.C.: Biotechnology Working Group, 1990).

Appendix A

1. This unofficial transcript, entitled "Malcolm Martin's Personal Notes on Enteric Bacteria Meeting, Tuesday, August 31, 1976," was circulated by the Office of the Director, NIH, in November 1977. A reference to its existence appears in an NIH document, "Background on the Proposed Revisions (9/27/77) of the NIH Guidelines for Research Involving Recombinant DNA Molecules, Compiled by the Office of the Director, NIH, November 1977." Speakers were not identified, but it is clear from the context that the first speaker was one of the cochairs of the meeting, Wallace Rowe or Malcolm Martin.

The complete document runs to ninety-three pages. These excerpts have been selected to give, first, a sense of the range of positions on the nature of possible recombinant DNA hazards and approaches to investigating them; second, a sense of the ways in which particular aspects of these problems were emphasized; third, a sense of the ways in which the controversy over recombinant DNA hazards was seen by scientists advising the NIH director.

The political significance of this meeting and its role in the reappraisal of possible recombinant DNA hazards that occurred in 1976–78 are discussed in chapter 5.

Appendix B

1. Four policy documents are cited in this appendix: Wallace Rowe, "Statement to the NIH Recombinant DNA Advisory Committee," May 1979, included in "Material Relevant to Risk Assessment," 4 May 1979, ORDAR, doc. 670, hereafter cited as Rowe; Ad Hoc Working Group on Risk Assessment, National Institute of Allergy and Infectious Diseases, transcript of meeting, 30 August 1979, ORDAR, hereafter cited as NIAID Ad Hoc Working Group; *E. coli* K-12 Host-Vector Systems Working Group, "Background Document for Proposed Exemption of *E.coli* K-12 Host-Vector Systems from the Recombinant DNA Guidelines," ca. September 1979, ORDAR, doc. 708 and supplement, distributed at RAC meeting, 6–7 September 1979, hereafter cited as RAC Background Document; and Department of Health, Education, and Welfare, National Institutes of Health, "Recombinant DNA Research: Proposed Actions under Guidelines," *Federal Register* 44 (30 November 1979): 69234–51, hereafter cited as NIH Decision Document.

BIBLIOGRAPHY

Unpublished Sources

Writing the history of a contemporary policy problem poses special problems of documentation because of the time elapsing between the period during which the issue was being addressed and the availability of relevant documents in government archives. In the United States, this time lag is generally at least five years and often considerably longer, depending on the practices of the government agency in question. In the United Kingdom, the fact that the government proceedings are covered by the Official Secrets Act of 1911 means that records of agency decisions and committee proceedings are unavailable for much longer—usually thirty years. Furthermore, waiting until material is officially archived by a government does not guarantee completeness of the official records. The sheer volume of material generated may mean that those responsible for sending material to archives are forced to be selective.

For these reasons, examining the history of a policy issue in progress often requires alternatives to reliance on government archives, particularly collecting material as events unfold. Such a strategy was used by historian of science Charles Weiner for documenting the issues associated with the development of genetic engineering in the mid-1970s. The rich collection of materials he assembled, including interviews with scientists and others prominent in the issue, personal correspondence, records of government committee meetings, conferences, and other meetings, which has been deposited at the Institute Archives of the Massachusetts Institute of Technology, provides a major resource for the early period of policymaking. In addition, the records maintained by the NIH Office of Recombinant DNA Activities provide an essential source for the decisions of the RAC. For British policy, archival resources are not as rich, largely because of the impact of the Official Secrets Act. There is also nothing comparable to the MIT collection for the United Kingdom. However, the development of the positions of trade unions is well documented by the material deposited in the Modern Records Centre at the University of Warwick by British unions, particularly the ASTMS.

Although the special collections mentioned above alleviate the problem of documentation, there are still many gaps in records of the calculations and decisions of prominent participants in policymaking, especially from the late 1970s onward and especially in the United Kingdom. The strategy adopted in this study to address these gaps was twofold. First, many of those I interviewed provided access to their personal files and to information from the files of the government office for which they were responsible. I am indebted to those who gave me access to material not available through other channels. Second, in the United States, the Freedom of Information Act provided a crucial tool for supplementing information that is readily accessible in the public domain, especially with respect to information concerning interactions between the government and the private sector.

The following collections of unpublished material have been cited.

Americans for Democratic Action papers, Washington, D.C.
Sydney Brenner personal papers
British Broadcasting Corporation Script Library, London
European Molecular Biology Organization files, Heidelberg, Germany
John Finklea files, National Institute of Occupational Safety and Health, Rockville, Md.
Friends of the Earth files, Washington, D.C.
J. Leslie Glick personal papers
Harlyn O. Halvorson personal papers
Robert Helling personal papers
Imperial Chemical Industries press files, Runcorn, United Kingdom
Jonathan King personal papers
Joshua Lederberg personal papers
Peter Lobban personal papers
London Weekend Television files, London
Michigan Historical Collections, Bentley Historical Library, University of Michigan, Ann Arbor
Modern Records Centre, University of Warwick, Coventry, United Kingdom
Office of Recombinant DNA Activities, National Institutes of Health records, Bethesda, Md.
Office of the Assistant to the Secretary of Defense for Atomic Energy files, Washington, D.C.
Office of the Director, National Institutes of Health files, Bethesda, Md.
Office of the Secretary, Department of Health, Education, and Welfare files, Washington, D.C.
Jerome Ravetz personal papers
Recombinant DNA History Collection, MC100, Institute Archives and Special Collections, MIT Libraries, Cambridge, Mass.
Robert L. Sinsheimer personal papers
John Subak-Sharpe personal papers
Robert Williamson personal papers
Burke K. Zimmerman personal papers

Published Sources

Abelson, John. 1979. "RNA Processing and the Intervening Sequence Problem." *Annual Review of Biochemistry* 48:1035–69.

———. 1980. "A Revolution in Biology." *Science* 209 (19 September): 1319–21.

Abelson, Philip. 1971. "Anxiety about Genetic Engineering." *Science* 173 (23 July): 225.

———. 1974. "Academic Science and Industry." *Science* 183 (29 March): 1251.

———. 1977. "Recombinant DNA." *Science* 197 (19 August): 721.

Abir-am, Pnina. 1980. "From Biochemistry to Molecular Biology: DNA and the Acculturated Journey of the Critic of Science Erwin Chargaff." *History and Philosophy of the Life Sciences* 2:3–60.

———. 1982. "The Discourse of Physical Power and Biological Knowledge in the 1930s: A Reappraisal of the Rockefeller Foundation's 'Policy' in Molecular Biology." *Social Studies of Science* 12: 341–82.

Adams, John. 1978. "Recombinant DNA Research: The Drug Industry Viewpoint." *Society for Industrial Microbiology News,* January, 3–5.

Agarwal, K. L., H. Büchi, M. H. Caruthers, N. Gupta, H. G. Khorana, K. Kleppe, A. Kumar, E. Ohtsuka, U. L. Rajbhandary, J. H. Van De Sande, V. Sgaramella, H. Weber, and T. Yamada. 1970. "Total Synthesis of the Gene for an Alanine Ribonucleic Acid from Yeast." *Nature* 227 (4 July): 27–34.

Allen, Garland. 1975. *Life Science in the Twentieth Century.* Cambridge: Cambridge University Press.

"All Safety Nets Failed, Says Shooter." 1979. *Nature* 277 (11 January): 78–79.

Anderson, E. S. 1974. "The Indiscriminate Use of Antibiotics Has Exerted More Pressure on the Bacterial Population Than Could Be Wielded by All Research Workers in the Field Put Together." *Nature* 250 (26 July): 279–80.

———. 1975. "Viability of, and Transfer of a Plasmid from *E. coli* K12 in the Human Intestine." *Nature* 255 (5 June): 502–4.

———. 1978. "Plasmid Transfer in *Escherichia coli. Journal of Infectious Diseases* 137 (May): 686–87.

Andreopoulos, Spyros. 1980. "Gene Cloning by Press Conference." *New England Journal of Medicine* 302 (27 March): 743–46.

Andrews, Richard. 1976. *Environmental Policy and Administrative Change: Implementation of the National Environmental Policy Act.* Lexington, Mass.: Lexington Books.

An-Na'im, Abdullahi Ahmed, ed. 1992. *Human Rights in Cross-Cultural Perspective: A Quest for Consensus.* Philadelphia: University of Pennsylvania Press.

Armstrong, Philip, Andrew Glyn, and John Harrison. 1984. *Capitalism since World War II.* London: Fontana.

Ashburner, Michael. 1976. "An Open Letter to the Health and Safety Executive." *Nature* 264 (4 November): 2.

Ashby, Eric. 1972. "Science and Antiscience." In *Sociological Review Monograph 18: The Sociology of Science,* ed. Paul Halmos, 209–26. Keele, U.K.: University of Keele.

Ashford, Nicholas. 1976. *Crisis in the Workplace: Occupational Disease and Injury.* Cambridge, Mass.: MIT Press.

Atkins, John. 1976. "Expression of a Eucaryotic Gene in *Escherichia coli.*" *Nature* 262 (22 July): 256–57.

Bachrach, Peter, and Morton Baratz. 1962. "Two Faces of Power." *American Political Science Review* 56:947–52.

———. 1963. "Decisions and Non-decisions." *American Political Science Review* 57:632–42.

———. 1970. *Power and Poverty: Theory and Practice.* New York: Oxford University Press.

Barfield, Claude E. 1982. *Science Policy from Ford to Reagan: Change and Continuity.* Washington, D.C.: American Enterprise Institute for Public Policy Research.

Barnes, Barry. 1981. "On the 'Hows' and 'Whys' of Cultural Change." *Social Studies of Science* 11:481–97.

Bartels, Ditta. 1983. "Oncogenes: Implications for the Safety of Recombinant DNA Work." *Search* 14 (April–May): 88–92.

———. 1986. "Genetische Manipulation an Krebsgenen." In Regine Kollek, Beatrix Tappeser, and Gunter Altner, eds., *Die ungeklarten Gefahrenpotentiale der Gentechnologie,* 70–83. Munich: J. Schweitzer Verlag.

———. 1987. "Escape of Cancer Genes?" *New Scientist,* 30 July, 52–54.

Bazell, Robert J. 1971. "Molecular Biology: Corporate Citizenship and Potential Profit." *Science* 174 (15 October): 275–76.

Beckwith, Jonathan. 1972. "The Scientist in Opposition in the United States." In Watson Fuller, ed., *The Biological Revolution: Social Good or Social Evil?* 295–304. London: Routledge and Kegan Paul.

Bennett, David, Peter Glasner, and David Travis. 1986. *The Politics of Uncertainty: Regulating Recombinant DNA Research in Britain.* London: Routledge and Kegan Paul.

Bennett, William, and Joel Gurin. 1977. "The Science That Frightens Scientists: The Great Debate over DNA." *Atlantic,* February, 43–62.

Berg, Paul, David Baltimore, Herbert W. Boyer, Stanley N. Cohen, Ronald W. Davis, David S. Hogness, Daniel Nathans, Richard Roblin, James D. Watson, Sherman Weissman, and Norton D. Zinder. 1974. "Potential Biohazards of Recombinant DNA Molecules." *Science* 185 (26 July): 303.

———. 1975. "Asilomar Conference on Recombinant DNA Molecules." *Science* 188 (6 June): 991–94.

Berg, Paul, Herbert W. Boyer, Stanley N. Cohen, Ronald W. Davis, David S. Hogness, Daniel Nathans, Richard Roblin, James D. Watson, Sherman Weissman, and Norton D. Zinder. 1974. "Potential Biohazards of Recombinant DNA Molecules." *Proceedings of the National Academy of Sciences of the United States of America* 71:2593–94.

Blake, Jules. 1980. "OSTP: The Last 4 Years" (letter). *Science* 210 (12 December): 1199.

Bloor, David. 1976. *Knowledge and Social Imagery.* London: Routledge and Kegan Paul.

Bok, Derek. 1981. "Business and the Academy." *Harvard Magazine,* May–June, 23–35.

Bolivar, Francisco, Raymond L. Rodriquez, Patricia J. Greene, Mary C. Betlach, Herbert L. Heyneker, and Herbert W. Boyer. 1977. "Construction and Characterization of New Cloning Vehicles: II. A Multipurpose Cloning System." *Gene* 2:95–113.

Bolton, Arthur P., ed. 1984. *Recombinant DNA Products: Insulin, Interferon, and Growth Hormone.* Boca Raton, Fla.: CRC Press.

Booth, Clive. 1984. "Higher Education and Research Training." In Maurice Goldsmith, ed., *U.K. Science Policy: A Critical Review of Policies for Publicly Funded Research,* 123–49. London: Longman.

Bouton, Katherine. 1983. "Academic Researchers and Big Business: A Delicate Balance." *New York Times Magazine,* 11 September.

Bowles, Samuel, David M. Gordon, and Thomas E. Weisskopf. 1984. *Beyond the Wasteland.* Garden City, N.Y.: Anchor Press.

Boyer, Herbert W. 1971. "DNA Restriction and Modification Mechanisms in Bacteria." *Annual Review of Microbiology* 25:153–76.

Boyer, Herbert W., and S. Nicosia, eds. 1978. *Genetic Engineering.* Proceedings of the International Symposium on Genetic Engineering: Scientific Developments and Practical Applications, Milan, 29–31 March 1978. Amsterdam: Elsevier.

Brenner, Sydney. 1974. "Evidence for the Ashby Working Party." Paper submitted to the Working Party on the Experimental Manipulation of the Genetic Composition of Micro-organisms. RDHC.

———. 1974. "New Directions in Molecular Biology." *Nature* 248 (26 April): 785–87.

———. 1978. "Six Months in Category IV." *Nature* 276 (2 November): 2–3.

Brill, Winston. 1985. "Safety Concerns and Genetic Engineering in Agriculture." *Science* 227 (25 January): 381–84.

Britten, Roy J., and Eric H. Davidson. 1969. "Gene Regulation for Higher Cells: A Theory." *Science* 165 (25 July): 349–57.

Brown, Donald. 1981. "Gene Expression in Eukaryotes." *Science* 211 (13 February): 667–74.

Bugler, Jeremy. 1972. *Polluting Britain: A Report.* Harmondsworth, U.K.: Penguin.

"Business and the Universities: A New Partnership." 1982. *Business Week,* 20 December, 58–62.

Bylinsky, Gene. 1974. "Industry Is Finding More Jobs for Microbes." *Fortune,* February, 96–102.

Callon, Michel, and John Law. 1982. "On Interests and Their Transformation: Enrolment and Counter-Enrolment." *Social Studies of Science* 12:615–25.

Capron, Alexander. 1974. Review of *Channeling Technology through Law,* by Laurence Tribe. *Southern California Law Review* 48:571–79.

Carson, Rachel. 1962. *Silent Spring.* Boston: Houghton Mifflin.

Cavalieri, Liebe. 1976. "New Strains of Life—or Death." *New York Times Magazine,* 22 August.

Chan, Hardy W., Mark A. Israel, Claude F. Garon, Wallace P. Rowe, and Malcolm A. Martin. 1979. "Molecular Cloning of Polyoma Virus DNA in *Escherichia coli:* Lambda Phage Vector System." *Science* 203 (2 March): 887–92.

Chang, Annie C. Y., and Stanley N. Cohen. 1974. "Genome Construction between Bacterial Species *in Vitro:* Replication and Expression of *Staphylococcus* Plasmid Genes in *Escherichia coli.*" *Proceedings of the National Academy of Sciences of the United States of America.* 71:1030–34.

Chang, Annie C. Y., Robert A. Lansman, David A. Clayton, and Stanley N. Cohen. 1975. "Studies of Mouse Mitochondrial DNA in *Escherichia coli*: Structure and Function of the Eucaryotic-Procaryotic Chimeric Plasmids." *Cell* 6:231–44.

Chang, Shing, and Stanley N. Cohen. 1977. "*In Vivo* Site-Specific Genetic Recombination Promoted by Eco R1 Restriction Endonuclease." *Proceedings of the National Academy of Sciences* 74 (November 1977): 4811–15.

Chargaff, Erwin. 1975. "Profitable Wonders." *Sciences* (August–September): 21.

———. 1976. "On the Dangers of Genetic Meddling." *Science* 192 (4 June): 938–40.

Chatigny, Mark A., Melvin T. Hatch, H. Wolochow, Tom Adler, John Hresko, Janet Macher, and Diana Besemer. 1979. "Studies on Release and Survival of Biological Substances Used in Recombinant DNA Laboratory Procedures." *Recombinant DNA Technical Bulletin* 2 (July): 62–67.

Chedd, Graham. 1981. "Genetic Gibberish in the Code of Life." *Science 81*, November, 50–55.

Clowes, Royston C. 1973. "The Molecule of Infectious Drug Resistance." *Scientific American*, April, 18–27.

Cohen, Stanley N. 1975. "The Manipulation of Genes." *Scientific American*, July, 20–33.

———. 1977. "Recombinant DNA: Fact and Fiction." *Science* 195 (18 February): 654–57.

Cohen, Stanley N., Annie C. Y. Chang, Herbert W. Boyer, and Robert B. Helling. 1973. "Construction of Biologically Functional Bacterial Plasmids *in Vitro*." *Proceedings of the National Academy of Sciences of the United States of America* 70:3240–44.

Cohen, Stanley N., and James Shapiro. 1980. "Transposable Genetic Elements." *Scientific American*, February, 40–49.

Collins, Harry. 1983. "An Empirical Relativist Programme in the Sociology of Scientific Knowledge." In Karin Knorr-Cetina and Michael Mulkay, eds., *Science Observed: Perspectives on the Social Study of Science*, 85–114. London: Sage Publications.

Colwell, Robert K., et al. 1985. "Genetic Engineering in Agriculture." *Science* 229 (12 July): 111–12.

Commoner, Barry. 1966. *Science and Survival*. New York: Viking.

Connally, William. 1972. "On 'Interests' in Politics," *Politics and Society* 2:459–77.

Cooke, Robert. 1973. "Scientists Worry That New Lab Organisms Could Escape, Do Harm." *Boston Globe*, 11 November, 66.

Cotgrove, Steven. 1973. "Anti-science." *New Scientist*, 12 July, 82–84.

Council of the European Communities. 1990. "Council Directive of 23 April 1990 on the Deliberate Release into the Environment of Genetically Modified Organisms." *Official Journal of the European Communities* 117 (8 May): 15–27.

Crampton, Roger C., and Barry B. Boyer. 1972. "Citizen Suits in the Environmental Field: Peril or Promise?" *Ecology Law Quarterly* 2, no. 3: 407–29.

Crenson, Matthew. 1972. *The Unpolitics of Air Pollution*. Baltimore: Johns Hopkins University Press.

Culliton, Barbara. 1976. "Recombinant DNA: Cambridge City Council Votes Moratorium." *Science* 193 (23 July): 300–301.

———. 1977. "Harvard and Monsanto: The $23-Million Alliance." *Science* 195 (25 February): 759–63.

———. 1978. "Recombinant DNA Bills Derailed: Congress Still Trying to Pass a Law." *Science* 199: (20 January): 274–77.

———. 1981. "Biomedical Research Enters the Marketplace." *New England Journal of Medicine* 304 (14 May): 1195–1201.

———. 1982. "The Hoechst Department at Mass General." *Science* 216 (11 June): 1200–1203.

Curtiss, Roy, III. 1978. "Biological Containment and Cloning Vector Transmissibility." *Journal of Infectious Diseases* 137 (May): 668–75.

Dahl, Robert. 1957. *A Preface to Democratic Theory.* Chicago: University of Chicago Press.

Danielli, James. 1972. "Artificial Synthesis of New Life Forms." *Bulletin of the Atomic Scientists* 28 (December): 20–24.

Darnell, James. 1982. "Variety in the Level of Gene Control in Eukaryotic Cells." *Nature* 297 (3 June): 365–70.

David, Edward. 1979. "Science Futures: The Industrial Connection." *Science* 203 (2 March): 837–40.

———. 1982. "To Prune, Promote, and Preserve." *Science* 217 (13 August): 589.

Davis, Bernard. 1970. "Prospects for Genetic Intervention in Man: Control of Polygenic Traits Is Much Less Likely than Cure of Monogenic Diseases." *Science* 170 (18 December): 1279–83.

———. 1976. "Discussion Forum: The Hazards of Recombinant DNA." *Trends in the Biological Sciences* 1 (August): N178–80.

———. 1976. "Evolution, Epidemiology, and Recombinant DNA. *Science* 193 (6 August): 442.

———. 1977. "The Recombinant DNA Scenarios: Andromeda Strain, Chimera, and Golem." *American Scientist* 65 (September/October): 547–55.

Day, Peter R. 1977. "Plant Genetics: Increasing Crop Yield." *Science* 197 (30 September): 1334–39.

———. 1985. "Engineered Organisms in the Environment: A Perspective on the Problem." In Harlyn O. Halvorson, David Pramer, and Marvin Rogul, eds., *Engineered Organisms in the Environment.* Washington, D.C.: American Society for Microbiology.

Dickson, David. 1978. "Friends of DNA Fight Back." *Nature* 272 (20 April): 664–65.

———. 1978. "GMAG: Stormy Weather Ahead?" *Nature* 271 (5 January): 5–6.

———. 1978. "NIH Confirms Violation of Recombinant DNA Research Guidelines." *Nature* 273 (4 May): 5.

———. 1978. "NIH May Loosen Recombinant DNA Research Guidelines." *Nature* 273 (18 May): 179.

———. 1978. "U.S. Extends Recombinant DNA Controls to Private Industry." *Nature* 276 (21/28 December): 744.

———. 1979. "DNA Critics Appointed to Advisory Committee." *Nature* 277 (11 January): 83.

———. 1979. "U.S. Drug Companies Push for Changes in Recombinant DNA Guidelines." *Nature* 278 (29 March): 385–86.

———. 1979. "U.S. Expected to Exempt Most Recombinant DNA Experiments from Federal Regulation." *Nature* 281 (13 September): 90.

———. 1980. "Backlash against DNA Ventures?" *Nature* 288 (20 November): 203.

———. 1980. "Genentech Makes a Splash on Wall Street." *Nature* 287 (23 October): 669–70.

———. 1984. *The New Politics of Science.* New York: Pantheon.

Dickson, David, and David Noble. 1981. "By Force of Reason: The Politics of Science and Technology Policy." In Thomas Ferguson and Joel Rogers, eds., *The Hidden Election: Politics and Economics in the 1980 Presidential Campaign,* 260–312. New York: Pantheon.

DiMento, J. 1977. "Citizen Environmental Litigation and Administrative Process." *Duke Law Journal* 22:409–52.

Dixon, Bernard. 1973. "Biological Research (1)." *New Scientist,* 25 October, 236.

"DNA Regulation Bill Hits Roadblock Again." 1978. *Congressional Quarterly,* 27 May, 1331–35.

"DoD Offers Dollars for Far-out Research Ideas in Molecular Detection." 1981. *McGraw-Hill's Biotechnology Newswatch* 1 (21 September): 1–2.

Douglas, John. 1978. "U.S. Geneticists Look to Europe for Research Facilities." *Nature* 275 (21 September): 170.

Douglas, Susan. 1990. "Technology and Society." *Isis* 81:80–83.

Drew, Elizabeth Brenner. 1967. "The Health Syndicate: Washington's Noble Conspirators." *Atlantic,* December, 73–82.

Drewry, Gavin. 1979. "Legislation." In S. A. Walkland and Michael Ryle, eds., *The Commons Today,* 87–117. Glasgow: Fontana.

Dutton, Diana B. 1988. *Worse Than the Disease: Pitfalls of Medical Progress.* Cambridge: Cambridge University Press.

Dutton, Diana B., and John L. Hochheimer. 1982. "Institutional Biosafety Committees and Public Participation: Assessing an Experiment." *Nature* 297 (6 May): 11–15.

Eagleton, Terry. 1986. *Against the Grain: Essays 1975–1985.* London: Verso.

Early, Stephen T., Jr., and Barbara B. Knight. 1981. *Responsible Government: American and British.* Chicago: Nelson-Hall.

Easlee, Brian. 1973. *Liberation and the Aims of Science.* London: Sussex University Press.

Eliot, Charles W. 1907. "Academic Freedom." *Science* 26 (5 July): 1–12.

"Evaluation of Risks Associated with Recombinant DNA Research." 1981. *Federal Register* 46 (4 December): 59385–92.

Fairfax, Sally. 1978. "A Disaster for the Environmental Movement." *Science* 199 (17 February): 743–49.

Falk, Richard. 1971. *This Endangered Planet: Prospects and Proposals for Human Survival.* New York: Random House.

———. 1990. "Inhibiting Reliance on Biological Weaponry: The Role and Relevance of International Law." In Susan Wright, ed., *Preventing a Biological Arms Race,* 241–66. Cambridge, Mass.: MIT Press.

————. 1992. "Cultural Foundations for the International Protection of Human Rights." In Abdullahi Ahmed An-Na'im, ed., *Human Rights in Cross-cultural Perspective: A Quest for Consensus*, 44–64. Philadelphia: University of Pennsylvania Press.

Ferguson, Thomas, and Joel Rogers. 1981. "The Reagan Victory: Corporate Coalitions in the 1980 Campaign." In Thomas Ferguson and Joel Rogers, eds., *The Hidden Election: Politics and Economics in the 1980 Presidential Campaign*, 13–17. New York: Pantheon.

"Flaws in GMAG's Guidelines." 1979. *Nature* 277 (15 February): 509–10.

Fogleman, Valerie. 1987. "Regulating Science: An Evaluation of the Regulation of Biotechnology Research." *Environmental Law* 17, no. 2: 183–273.

"Forever Amber on Manipulating DNA Molecules?" 1975. *Nature* 256 (17 July): 155.

Formal, Samuel B., and Richard B. Hornick. 1978. "Invasive *Escherichia coli*." *Journal of Infectious Diseases* 137 (May): 641–44.

Foucault, Michel. 1980. *Power/Knowledge: Selected Interviews and Other Writings, 1972–1977*. New York: Pantheon Books.

Fox-Genovese, Elizabeth. 1982. "Placing Women's History in History." *New Left Review* 133:6–19.

Fredrickson, Donald. 1977. "Health and the Search for New Knowledge." *Daedalus* 106 (Winter): 159–70.

————. 1986. "The Recombinant DNA Controversy: The NIH Viewpoint." In Raymond A. Zilinskas and Burke K. Zimmerman, eds., *The Gene-Splicing Wars: Reflections on the Recombinant DNA Controversy*, 13–26. New York: Macmillan.

Freedman, Robert. 1978. "Gene Manipulation: A New Climate." *New Scientist*, 27 July, 268–69.

Freifelder, David, ed. 1978. *Recombinant DNA*. San Francisco: Freeman.

Freter, Rolf, Howard Brickner, Janet Fekete, Patricia C. M. O'Brien, and Mary M. Vickerman. 1978. "Possible Effects of Foreign DNA on Pathogenic Potential and Intestinal Proliferation of *Escherichia coli*." *Journal of Infectious Diseases* 137 (May): 624–29.

————. 1979. "Testing of Host-Vector Systems in Mice." *Recombinant DNA Technical Bulletin* 2 (July): 68–76.

Frey, Frederick. 1971. "Comment: On Issues and Non-issues in the Study of Power." *American Political Science Review* 65:1081–1101.

Friedmann, Theodore, and Richard Roblin. 1972. "Gene Therapy for Human Genetic Disease? Proposals for Genetic Manipulation in Humans Raise Difficult Scientific and Ethical Problems." *Science* 175 (3 March): 949–55.

Fudenberg, H. H., ed. 1982. *Biomedical Institutions, Biomedical Funding, and Public Policy*. New York: Plenum.

Fujimura, Joan. 1988. "The Molecular Bandwagon in Cancer Research: Where Social Worlds Meet." *Social Problems* 35:261–83.

Geertz, Clifford. 1973. *The Interpretation of Cultures*. New York: Basic Books.

"Genetic Engineering." 1978. *Medical World* November–December, 10.

"Genetic Engineering Gets British Go-Ahead." 1976. *New Scientist*, 2 September, 475.

Gilbert, Walter. 1977. "Recombinant DNA Research: Government Regulation." *Science* 197 (15 July): 206.

———. 1980. "Useful Proteins from Recombinant Bacteria." *Scientific American*, April, 74–94.

———. 1981. "DNA Sequencing and Gene Structure." *Science* 214 (18 December): 1305–12.

Gilpin, Robert. 1962. *American Scientists and Nuclear Weapons Policy*. Princeton, N.J.: Princeton University Press.

Glaucon. 1980. "Biology Loses Her Virginity." *New Scientist*, 18/25 December, 826.

Glendon, A. Ian, and Richard T. Booth. 1982. "Worker Participation in Occupational Health and Safety in Britain." *International Labour Review* 121 (July–August): 399–416.

Glick, J. Leslie. 1976. "Reflections and Speculations on the Regulation of Molecular Genetic Research." *Annals of the New York Academy of Sciences* 265: 178–92.

———. 1982. "Impact of Recombinant DNA Technology on the Economy." In H. H. Fudenberg, ed., *Biomedical Research Institutions, Biomedical Funding, and Public Policy*, 640–73. New York: Plenum.

Goldburg, Rebecca, Jane Rissler, Hope Shand, and Chuck Hassebrook. 1990. *Biotechnology's Bitter Harvest: Herbicide Tolerant Crops and the Threat to Sustainable Agriculture*. Washington, D.C.: Biotechnology Working Group.

Golinski, Jan. 1990. "The Theory of Practice and the Practice of Theory: Sociological Approaches to the History of Science." *Isis* 81: 492–505.

Goodell, Ray. 1980. "The Gene Craze." *Columbia Journalism Review* 19 (November–December): 41–45.

Gorbach, Sherwood L., ed. 1978. "Risk Assessment of Recombinant DNA Experimentation with *Escherichia coli* K12" (proceedings from a workshop held at Falmouth, Mass., 20–21 June 1977). *Journal of Infectious Diseases* 137: 611–714.

———. 1978. "Recombinant DNA: An Infectious Disease Perspective." *Journal of Infectious Diseases* 137 (May): 615–23.

———. 1978. "Risk Assessment Protocols for Recombinant DNA Experimentation." *Journal of Infectious Diseases* 137 (May): 704–8.

Greenberg, Daniel S. 1967. *The Politics of Pure Science*. New York: New American Library.

Grossman, Lawrence I., and Gordon P. Morre. 1983. "Recombinant DNA and Basic Research: What We Have Learned." *LSA* (bulletin of the University of Michigan College of Literature, Science, and the Arts) 6, no. 2: 15–20.

Grunstein, Michael, and David S. Hogness. 1975. "Colony Hybridization: A Method for the Isolation of Cloned DNAs That Contain a Specific Gene." *Proceedings of the National Academy of Sciences of the United States of America* 72: 3961–65.

"Guide to Biotechnology Companies." 1982. *Genetic Engineering News*, November–December, 6–19.

Gummett, Philip. 1980. *Scientists in Whitehall*. Manchester: Manchester University Press.

Gwynne, Peter, Stephen G. Michaud, and William J. Cook. 1976. "Politics and Genes." *Newsweek*, 12 January, 50–52.

Habermas, Jurgen. 1970. "Toward a Theory of Communicative Competence." *Inquiry* 13:360–75

Hall, B. D. 1979. "Mitochondria Spring Surprises." *Nature* 282 (8 November): 129–30.

Hall, John. 1978. "The Union Boss." *New Scientist,* 26 October, 266–67.

Halsey, A. H., and M. A. Trow. 1971. *The British Academics.* Cambridge, Mass.: Harvard University Press.

Halvorson, Harlyn O. 1977. "ASM on Recombinant DNA." *Science* 196 (10 June): 1154.

———. 1977. "Recombinant DNA Legislation: What Next?" *Science* 198 (28 October): 357.

Halvorson, Harlyn O., David Pramer, and Marvin Rogul, eds. 1985. *Engineered Organisms in the Environment.* Washington, D.C.: American Society for Microbiology.

Hanashan, Douglas, and Matthew Meselson. 1980. "Plasmid Screening at High Colony Density." *Gene* 10:63–67.

Handler, Philip. 1969. "Sombre Greeting from Abroad." *Nature* 224 (27 December): 1250.

———, ed. 1970. *Biology and the Future of Man.* New York: Oxford University Press.

———. 1970. "Science and the Federal Government: Which Way To Go?" *Federation Proceedings* 29 (May–June): 1089–97.

———. 1970. "Science's Continuing Role." *Bioscience* 20 (15 October): 1101–6.

Haraway, Donna. 1991. *Simians, Cyborgs, and Women: The Reinvention of Nature.* New York: Routledge, Chapman, and Hall.

Harding, Sandra. 1986. *The Science Question in Feminism.* Ithaca: Cornell University Press.

Hartley, Brian. 1980. "The Biology Business: The Bandwagon Begins to Roll." *Nature* 283 (10 January): 122.

Hartsock, Nancy. 1990. "Foucault on Power." In Linda J. Nicholson, ed., *Feminism/Postmodernism,* 157–75. New York: Routledge, Chapman, and Hall.

Hawkes, Nigel. 1979. "Science in Europe: Smallpox Death in Britain Challenges Presumption of Laboratory Safety." *Science* 203 (2 March): 855–56.

Haynes, Robert H., and Philip C. Hanawalt, eds. 1968. *The Molecular Basis of Life.* San Francisco: Freeman.

Hedgpeth, Joe, Howard M. Goodman, and Herbert W. Boyer. 1972. "DNA Nucleotide Sequence Restricted by the RI Endonuclease." *Proceedings of the National Academy of Sciences of the United States of America* 69:3448–52.

Heirich, Max. 1979. "Why We Avoid the Key Questions: How Shifts in the Funding of Scientific Inquiry Affect Decision Making about Science." In Steven P. Stich and David A. Jackson, eds., *The Recombinant DNA Debate,* 234–60. Englewood Cliffs, N.J.: Prentice Hall.

Hellman, Alfred, Michael N. Oxman, and Robert Pollack. 1973. *Biohazards in Biological Research.* Cold Spring Harbor, N.Y.: Cold Spring Harbor Laboratory.

Herrmann, Robert O. 1971. "The Consumer Movement in Historical Perspective." In David A. Aaker and George S. Day, eds., *Consumerism: Search for Consumer Interest.* New York: Free Press.

Hershfield, Vickers, Herbert W. Boyer, Charles Yanofsky, Michael A. Lovett, and Donald R. Helsinki. 1974. "Plasmid Co1E1 as a Molecular Vehicle for Cloning and Application of DNA." *Proceedings of the National Academy of Sciences of the United States of America* 71:3455–59.

Heyneker, Herbert L., John Shine, Howard M. Goodman, Herbert W. Boyer, John Rosenberg, Richard E. Dickerson, Saran A. Narang, Keiichi Itakura, Syryaung Lin, and Arthur D. Riggs. 1976. "Synthetic *lac* Operator DNA Is Functional *in Vivo*." *Nature* 263 (28 October): 748–52.

Holden, Constance. 1971. "Corporate Responsibility Movement Is Alive and Well." *Science* 172 (28 May): 920.

Hotchkiss, Rollin D. 1965. "Portents for a Genetic Engineering." *Journal of Heredity* 56:197–202.

Hunter, J. R., D. C. Ellwood, A. Atkinson, and P. J. Greenaway. 1978. "Bile Salt Sensitivity of *Escherichia coli* χ1776." *Nature* 275 (7 September): 70–71.

"Ignorance Is Never Bliss." 1979. *Nature* 277 (11 January): 75–76.

Ince, Martin. 1986. *The Politics of British Science*. Brighton, Sussex: Wheatsheaf Books.

Israel, Mark A., Hardy W. Chan, Malcolm A. Martin, and Wallace P. Rowe. 1979. "Molecular Cloning of Polyoma Virus DNA in *Escherichia coli:* Oncogenicity Testing in Hamsters." *Science* 205 (14 September): 1140–42.

Israel, Mark A., Hardy W. Chan, Wallace P. Rowe, and Malcolm A. Martin. 1979. "Molecular Cloning of Polyoma Virus DNA in *Escherichia coli:* Plasmid Vector System." *Science* 203 (2 March): 883–87.

Israel, Mark A., Daniel T. Simmons, Sara L. Hourihan, Wallace P. Rowe, and Malcolm A. Martin. 1979. "Interrupting the Early Region of Polyoma Virus DNA Enhances Tumorigenicity." *Proceedings of the National Academy of Sciences.* 76 (August): 3713–16.

Itakura, Keiichi, Tadaaki Hirose, Roberto Crea, Arthur D. Riggs, Herbert L. Heyneker, Francisco Bolivar, and Herbert W. Boyer. 1977. "Expression in *Escherichia coli* of a Chemically Synthesized Gene for the Hormone Somatostatin." *Science* 198 (9 December): 1056–63.

"It Takes a Death: We Publish Professor Shooter." 1979. *Medical World,* January 1979, 2–13.

Jackson, David, Robert Symons, and Paul Berg. 1972. "Biochemical Method for Inserting New Genetic Information in DNA of Simian Virus 40: Circular SV40 Molecules Containing Lambda Phage Genes and the Galactose Operon of *Escherichia coli*." *Proceedings of the National Academy of Sciences of the United States of America* 69:2904–9.

Jacobus, Mary, Evelyn Fox Keller, and Sally Shuttleworth, eds. 1990. *Body/Politics: Women and the Discourses of Science*. New York: Routledge, Chapman, and Hall.

Jameson, Frederic. 1972. *The Prison-House of Language*. Princeton, N.J.: Princeton University Press.

Jelsma, Japp, and W. A. Smit. 1986. "Risks of Recombinant DNA Research: From Uncertainty to Certainty." In Henk A. Becker and Alan L. Porter, eds., *Impact Assessment Today,* 715–40. Utrecht: Uitgeverij Jan van Arkel.

Johnson, Judith. 1982. "Biotechnology: Commercialization of Academic Research." Issue Brief IB81160. Congressional Research Service, Library of Congress, July.

Johnson, Stanley. 1973. *The Politics of the Environment: The British Experience*. London: Tom Stacey.

Johnston, Ron, and Philip Gummett, eds. 1979. *Directing Technology: Policies for Promotion and Control*. London: Croom Helm.

Jordanova, Ludi. 1989. *Sexual Visions: Images of Gender in Science and Medicine between the Eighteenth and Twentieth Centuries*. Brighton: Harvester Wheatsheaf.

Judson, Horace F. 1975. "Fearful of Science: Who Shall Watch the Scientists?" *Harper's*, June, 70–76.

Kass, Leon. 1971. "The New Biology: What Price Relieving Man's Estate?" *Science* 174 (19 November): 779–88.

Kay, Lily E. 1993. *The Molecular Vision of Life: Caltech, the Rockefeller Foundation, and the Rise of the New Biology*. Oxford: Oxford University Press.

Keatley, Anne. 1983. "Knowledge as Real Estate." *Science* 222 (18 November): 717.

Keller, Evelyn Fox. 1981. "McClintock's Maize." *Science 81*, October, 55–58.

———. 1985. *Reflections on Gender and Science*. New Haven: Yale University Press.

Kennedy, Donald. 1980. "Health Research: Can Utility and Quality Co-exist?" Speech given at University of Pennsylvania, December.

"Kennedy Opens Door to Discussion of His Bill to Regulate Recombinant DNA Research." 1977. *Federation of American Societies for Experimental Biology Newsletter* 10 (July): 1–2.

Kenney, Martin. 1986. *Biotechnology: The University-Industrial Complex*. New Haven: Yale University Press.

Kevles, Daniel. 1977. "The National Science Foundation and the Debate over Postwar Research Policy, 1942–1945: A Political Interpretation of Science: The Endless Frontier." *Isis* 68:5–26.

———. 1979. *The Physicists: The History of a Scientific Community in Modern America*. New York: Vintage.

Kidd, Charles. 1973. "The NIH Phenomenon." Review of *Politics, Science, and Dread Disease* by Steven P. Strickland. *Science* 179 (19 January): 270–72.

King, Jonathan. 1978. "Recombinant DNA and Autoimmune Disease." *Journal of Infectious Diseases* 137 (May): 663–66.

Klaw, Spencer. 1968. *The New Brahmins: Scientific Life in America*. New York: William Morrow.

Knorr-Cetina, Karin. 1981. *The Manufacture of Knowledge: An Essay on the Constructivist and Contextual Nature of Science*. Oxford: Pergamon Press.

Knorr-Cetina, Karin, and Michael Mulkay, eds. 1983. *Science Observed: Perspectives on the Social Study of Science*. London: Sage Publications.

Kohler, Robert E. 1990. *Partners in Science: Foundations and Natural Scientists, 1900–1945*. Chicago: Chicago University Press.

Kolata, Gena Bari. 1976. "DNA Sequencing: A New Era in Molecular Biology." *Science* 192 (14 May): 645–47.

Krimsky, Sheldon. 1979. "A Comparative View of State and Municipal Laws Regulating the Use of Recombinant DNA Molecules Technolgy." *Recombinant DNA Technical Bulletin* 2 (November): 121–33.

————. 1982. *Genetic Alchemy*. Cambridge, Mass.: MIT Press.

————. 1988. "Release of Genetically Engineered Organisms into the Environment." In Sheldon Krimsky and Alonzo Plough, *Environmental Hazards: Communicating Risks as a Social Process*, 75–129. Dover, Mass.: Auburn House Publishing Company.

Krimsky, Sheldon, Anne Baeck, and John Bolduc. 1982. *Municipal and States Recombinant DNA Laws: History and Assessment*. Report prepared for the Boston Neighborhood Network. June.

Krimsky, Sheldon, and Alonzo Plough. 1988. *Environmental Hazards: Communicating Risks as a Social Process*. Dover, Mass.: Auburn House Publishing Company.

Kuhn, Thomas S. 1962. *The Structure of Scientific Revolutions*. Chicago: Chicago University Press.

Lake, Gordon. 1991. "Scientific Uncertainty and Political Regulations: European Legislation on the Contained Use and Deliberate Release of Genetically Modified (Micro) Organisms." *Project Appraisal* 6 (March): 7–15.

Landrigan, Philip J., John M. Harrington, and Larry J. Elliott. 1984. "The Biotechnology Industry." In John M. Harrington, ed., *Recent Advances in Occupational Health* 2:3–13. Edinburgh: Churchill Livingstone.

Lass, Jonathan, Katherine Gillman, and David Sheridan. 1984. *A Season of Spoils: The Story of the Reagan Administration's Attack on the Environment*. New York: Pantheon.

Lawrence, Eleanor. 1976. "Genetic Manipulation: Guidelines Out." *Nature* 263 (2 September): 4–5.

————. 1976. "Nuts and Bolts of Genetic Engineering. *Nature* 263 (28 October): 726–27.

————. 1979. "Bacteriologists Lobby GMAG's First Public Meeting." *Nature* 277 (4 January): 3.

Lear, John. 1978. *Recombinant DNA: The Untold Story*. New York: Crown.

Leder, Philip. 1982. "The Genetics of Antibody Diversity." *Scientific American*, May, 102–15.

Lederberg, Joshua. 1968. "Genetics. In *Encyclopaedia Britannica Yearbook of Science and the Future 1969*, 318–21. Chicago: Encyclopaedia Britannia.

Leopold, John W., and P. B. Beaumont. 1982. "Joint Health and Safety Committees in the United Kingdom: Participation and Effectiveness—A Conflict?" *Economic and Industrial Democracy* 3:263–84.

Levidov, Les, and Joyce Tait. 1992. "Risk Regulation: Release of Genetically Modified Organisms: Precautionary Legislation." *Project Appraisal* 7 (June): 93–105.

Levy, Stuart, and Bonnie Marshall. 1979. "Survival of *E. coli* Host-Vector Systems in the Human Intestinal Tract." *Recombinant DNA Technical Bulletin* 2 (July): 77–80.

Lewin, Roger. 1974. "Ethics and Genetic Engineering." *New Scientist*, 17 October, 16.

————. 1974. "The Future of Genetic Engineering." *New Scientist*, 17 October, 166.

————. 1976. "Genetic Engineers Ready for Stage Two." *New Scientist,* 14 October, 86–87.

————. 1977. "Total Ban Sought on Genetic Engineering." *New Scientist,* 10 March, 571.

————. 1977. "US Genetic Engineering in a Tangled Web." *New Scientist,* 17 March, 640–41.

————. 1978. "Modern Biology at the Industrial Threshold." *New Scientist,* 5 October, 18–19.

————. 1981. "Do Jumping Genes Make Evolutionary Leaps?" *Science* 213 (7 August): 634–36.

————. 1981. "Tale of the Orphaned Genes." *Science* 212 (1 May): 530.

————. 1982. "Repeated DNA Still in Search of a Function." *Science* 217 (13 August): 621–23.

Leys, Colin. 1989. *Politics in Britain: From Labourism to Thatcherism.* London: Verso.

Lobban, Peter. 1969. "The Generation of Transducing Phage *in vitro.*" Essay for third Ph.D. examination, Stanford University. 6 November.

Lobban, Peter, and A. Dale Kaiser. 1973. "Enzymatic End-to-End Joining of DNA Molecules." *Journal of Molecular Biology* 78:453–71.

Locke, John. 1976. "An Open Reply from the Director of the Executive." *Nature* 264 (4 November): 3.

Lukes, Steven. 1974. *Power: A Radical View.* London: Macmillan.

————. 1977. *Essays in Social Theory.* New York: Columbia University Press.

————. 1991. *Moral Conflict and Politics.* Oxford: Clarendon Press.

Luria, Salvador. 1965. "Directed Genetic Change: Perspectives from Molecular Genetics." In T. M. Sonneborn, ed., *The Control of Human Heredity and Evolution,* 1–19. New York: Macmillan.

————. 1969. "Modern Biology: A Terrifying Power." *Nation,* 20 October. 406–9.

McDonald, Kim. 1982. "North Carolina Governor Urges States to Support Scientific Research at Academic Institutions." *Chronicle of Higher Education,* 13 January, 9.

McDougall, Walter A. 1985. *The Heavens and the Earth: The Rationale of the Space Program.* New York: Basic Books.

McGinty, Lawrence. 1979. "Smallpox Laboratories, What Are the Risks? *New Scientist,* 4 January, 8–14.

MacKenzie, Donald A. 1981. "Interests, Positivism, and History." *Social Studies of Science* 11:498–503.

————. 1981. *Statistics in Britain, 1865–1930: The Social Construction of Scientific Knowledge.* Edinburgh: Edinburgh University Press.

————. 1987. "Missile Accuracy: A Case Study in the Social Processes of Technological Change." In Wiebe E. Bijker, Thomas P. Hughes, and Trevor J. Pinch, eds., *The Social Construction of Technological Systems,* 195–22. Cambridge, Mass.: MIT Press.

————. 1990. *Inventing Accuracy: A Historical Sociology of Nuclear Missile Guidance.* Cambridge, Mass.: MIT Press.

McKie, Robin. 1979. "Britain's Shadow Minister Believes in Experts." *Nature* 178 (29 March): 63.

Maddox, John. 1969. "On Which Side Are the Angels?" *Nature* 224 (27 December): 1241–42.

———. 1979. "GMAG: NIH Guidelines Are Not the Way." *Nature* 278 (1 March): 10.

Marchant, Gary. 1988. "Modified Rules for Modified Bugs: Balancing Safety and Efficiency in the Regulation of Deliberate Release of Genetically Engineered Microorganisms." *Harvard Journal of Law and Technology* 1 (Spring): 163–208.

Marshall, Eliot. 1979. "Gene Splicers Simulate a 'Disaster,' Find No Risk." *Science* 203 (23 March): 1223.

Martin, Brian. 1993. "The Critique of Science Becomes Academic." *Science, Technology, and Human Values* 18 (Spring): 247–59.

Marx, Jean L. 1976. "Molecular Cloning: Powerful Tool for Studying Genes." *Science* 191 (19 March): 1160–62.

———. 1977. "Nitrogen Fixation: Prospects for Genetic Manipulation." *Science* 196 (6 May): 638–41.

———. 1981. "Antibodies: Getting Their Genes Together." *Science* 212 (29 May): 1015–17.

———. 1981. "Gene Control Puzzle Begins to Yield." *Science* 212 (8 May): 653–55.

———. 1982. "Agricultural Applications of Genetic Engineering." *Science* 216 (18 June): 1306.

Maxam, Allan M., and Walter Gilbert. 1977. "A New Method for Sequencing DNA." *Proceedings of the National Academy of Sciences of the United States of America* 74:560–64.

Medawar, Peter. 1969. "On 'the Effecting of All Things Possible.'" *New Scientist,* 4 September, 467.

———. 1977. "The DNA Scare: Fear and DNA." *New York Review of Books,* 27 October, 15–20.

Mellon, Margaret. 1988. *Biotechnology and the Environment.* Washington, D.C.: National Wildlife Federation.

Merchant, Carolyn. 1980. *The Death of Nature: Women, Ecology, and the Scientific Revolution.* San Francisco: Harper and Row.

Merton, Robert K. 1973. *The Sociology of Science.* Chicago: University of Chicago Press.

Mertz, Janet E., and Ronald W. Davis. 1972. "Cleavage of DNA by R Restriction Endonuclease Generates Cohesive Ends." *Proceedings of the National Academy of Sciences of the United States of America* 69:3370–74.

Milewski, Elizabeth A. 1984. "The NIH Guidelines for Research Involving Recombinant DNA Molecules." In Arthur P. Bolton, ed., *Recombinant DNA Products: Insulin, Interferon and Growth Hormone,* 155–69. Boca Raton, Fla.: CRC Press.

Mitchell, Robert Cameron, and J. Clarence Davies, III. 1978. "The United States Environmental Movement and Its Political Context: An Overview." Resources for the Future Discussion Paper. May.

"Molecular Biology: Suffering from Shock." 1979. *Nature* 278 (12 April): 587.

Morgan, Joan, and William J. Whelan, eds. 1979. *Recombinant DNA and Genetic Experimentation*. Oxford: Pergamon Press.

Morrow, John F., Stanley N. Cohen, Annie C. Y. Chang, Herbert W. Boyer, Howard M. Goodman, and Robert B. Helling. 1974. "Replication and Transcription of Eucaryotic DNA in *Escherichia coli*." *Proceedings of the National Academy of Sciences of the United States of America* 71:1743–47.

Muller, Hermann. 1965. "Means and Aims in Human Genetic Betterment." In T. M. Sonneborn, ed., *The Control of Human Heredity and Evolution*, 101–22. New York: Macmillan.

Murray, Kenneth. 1974. "Alternative Experiments?" *Nature* 250 (26 July): 279.

Murray, Noreen E., and Kenneth Murray. 1974. "Manipulation of Restriction Targets in Phage to Form Receptor Chromosomes for DNA Fragments." *Nature* 251 (11 October): 476–81.

Nader, Ralph. 1965. *Unsafe at Any Speed*. New York: Grossman.

National Academy of Sciences. 1969. *Technology: Processes of Assessment and Choice*. Washington, D.C.: National Academy of Sciences.

———. 1977. *Research with Recombinant DNA*. Washington, D.C.: National Academy of Sciences.

National Academy of Sciences. Ad Hoc Committee. 1969. *Applied Science and Technological Progress*. Washington, D.C.: National Academy of Sciences.

National Academy of Sciences. Committee on Scientific Evaluation of the Introduction of Genetically Modified Microorganisms and Plants into the Environment. 1989. *Field Testing Genetically Modified Organisms: Framework for Decisions*. Washington, D.C.: National Academy Press.

National Academy of Sciences. Committee on the Introduction of Genetically Engineered Organisms into the Environment. 1987. *Introduction of Recombinant DNA–Engineered Organisms into the Environment: Key Issues*. Washington, D.C.: National Academy Press.

National Science Board. 1981. *Science Indicators 1980*. Washington, D.C.: U.S. Government Printing Office.

———. 1985. *Science Indicators: The 1985 Report*. Washington, D.C.: U.S. Government Printing Office.

Neilands, J. B. 1972. *Harvest of Death: Chemical Warfare in Vietnam and Cambodia*. New York: Free Press.

Nelkin, Dorothy. 1977. "Public Participation in the Decisions about Science and Technology in the United States." Report to the OECD Science Policy Directorate, December.

———. 1989. "Science Studies in the 1990s." *Science, Technology, and Human Values* 14 (Summer): 305–11.

Nelson, Gaylord. 1977. "Remarks on Recombinant DNA Act, S. 1217 (Amendment No. 754)." *Congressional Record*, 2 August, S.13312–19.

Newman, Stuart. 1982. "The 'Scientific' Selling of rDNA." *Environment* 24 (July–August): 21–23, 53–57.

Newmark, Peter. 1978. "European Molecular Biology." *Nature* 273 (18 May): 182–83.

————. 1978. "WHO Looks for Benefits from Genetic Engineering." *Nature* 272 (20 April): 663–64.

————. 1980. "France: Biotechnology Company Planned." *Nature* 285 (15 May): 124.

Nichols, David. "Associated Interest Groups of American Science." In Albert Teich, ed., *Scientists and Public Affairs.* Cambridge, Mass.: MIT Press.

Nichols, Theo, and Pete Armstrong. 1973. *Safety or Profit: Industrial Accidents and the Conventional Wisdom.* Bristol, U.K.: Falling Wall Press.

Nicholson, Linda J., ed. 1990. *Feminism/Postmodernism.* New York: Routledge, Chapman, and Hall.

NIH Factbook: Guide to National Institutes of Health Programs and Activities. 1976. Chicago: Marquis Who's Who.

Noble, Charles. 1986. *Liberalism at Work: The Rise and Fall of OSHA.* Philadelphia: Temple University Press.

Noble, David. 1977. *America by Design: Engineers as Agents and Victims of Capital.* New York: Knopf.

Noll, Roger G., and Paul A. Thomas. 1977. "The Economic Implications of Regulation by Expertise: The Guidelines for Recombinant DNA Research." In National Academy of Sciences, *Research with Recombinant DNA,* 262–77. Washington, D.C.: National Academy of Sciences.

Norman, Colin. 1975. "Berg Conference Favors Use of Weak Strains." *Nature* 254 (6 March): 6–7.

————. 1976. "Genetic Manipulation: Guidelines Issued." *Nature* 262 (1 July): 2.

————. 1981. "Is Reaganomics Good for Technology?" *Science* 213 (21 August): 843–45.

"No Sci-Fi Nightmare, After All." 1977. *New York Times,* 24 July, 18.

Novick, Peter. 1989. *That Noble Dream: The Objectivity Question and the American Historical Profession.* Cambridge: Cambridge University Press.

"Now Reason Can Prevail." 1978. *Nature* 276 (9 November): 103.

Omenn, Gilbert S. 1981. "Government as a Broker between Private and Public Institutions in the Development of Recombinant DNA Applications." In Melissa Kenberg, ed., *Proceedings of the Battelle Conference on Genetic Engineering* 1:34–45. Seattle: Battelle Seminars and Studies Program.

Organization for Economic Cooperation and Development. 1968. *Reviews of National Science Policy: United States.* Paris: OECD.

————. 1974. *The Research System: Comparative Survey of the Organization and Financing of Fundamental Research.* Paris: OECD.

Orgel, L. E., and F. H. C. Crick. 1980. "Selfish DNA: The Ultimate Parasite." *Nature* 284 (17 April): 604–7.

O'Riordan, Timothy. 1976. *Environmentalism.* London: Pion.

Orlans, Harold. 1962. *The Effects of Federal Programs on Higher Education: A Study of 36 Universities and Colleges.* Washington, D.C.: The Brookings Institution.

Perlman, David. 1977. "Scientific Announcements" (letter). *Science* 198 (25 November): 782

Petrocheilou, V., and M. H. Richmond. 1977. "Absense of Plasmid or *Escherichia coli* K12 Infection among Laboratory Personnel Engaged in R-Plasmid Research." *Gene* 2:323–27.

Phillips, Don I., Catherine C. Cleare, and Maria Patterson, eds. 1977. *Research and Development in the Federal Budget: Colloquium Proceedings, June 15–16, 1977.* Washington, D.C.: American Association for the Advancement of Science.

Philp, Mark. 1985. "Foucault." In Quentin Skinner, ed., *The Return of Theory in the Human Sciences*, 65–82. Cambridge: Cambridge University Press.

Pinch, Trevor J., and Wiebe E. Bijker. 1987. "The Social Construction of Facts and Artifacts; Or, How the Sociology of Science and the Sociology of Technology Might Benefit Each Other." In Wiebe E. Bijker, Thomas P. Hughes, and Trevor J. Pinch, eds., *The Social Construction of Technological Systems*, 17–50. Cambridge, Mass.: MIT Press.

"Plenty for GMAG to do." 1976. *Nature* 276 (2 November): 1.

Pollard, Sidney. 1984. *The Wasting of the British Economy.* 2d ed. Beckenham, Kent: Croon Helm.

Press, Frank. 1981. "Science and Technology in the White House, 1977–1980: Part I." *Science* 211 (9 January): 139–45.

Pritchard, R. H. 1978. "Recombinant DNA Is Safe." *Nature* 273 (29 June): 696.

Rabinow, Paul, and William Sullivan. 1987. "The Interpretive Turn: A Second Look." In Paul Rabinow and William Sullivan, eds., *Interpretive Social Science: A Second Look*, 1–30. Berkeley: University of California Press.

Rambach, Alain, and Pierre Tiollais. 1974. "Bacteriophage Having Eco RI Endonuclease Sites Only in the Nonessential Region of the Genome." *Proceedings of the National Academy of Sciences of the United States of America* 71:3927–80.

"Rational Risk Assessment for Genetic Engineering." 1978. *New Scientist*, 9 November.

Ravetz, Jerome. 1971. *Scientific Knowledge and Its Social Problems.* Oxford: Oxford University Press.

———. 1990. "Recombinant DNA Research: Whose Risks?" In *The Merger of Knowledge and Power: Essays in Critical Science*, 63–80. London: Mansell Publishing.

"Recombinant DNA Debate Three Years On." 1977. *Nature* 268 (21 July): 185.

Redfearn, Judy. 1979. "UK Genetic Engineering Regulations Relax Gradually." *Nature* 282 (20/27 December): 769.

———. 1980. "Occupational Safety: Cuts Ahead." *Nature* 286 (21 August): 751.

Restivo, Sal. 1988. "Modern Science as a Social Problem." *Social Problems* 35 (June): 206–25.

Rettig, Richard. 1977. *Cancer Crusade: The Story of the National Cancer Act of 1971.* Princeton: Princeton University Press.

"Risk Assessment Protocols for Recombinant DNA Experimentation." 1978. *Journal of Infectious Diseases* 137 (May): 704–8.

Roark, Ann. 1979. "Easing of DNA Rules Draws Fire from Scientists." *Chronicle of Higher Education*, 29 October 29, 15.

Roberts, Thomas M., and Gail D. Lauer. 1979. "Maximizing Gene Expression on a Plasmid Using Recombination *in Vitro*." *Methods in Enzymology* 68:473–82.

Robertson, Miranda. 1974. "ICI Puts Money on Genetic Engineering." *Nature* 251 (18 October): 564–65.

————. 1981. "Gene Families, Hopeful Monsters, and the Selfish Genetics of DNA." *Nature* 293 (1 October): 333–34.

————. 1983. "Clues to the Genetic Basis of Cancer." *New Scientist,* 9 June, 688–91.

Rockefeller, David. 1972. "Corporate Tasks for 70's: Social Action Not Words." *Commercial and Financial Chronicle,* 6 January, (69)21.

Rogers, Michael. 1975. "The Pandora's Box Congress." *Rolling Stone,* 19 June, 15–19, 37–38.

————. 1977. *Biohazard.* New York: Knopf.

Roland, Alex. 1986. "Science and War." *Osiris* 2:247–72.

Rolleston, Frances. 1979. "Commentary on Risk Assessment Scheme." *Nature* 281 (25 October): 626–27.

Rose, Hilary, and Stephen Rose. 1970. *Science and Society.* Harmondsworth, Middlesex: Penguin Books.

————, eds. 1976. *The Political Economy of Science: Ideology of/in the Natural Sciences.* London: Macmillan.

————, eds. 1976. *The Radicalization of Science.* London: Macmillan.

Rose, Richard. 1983. "Still the Era of Party Government." *Parliamentary Affairs* 36 (Summer): 282–99.

Rose, Stephen, ed. 1968. *Chemical and Biological Warfare.* London: George G. Garrup.

Rosenberg, Barbara, and Lee Simon. 1979. "Recombinant DNA: Have Recent Experiments Assessed All the Risk?" *Nature* 282 (20/27 December): 773.

Rossiter, Margaret. 1982. *Women Scientists in America: Struggles and Strategies to 1940.* Baltimore: Johns Hopkins University Press.

————. 1985. "Science and Public Policy since World War II." *Osiris* 1:273–94.

Rothman, Harry, and Zbigniew Towalski. 1984. "British Biotechnology Policy." In Maurice Goldsmith, ed., *U.K. Science Policy,* 44–45. London: Longman.

Rothschild, Lord. 1971. *A Framework for Government Research and Development.* Cmnd. 4814.

Royal Society. 1981. "Biotechnology and Education: Report of a Working Group." London: Royal Society.

Rudwick, Martin. 1985. *The Great Devonian Controversy.* Chicago: University of Chicago Press.

"Safety Scene: TUC Warning on Risk to Health and Safety." 1983. *Occupational Safety and Health,* July.

Sagik, B. P., and C. A. Sorber. 1979. "The Survival of Host-Vector Systems in Domestic Sewage Treatment Plants." *Recombinant DNA Technical Bulletin* 2 (July): 55–61.

Sambrook, J. 1977. "Adenovirus Amazes at Cold Spring Harbor." *Nature* 268 (14 July): 101–4.

Sandbach, Francis. 1980. *Environmental Ideology and Policy.* Oxford: Basil Blackwell.

Sanger, F., and A. R. Coulson. 1975. "A Rapid Method for Determining Sequences in DNA by Primed Synthesis with DNA Polymerase." *Journal of Molecular Biology* 94:441–48.

Sansonetti, Philippe J., Dennis J. Kopecko, and Samuel B. Formal. 1981–82.

"*Shigella sonnei* Plasmids: Evidence That a Large Plasmid Is Necessary for Virulence." *Infection and Immunity* 34:75–83, 35:852–60.

Sargeant, K., and C. G. T. Evans. 1979. *Hazards Involved in the Industrial Use of Micro-organisms.* Brussels-Luxembourg: Commissions of European Communities.

Saunders, Peter. 1979. *Urban Politics: A Sociological Interpretation.* London: Hutchinson.

Scarrow, Howard. 1972. "The Impact of British Domestic Air Pollution Legislation." *British Journal of Political Science* 2:261–82.

Schattschneider, E. E. 1960. *The Semi-sovereign People: A Realist's View of Democracy in America.* New York: Holt, Rinehart, and Wilson.

Scheller, Richard H., Richard E. Dickerson, Herbert W. Boyer, Arthur D. Riggs, and Keiichi Itakura. 1977. "Chemical Synthesis of Restriction Enzyme Recognition Sites Useful for Cloning." *Science* 196 (8 April): 177–80.

Schnaiberg, Allan. 1973. "Politics, Participation, and Pollution: The 'Environmental Movements.'" In J. Walton and D. E. Carns, eds., *Cities in Change: Studies on the Urban Condition,* 605–27. Boston: Allyn and Bacon.

———. 1980. *The Environment from Surplus to Scarcity.* New York: Oxford University Press.

Schwartz, Charles. 1970. "The Movement vs. the Establishment." *Nation,* 22 June, 747–51.

Scott, Joan Wallach. 1989. "History in Crisis? The Others' Side of the Story." *American Historical Review* 94:680–92.

Shannon, James. 1967. "The Advancement of Medical Research: A Twenty-Year View of the Role of the National Institutes of Health." *The Journal of Medical Education* 42:97–108.

Shapin, Steven. 1982. "History of Science and Its Sociological Reconstructions." *History of Science* 20:157–211.

———. 1988. "Following Scientists Around." *Social Studies of Science* 18:533–50.

Shapiro, James, Larry Eron, and Jonathan Beckwith. 1969. "More Alarms and Excursions." *Nature* 224 (27 December): 1337.

Shapiro, Sidney. 1990. "Biotechnology and the Design of Regulation." *Ecology Law Quarterly* 17, no. 1: 1–70.

Sherwin, C. W., and R. W. Isenson. 1967. "Project Hindsight: A Defense Department Study of the Utility of Research." *Science* 156 (23 June): 1571–77.

Silk, Leonard, and David Vogel. 1976. *Ethics and Profits: The Crisis of Confidence in American Business.* New York: Simon and Schuster.

Sills, David. 1975. "The Environmental Movement and Its Critics." *Human Ecology* 3, no. 1: 1–41.

Simring, Francine Robinson. 1977. "The Double Helix of Self-interest." *The Sciences,* May–June, 10–13, 27.

Singer, Maxine. 1979. "Spectacular Science and Ponderous Progress." *Science* 203 (5 January): 9.

Singer, Maxine, and Dieter Soll. 1973. "Guidelines for DNA Hybrid Molecules." *Science* 181 (21 September): 1114.

Sinsheimer, Robert L. 1966. "The End of the Beginning." *Engineering and Science* 33 (December): 7–10.

———. 1970. "Genetic Engineering: The Modification of Man." *Impact of Science on Society* 20:279–90.

———. 1975. "Troubled Dawn for Genetic Engineering." *New Scientist,* 16 October, 148–51.

———.1976. "Discussion Forum: The Hazards of Recombinant DNA." *Trends in the Biological Sciences* 1 (August): B178–N180.

———. 1976. "Recombinant DNA: On Our Own." *BioScience* 26, no. 10:599.

———. 1976. "Recombinant DNA Research (Genetic Engineering): Social Implications and Potential Hazards." UCLA Distinguished Lecture Series, 1976–77, 22 November.

———. 1977. "An Evolutionary Perspective for Genetic Engineering." *New Scientist,* 20 January, 150–52.

———. 1977. "Recombinant DNA." *Annual Review of Biochemistry,* 46:415–38.

Smith, Cecil R., and Malcolm A. Martin. 1982. "In Vivo Biological Activity of Polyoma-pBR322 Recombinants Cloned in WT *Escherichia coli* or *Escherichia coli* K12."

Smith, H. R., and B. Rowe. 1982. "Testing of Disabled Strains of *Escherichia coli* K12." In United Kingdom, Genetic Manipulation Advisory Group, *Third Report of the Genetic Manipulation Advisory Group.* Cmnd. 8665 (September), 132–35.

Smith, H. Williams. 1975. "Survival of Orally Administered *E. coli* K12 in Alimentary Tract of Man." *Nature* 255 (5 June): 500–502.

———. 1978. "Is It Safe to Use *Escherichia coli* K12 in Recombinant DNA Experiments?" *Journal of Infectious Diseases* 137 (May): 655–60.

Smith, Keith. 1984. *The British Economic Crisis: Its Past and Future.* Harmondsworth, Middlesex: Penguin Books.

Sonneborn, T. M., ed. 1965. *The Control of Human Heredity and Evolution.* New York: Macmillan.

Spero, Joan. 1988–89. "Guiding Global Finance." *Foreign Policy* 73:114–34.

Stent, Gunther. 1969. *The Coming of the Golden Age: A View of the End of Progress.* New York: Natural History Press.

———. 1970. "DNA." *Daedalus* 99:909–37.

Stetten, DeWitt. 1982. "The DNA Disease" (letter). *Nature* 297 (27 May): 260.

Stevenson, Adlai. 1977. "Recombinant DNA Legislation." *Congressional Record,* 22 September, 30457–60.

Stich, Steven P., and David A. Jackson, eds. 1979. *The Recombinant DNA Debate.* Englewood Cliffs, N.J.: Prentice Hall.

"Still Looser UK Guidelines." 1980. *Nature* 287 (25 September): 265–66.

Stoker, Michael. 1974. "Molecular Dirty Tricks Ban." *Nature* 250 (26 July): 278.

Strickland, Steven P. 1971. "Integration of Medical Research and Health Policies." *Science* 173 (17 September): 1093–1103.

———. 1972. *Politics, Science, and Dread Disease: A Short History of United States Medical Research Policy.* Cambridge, Mass.: Harvard University Press.

Struhl, Kevin, John R. Cameron, and Ronald W. Davis. 1976. "Functional Genetic Expression of Eucaryotic DNA in *Escherichia coli.*" *Proceedings of the National Academy of Sciences of the United States of America* 73:1471–75.

Sun, Marjorie. 1980. "Insulin Wars: New Advances May Throw Market into Turbulence." *Science* 210 (12 December): 1225–28.

Swain, Donald. 1962. "The Rise of a Research Empire: NIH, 1930–1950." *Science* 138 (14 December): 1233–37.

Szabo, G. S. A. 1977. "Patents and Recombinant DNA." *Trends in the Biochemical Sciences* 2 (November): N246–49.

Szekely, Maria. 1976. "Two Approaches to Gene Synthesis." *Nature* 263 (23 September): 277–78.

Tatum, Edward L. 1959. "A Case History in Biological Research." *Science* 129 (26 June): 1711–15.

Teich, Albert H., ed. 1974. *Scientists and Public Affairs.* Cambridge, Mass.: MIT Press.

———, ed. 1981. *R&D and the New National Agenda: Federal R&D, Issues in Defense R&D, R&D in the FY Budget, Impacts.* Washington, D.C.: American Association for the Advancement of Science.

Teich, Albert H., Gail J. Breslow, and Ginger F. Payne, eds. 1980. *R&D in an Inflationary Environment: Federal R&D, Industry and the Economy, Universities, Intergovernmental Science; Colloquium Proceedings, June 19–20, 1980.* Washington, D.C.: American Association for the Advancement of Science.

Teich, Albert H., and Ray Thornton. 1982. *Science, Technology and the Issues of the Eighties: Policy Outlook.* Boulder, Colo.: Westview Press.

Thomas, Lewis. 1977. "Notes of a Biology Watcher: The Hazards of Science." *New England Journal of Medicine* 296 (10 February): 324–27.

Thomas, Marjorie, John R. Cameron, and Ronald W. Davis. 1974. "Viable Molecular Hybrids of Bacteriophage Lambda and Eucaryotic DNA." *Proceedings of the National Academy of Sciences of the United States of America* 71:4579–83.

Tiedje, James M., et al. 1989. "The Planned Introduction of Genetically Engineered Organisms: Ecological Considerations and Recommendations." *Ecology* 70:298–315.

Toews, John E. 1987. "Intellectual History after the Linguistic Turn: The Autonomy of Meaning and the Irreducibility of Experience." *American Historical Review* 92:879–907.

Tooze, John. 1978. "Harmonizing Guidelines: Theory and Practice." In Herbert W. Boyer and S. Nicosia, eds., *Genetic Engineering,* 279–85. Amsterdam: Elsevier.

Triano, Christine, and Nancy Watzman. 1991. *All the Vice President's Men.* Washington, D.C.: OMB Watch and Public Citizen's Congress Watch.

Tuana, Nancy, ed. 1989. *Feminism and Science.* Bloomington: Indiana University Press.

"A Two-Edged Sword." 1972. *Nature* 240 (10 November): 73–74.

"UK Plant Biotechnology: ARC Joins In." 1982. *Nature* 298 (29 July): 412.

United Kingdom. 1970. *The Protection of the Environment.* Cmnd. 4373. May.

———. 1974. *Report of the Committee of Inquiry into the Smallpox Outbreak in London in March and April 1973.* (the Cox Report). Cmnd. 5626. June.

————. 1975. *Report of the Working Party on the Experimental Manipulation of the Genetic Composition of Micro-organisms*. Cmnd. 5880. January.

————. 1976. *Report of the Working Party on the Practice of Genetic Manipulation*. Cmnd. 6600. August.

————. 1980. *Report on Non-departmental Public Bodies*. Cmnd. 7797. January.

————. 1981. *Biotechnology*. Cmnd. 8177. March.

United Kingdom. Advisory Board for the Research Councils. 1976. *Second Report for the Research Councils*. Cmnd. 6430. March.

————. 1979. *First Report for the Research Councils*. Cmnd. 7367. February.

————. 1979. *Third Report of the Advisory Board for the Research Councils, 1976–1978*. Cmnd. 7467. February.

United Kingdom. Advisory Board for the Research Councils/University Grants Committee. 1982. *Report of a Joint Working Party on the Support of University Scientific Research*. Cmnd. 8567. June.

United Kingdom. Advisory Council for Applied Research and Development. 1978. *Industrial Innovation*. London: HMSO.

————. 1979. *Technological Change: Threats and Opportunities for the United Kingdom*. London: HMSO, December.

United Kingdom. Advisory Council for Applied Research and Development/Advisory Board for the Research Councils. 1980. *Biotechnology: Report of a Joint Working Party*. London: HMSO.

————. 1983. *First Joint Report*. Cmnd. 8957. July.

————. 1983. *Improving Research Links between Higher Education and Industry*. London: HMSO, June.

United Kingdom. Committee on Safety and Health at Work. 1972. *Safety and Health at Work: Report of the Committee, 1970–1972*. Cmnd. 5034. July.

United Kingdom. Council for Science Policy. 1966. *First Report*. Cmnd. 3007. May.

United Kingdom. Department of Health and Social Security. 1975. *Report of the Working Party on the Laboratory Use of Dangerous Pathogens*. Cmnd. 6054. May.

United Kingdom. Genetic Manipulation Advisory Group. 1978. *First Report of the Genetic Manipulation Advisory Group*. Cmnd. 7215. May.

————. 1978. "Genetic Manipulation: New Guidelines for UK." *Nature* 276 (9 November): 104–8.

————. 1979. *Second Report of the Genetic Manipulation Advisory Group*. Cmnd. 7785. December.

————. 1982. *Third Report of the Genetic Manipulation Advisory Group*. Cmnd. 8665. September.

United Kingdom. Health and Safety Commission. 1976. *Consultative Document: Compulsory Notification of Proposed Experiments in the Genetic Manipulation of Micro-organisms*. London: HMSO.

————. 1978. *Genetic Manipulation: Regulations and Guidance Notes*. London: HMSO.

————. 1978. *Report 1977–1978*. London: HMSO.

————. 1979. *Report 1978–1979*. London: HMSO.

————. 1980. *Report 1979–1980*. London: HMSO.

United Kingdom. House of Commons. Education, Science, and Arts Committee. 1982. *Biotechnology: Interim Report on the Protection of the Research Base in Biotechnology*. Session 1981–82. 27 July.

United Kingdom. House of Commons. Select Committee on Science and Technology. 1979. *Recombinant DNA Research: Interim Report*. London: HMSO, 3 April.

United Kingdom. House of Lords. Select Committee on the European Communities. 1980. *Genetic Manipulation (DNA)*. Session 1979–80. 4 March.

United Kingdom. Medical Research Council. 1974. *Annual Report, 1973–1974*. London: HMSO.

United Kingdom. Royal Commission on Environmental Pollution. 1971. *First Report*. Cmnd. 4584. February.

————. 1972. *Second Report: Three Issues in Industrial Pollution*. Cmnd. 3894. March.

————. 1972. *Third Report: Pollution in Some British Estuaries and Coastal Waters*. Cmnd. 5054. September.

————. 1976. *Fifth Report: Air Pollution Control: An Integrated Approach*. Cmnd. 6371. January.

United Kingdom. University Grants Committee. 1970. *Annual Survey, Academic Year 1968–1969*. Cmnd. 4261.

United States. Army Medical Research Institute of Infectious Diseases. 1985. "Biological Defense: Functional Area Assessment: Overview." 4 January.

United States. Chemical Warfare Review Commission. 1985. *Report of the Chemical Warfare Review Commission*. Washington, D.C.: U.S. Government Printing Office.

United States. Congress. House. Committee on Appropriations. 1963. *Hearings: Department of Defense Appropriations for 1963*. 87th Cong. 2d Sess. 10 January–13 October.

————. 1970. *Hearings: Department of Defense Appropriations for 1970*. 91st Cong. 1st Sess. 3 January–23 December.

United States. Congress. House. Committee on Foreign Affairs. 1969. *Hearings on Chemical-Biological Warfare: U.S. Policies and International Effects*. 91st Cong. 1st Sess.

United States. Congress. House. Committee on Interstate and Foreign Commerce. 1978. *Report: Recombinant DNA Act*. 95th Cong. 2d Sess.

United States. Congress. House. Committee on Science and Technology. Subcommittee on Science, Research and Technology. 1977. *Hearings: Science Policy Implications of DNA Recombinant Molecule Research*. 95th Cong. 1st Sess.

————. 1978. *Report: Science Policy Implications of DNA Recombinant Molecule Research*. 95th Cong. 2d Sess.

United States. Congress. Office of Technology Assessment. 1981. *Impacts of Applied Genetics: Micro-organisms, Plants, and Animals*. Washington, D.C.: U.S. Government Printing Office.

————. 1984. *Commercial Biotechnology: An International Analysis*. Washington, D.C.: U.S. Government Printing Office.

————. 1988. *New Developments in Biotechnology: 3. Field-Testing Engineered Organisms: Genetic and Ecological Issues.* Washington, D.C.: U.S. Government Printing Office.

————. 1988. *New Developments in Biotechnology: 4. U.S. Investment in Biotechnology.* Washington, D.C.: U.S. Government Printing Office.

————. 1991. *Biotechnology in a Global Economy.* Washington, DC: U.S. Government Printing Office.

United States. Congress. Senate. Committee on Commerce, Science, and Transportation. Subcommittee on Science, Technology, and Space. 1977. *Hearings: Regulation of Recombinant DNA Research.* 95th Cong. 1st Sess.

————. 1978. *Report: Recombinant DNA Research and Its Applications.* 95th Cong. 2d Sess.

————. 1980. *Hearing: Industrial Applications of Recombinant DNA Techniques.* 96th Cong, 2d Sess.

United States. Congress. Senate. Committee on Human Resources. Subcommittee on Health and Scientific Research. 1977. *Hearing: Recombinant DNA Regulation Act, 1977.* 95th Cong. 1st Sess.

United States. Congress. Senate. Committee on Labor and Public Welfare. Subcommittee on Health. 1975. *Hearing: Genetic Engineering, 1975.* 94th Cong. 1st Sess.

United States. Department of Defense. 1980. "Annual Report on Chemical Warfare and Biological Research Programs (October 1, 1978–September 30, 1979), 30 November 1979." *Congressional Record,* 5 August, S10852–68.

————. 1990. *Critical Technologies Plan.* Washington, D.C.: U.S. Government Printing Office.

————. 1991. *Report of the National Critical Technologies Panel.* Washington, D.C.: U.S. Government Printing Office, March.

United States. Department of Health and Human Services. National Institutes of Health. 1981. "Guidelines for Research Involving Recombinant DNA Molecules." *Federal Register* 46 (1 July): 34462–87.

————. 1981. "Recombinant DNA Research: Actions under Guidelines." *Federal Register* 46 (30 October): 53980–85.

————. 1982. "Guidelines for Research Involving Recombinant DNA Molecules." *Federal Register* 47 (21 April): 17180–98.

————. 1982. "Guidelines for Research Involving Recombinant DNA Molecules." *Federal Register* 47 (27 August): 38048–68.

————. 1982. "Program to Assess the Risks of Recombinant DNA Research: Proposed Second Annual Update," *Federal Register* 47 (7 December): 55104–9.

United States. Department of Health, Education, and Welfare. 1976. *Report of the President's Biomedical Research Panel.* Washington, D.C.: U.S. Government Printing Office. 30 April.

United States. Department of Health, Education, and Welfare. Education Office. 1968. *Higher Education Finances: Selected Trends and Summary Data.* Washington, D.C.: U.S. Government Printing Office, June.

United States. Department of Health, Education, and Welfare. Food and Drug Administration. 1978. "Recombinant DNA: Intent to Propose Regulations." *Federal Register* 43 (22 December): 60134–35.

United States. Department of Health, Education, and Welfare. National Institutes of Health. 1974. "Advisory Committee: Establishment of Committee." *Federal Register* 39 (6 November): 39306.

————. 1976. "Decision of the Director, National Institutes of Health, to Release Guidelines for Research on Recombinant DNA Molecules." 23 June. *Federal Register* 41 (7 July): 27902–29811.

————. 1976. "Guidelines for Research Involving Recombinant DNA Molecules." *Federal Register* 41 (7 July): 27911–43.

————. 1976–90. *Recombinant DNA Research* (Documents Relating to NIH Guidelines for Research Involving Recombinant DNA Molecules). 12 vols. Bethesda, Md.: NIH Office of Recombinant DNA Activities.

————. 1977. "Recombinant DNA Research: Proposed Revised Guidelines." *Federal Register* 42 (27 September): 49596–49609.

————. 1978. "Guidelines for Research Involving Recombinant DNA Molecules." *Federal Register* 43 (22 December): 60108–31.

————. 1978. "Recombinant DNA Research: Proposed Revised Guidelines." *Federal Register* 43 (28 July): 33042–33178.

————. 1978. "Recombinant DNA Research: Revised Guidelines." *Federal Register* 43 (22 December): 60080–105.

————. 1978. "U.S.-EMBO Workshop to Assess Risks for Recombinant DNA Experiments Involving the Genomes of Animal, Plant, and Insect Viruses." *Federal Register* 43 (31 March): 13748–55.

————. 1979. "Proposed Guidelines for Research Involving Recombinant DNA Molecules." *Federal Register* 44 (30 November): 69210–34.

————. 1979. "Proposed Supplement to the NIH Guidelines for Recombinant DNA Research." *Federal Register* 44, No. 151 (3 August): 45868–69.

————. 1979. "Recombinant DNA Research: Proposed Actions under Guidelines." *Federal Register* 44 (13 April): 22314–16.

————. 1979. "Recombinant DNA Research: Proposed Actions under Guidelines." *Federal Register* 44 (30 November): 69234–51.

————. 1979. "Recombinant DNA Research: Proposed Actions under Guidelines: 1. Proposed Exemption for *E. coli* K12 Host-vector Systems." *Federal Register* 44 (31 July): 145.

————. 1980. "Guidelines for Research Involving Recombinant DNA Molecules." *Federal Register* 45 (29 January): 6724–49.

————. 1980. "Recombinant DNA Research: Actions under Guidelines." *Federal Register* 45 (21 November): 77372–81.

————. 1980. "Recombinant DNA Research: Physical Containment Recommendations for Large-Scale Uses of Organisms Containing Recombinant DNA Molecules." *Federal Register* 45 (11 April): 24968–71.

————. 1981. "Recombinant DNA Research: Proposed Actions under Guidelines." *Federal Register* 46 (20 March): 17995.

————. 1981. "Recombinant DNA Research: Proposed Actions under Guidelines." *Federal Register* 46 (7 December): 59734–37.

————. 1981. "Recombinant DNA Research: Proposed Revised Guidelines." *Federal Register* 46 (4 December): 59368–59425.

United States. Library of Congress. Congressional Research Service. 1972. "Genetic Engineering: Evolution of a Technological Issue." Report prepared for the House Subcommittee on Science and Astronautics, 92d Cong, 2d Sess. November.

United States. National Science Foundation. 1959–67. *Federal Funds for Science.* Surveys of Science Resources Series, vols. 8–15. Washington, D.C.: U.S. Government Printing Office.

United States. National Science Foundation. Division of Science Resource Studies. N.d. "Federal Funds for Research and Development: Detailed Historical Tables, Fiscal Years 1955–1987."

United States. Office of Science and Technology Policy. 1984. "Proposal for a Coordinated Framework for Regulation of Biotechnology." *Federal Register* 49 (31 December): 50856.

———. 1985. "Coordinated Framework for Regulation of Biotechnology: Establishment of the Biotechnology Science Coordinating Committee." *Federal Register* 50 (14 November): 47, 174–75.

———. 1986. "Coordinated Framework for Regulation of Biotechnology." *Federal Register* 51 (26 June): 23302–93.

United States. President's Council on Competitiveness. 1991. *Report on National Biotechnology Policy.* February.

Vaquin, Madeleine. 1981. "Biotechnology in Britain." Draft contract report prepared for the U.S. Office of Technology Assessment, October.

Vickers, Tony. 1978. "Flexible DNA Regulation: The British Model." *Bulletin of the Atomic Scientists* 34 (January): 4–5.

Vig, Norman. 1968. *Science and Technology in British Politics.* Oxford: Pergamon Press.

Vogel, David. 1986. *National Styles of Regulation: Environmental Policy in Great Britain and the United States.* Ithaca: Cornell University Press.

Wade, Nicholas. 1973. "Microbiology: Hazardous Profession Faces New Uncertainties." *Science* 182 (9 November): 566–67.

———. 1975. "Genetics: Conference Sets Strict Controls to Replace Moratorium." *Science* 187 (4 March): 931–35.

———. 1975. "Recombinant DNA: NIH Group Stirs Storm by Drafting Laxer Rules." *Science* 190 (21 November): 767–69.

———. 1975. "Recombinant DNA: NIH Sets Strict Rules." *Science* 190 (19 December): 1175–79.

———. 1976. "Guidelines Extended but EPA Balks." *Science* 194 (15 October): 304.

———. 1976. "Recombinant DNA: The Last Look Before the Leap." *Science* 192: (16 April): 236–38.

———. 1977. "Gene Splicing: Senate Bill Draws Charges of Lysenkoism." *Science.* 197 (July 22): 348–50.

———. 1977. "Recombinant DNA: NIH Rules Broken in Insulin Gene Project." *Science* 197 (30 September): 1342–45.

———. 1977. *The Ultimate Experiment: Man-Made Evolution.* New York: Walker and Company.

———. 1978. "Gene-Splicing Rules: Another Round of Debate." *Science* 199 (6 January): 30–33.

————. 1978. "New Smallpox Case Seems Lab Caused." *Science* 201 (8 September): 893.

————. 1978. "A Tangled Tale from the Biology Classroom." *Science* 200 (5 May): 516–17.

————. 1979. "Major Relaxations in DNA Rules." *Science* 205 (21 September): 1238.

————. 1979. "Recombinant DNA: Warming Up for Big Pay-off." *Science* 206 (9 November): 663–65.

————. 1980. "Cloning Gold Rush Turns Basic Biology into Big Business." *Science* 208 (16 May): 688–93.

————. 1980. "Gene Gold Rush Splits Harvard, Worries Brokers." *Science* 210 (21 November): 878-9.

————. 1980. "Gene Splicing Company Wows Wall Street." *Science* 210 (31 October): 506–7.

————. 1980. "University and Drug Firm Battle over Billion-Dollar Gene." *Science* 209 (26 September): 1492–94.

————. 1982. "The Roles of God and Mammon in Molecular Biology." In William J. Whelan and Sandra Black, eds., *From Genetic Experimentation to Biotechnology: The Critical Transition*, 203–11. New York: John Wiley.

Walgate, Robert. 1978. "GMAG Wants to Stay On." *Nature* 273 (25 May): 25.

————. 1979. "Genetic Manipulation: Britain May Exempt 'Self-Cloning.'" *Nature* 277 (22 February): 589.

————. 1980. "Brussels Asked to Spend £16M on Biotechnology." *Nature* 283 (10 January): 125–26.

————. 1980. "United Kingdom: Biotechnology Report Urges £10 Million Programme to Match Competitors'." *Nature* 283 (24 January): 324–25.

Walker, Jack. 1977. "Setting the Agenda in the U.S. Senate: A Theory of Problem Selection." *British Journal of Political Science* 7 : 423–45.

Walkland, S. A., and Michael Ryle, eds. 1979. *The Commons Today*. Glasgow: Fontana.

Walsh, John. 1981. "Biotechnology Boom Reaches Agriculture." *Science* 213 (18 September): 1339–41.

Watanabe, Tsutomu. 1967. "Infectious Drug Resistance." *Scientific American*, December, 19–27.

Watson, James. 1971. "Moving toward the Clonal Man." *Atlantic*, May, 50–53.

————. 1977. "An Imaginary Monster." *Bulletin of the Atomic Scientists*, May, 19–20.

————. 1977. "In Defense of DNA." *New Republic*, 25 June, 11–14.

————. 1977. "Remarks on Recombinant DNA." *CoEvolution Quarterly* 14 (Summer): 40–41.

————. 1978. "Trying to Bury Asilomar." *Clinical Research* 26 (April): 113–15.

Watson, James D., and John Tooze. 1981. *The DNA Story: A Documentary History of Gene Cloning*. San Francisco: Freeman.

Weart, Spencer. 1976. "Scientists with a Secret." *Physics Today*, February, 23–30.

Webster, Andrew. 1991. *Science, Technology, and Society*. London: Macmillan.

Wehr, Elizabeth. 1978. "DNA Regulation Bill Hits Roadblock Again." *Congressional Quarterly*, 27 May, 1331–35.

Weinberg, Alvin. 1970. "In Defense of Science." *Science* 167 (9 January): 141–145.

Weinberg, Janet. 1975. "Asilomar Decision: Unprecedented Guidelines for Gene-transplant Research." *Science News,* 8 March, 148–49, 156.

———. 1975. "Decision at Asilomar." *Science News,* 22 March, 194–96.

Weinberg, Robert A. 1983. "A Molecular Basis of Cancer." *Scientific American,* November, 126–42.

Weiner, Charles. 1982. "Relations of Science, Government, and Industry: Historical Precedents and Problems." In Albert H. Teich and Ray Thornton, eds., *Science, Technology, and the Issues of the Eighties: Policy Outlook,* 71–97. Boulder, Colo.: Westview.

———. 1982. "Science in the Marketplace: Historical Precedents and Problems." In William J. Whelan and Sandra Black, eds., *From Genetic Experimentation to Biotechnology: The Critical Transition,* 123–31. New York: John Wiley and Sons.

———. 1986. "Universities, Professors, and Patents: A Continuing Controversy." *Technology Review,* February–March, 33–43.

Weiss, Edith Brown. 1979. "International Legal Aspects of Recombinant DNA Research." In Joan Morgan and William J. Whelan, eds., *Recombinant DNA and Genetic Experimentation,* 245–54. Oxford: Pergamon Press.

Weiss, Robin. 1975. "Virological Hazards in Routine Procedures." *Nature* 255 (5 June): 445–47.

Weissmann, Charles. 1981. "The Cloning of Interferon and Other Mistakes." In I. Gresser, ed., *Interferon 3,* 101–34. New York: Academic Press.

Wheelwright, Jeff. 1980. "Boom in the Biobusiness." *Life,* May, 57–58.

Whelan, William J., and Sandra Black, eds. 1982. *From Genetic Experimentation to Biotechnology: The Critical Transition.* New York: John Wiley.

"Why Are We Waiting?" 1976. *Nature* 262 (22 July): 244.

Williams, Roger. 1973. "Some Political Aspects of the Rothschild Affair." *Science Studies* 3, no. 1:31–46.

———. 1977. *Government Regulation of the Occupational and General Environments in the United Kingdom, the United States and Sweden.* Science Council of Canada Background Study no. 40. October. Ottawa: Science Council of Canada.

Williamson, Alan R. 1979. "GMAG Should Look to NIH." *Nature* 277 (1 February): 346.

Williamson, Robert. 1976. "First Mammalian Results with Genetic Engineering." *Nature* 260 (18 March): 189–90.

"Will the DNA Guidelines Wither Away?" 1980. *Nature* 287 (9 October): 474.

Winner, Langdon. 1993. "Upon Opening the Black Box and Finding It Empty: Social Constructivism and the Philosophy of Technology." *Science, Technology, and Human Values* 18 (Summer): 362–78.

Wolfinger, Raymond. 1971. "Nondecisions and the Study of Local Politics." *American Political Science Review* 65:1063–80.

Wolfle, Dael. 1972. *The Home of Science: The Role of the University.* New York: McGraw-Hill.

Wolstenholme, Gordon E. W., and Maeve O'Connor, eds. 1972. *Civilization and Science: In Conflict or Collaboration?* Amsterdam: Elsevier.

Woolgar, Steve. 1981. "Interests and Explanation in the Social Study of Science." *Social Studies of Science* 11:365–94.

———. 1988. *Science: The Very Idea*. London: Tavistock Publications.

Wright, Susan. 1979. "The Recombinant DNA Advisory Committee." *Environment* 21 (April): 2–5.

———. 1979. "Recombinant DNA Policy: From Prevention to Crisis Intervention." *Environment* 21 (November): 34–37, 42.

———. 1990. "Biotechnology and the Military." In Steven M. Gendel, A. David Kline, D. Michael Warren, and Faye Yates, *Agricultural Bioethics: Implications of Agricultural Biotechnology*, 76–96. Ames: Iowa State University Press.

———. 1990. "Evolution of Biological Warfare Policy, 1945–1990." In Susan Wright, ed., *Preventing a Biological Arms Race*, 26–68. Cambridge, Mass.: MIT Press.

———, ed. 1990. *Preventing a Biological Arms Race*. Cambridge, Mass.: MIT Press.

———. 1992. "Prospects for Biological Disarmament." *Transnational Law and Contemporary Problems* 2 (Fall): 453–92.

Wright, Susan, and Stuart Ketcham. 1990. "The Problem of Interpreting the U.S. Biological Defense Research Program." In Susan Wright, ed., *Preventing a Biological Arms Race*, 169–96. Cambridge, Mass.: MIT Press.

Wright, Susan, and Robert L. Sinsheimer. 1983. "Recombinant DNA and Biological Warfare." *Bulletin of the Atomic Scientists* 39 (November): 20–26.

Yanchinski, Stephanie. 1981. "DNA: Ignorant, Selfish, and Junk." *New Scientist*, 16 July, 154–55.

———. 1982. "France Entices Its Biotechnologists into Industry." *New Scientist*, 25 March, 767.

———. 1983. "Oncogene Researchers May Run Cancer Risk." *New Scientist*, 25 August, 530.

Young, Robert. 1977. "Science *Is* Social Relations." *Radical Science Journal* 5:65–129.

Yoxen, Edward. 1979. "Regulating the Exploitation of Recombinant Genetics." In Ron Johnston and Philip Gummett, eds., *Directing Technology: Policies for Promotion and Control*, 225–44. London: Croom Helm.

———. 1981. "Life as a Productive Force: Capitalizing the Science and Technology of Molecular Biology." In Les Levidow and Bob Young, eds., *Science, Technology, and the Labor Process*, 66–122. London: CSE Books.

———. 1982. "Giving Life a New Meaning: The Rise of the Molecular Biology Establishment." In N. Elias, H. Martins, and R. Whitley, eds., *Scientific Establishments and Hierarchies: Sociology of the Sciences*, 123–43. Vol. 6 of *Sociology of the Sciences: A Yearbook*. Dordrecht: D. Reidel.

———. 1983. *The Gene Business: Who Should Control Biotechnology?* New York: Harper and Row.

———. 1984. "Assessing Progress with Biotechnology." In Michael Gibbons, Philip Gummett, and Bhalchandar Udgaonkar, eds., *Science and Technology Policy in the 1980s and Beyond*. London: Longman.

Ziff, Edward. 1973. "Benefits and Hazards of Manipulating DNA." *New Scientist*, 25 October, 274–75.

Zilinskas, Raymond A., and Burke K. Zimmerman, eds. 1986. *The Gene-Splicing Wars: Reflections on the Recombinant DNA Controversy.* New York: Macmillan.

Zimmerman, Burke K. 1986. "Science and Politics: DNA Comes to Washington." In Raymond A. Zilinskas and Burke K. Zimmerman, eds., *The Gene-Splicing Wars: Reflections on the Recombinant DNA Controversy,* 33–54. New York; Macmillan.

INDEX

Page numbers followed by T or F refer to tables or figures, respectively.

ACS-4561   1/19/95

QH
442
W75
1994